First CHAMP Mission Results for Gravity, Magnetic and Atmospheric Studies

Springer
Berlin
Heidelberg
New York
Hong Kong
London
Milan
Paris
Tokyo

C. Reigber · H. Lühr · P. Schwintzer (Eds.)

First CHAMP Mission Results for Gravity, Magnetic and Atmospheric Studies

With 302 Figures and 49 Tables

Springer

Professor Dr. Christoph Reigber
Professor Dr. Hermann Lühr
Dr. Peter Schwintzer

Telegrafenberg A 17
14473 Potsdam, Germany

ISBN 3-540-00206-5 Springer-Verlag Berlin Heidelberg New York

Cataloging-in-Publication Data applied for

A catalog record for this book is available from Library of Congress.

Bibliographic information published by Die Deutsche Bibliothek.
Die Deutsche Bibliothek lists this publication in the Deutsche Nationalbibliografie;
detailed bibliographic data is available in the Internet at http://dnb.ddb.de

Springer-Verlag Berlin Heidelberg New York
a member of BertelsmannSpringer Science+Business Media GmbH

http://www.springer.de

Printed in Germany

Production: PRO EDIT GmbH, Heidelberg, Germany
Cover Design: Erich Kirchner, Heidelberg, Germany
Typesetting: Camera-Ready by Authors

Printed on acid-free paper 32/3141/Di 5 4 3 2 1 0

PREFACE

In 1995, the German Space Agency DARA selected the *CHA*llenging *M*inisatellite *P*ayload (CHAMP) mission for development under a special support programme for the space industry in the new states of the unified Germany, with the Principal Investigator and his home institution GFZ Potsdam being ultimately responsible for the success of all mission phases. After three years of spacecraft manufacturing and testing, the satellite was injected successfully into its final, near circular, almost polar and low altitude (450 km) orbit from the cosmodrome Plesetsk in Russia on July 15, 2000. After a nine month commissioning period during which all spacecraft systems and instruments were checked , calibrated and validated, the satellite has been delivering an almost uninterrupted flow of science data since May 2001. Since this date, all science data have been made available to the more than 150 selected co-investigator teams around the globe through an international Announcement of Opportunity.

The scientific goals of the CHAMP mission are to gain a better understanding of dynamic processes taking place in the Earth's interior and in the space near Earth. These goals can be achieved by improved observation of the Earth's gravity and magnetic fields and their time variability with high-performance on-board instrumentation and by exploring the structure of the Earth's atmosphere and ionosphere through radio occultation measurements. CHAMP carries a Global Positioning System (GPS) receiver, a high-precision accelerometer and a laser retro-reflector for the orbit, gravity and occultation tasks, and vector and scalar magnetometers and an ion drift meter for the magnetic field observations. Two pairs of star sensors provide the orientation for the vector instruments. This instrumentation, partially flown for the first time is innovative in its combination for scientific and operational use. Never before in the space age was it possible to provide simultaneously almost continuous, precisely time-tagged and long-duration measurement records of the gravity (orbit) and magnetic field variations and of atmospheric state parameters.

To discuss the status, challenges and possible directions of this first combined gravity/magnetic field and atmosphere/ionosphere remote sensing mission, a *First CHAMP Science Meeting* was held in Potsdam, Germany, from 21 until 24 January, 2002. About 220 scientists from 18 countries participated in the meeting. This international gathering was organized in the form of three panels, under themes corresponding approximately to the main chapters in this book. Each of the panel themes was addressed through invited lectures by experts for mission and science aspects. Oral presentation sessions were complemented by poster sessions and by a series of working group sessions, focussing on specific problems or planning issues.

We would like to thank all the participants of the 1st CHAMP Science Meeting for their contributions to the realization of this book. In particular, we acknowledge the panel chairmen for their contributions in structuring the panel programmes and the reviewers for their work on the contributions which went into this book. Special thanks are due to Mrs. Katrin Weisse and Mr. Philip Horsfield for the collection, formatting and editing of the papers. Lastly, the financial and operational support by the Federal Ministry of Education and Research (BMBF), the GFZ Potsdam and the German Aerospace Center (DLR) is gratefully acknowledged, without which neither the CHAMP satellite nor its operation would have been feasible.

We hope that the publication of this book illustrates the manifold possibilities of the use of CHAMP data and further widens the circle of CHAMP users for investigations on the System Earth in the course of the current *Decade of Geopotential Research* as adopted by the International Union of Geophysics and Geodesy (IUGG).

Potsdam, December 2002

Christoph Reigber
Hermann Lühr
Peter Schwintzer

Table of Contents

Orbit and Earth Gravity Field

Instruments, Orbit Determination and Gravity Field Recovery

CHAMP Orbit and Gravity Instrument Status

Ludwig Grunwaldt[1] and Thomas K. Meehan[2]

[1]GeoForschungsZentrum Potsdam (GFZ), Division 1, Kinematics & Dynamics of the Earth
[2]Jet Propulsion Laboratory/NASA, Mail Stop 238-540, 4800 Oak Grove Drive, Pasadena, CA 91109-8099, USA

Summary. The status of the instruments from the CHAMP Orbit and Gravity Package (GPS Receiver, STAR Accelerometer and Laser Retro Reflector) is reported with respect to technical performance, data availability and possible operational peculiarities. An outlook of upcoming software development and / or modifications in operational modes for both the GPS Receiver and the electrostatic accelerometer is presented.

Key words: CHAMP, GPS Receiver, STAR Accelerometer, Laser Retro Reflector

1 Introduction

The CHAMP Orbit and Gravity Package consists of NASA's „BlackJack" GPS Receiver, the electrostatic STAR Accelerometer (provided by CNES) and the passive Laser Retro Reflector (LRR) made by GFZ Potsdam. All 3 instruments support the CHAMP Precision Orbit Determination (POD); the GPS Receiver is also used for satellite bus operations as real-time navigation supporting the CHAMP Attitude and Orbit Control System (AOCS) and precision timing of the onboard experiments by supplying a 1pps reference pulse to the Onboard Data Handling System (OBDH).

2 Instrument Status and Performance

2.1 The CHAMP Laser Retro Reflector

The CHAMP retro reflector was designed with special emphasis on a high link budget in combination with negligible target signature in order to serve as an external calibration device for the onboard GPS receiver and to support CHAMP POD. It is formed of an array consisting of four cube corner prisms mounted in a compact aluminium frame (Fig. 1, left panel). This special design ensures that only one prism contributes to the return signal in most of the cases, except for short periods near the culmination of the satellite where the signals of two prisms start to interfere.

The cube corner prisms of 38 mm clear aperture produce two-spot far field diffraction patterns (Fig. 1, right panel) separated by about 20" in order to correct for

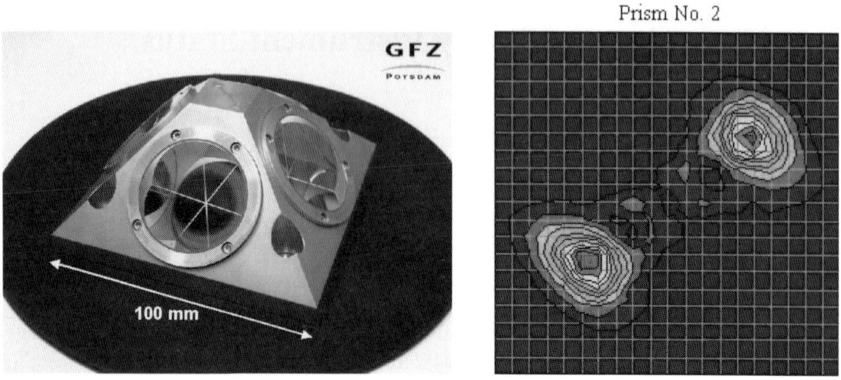

Fig. 1. (Left Panel) The CHAMP Laser Retro Reflector (LRR); (Right Panel) Two-spot far-field diffraction pattern of the CHAMP LRR prism No.2

the effect of velocity aberration of the returned signal. Proper orientation of the prisms within the aluminium frame ensures that one of the far field lobes is directed to the apparent position of the station as seen from the reflector. The symmetric location of these lobes also allows for laser tracking of the CHAMP satellite both in nominal flight orientation and with the magnetometry boom oriented in anti-velocity direction. A detailed description of the CHAMP LRR design and early in-orbit performance can be found in Grunwaldt, Neubert and Lapushka (2000).

By the end of January 2002, a total of more than 2500 CHAMP laser passes was tracked by the worldwide Satellite Laser Ranging (SLR) network indicating a good health of the reflector array. The return signal is generally reported to be very strong (comparable with returns from SLR satellites as TOPEX and AJISAI) under most orientations between tracking station and CHAMP. In connection with precise GPS-derived tracking predictions available from GFZ, this LRR feature supports fast target acquisition both under day- and nighttime conditions.

Because of the small size of the array, the target signature cannot be resolved by most of the existing SLR ground stations. This allows for highly accurate laser ranging to the CHAP LRR. The tracking station 7839 Graz (Austria) reports a single-shot RMS for CHAMP as low as 3.2 mm averaged over all 132 passes obtained in 2001. This is the lowest single-shot RMS obtained for any SLR satellite so far.

2.2 The STAR Accelerometer

The STAR instrument (Space Triaxial Accelerometer for Research Missions) uses electrostatic forces to keep a proof-mass at the center of an electrode-equipped cage. The forces necessary to keep this mass motionless in its equilibrium position provide a measure of the acceleration acting on the cage which is hardmounted to the CHAMP satellite body. A total of 6 pairs of electrodes (3 for X, 2 for Y, 1 for

Z) is used to control the 6 degrees of freedom (3 translations, 3 rotations). Fig. 2 shows the arrangement of the accelerometer electrodes and axes with respect to the main axes of the satellite.

The accelerometer Sensor Unit (SU) delivers linear and angular accelerations as output voltages which are linear combinations of the related electrode voltages. These voltages are digitized together with a set of housekeeping parameters within the Interface and Control Unit (ICU) and reported to the CHAMP OBDH.

The main specifications of the STAR accelerometer are summarized in Table 1. It has to be noted that - according to the design of the STAR SU - not all axes display an equal degree of sensitivity. The high-sensitive accelerometer axes (considering linear accelerations) are arranged about parallel to the spacecraft's along-track and cross-track axes, resp. A pair of star sensor heads is hardmounted to the

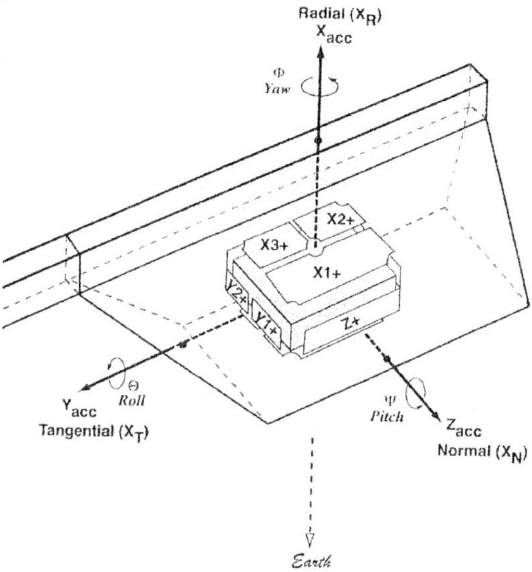

Fig. 2. Relative orientation between CHAMP and ACC sensor main axes

Measurement bandwidth	$10^{-4} \ldots 10^{-1}$ Hz
Measurement range	$\pm 10^{-4}\,ms^{-2}$
Resolution of high-sensitive axes (ACCY, ACCZ) *)	$< 3 \cdot 10^{-9}\,ms^{-2}$
Resolution of lower sensitive axis (ACCX) *)	$< 3 \cdot 10^{-8}\,ms^{-2}$
Temperature coefficient of bias (ACCY, ACCZ)	$5 \cdot 10^{-3}\,°C^{-1}$
Temperature coefficient of bias (ACCX)	$2 \cdot 10^{-3}\,°C^{-1}$
*) within the specified measurement bandwidth	

Table 1. Specifications of the STAR accelerometer for linear accelerations

thermal housing of the accelerometer sensor. The quaternions from these star sensor heads determine the exact instantaneous orientation of the accelerometer axes in space.

The status of the CHAMP accelerometer can be shortly summarized as follows:

- The STAR accelerometer has been powered-on continuously since the day 201_2000 and is delivering a permanent stream of science and housekeeping data with nearly 100% availability. The very few data gaps are related rather to spacecraft events than to accelerometer problems.
- The obtained values for accelerometer noise level in linear accelerations are close to the specifications, the sensitivity is even better than specified.
- The long-term bias and temperature stability of the high-sensitive axes are excellent and meet all expectations.

There is observed a variety of spurious signals within the accelerometer raw data:

- 1-second long spikes triggered by the activation / de-activation of dedicated heaters within the CHAMP satellite.
- Longer periodic signal (10-20 minutes) in all accelerations containing the signal from the X3 electrode pair, correlated to the activation of the 4 heaters on the thermal housing of the accelerometer sensor.
- Large, short peaks in all accelerations containing the signal from the X3 electrode pair, appearing once per orbit (the „orbit-counter spikes").
- Spikes going both positive and negative within 2 seconds („up-down spikes") showing up in all accelerometric channels.
- Infrequent discontinuities within the bias of an accelerometric channel („bias jumps"), following ICU reboots and correlated to the „orbit counter spikes".

There is good reason to assume that all those spurious signals have an electronic origin and do not exhibit real accelerations. With the exception of the rare bias discontinuities (which require a bias re-calibration after their occurrence), it appears possible to eliminate these noise signals in the process of accelerometer data processing [see Förste (2002) and Oberndorfer and Müller (2002)]. Fig. 3 shows an example of several different types of spurious signals within the STAR Level-1 data for the along-track acceleration during one orbit.

Generally, there appears to be a hyper-sensitivity to both temperature variations in the accelerometer cage and external noise signals by the X3 electrode pair. This gave rise to the assumption that probably a drive-voltage amplifier controlling one of the X3 electrodes shows a malfunction. Some activity is under way in order to overcome this unwanted effect and to provide accelerations free of the X3-influence by a different post-processing strategy [see Perret et al. (2002)].

The hyper-sensitivity to temperature variations within the accelerometer sensor shows noticeable effects only during the rare cases of a disturbed thermal regime within the spacecraft. While the specification for the maximum allowed orbit variations of the cage temperature was ±1 K, the satellite manufacturer managed to obtain a tenfold better thermal control of ±0.1 K per orbit which minimizes the

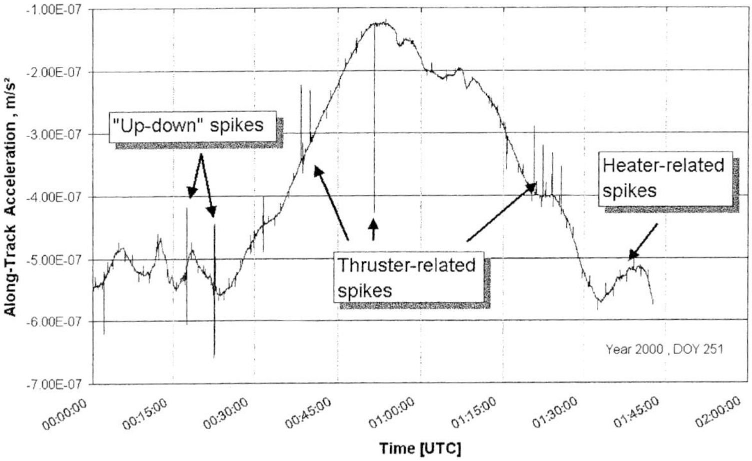

Fig. 3. Various types of spurious signals within the ACC Level-1 data

Date	Event
254_2000	OBDH redundancy switch, CHAMP temporarily in thermal survival mode
276 - 307_2000	CHAMP in full Sun orbit; deficiencies in thermal control strategy of the accelerometer sensor
035 – 036_2001	OBDH redundancy switch, CHAMP temporarily in thermal survival mode
328_2001	OBDH software failure, thermal control halted with accelerometer heaters enabled for several hours

Table 2. Major events in the thermal regime of the STAR accelerometer on CHAMP

problems. The few events in which the thermal regime within CHAMP was temporarily disturbed to a larger degree are listed in Table 2.

2.3 The „BlackJack“ GPS Receiver

The CHAMP GPS receiver (NASA/JPL's „BlackJack“ instrument) performs dual frequency phase and range measurements and serves 4 antennas: one zenith-viewing POD, two limb-viewing and one on the nadir side of the spacecraft. The arrangement of POD and limb-viewing antennas is shown in Fig. 4. The receiver architecture employs redundant digital electronics and relies on commercial grade, non radiation-hardened components. The receiver firmware is re-programmable in orbit and displays a high level of flexibility. It allows for tracking of up to 12 GPS satellites in parallel for POD.

Besides the above mentioned role within the satellite bus operations , the „BlackJack“ GPS receiver serves for precise radiometric observations of phase and range for post-processed orbit determination in the order of few centimeters. Simultaneously with POD observations, the medium gain, limb viewing RHCP (Right-Hand Circular Polarization) helix antenna allows high-frequency occultation measurements of the ionosphere and the neutral atmosphere. The nadir-viewing LHCP helix antenna will serve for a GPS altimetry experiment. A short list of the main specifications for the receiver can be found in Table 3.

The status of the instrument can be summarized as follows:

Fig. 4. CHAMP GPS antenna array of limb-viewing elements (left) and the zenith-looking POD antenna on a choke ring

Accuracy of the real-time navigation solution	< 10 m (SA off)
Time calibration accuracy	< 1 μs
Dual-frequency range and integrated carrier for POD	
Phase (ionosphere-free)	< 0.2 cm
Range (inosphere-free)	< 30 cm
Limb-sounding observables	
L1 carrier phase	<0.005 cm (1s)
L2 carrier phase	<0.03 cm (1s)

Table 3. Specifications of the „BlackJack“ GPS Receiver on CHAMP

- The GPS receiver was powered-on one day after the launch of CHAMP and delivers real-time navigation solutions and POD data on a continuous base. The overall data availability for these types of data is about 98% (only limited by occasional resets of the receiver software and periods of software uploads); data quality is well within the specified values.
- Several in-orbit software modifications have been performed since November 2000 in order to upgrade the tracking performance and to add improved occultation measurement capabilities.
- Radio occultation measurements have been carried out periodically since February 2001 and are enabled continuously since August 2001. The precision of limb-sounding observables exceeds any previous orbiting GPS receiver while scheduling over 200 observations per day.

There are a few operational issues remaining (as of January 2002):

- The receiver software resets at higher rate when configured for occultation measurements. The number of such resets (causing an interruption of tracking for several minutes) increases from less than 1 per day to about 2-3 per day in this case.
- Due to the sensitivity of gravity field recovery to GPS data gaps, the maximum number of simultaneous POD tracks has to bet set to no more than 8 in order to keep the number of resets low. This reduces the potential orbit determination accuracy.
- High frequency oscillator disturbance shows up as periodic phase spikes in the 50 Hz calibrated occultation data.
- The overall occultation data production is reduced due to problems with the tracking of the reference satellite.

The combined ionospheric/atmospheric occultation mode causes a frequent loss of reference satellite data and thus limits the number of successfully processed atmospheric profiles.

3 Upcoming activities and modifications in operational modes

3.1 STAR Accelerometer

The excellent performance of the highly sensitive axes suggests a test for a possible increase in the resolution of the accelerometer. The science telemetry per accelerometric channel consists of 20 bits (including one sign bit which is always transmitted). The „bit window“ of the remaining 19 bits can be shifted by 1 to 4 bits using a dedicated telecommand. A 1-bit shift will increase the resolution by a factor of 2. This experiment requires a few modifications in the on-ground processing software and will be carried out on announcement.

3.2 GPS Receiver

A list of possible software fixes has been discussed in order to overcome the remaining operational issues:

- Recent updates to the OS libraries in GRACE (which displays mostly common GPS hardware and software) have been ported to the CHAMP / SAC-C software and tested on a simulator. These software changes will allow for the specified maximum number of 12 POD satellites while reducing the number of software resets considerably. The amount of software to be uploaded amounts to about 400 kB which cannot be performed within the regular ground station contacts. In order to perform the uplink within about one day (the receiver does not produce tracking data during this operation), support from NASA's Polar Ground Network will be required.
- Pending on the successful upload of the upgraded OS libraries, the improvement of the occultation software can be performed by uploading another two small libraries. These libraries will contain L2 tracking loop modifications (in order to overcome the primary reason for the reduced occultation data production) and software updates for the combined ionospheric/atmospheric occultation mode.
- Further experimental science opportunities (use of the nadir antenna for GPS altimetry studies) can be considered in the coming months.

Acknowledgement. The GFZ part of this study was carried out under the grant of the German Federal Ministry of Education and Research (BMBF) No. 03F0333A.

References

Grunwaldt L, Neubert R, Lapushka K (2000) First Results of Laser Ranging to the CHAMP Retroreflector. Proceedings of the 12th Workshop on Laser Ranging, Matera, http://geodaf.mt.asi.it/html/news/iwlr/Grunwaldt_et_al_CHAMP%20Retroreflector.pdf

Perret A, Biancale R, Perosanz F, Lemoine J M, Fayard T, Vales N Sarrailh M, Loyer S, Toubouls P, Foulon B, Pradels G, Grunwaldt L, Duvivier E (2002) On-board evalution of the STAR accelerometer. This issue

Förste C (2002) CHAMP accelerometer data processing – level-2-data generation at GFZ Potsdam. This issue

Oberndorfer H, Müller J (2002) CHAMP-Accelerometer and star sensor data combination. This issue

On Board Evaluation of the STAR Accelerometer

Félix Perosanz[1], Richard Biancale[1], Sylvain Loyer[2], Jean-Michel Lemoine[1], Alain Perret[1],Pierre Touboul[3], Bernard Foulon[3], Gregory Pradels[3], Ludwig Grunwald[4], Thierry Fayard[1], Nicole Vales[1], Michel Sarrailh[1]

[1] CNES, 18 avenue Edouard Belin, 31401 Toulouse, France
[2] Noveltis, 2 avenue de l'Europe, 31520 Ramonville, France
[3] ONERA, 29 avenue de la Division Leclerc, 92322 Chatillon, France
[4] GeoForschungsZentrum Potsdam (GFZ), Division 1, Kinematics & Dynamics of the Earth

Summary. The main results of the on-board evaluation of the STAR accelerometer are presented after 18 months of mission. The instrument demonstrates high performances in terms of resolution and reliability and its contribution to dynamic orbit determination is clear. However, unexplained signal jumps have been detected and analysed. In addition, an anomalous behaviour of the X3 electrode of the accelerometer, affects the Radial, Roll and Pitch accelerations. Nevertheless, corrected observations can be recovered from a new combination of the electrode voltages. The in-orbit calibration will also benefit from the new EIGEN gravity field model that includes CHAMP data.

Key words: CHAMP, STAR-accelerometer, calibration, dynamic orbit

1 Introduction

This study is based on the analysis of 18 months of raw data (level-1) of the STAR accelerometer flying on board the CHAMP German satellite launched the 15th of July 2000 (figure 1). The checking of the instrument measurements and associated house-keeping parameters, the detection and correction of possible malfunctions are the goals of the common and continuous effort of the different teams involved. In addition a calibration strategy has to be defined and refined in order to provide users with the best possible level-2 data. STAR accelerometer clearly shows a high in-orbit performance. The main results of the « commissioning phase » are described in Perret et al. (2001) and summarized in Table 1. However, two critical anomalies concerning the « X3 » electrode behaviour and unexpected « signal jumps » have been detected and are analysed here. The present status and the perspectives of the accelerometer data calibration strategy are also presented.

2 Signal jumps analysis

Jumps are defined as rapid (1 or 2 seconds) variation of the mean signal level. As expected, accelerometer operations, like ICU (Interface and Control Unit) reboot or change of active board, modify the calibration parameters and consequently

Contribution to measurement	**Origin**	**Amplitude**	**Status**
Dynamical	atmospheric drag	10^{-5} ms^{-2}	observation
	radiation pressure	10^{-8} ms^{-2}	
Attitude Control	linear acceleration resulting from thruster firing	up to 10^{-6} ms^{-2}	measured or modeled or adjusted
Magnetic	Lorentz force	< 5 10^{-8} ms^{-2}	neglected
Geometrical	proof mass eccentricity	< 2 mm (< 3 10^{-9} ms^{-2})	neglected
	boom oscillations (2.8 Hz)	10^{-7} ms^{-2}	filtered
STAR + ICU calibration parameter	bias R,T,N	< 10^{-5} ms^{-2}	a priori corrected + adjusted
	scale factor R,T,N	1± < 20 %	
	unexpected drift (Radial)	***7 10^{-8} ms^{-2} /day***	
	out of spec. temperature sensitivity (Radial)	***-5 10^{-6} ms^{-2} /°C*** ***-2 10^{-7} ms^{-2} /°C/day***	
Peaks	"X3" (shadow)	up to 3 10^{-5} ms^{-2}	filtered
	"heater switching"	5 10^{-8} ms^{-2}	
	"up-down" (unexplained)	10^{-7} ms^{-2}	
Jumps	ICU slot permutation	< 10^{-7} ms^{-2}	define processing period
	ICU reboot	< 10^{-7} ms^{-2}	
	"unexplained"	***1-5 10^{-8} ms^{-2}***	
Observed Noise	instrumental	R: 8 10^{-9} ms^{-2}Hz$^{1/2}$ T: 2 10^{-9} ms^{-2}Hz$^{1/2}$ N: 4 10^{-9} ms^{-2}Hz$^{1/2}$	limited by digitization

Table 1. Summary of the different accelerometer level-1 signal component. The comparison to the expected performances of the instrument shows two problems: "unexplained" jumps of the 6 measurement channels and an anomalous behaviour of the radial linear component

$$V_{X1+} = V_X + V_\theta$$
$$V_{X2+} = V_X - V_\theta + V_\psi$$
$$V_{X3+} = V_X - V_\theta - V_\psi$$
$$V_{Y1+} = V_Y + V_\theta$$
$$V_{Y2+} = V_Y - V_\theta$$
$$V_{Z+} = V_Z$$

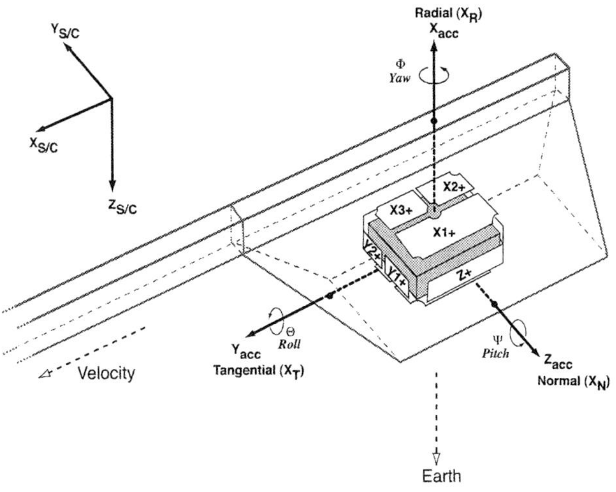

Fig. 1. Illustration of the STAR accelerometer at the CHAMP's center-of-mass. The charged proof-mass (in gray) is maintained at the center of the electrostatic cage by applying variable voltages to the 6 couples of electrodes X1, X2, X3, Y1, Y2, Z. A linear combination of the measured voltages gives the non-gravitational linear and angular accelerations of the satellite.

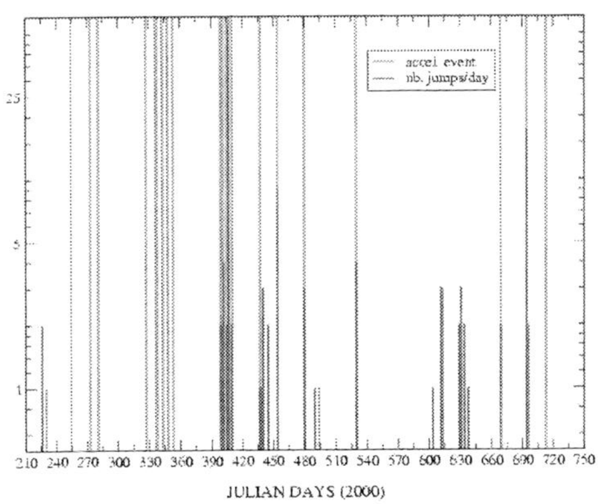

Fig. 2. Number of "unexplained" jumps per day and their correlation with on-board accelerometer operations.

produce a signal jump. However numerous jumps of unknown origin with amplitude of a few 10^{-8} m/s/s affect the 6 measurement channels.

Figure 2 gives the number of these unexplained jumps per day and the occurrence of on-board accelerometer events. Periods of one or two months without any jumps exist while up to 10 jumps can appear during the same day. Frequently and without any explanation, series of jumps appear within the few hours of an ICU operation. The understanding of these signals is still under investigation. They remain a critical limitation to the accelerometer calibration precision at the level of 10^{-8}m/s/s.

3 Mean signal analysis

The voltage applied to the 6 couple of electrodes can be recovered from the 6 acceleration measurements by applying a transfer matrix. A day-by-day averaging of the X1, X2, X3, Y1, Y2, Z electrode voltages are plotted on figure 3.

The signal variation include: (i) the seasonal evolution of the physical signal; (ii) the ICU calibration parameters change due to on-board operations (reboots, change of active board...); (iii) unexplained signal jumps.

All the electrode signals have a nominal behavior except the X3 electrode that clearly shows anomalous temperature sensitivity (100 times larger than expected)

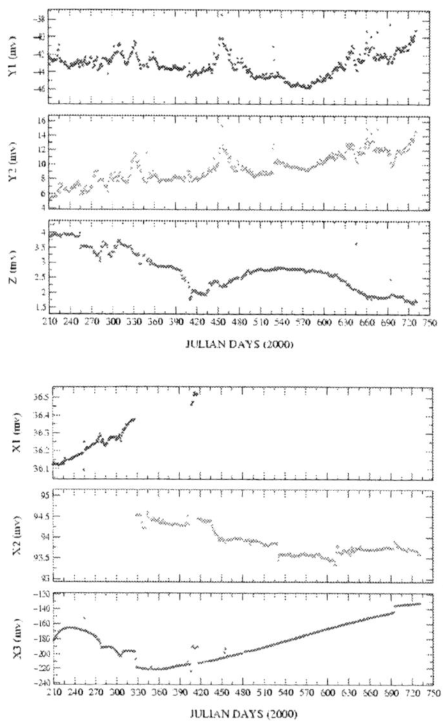

Fig. 3. Daily average of the 6 electrode voltages. A large (up to 100 mV) secular drift and a hypersensitivity to the accelerometer temperature variations affect the X3 electrode. The discontinuities of the signal correspond to (expected) ICU bias changes due to on-board operations

and secular drift (100 mV ~ 10^{-5} m/s/s amplitude). As the X3 electrode contributes to the radial, roll and pitch acceleration measurements, these observations are critically affected.

4 Technical analysis of the X3 electrode behaviour

Several hypothesis of malfunction of different components of the SU have been evaluated. The failure of the Drive Voltage Amplifier (DVA) of one of the two X3 electrodes (X3+ or X3-) is very likely. Such hypothesis remains compatible with the observed

- levitation of the proof mass (the SU is still working),
- nominal measurements on the 3 channels independent of X3,
- nominal signal noise on the 6 channels,
- high sensibility to temperature variation,
- amplification of unexpected signals (peaks, jumps…).

In addition, it can be demonstrated that the 6 degrees of freedom of the proof-mass can be recovered from the 11 (out of 12) electrodes working properly. However a new transfer matrix must be applied in order to get "corrected" radial, Roll and Pitch accelerations:

Definitions:

Γ_X: radial acceleration

$\ddot{\theta}$: roll acceleration

$\ddot{\psi}$: pitch acceleration

Φ : sensitivity of the corresponding channel

In a nominal mode (figure 1):

$$V_{X3+} = V_X - V_\theta - V_\Psi$$
$$V_{X3-} = - V_{X3+}$$

and:

$$\begin{bmatrix} \Gamma_X \\ l.\ddot{\theta} \\ l.\ddot{\psi} \end{bmatrix}_{real} = \begin{bmatrix} 1 & 0 & 0 \\ 0 & 1 & 0 \\ 0 & 0 & 1 \end{bmatrix}_{acc} . \begin{bmatrix} \Phi_X . V_X \\ \Phi_\theta . V_\theta \\ \Phi_\Psi . V_\Psi \end{bmatrix}$$

In case of a X3 DVA failure:

$$V_{X3+} = V_X - V_\theta - V_\Psi$$
$$V_{X3-} = V_\varepsilon$$

where V_ε is the perturbing voltage (including drift, temperature sensitivity…)
After some algebra we get:

$$\begin{bmatrix}\Gamma_X\\ I.\ddot{\theta}\\ I.\ddot{\Psi}\end{bmatrix}_{real} = \begin{bmatrix} \frac{7}{8}.\Phi_X & \frac{1}{8}.\Phi_X & \frac{1}{8}.\Phi_X \\ \frac{1}{8}.\Phi_\theta & \frac{7}{8}.\Phi_\theta & -\frac{1}{8}.\Phi_\theta \\ \frac{1}{4}.\Phi_\Psi & -\frac{1}{4}.\Phi_\Psi & \frac{3}{4}.\Phi_\Psi \end{bmatrix} . \begin{bmatrix}V_X\\ V_\theta\\ V_\Psi\end{bmatrix}_{acc} - \frac{1}{8}V_\varepsilon . \begin{bmatrix}-\Phi_X\\ \Phi_\theta\\ \Phi_\Psi\end{bmatrix}$$

According to Touboul and Foulon (2001) and considering that the roll and pitch accelerations given by the accelerometer are dominated by their biases (the real angular accelerations should remain close to zero in a nominal attitude mode):

$$\ddot{\theta}_{real} \approx \ddot{\Psi}_{real} << \ddot{\theta}_{acc}, \ddot{\Psi}_{acc}$$
$$\Rightarrow V_\Psi \approx 2.V_\theta$$
$$\Rightarrow V_\varepsilon \approx V_X + 5.V_\theta$$

And the new transfer matrix becomes:

$$\Rightarrow \begin{bmatrix}\Gamma_X\\ 1.\ddot{\theta}\\ 1.\ddot{\Psi}\end{bmatrix}_{real} = \begin{bmatrix} 1.\Phi_X & \frac{3}{4}.\Phi_X & \frac{1}{8}.\Phi_X \\ 0 & \frac{2}{8}.\Phi_\theta & -\frac{1}{8}.\Phi_\theta \\ 0 & -\frac{3}{2}.\Phi_\Psi & \frac{3}{4}.\Phi_\Psi \end{bmatrix} . \begin{bmatrix}V_X\\ V_\theta\\ V_\Psi\end{bmatrix}_{acc}$$

Figure 4 shows how the signal is improved on both short and long time scales when this new transfer matrix is considered.

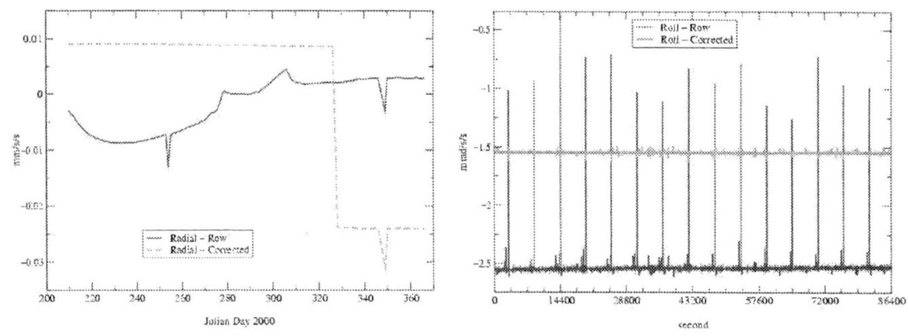

Fig. 4. Comparison of the corrected accelerations (new transfer matrix applied) to the raw signal. a) The anomalous secular drift and sensitivity to temperature variation (see figure 2) do not appear anymore in the long-term (daily average) radial acceleration signal. b) The large (non dynamical) spikes affecting the short-term roll accelerations are eliminated.

5 Accelerometer calibration

An a posteriori estimation of calibration parameters (bias+scale factor) of the accelerometer (SU+ICU) is necessary. These parameters are adjusted simultaneously with the orbital parameters through a classical least squares solution (inverse method) and a dynamic approach.

Calibration parameters are cross-correlated and also correlated with the spherical harmonics of the gravity field model. An iterative procedure is needed in order to successively improve the two sets of unknowns. In addition the current decimetre orbital accuracy gives a significant solution only for the tangential bias and scale factor. Furthermore, ICU on-board operations (transmitted from ground) can modify the accelerometer calibration parameters:

- reboot (16 events)
- in orbit tests (3 events)
- change of active board (between spare and main) (3 events)

Unexplained signal jumps also affect the accelerometer calibration. Many of them have been detected (Figure 2) and they limit the calibration accuracy at the level of few 10^{-8} m/s/s.

The present calibration strategy consists in estimating a constant scale factor (between two « events ») and a daily bias. An elaborated procedure is needed to correct the radial acceleration component from the « X3 electrode problem ». See Perret et al. (2001) for more details.

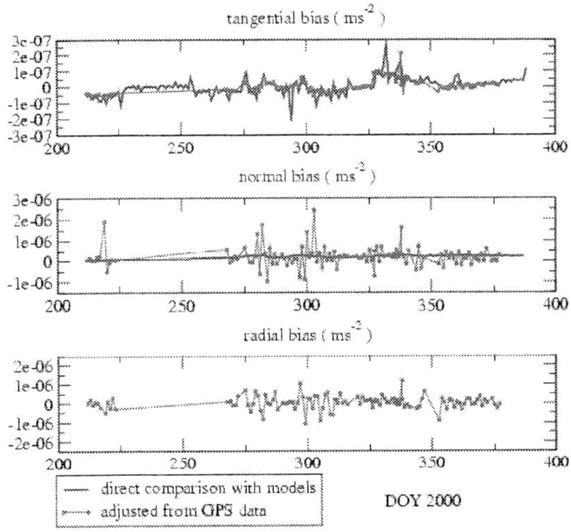

Fig. 5. Radial, tangential and normal accelerometer biases estimated by "external" (comparison to models) and dynamic (using GPS tracking data) calibration methods.

The accelerometer biases obtained by two different methods are shown on figure 5 for the three directions. The first solution is obtained by a direct comparison with non-gravitational models and the adjustment of daily biases (external calibration). The second solution is the result of a GPS orbit processing in which we solve for gravity field, orbit and calibration parameters (GPS calibration). In the two cases, the scale factors in each direction were fixed over the entire period. The estimated biases are clearly better determined in the tangential direction by the "GPS calibration". Nevertheless, the "external calibration" remains a good method to calibrate the radial and normal axes on which a dynamic approach has a lack of sensitivity.

6 Conclusion

The STAR accelerometer demonstrates a high performance in terms of resolution and reliability. Its contribution to dynamic orbit determination and gravity field model improvement has been demonstrated (Loyer et al., this issue). Nominal house-keeping parameters have been observed since the beginning of the mission. In addition, several commands from ground have been successfully executed by the accelerometer (resolution change, ICU self calibration, ICU reboot, change of active board).

Unexplained signal jumps ($\sim 10^{-8}$ m/s/s) affect all the accelerometer observations. A signal pre-processing tool detecting and repairing these discontinuities is under development.

The erroneous behaviour of the X3 electrode of the SU, affecting the radial, roll and pitch accelerations is very likely due to the failure of a Drive Voltage Amplifier (DVA). Nevertheless, corrected observations can be recovered from a new transfer matrix.

The in-orbit calibration strategy and accuracy will benefit from the

- detection/reparation of the « unexplained signal jumps »,
- corrected Radial acceleration observation,
- new EIGEN gravity field model that includes CHAMP data (Reigber et al., this issue).

Consequently, improved accelerometer level-2 products will be available for the community.

References

Perret A, Biancale R, Camus AL, Lemoine JM, Fayard T, Loyer S, Perosanz F, Sarrailh M (2001) STAR commissioning phase calibration validation activities by CNES. Vol 1& 2, DSO/ED/TU/EI 65-2001, CNES, France.

Touboul P, Foulon B (2001) Accelerometre STAR – Analyse du comportement de l'electrode X3. Rapport Technique 1/05559DMPH, ONERA, France.

Determination of CHAMP Accelerometer Calibration Parameters

Zhigui Kang, Srinivas Bettadpur, Byron Tapley, Minkang Cheng, John Ries

Center for Space Research, 3925 W Braker Lane Suite 200, Austin TX 78759 USA

Summary. One of the key science instruments for gravity science aboard CHAMP is a three-axis STAR accelerometer, which measures the non-gravitational accelerations acting on the spacecraft. These measurements are used in precise orbit determination and for gravity field estimation. Usually, the accelerometer output is not an absolute value of non-gravitational accelerations and has to be calibrated by applying a bias and a scale factor for each axis. Since the bias and scale factor parameters cannot be obtained from ground tests, they have to be determined using the flight data. This paper describes different approaches for determination of the CHAMP accelerometer calibration parameters: comparison of observed accelerations with accelerations computed using models, and determination of the parameters in orbit determination and gravity field recovery schemes. The impact of these calibration parameters on the gravity field recovery are also discussed.

Key words: Accelerometry, orbit determination, gravity recovery, STAR accelerometer, calibration parameter

1 Introduction

CHAMP (Challenging Mini-Satellite Payload for Geoscience and Application) is a German small satellite mission (Reigber et al., 2000) designed for:

- global long- to medium-wavelength Earth gravity and magnetic field recovery and its temporal variations,
- atmosphere/ionosphere GPS limb sounding

Launched in July 2000, the characteristic orbit parameters are near-polar (inclination=87 degree), near-circular, with an initial altitude of 450 km. The design lifetime is 5 years.

For precise orbit determination and global gravity recovery, CHAMP carries an accelerometer to directly measure the hard-to-model non-gravitational accelerations, mainly air drag, which then can be applied in CHAMP's dynamic orbit determination with GPS satellite-to-satellite tracking (SST) to isolate the gravitational orbit perturbations. A central problem in using those measurements is the determination of the instrument's calibration parameters, i.e. the bias and scale factor for each of the three axes. The calibration parameters have to be adjusted simultaneously with the other parameters during the orbit determination and gravity recovery process because these parameters cannot be obtained on the ground under the 1-g environment. Therefore, the goals of this study are:

- to investigate how well the accelerometer calibration parameters can be estimated
- to study the impact of the calibration parameters on the gravity recovery
- to develop an optimal accelerometer data processing approach

2 CHAMP data processing strategy

The baseline tracking system for CHAMP is GPS satellite-to-satellite tracking, which generally consists of the GPS satellite constellation, the GPS ground receiver network and the CHAMP flight GPS receiver. Measurements are taken between the high flying GPS satellite and the low flying CHAMP spacecraft, and between the GPS satellites and the ground receivers. The data processing is carried out using double-differenced carrier phase observations. In orbit determination, the accelerometer measurements replace the non-gravitational accelerations. The accelerometer observation equation is as follows (Schwintzer et al. 1998):

$$a_{ci} = a_i + k_i a_{mi} \quad \text{i = 1, 2, 3 (x-, y-, z-axis)}$$

where a_{ci} is calibrated acceleration in i-direction, a_i accelerometer bias in i-direction, k_i scale factors in i-direction, and a_{mi} measured acceleration in i-direction.

To determine CHAMP accelerometer calibration parameters, we have used three different approaches:

1. comparison of observed accelerations with accelerations computed using models
2. estimation of accelerometer biases in CHAMP orbit determination
3. determination of accelerometer calibration parameters during the gravity field estimation

In this study, the following processing strategy was used. First, orbits were dynamically determined using only GPS data to get non-gravitational accelerations. Then the accelerometer data were compared with the non-gravitational accelerations to obtain the initial accelerometer bias parameters. After that, only three accelerometer biases were estimated in the orbit determination (one per day for each axis) due to the large correlation between bias and scale factor. Finally, different combinations of accelerometer calibration parameters were estimated in the gravity field recovery.

3 Test cases

Six experiments were conducted using 28 one-day arcs between May 14 and June 10, 2001. The first test case used biases determined by comparison of the accelerometer data with the non-gravitational accelerations. The second case used accelerometer biases which were estimated in the orbit determination without gravity field estimation. The remaining cases were the following different combinations of accelerometer parameters determined as part of the gravity field recovery: (1) 3B: three daily biases, (2) 3B3S: three daily biases and three global scale factors, (3) 3B2STN: three daily biases and two global scale factors in the along-track and cross-track directions, (4) 3B2SRT: three daily biases and two global scale factors in the radial and along-track directions. In addition, two gravity solutions were performed: CHAMP-only and combined solution with TEG4S8 (Tapley et al., 2000), which is one of the CSR satellite-only gravity solutions without CHAMP data.

4 Results and discussion

The accelerometer calibration parameters are determined based on the strategy and test cases described above. The accelerometer calibration parameters determined in different test cases are summarized in Tables 1 and 2. It is observed that a large drift exists in the radial direction, and the bias for each axis depends on whether the corresponding scale factor is estimated or not. This is because the estimated scale factors differ significantly from unity, and there is large correlation between the bias and scale factor for each axis. Since the models are not sufficiently accurate, the biases derived from comparison of measured accelerations with computed accelerations can be used only as initial values in the orbit determination and for checking the data quality.

Table 2 gives the estimated scale factors and their formal errors. All three scale factors appear to be well determined. The values in the along-track and across-track are closer the unity.

	Radial		Along-track		Cross-track	
Case	Bias	Drift	Bias	Drift	Bias	Drift
Data vs Model	8.294	0.097	-3.635	0.002	0.495	-0.001
POD	8.221	0.101	-3.635	0.0003	0.747	0.013
3B	8.381	0.099	-3.648	0.0002	0.552	-0.001
3B3S	7.936	0.051	-3.715	-0.0003	0.531	-0.002
3B2S(TN)	8.343	0.098	-3.707	-0.0002	0.530	-0.001
3B2S(RT)	7.939	0.051	3.714	0.0003	0.546	0.002

Table 1. CHAMP mean accelerometer biases (μms^{-2}) and drifts (μms^{-2}/day)

	Radial		Along-track		Cross-track	
Case	Value	Error	Value	Error	Value	Error
3B3S	0.51003	0.00154	0.80840	0.00028	0.93084	0.00107
3B2S(TN)			0.83055	0.00028	0.93962	0.00109
3B2S(RT)	0.51283	0.00154	0.80891	0.00028		

Table 2. CHAMP Accelerometer scale factors

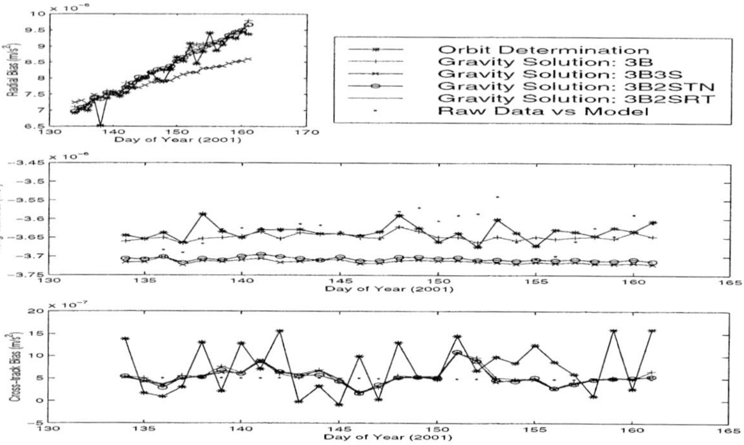

Fig. 1. CHAMP accelerometer biases

Figure 1 is a time-series plot of estimated CHAMP daily accelerometer biases using different approaches. It also demonstrates that there is a large drift in the radial direction. The values for the along-track bias from the gravity solution cases are very stable. The other biases are apparently not as well determined.

Figure 2 is the CHAMP-only gravity model comparison which presents the relative geopotential coefficient differences. The differences between CHAMP-ACC 3B3S and 3B are largest. The cumulative geoid error (70X70) is about 100 cm. The second largest effect on the CHAMP-only solution is the radial scale factor. The cumulative geoid error (70X70) is about 50 cm. There are relatively small effects (10 cm) resulting from estimation the cross-track scale factor or not. Figure 3 shows the combination gravity model comparison by solving for different parameters. In this case, there are very small impacts of different parameter combinations on gravity recovery. This is due to the strong gravity information from TEG4S8. The largest effect on the geoid is only about 1 cm (70X70).

Figure 4 shows the comparison of EGM96 with CHAMP-only gravity models. This is an external gravity comparison. We assume that the EGM96 model is probably better than the CHAMP-only gravity models. Therefore, the CHAMP-only solution with scale factors estimated is better than the solution without scale

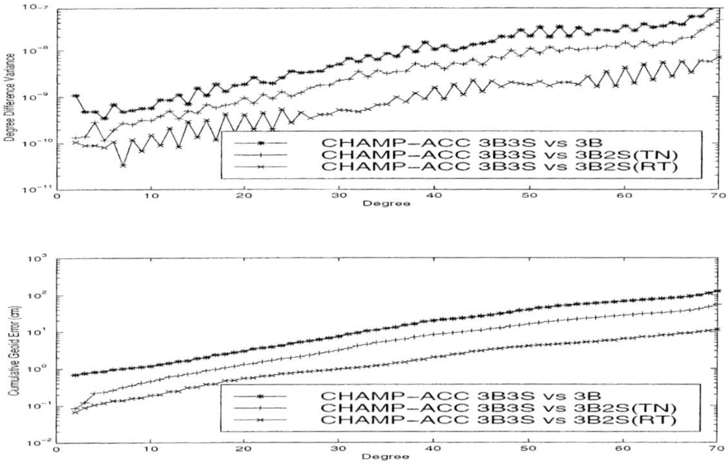

Fig. 2. CHAMP-only gravity model comparison

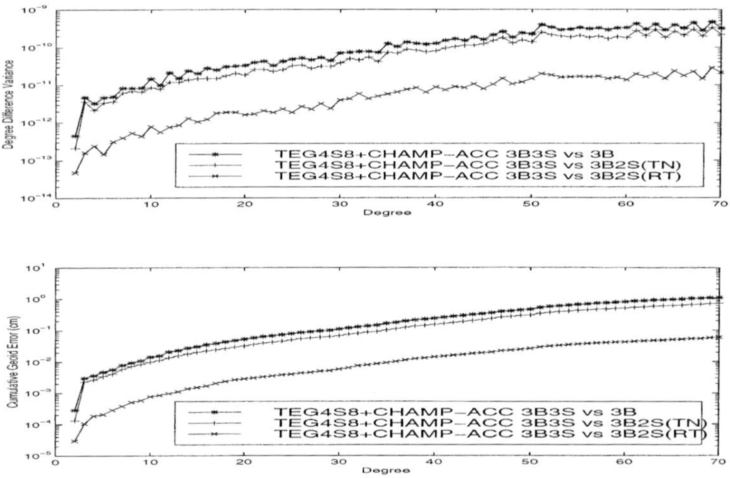

Fig. 3. Combination model (TEG4S8+CHAMP-ACC) comparison

factors. Table 3 also shows the same conclusion. In addition, the cross-track scale factor has nearly no impact on the gravity solution. From Table 3, we can see that CHAMP-only gravity models for some satellites (LAGEOS, TOPEX) are comparable with EGM96. For the CHAMP satellite, the CHAMP-only gravity models are even better than EGM96 model, as would be expected since the EGM96 has no CHAMP tracking data included.

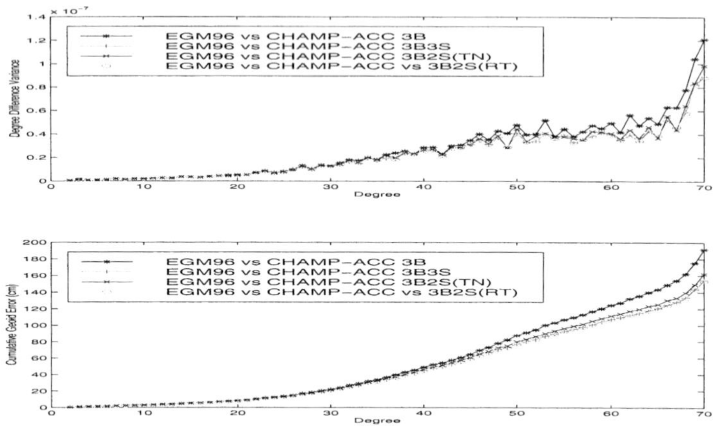

Fig. 4. Comparison of EGM96 with CHAMP-only gravity models

	EGM96	CHA 3B	CHA 3B3S	CHA 3B2S(TN)	CH A 3B2S (RT)
CHAMP	6.29	1.42	1.33	1.33	1.33
GFO-1	11.99	23.10	20.43	22.16	20.39
LAGEOS-1	1.18	2.32	1.41	1.40	1.44
LAGEOS-2	1.07	1.99	1.63	1.39	1.66
STELLA	6.37	65.33	42.95	60.79	44.68
ERS-2 (SLR)	6.81	30.70	14.96	13.93	14.58
TOPEX(SLR)	2.35	4.10	3.81	3.61	3.97

Table 3. Orbit fit comparison (cm)

5 Conclusions

The CHAMP accelerometer calibration parameters were estimated from acceleration comparisons, dynamic orbit adjustment and gravity solution using GPS and accelerometer observations. The following conclusions can be drawn:

- The initial accelerometer biases can be obtained through comparison of observed accelerations with the accelerations computed using models. It can be used for checking data quality.
- The accelerometer biases are well determined in the gravity solution. From the solutions, a large drift (about 0.1 μms^{-2} per day) in the radial direction is observed.
- The along-track and cross-track scale factor can be relatively well determined.

- The impacts of the accelerometer scale factors on CHAMP-only gravity solution are large. This is likely due to the fact that the scale factors for each axis are significantly different from unity. The cumulative geoid difference with and without the scale factors is about 100 cm (70X70) for a 28 day solution. But there is only about 10 cm geoid difference for gravity model complete to spherical harmonic degree and order 30.
- Through the orbit fit and gravity model comparison, the CHAMP-only solution with scale factors is better than the solution without the scale factors, as would be expected.

References

Reigber Ch, Luehr H, Schwintzer P (2000) CHAMP Mission Status and Perspectives, Suppl To EOS, Transactions, AGU, 81, 48, F307

Schwintzer P, Kang Z, Perosanz F (2000) Accelerometer Aboard CHAMP, International association of Geodesy Symposia, Vol 120, pp 197-200

Tapley B, Chambers D, Cheng M, Kim M, Poole S, Ries J (2000) The TEG-4 Earth Gravity Model, 25th European Geophysical Society General Assembly, Nice, France

CHAMP Accelerometer and Star Sensor Data Combination

Helmut Oberndorfer[1] and Jürgen Müller[2]

[1] Institut für Astronomische und Physikalische Geodäsie (IAPG), Technische Universität München, Germany. *helmut@alpha.fesg.tu-muenchen.de*
[2] Institut für Erdmessung (IFE), Universität Hannover, Germany. *mueller@ife.uni-hannover.de*

Summary. The STAR accelerometer on board CHAMP measures the non-gravitational accelerations, which are needed to separate them from the gravitational ones when determining the Earth gravity field from orbit perturbations. Besides translational accelerations the test mass of the accelerometer is also sensitive to angular accelerations of the spacecraft. Therefore it is possible to combine the accelerometer data with the star sensor data, which provide the orientation of the spacecraft. The combination of both measurement data was performed in the frequency domain and by applying Kalman filter techniques. As a result, the reduction of the star sensor measurement noise was achieved and the estimation of accelerometer parameters like scale factor and bias became possible.

Key words: accelerometer, star sensor, Kalman filter

1 Introduction and Analysis Strategy

The STAR accelerometer on board CHAMP measures the non-gravitational accelerations. They have to be known in order to separate them from the gravitational ones when determining the Earth gravity field from orbit perturbations. Apart from the measurement of the translational accelerations, the STAR accelerometer is capable to measure the angular motion of the spacecraft. Therefore it is possible to combine the accelerometer data with the star sensor data, which provide the orientation of the spacecraft. In addition the housekeeping data contain a record of thruster firing events which can be used for the calculation of the angular accelerations, too. Here, we discuss measurements of the accelerometer and of the star sensor from day 324 in 2001. The sensor data and satellite housekeeping data were compared in the time and frequency domain. The data were processed in different steps. First we remove peaks in the accelerometer data which are not caused by thruster events. After that we show the strong coupling of the radial accelerometer circuits with accelerometer temperature control and remove this effect from the data. In the next step, we estimate the angular acceleration scale factors of the accelerometer by comparing them with the thruster housekeeping data. Finally, prefiltered and scaled accelerometer and star sensor measurements

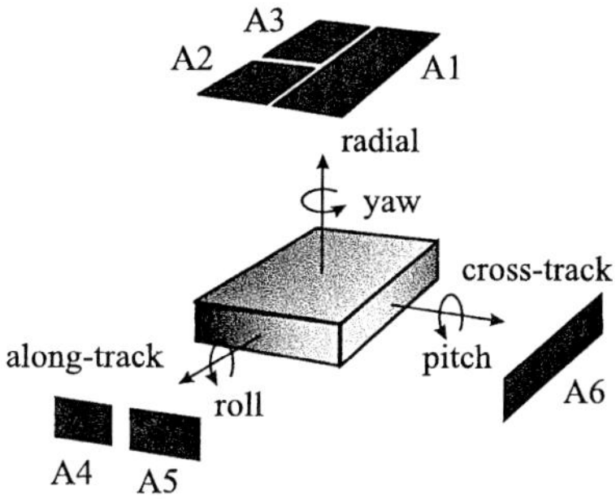

Fig. 1. Accelerometer model

ments were combined by applying KALMAN filter techniques in order to get optimal attitude information.

2 The Accelerometer

The accelerometer [1] consists of six electrode pairs (denoted X1 to X6) which control the motion of the test mass in six degrees of freedom. The measurement signal is a linear combination of the voltage applied to the electrodes. For example, the sum of the voltages of X4 and X5 multiplied by a scale factor which depends on the accelerometer parameters represents the along track acceleration of center of mass of the satellite. The difference of X4 and X5 represents the rotational acceleration about the yaw axis of the spacecraft.

Electrode voltages of the accelerometer with an arbitrary bias are shown in Fig. 2. As expected, the highest amplitudes are observed in the along track direction (X4, X5). Spikes in the signal are caused by thruster activities.

3 Accelerometer Coupling with Heater Activities

All three accelerometer control circuits in radial direction are affected by the accelerometer temperature control. As example the pitch angular acceleration signal is discussed. The heater activities around the accelerometer are shown in the upper part of Fig. 3. A strong correlation between the accelerometer temperature control and the accelerometer angular acceleration output was detected. The coupling of heater activities and angular acceleration signal was estimated in the frequency domain using a first order filter. The modeled disturbance is shown in the lower curve. It has to be subtracted from the measurements which results in corrected angular accelerations. In the control circuits for along and normal direction, a similar effect has not been detected.

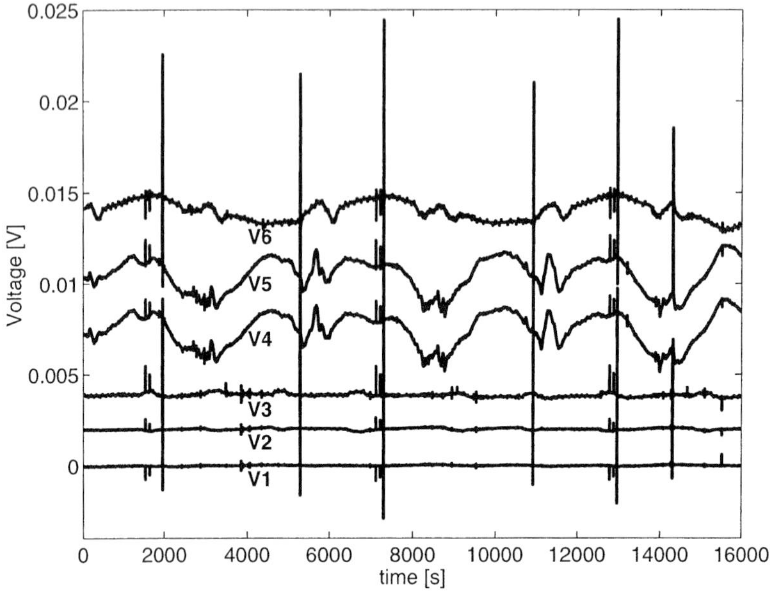

Fig. 2. Electrode voltages of the accelerometer with an arbitrary bias

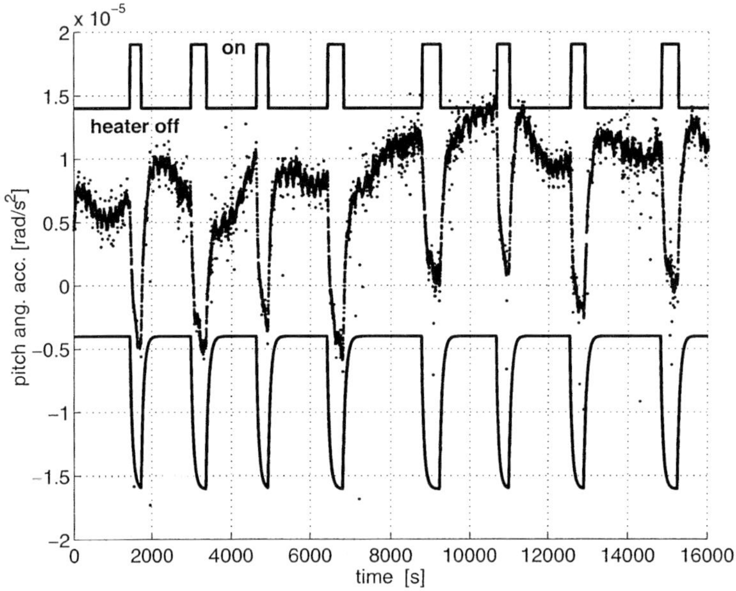

Fig. 3. Comparison of heater switch on/off, pitch acceleration signal (dots) and estimated correction

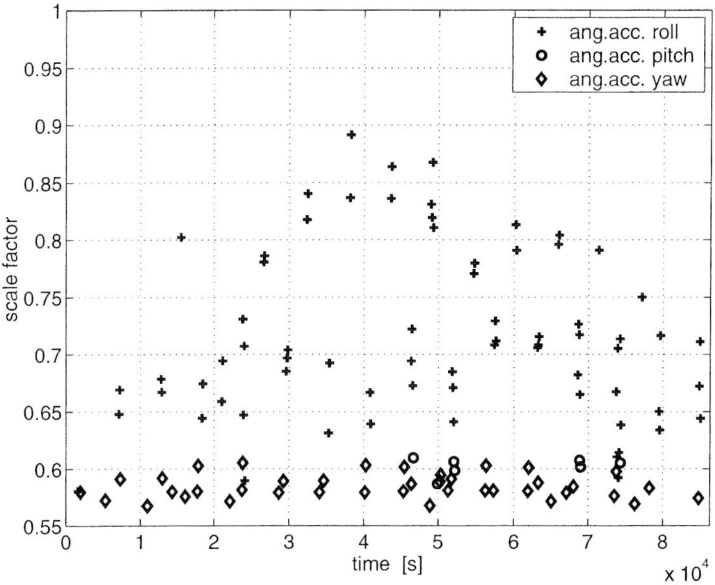

Fig. 4. Estimated scale factors

4 Scale Factor Estimation

The scale factors of the rotational circuit of the accelerometer are estimated by comparing the measured angular acceleration with the S/C angular moments during each manoeuvre. The S/C angular moment is calculated from the thruster-on duration and the S/C inertia moments. Fig. 4 shows the estimated scale factors for the three rotational circuits. Using the heater corrected output, the estimated scale factors become about 0.6 for the pitch and yaw angular acceleration and about 0.75 for the roll angular acceleration. It should be noted that these values differ much from the expected value of one.

5 Kalman filter

The corrected and scaled accelerometer data are compared with the star sensor data using a kalman filter. The measured attitude angles, relative to the nominal satellite orientation are shown in the left side of Fig. 5. The star sensor (ASC = advanced stellar compass) [2] consists of two sensor heads and a unit where the signals of both sensor heads are combined. For a single sensor head, the rotation angle about the line of sight is five times poorer than in the transverse direction. For time spans where no measurement is

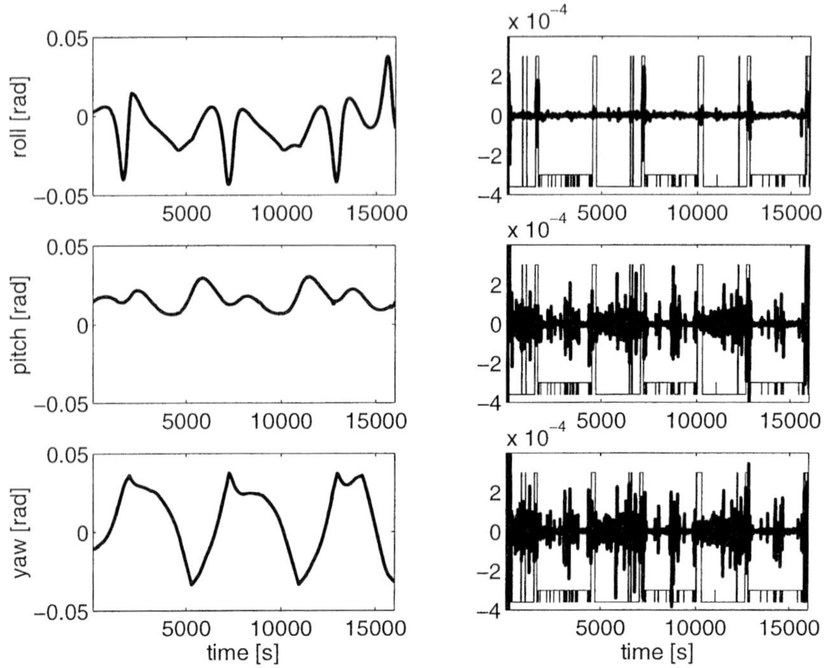

Fig. 5. Star sensor measurement and errors derived by kalman filtering

available, the star sensor pointing angles were interpolated by a polynomial fit in order to obtain a continuous measurement signal.

Differences of pointing angles measured by the star sensor and the estimated pointing angle by the KALMAN filter are shown at the right side of Fig. 5. They represent the star sensor error at higher frequencies. The derived star sensor measurement noise is given in Table 1.

Table 1. Star sensor measurement noise [rad/$\sqrt{Hz}$]

attitude angle	roll	pitch	yaw
1 ASC head	1e-5	5e-5	5e-5
2 ASC head	8e-6	1e-5	1e-5

Additionally in Fig. 5, the thin lines indicate the status of the sensor heads. A level of -3·10^{-4} indicates that two sensor heads measure, a level of -3.5·10^{-4} means that only one sensor head is active and +3·10^{-4} that no measurement is available. Time periods, where no measurements are available are short. Mostly the sensor switches between two and one sensor heads.

6 Conclusion

The comparison of the accelerometer data with housekeeping and star sensor data enables a better understanding of the accelerometer behaviour. Especially the angular acceleration measurement of the accelerometer is sensitive to such investigations. Remarkable is the strong coupling of the accelerometer measurement with the heater activities. On the other hand, the estimation of accelerometer parameters like scale factors becomes possible. Combining accelerometer and star sensors measurements leads to a better attitude information. Moreover, phases where no star sensor information is available can better be interpolated.

Acknowledgment

This work has been funded by the Deutsche Forschungsgemeinschaft (contract No. Mu 1141/2-1, 2-2). We are mostly indebted to Dr. Grunwaldt from the GeoForschungsZentrum Potsdam.

References

1. Touboul P, Foulon B, Le Clerc G-M (1998) STAR, the accelerometer of the geodesic mission CHAMP. 49th International Astronautical Congress Melbourne (Australia)
2. Jørgensen JL (1999) In orbit performance of a fully autonomous star tracker. 4th ESA International Conference on Spacecraft Guidance, Navigation and Control Systems, ESTEC, Noordwijk (Netherlands)

CHAMP Clock Error Characterization

Rolf König, Karl Hans Neumayer, Grzegorz Michalak, and Ludwig Grunwaldt

GeoForschungsZentrum Potsdam, Telegrafenberg, D-14473 Potsdam, Germany, *koenigr@gfz-potsdam.de*

Summary. The behaviour of the steered quartz clock of the "BlackJack" GPS receiver on-board CHAMP is characterized. The receiver clock estimates from on-ground orbit determination and clock offsets as reported within the CHAMP navigation solution are presented, compared, and discussed. No long term trend can be detected in the first 20 months of the mission. In the short term the clock estimates oscillate at orbital frequencies and at standard deviations around the μs.

Key words: CHAMP, BlackJack, GPS receiver, clock error

1 Introduction and Data

The CHAMP satellite carries an instrumentation for the evaluation of the gravity field and of the magnetic field and for the sounding of the neutral atmosphere and the ionosphere [1]. All on-board measurements are time stamped by the time signal out of the on-board GPS receiver called BlackJack [2]. In the following it is tried to evaluate the on-board clock by analyzing time series of estimates of the clock offset determined by off-line and on-ground POD (Precise Orbit Determination) and by the real-time NAV (navigation) solution of the BlackJack receiver itself.

For the CHAMP mission, GFZ installed the so-called OG (Orbit and Gravity) system for the generation of routine orbits for use in the various chains of mission data processing and also for external users. Within the OG processing, clock estimates are generated inherently in the POD processes where GPS data are evaluated. Different orbit products are being made available with different features as latencies and accuracies due to the intention of the particular product. Namely the so-called PSO (Postprocessed Science Orbit) and the RSO (Rapid Science Orbit) are produced.

The PSO results as an intermediate step in the course of the generation of gravity field solutions. For the evaluation of the EIGEN-1S gravity model [3], the period DoY (Day of Year) 212 in 2000 (denoted day 212 since 2000.0 in the graphics throughout this paper) to DoY 10 in 2001 (or day 376 since 2000.0) was evaluated. This period however is not fully covered but gaps from hours to weeks are included. So the data cover approximately 90 days of this half year of the mission.

The RSO is intended for use in the processing of the atmospheric occultations mainly. Its production started with the enabling of the occultation measurements in March 2001. For the purpose of this analysis, three periods were selected from DoY 107 to 137 in 2001 (day 473 to day 503 since 2000.0), from DoY 182 to 213 in 2001 (day 548 to 579 since 2000.0), and a recent one from DoY 30 to 54 in 2002 (day 760 to 784 since 2000.0).

2 Clock Parameter Estimates

From the PSO the time series as depicted in Fig. 1 results. There are obviously some events where large clock offsets are computed. F.i. the event at day 318 can be related to a receiver problem which occured fortunately only once in the mission so far. Following a bit error after a receiver reset the clock offset was erraneously set to 25 ms, the navigation algorithm tried then to correct it. The operator on-ground noticed this and solved the problem by resetting the receiver once more. A blow up of this event is shown in Fig. 2 where also the BlackJack NAV solution is given for comparison and verification. Both solutions are in full agreement.

In Fig. 3 the time series of both POD clock solutions, i.e. from the PSO and from the RSO, are compiled. The range of the clock offset axis is restricted to ± 20 μs for a closer look into the stochastic behaviour of the clock

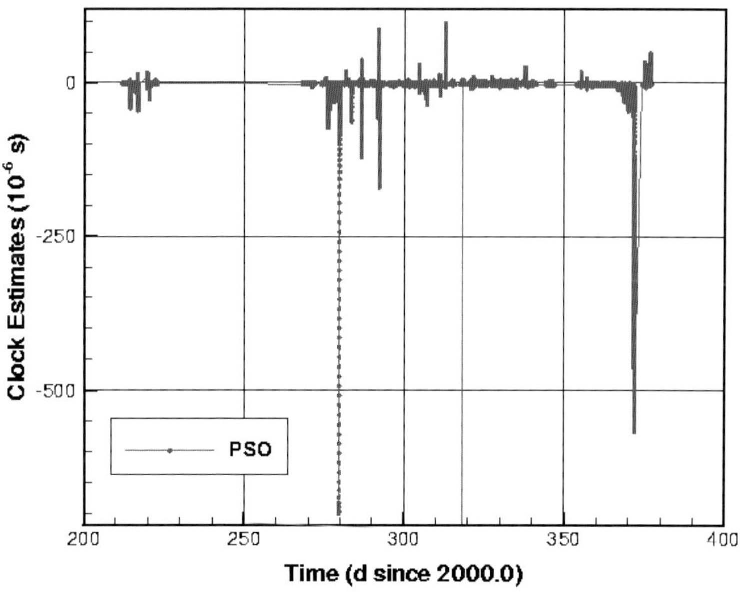

Fig. 1. Postprocessed Science Orbit clock estimates.

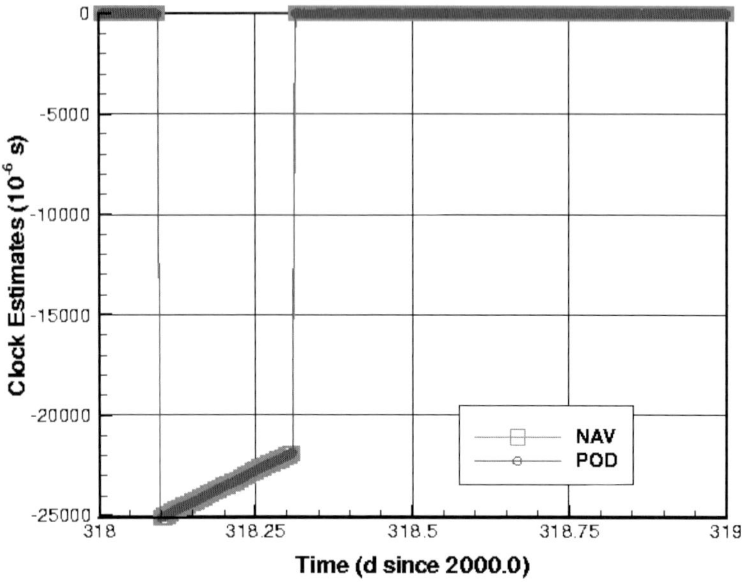

Fig. 2. Exceptional clock offset.

estimates. No long term trend can be detected in the time series over a period of approximately 20 months since launch. There is evidence for different stochastic behaviour for the PSO and the RSO time series. The reason is that in the RSO case the GPS range observations exceeding a certain limit are eliminated before the solution starts. So the range is not yet corrected for the CHAMP clock offset. This results in the sharp upper and lower bounds of the RSO clock offset estimates at around 10 μs.

Fig. 4 looks even closer into the time series at the example of day 212, the first day of the analysis period. As with Fig. 2 also the clock estimates by the BlackJack receiver are given. Both time series, from POD and from NAV, behave quite similar. The NAV solution shows some outliers because it comes in real time and blunders may therefore not be detected that efficient.

An offset between the NAV and the POD time series can be seen in Fig. 4. This is the offset of the reference clocks of the NAV and the POD solutions. In the NAV case, the receiver clock is steered with respect to the GPS broadcast clocks. In the POD case, the on-board receiver clock is determined in two steps. First the GPS satellite clocks are estimated from on-ground undifferenced GPS observations with respect to a selected GPS ground station equipped with a stable H-maser clock. In the second step the GPS clocks are fixed and then the CHAMP clock parameters estimated from on-board undifferenced GPS observations. So the POD CHAMP clock is estimated relative to the selected ground station clock.

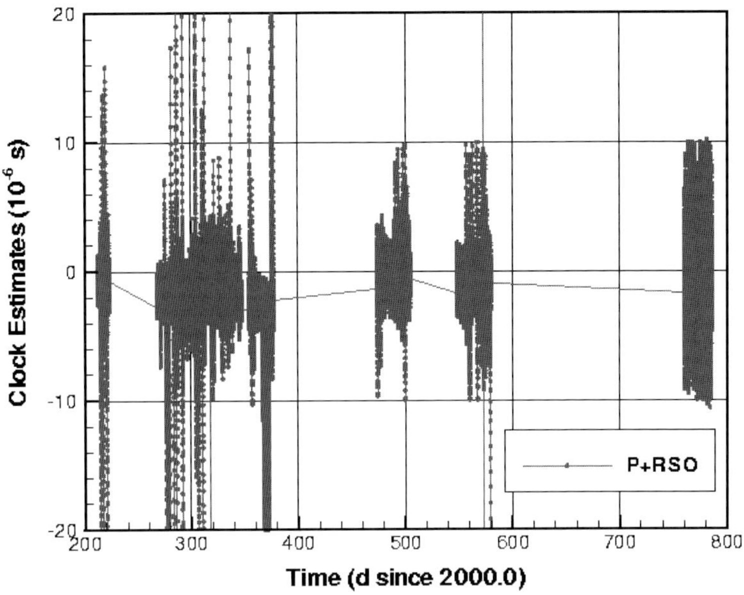

Fig. 3. Postprocessed and Rapid Science Orbit clock estimates.

In order to show that the scatter of the clock estimates is independent of the orbit determination approach, once more POD (this time from the RSO) and NAV clock estimates are plotted in Fig. 5 for day 784, the last day of the analysis period. Again the agreement in the stochastic behaviour is obvious despite of a systematic offset of approximately 0.5 μs as for day 212 above.

The standard deviation of the POD and the NAV time series for day 212 is 0.6 μs, for day 784 a larger standard deviation of 1.7 μs results. The data coverage of the full analysis period does not allow to deduce this phenomenon. Therefore a detailed analysis must be left for future investigations once the gaps in the time series can be filled.

A Fourier analysis reveals dominant frequencies close to once, twice, and three times per orbital revolution of the satellite for both time series, POD and NAV, and for both days, 212 and 784. These fequencies could likely stemm from the orbit determination procedures.

3 Conclusions

The CHAMP BlackJack receiver clock offsets were estimated from on-ground POD solutions for particular periods of the first 20 months of spaceborn operation. In the long term the time series of the clock estimates shows no linear or harmonic systematics. In the short term a comparison with the clock offsets reported by the receiver itself showed besides a physical bias,

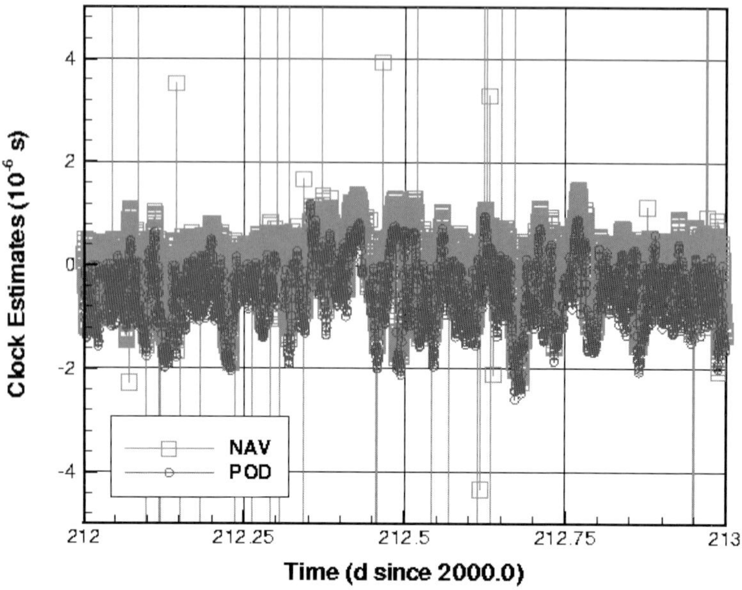

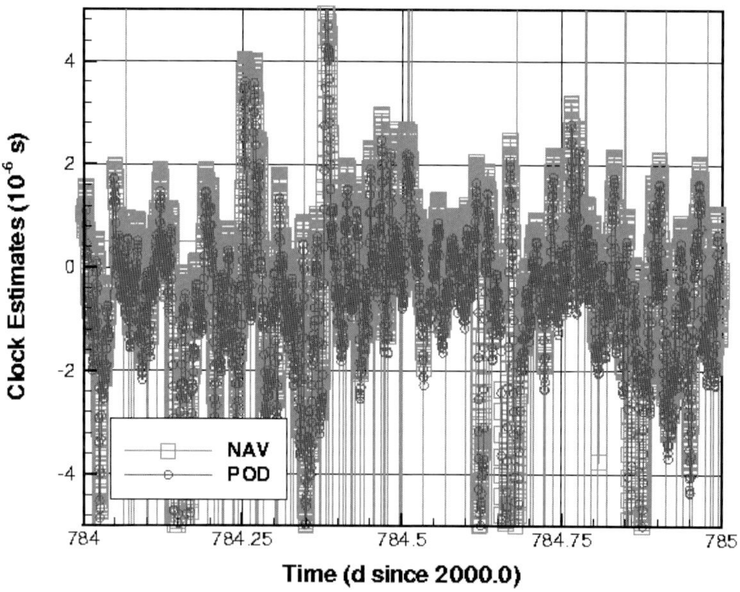

Fig. 5. Clock offsets at the last day of the analysis period.

similar harmonic and stochastic properties. Dominant frequencies related to the orbital revolution of the satellit can be observed. Despite some large exceptional deviations, the standard deviations of the clock estimates size for normal days at around 0.6 μs and for a recent period around 1.7 μs.

Acknowledgement. This study was carried out under the grant of the German Federal Ministry of Education and Research (BMBF) No. 03F0333A.

References

1. Reigber Ch, Schwintzer P, Lühr H (1999) The CHAMP geopotential mission. Boll Geof Theor Appl 40:285–289
2. Kuang D, Bar-Sever Y, Bertiger W, Desai S, Haines B, Iijima B, Kruizinga G, Meehan Th, Romans L (2001) Precise Orbit Determination for CHAMP using GPS Data from BlackJack Receiver. In: 2001 ION National Technical Meeting Proceedings. ION Institute of Navigation, Fairfax, VA, USA, pp 762–770
3. Reigber Ch, Balmino G, Schwintzer P, Biancale R, Bode A, Lemoine J-M, König R, Loyer S, Neumayer H, Marty J-C, Barthelmes F, Perosanz F, Zhu SY (2002) A High-Quality Global Gravity Field Model from CHAMP GPS Tracking Data and Accelerometry (EIGEN-1S). Geophysical Research Letters, 29 (14), 10.1029/ 2002 GL015064

Determination of the Offset of CHAMP GPS Antenna with Respect to Satellite's Mass Center

Chuang Shi, Ludwig Grunwaldt, Jean-Claude Raimondo, Franz-Heinrich Massmann and Sheng Yuan Zhu

GeoForschungsZentrum Potsdam (GFZ), Telegrafenberg 14473 Potsdam, Germany, Div. 1: Kinematics and Dynamics of the Earth

Summary. It is well known that there are systematic differences between the TOPEX/POSEIDON orbits determined by GPS and by SLR plus DORIS. The residuals of laser tracking data to GPS 35 and GPS 36 also show significant bias. One of the possible reasons of such systematic differences is the uncertainty in the phase center offset of the GPS instrument. The offset of the CHAMP GPS antenna has been measured by the hardware team, and the related numbers are released. But the actual electronic phase center is difficult to be measured accurately. Therefore we try to use the real CHAMP data to estimate the offset again in order to check the validity of the numbers. Our preliminary estimations of the offset have an uncertainty of about 1 cm or less. The results indicate that the official numbers of the offset might need to be corrected by about 1.7 cm in the radial direction.

Key words: GPS, antenna phase center offset, SLR

1 Motivation

Data from various tracking techniques are used for accurate satellite orbit restitution, e.g. GPS, SLR, DORIS, PRARE. For example, GPS (SST), laser and DORIS data types are available for T/P. Comparing the orbits computed from GPS data and those from SLR/DORIS data, one has found a significant bias at 2-3 cm, although the orbit accuracy at the radial direction has already reached the same level, see Schwintzer et. al. (1995). By altimeter data processing, this orbit bias will produce an ocean surface height bias. Therefore for high accuracy scientific application this systematic error is harmful. Another example is the GPS satellite. When one computes the SLR residual (to PRN 35 and 36) by fixing IGS orbits, a few cm systematic difference is clearly seen, see Zhu et. al. (1997).

In order to properly use the scientific data from CHAMP, one may ask the question: whether there are also such systematic errors in the CHAMP orbits? It is well known that one of the possible reasons for such kind of systematic errors is the uncertainty in the offset of the GPS antenna with respect to the mass-center of the satellite. How accurate (or uncertain) are the numerical values of the CHAMP GPS-POD antenna's offset? This is the question we are dealing with in this study.

2 Methods

The phase center of the GPS antenna is an electronic one. It is difficult to measure it accurately by a mechanical method alone. Even when one measures it on the ground, due to the difference in space environment, the electronic center may change its position. A meaningful way is to use the tracking data to estimate (the correction of) the antenna phase center.

The first method uses only the on-board GPS data. The offset and CHAMP orbit are determined simultaneously. Due to the high correlation with ambiguities and clock corrections, the flight-direction and normal-direction components of the offset cannot be estimated reasonably. Therefore, only the radial component is estimated. This is a "direct method".

The second method is "indirect". The CHAMP orbits (without solving the offset) are held fixed. The SLR data are used to estimate the correction of the offset of the laser reflector. This is interpreted as the opposite sign of GPS antenna phase center offset correction.

3 Preliminary results

Ten days CHAMP data from 21.05.01 to 30.05.01 were used to get the following preliminary results. The solution strategy was described in Zhu et. al. (2002). GFZ.a denotes a 10 one-day arc solution, in which the GPS orbits and clocks computed by GFZ CHAMP team were fixed. GFZ.b is a 5 times two-day arc solutions with same GPS clocks and orbits. GFZ.c is again a 10 times one-day arc results, but the GPS orbits and clocks were from IGS. At first the offset is not solved. We simply fix the CHAMP orbits for each case and compute the rms and mean residual of the laser data. Table 1 gives the results. Note the sigma of the mean is a 'formal' one, the actual uncertainty may be a few times larger. From this table one can see that: 1) the mean residual (systematic error) is not very large (not as in the case of GPS 35 and 36, which is around 5 cm); 2) no matter what arc length was used, which GPS orbits were fixed, the mean residuals have no substantial differences.

Orbits	rms laser residuals	Mean SLR residual
GFZ.a	6.89	-1.1±0.16
GFZ.b	6.60	-1.2±0.15
GFZ.c	6.55	-1.0±0.15

Table 1. Mean residual of SLR fit to CHAMP orbits (in cm)

The daily (or each 2-d arc) estimates of the antenna offset in the radial direction are listed in Table 2. The mean value is listed in the last row. The fluctuation of the 2-d arc solutions is smaller than that of one-day arc, and the mean value of the former is larger. The sigma given in the last row is computed based on the fluctuations of the individual estimated values. The results computed from the second method are given in table 3. Individual CHAMP orbits are linked together to form a 10-d arc orbit. Then use the laser data in this 10 d interval to estimate the satellite mass-center correction with respect to the laser reflector, and interpret it (with opposite sign) as the GPS antenna phase center offset correction. The sigma is again a formal one. The major difference from the first method is that all three components of the offset can be estimated with reasonable precision. Note that the results in the flight or transversal direction have the biggest discrepancy. The corrections in the radial are similar as those solved from method one. Form these preliminary results one can see that only the correction in the radial is somehow statistically significant. The average value of it is about 1.7 cm with a uncertainty of a few mm.

<table>
<tr><th>DOY of 2001</th><th>GFZ.a</th><th>GFZ.b</th><th>GFZ.c</th></tr>
<tr><td>d141</td><td>1.2</td><td rowspan="2">1.5</td><td>1.0</td></tr>
<tr><td>d142</td><td>1.6</td><td>1.7</td></tr>
<tr><td>d143</td><td>1.7</td><td rowspan="2">2.7</td><td>1.8</td></tr>
<tr><td>d144</td><td>0.3</td><td>-0.3</td></tr>
<tr><td>d145</td><td>0.9</td><td rowspan="2">1.9</td><td>1.5</td></tr>
<tr><td>d146</td><td>2.8</td><td>2.0</td></tr>
<tr><td>d147</td><td>1.4</td><td rowspan="2">1.7</td><td>2.4</td></tr>
<tr><td>d148</td><td>1.7</td><td>2.3</td></tr>
<tr><td>d149</td><td>1.4</td><td rowspan="2">1.8</td><td>1.5</td></tr>
<tr><td>d150</td><td>1.2</td><td>1.3</td></tr>
<tr><td>Mean</td><td>1.4±0.6</td><td>1.9±0.3</td><td>1.5±0.7</td></tr>
</table>

Table 2. The estimated radial component of the offset by method 1 (in cm)

	Radial	Normal	Transversal
GFZ.a	1.7 ±0.3	-1.1 ±0.3	1.9 ±0.4
GFZ.b	2.0 ±0.3	-0.2 ±0.3	-0.8 ±0.4
GFZ.c	1.6 ±0.3	-1.4 ±0.3	-1.8 ±0.4

Table 3. SLR estimated GPS phase center corrections (in cm)

4 Summarizing Conclusion

1. The information concerning the offset of on-board GPS antenna phase center is correct at about 1 to 2 cm. This can be confirmed through the laser tracking (to CHAMP) data.
2. One can use on-board GPS data and/or SLR data to estimate the phase center corrections.
3. Preliminary results have shown that for the radial component both methods are consistent with each other at a few mm level. The mean value from these tests is 1.7 cm. The estimates of the other two components are statistically not significant (from zero).
4. The original values of the GPS POD antenna offset are: along-track = −148.8 cm, cross-track = 0 cm, radial = −39.28 cm. According to our preliminary results, the corrections of them are: along-track = 0cm±1cm, cross-track = 0cm±1cm, radial = 1.7cm±0.5cm.
5. In the future more data should be used to estimate the corrections with an uncertainty of 1 mm or better.

Acknowledgement. This study was carried out under the grant of the German Federal Ministry of Education and Research (BMBF) No. 03F0333A.

References

Schwintzer P, Kang Z, Reigber Ch, Zhu S (1995) GPS satellite-to-satellite tracking for TOPEX/POSEIDON precise orbit determination and gravity field model improvement. J Geodynamics vol 20, No 2 pp 155-166.

Zhu SY, Reigber Ch, Kang Z (1997) Apropos laser tracking to GPS satellite. Journal of Geodesy, vol 71, pp 423-431

Zhu S, Neumayer H, Massmann FH, Shi C, and Reigber Ch (2002) Impact of different data combinations on the CHAMP orbit determination. This issue

Spaceborne GPS for POD and Earth Science

Thomas P. Yunck

Jet Propulsion Laboratory, California Institute of Technology, 4800 Oak Grove Drive, Pasadena, CA 91109-8099, USA

Summary. With the emergence of spaceborne geodetic receivers, GPS is rapidly invading the realm of Earth remote sensing. Among the current science applications are centimeter-level precise orbit determination; gravity model improvement; high resolution ionospheric imaging; and atmospheric limb sounding to recover precise profiles of atmospheric density, pressure, temperature, and water vapor distribution. In the near future we may also see spaceborne GPS used for direct ocean altimetry and scatterometry. Almost uniquely among science sensors, GPS receivers lend themselves to incorporation into highly miniaturized, low cost, and fully autonomous sensorcraft.

Key words: GPS, atmospheric occultation, precise orbit determination, ocean reflections, ionospheric imaging, reduced dynamic tracking

Introduction. In the 1980's, the GPS techniques then being devised for centimeter-level global geodesy were adapted to provide sub-decimeter orbit accuracy for low satellites (Yunck et al., 1990), inaugurating the era of spaceborne GPS geodesy. This has now blossomed into a profusion of Earth science pursuits exploiting spaceborne GPS. Current applications include precise orbit determination (POD), gravity recovery, ionospheric imaging, indirect enhancements to global geodesy, atmospheric occultation, and ocean reflection.

1 Precise Orbit Determination & Space Altimetry

Since the 1960's the needs of Earth remote sensing, and in particular of space-based ocean altimetry, have stimulated advances in precise orbit determination. Examples of ocean altimetric satellites over the years include the US Seasat (1978) and Geosat (1985-1990), the European Remote Sensing satellites (1991-); the US-French TOPEX/ Poseidon (1992-); and the French-US Jason-1 mission (2001-).

Figure 1 illustrates a central problem in the study of ocean circulation by space-based altimetry. Of principal interest are the systematic departures of the ocean surface from the geoid—the surface of constant gravitational potential at sea level—caused by the forces that drive ocean circulation. Modern satellite altimeters can measure the height of the sea surface with a precision of better than 2 cm, averaged over a footprint of several kilometers. But the height of the altimeter can vary by hundreds of meters around its orbit, confounding the interpretation of the

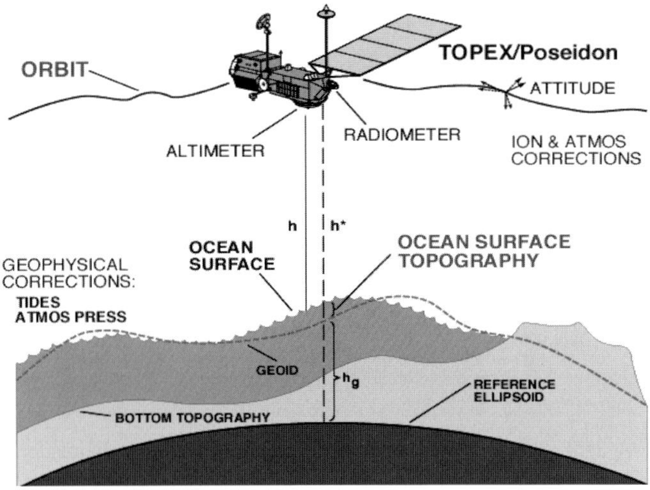

Fig. 1. Elements of ocean altimetry from space

measurements. We must therefore track the satellite's absolute geocentric altitude continuously to a few centimeters as well. In 1991 an international science panel recommended the pursuit of "technology developments that lead to...orbits of altimetric satellites with no more than 1 cm RMS error" (NASA, 1991).

1.1 Classical Dynamic Orbit Determination

Conventional ground-based tracking systems generally provide coverage from only one direction at a time, and often provide no coverage. To supply the missing information for orbit determination, an orbit model must be fit to the tracking data. The most precise strategies employ orbit models derived from detailed models of the forces acting on the satellite, a technique known as dynamic orbit determination. The technique begins with a set of tracking measurements along with models of the forces and satellite mass. Those models yield a model of satellite acceleration over time, from which, by double integration, one can form a nominal or a priori trajectory. To produce the orbit solution we need only estimate the two constants of integration—the initial position and velocity—also known as the epoch state. To improve the fit, we can also adjust various model parameters, such as drag, solar radiation, or gravity coefficients. The solution accuracy depends on how well the force models, fixed or adjusted, describe the real forces acting on the satellite. To reach centimeter accuracy over a fitting interval of days, those models must be surpassingly fine.

1.2 Reduced Dynamic Orbit Estimation

Continuous 3D coverage from GPS allows us to discard detailed force models in a kinematic orbit solution, illustrated in Fig. 2. The dashed line represents the true

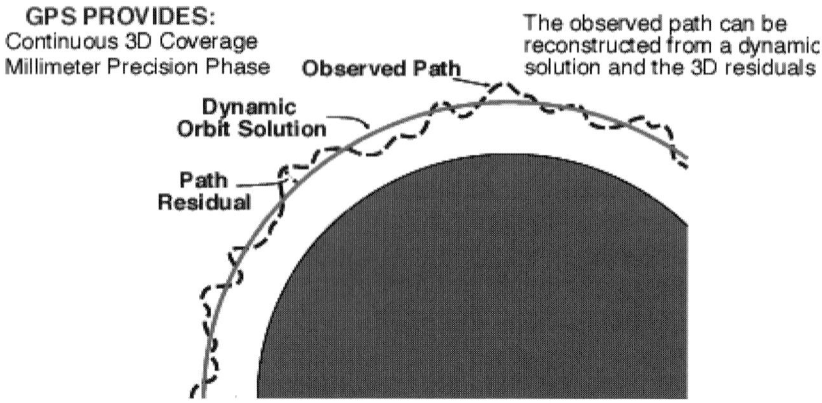

Fig. 2. Elements of kinematic and reduced dynamic orbit estimation

trajectory of a low orbiter subject to variable forces, while the smooth line represents a simple dynamic orbit solution. Continuous data from multiple GPS satellites permit us to construct at every time point a full 3D residual vector between the dynamic solution and a position estimate based only on the measurements. We can thus assemble a complete observed trajectory, geometrically, from a dynamic solution and its post-fit residuals.

In practice we can build the solution by solving for a local 3D correction at each time step. Each correction represents the incremental force needed to explain the observed departure from the model orbit over the last time step (Yunck, 1996). We can benefit by placing constraints on the estimated corrections, consistent with the force model errors. And if the model errors are correlated between time points, we can benefit further by representing that correlation in the stochastic adjustment. More generally, we can model the force corrections as colored (correlated) noise with a variance that reflects the expected force model errors. If the variance and correlation model are chosen carefully, we can fully exploit the dynamic information while minimizing the consequences of dynamic model error. This *reduced dynamic* solution will in general yield the lowest overall error.

1.3 GPS Orbit Improvement and TOPEX/Poseidon

In the early days of GPS POD, a global differential technique was used to maximize accuracy. A small global network of ground receivers at well-known locations continuously tracked the GPS satellites. Data from the orbiter and ground were then combined in a grand solution for GPS and user orbits, some or all ground positions, phase biases, receiver and satellite clocks, and atmospheric delays. All positions are recovered in a consistent frame and GPS orbits and clocks are removed as a direct source of error.

This technique was first demonstrated in the early 1990's on TOPEX/ Poseidon. Initial dynamic orbit solutions gave altitude errors of about 6 cm, limited primarily by the quality of the gravity model, while reduced dynamic solutions were accurate to 3 or 4 cm. Eventually the errors for all solutions, dynamic and reduced dynamic, fell to about 2 cm after the gravity model was tuned with GPS flight data by investigators at the University of Texas (Schutz et al., 1994).

Today, with the high accuracy of the GPS orbit and clock solutions distributed by the International GPS Service (IGS), this grand strategy is not needed. IGS solutions can be adopted without adjustment and precise LEO orbits computed directly from the flight data alone, with little or no performance penalty, vastly speeding POD solutions. Rapid service IGS orbits are now available within hours of real time, permitting comparably fast precise LEO orbit solutions. Moreover, JPL now distributes precise GPS orbits and clocks on an experimental basis over the Internet in real time (few-second delay). We can expect that within a year a low earth satellite will be able to compute onboard a true real time orbit with sub-decimeter accuracy.

2 The Geoid and Gravity Estimation

There is a second element to confound ocean altimetry: the geoid. Undulations of the geoid are on the order of 100 m, or 100 times those of the dynamic topography. Accurate knowledge of the geoid is therefore essential to expose the permanent component of ocean circulation. The more precisely we can observe orbit variations, the more precisely we can reconstruct the field. It is the nature of the gravitational potential that precise knowledge of the gravity signal for one orbit geometry does not imply comparable knowledge for other geometries. We therefore require tracking data from a variety of orbits.

The geopotential is commonly represented as an expansion of solid spherical harmonics (e.g. Lambeck, 1988). The perturbing effects of geopotential irregularities on a satellite's motion attenuate sharply with altitude and with decreasing wavelength. This attenuation goes approximately as exp(–2□z/□), where z is the distance from the source of the anomaly (e.g., height of the satellite) and □ is the wavelength. Accurate recovery of the geopotential through observation of the absolute motion of a satellite is therefore limited to wavelengths down to about twice the satellite altitude, at which point the attenuation is more than a factor of 20. This can be considerably surpassed for selected harmonics in resonance with a particular satellite orbit.

It was recognized early on that the high precision and continuous 3D coverage from GPS would provide a powerful means of improving gravity models through dynamic orbit estimation. As we have seen, it was the use of GPS data from TOPEX/Poseidon in the gravity adjustment process that first pushed the dynamic modeling errors down to the 2 cm level. But the resulting gravity improvements are specialized to that orbit and can be markedly less pronounced for others. Happily, we now have more to choose from. BlackJack GPS receivers on CHAMP, SAC-C, and the twin GRACE spacecraft will permit useful improvement of current gravity models, such as EGM96 (Lemoine et al, 1998), down to wavelengths below 1000 km.

Of course, GRACE (Gravity Recovery And Climate Experiment) itself is designed for high precision gravity mapping by quite different means. As the twin GRACE spacecraft fly a few hundred kilometers apart, a microwave crosslink will tracks changes in their relative distance with an accuracy of microns. This will enable recovery of the gravity field down to wavelengths of a few hundred kilometers. The onboard GPS data, while providing few-centimeter orbit determination, will provide little direct enhancement to the gravity solutions.

3 Ionospheric Imaging

A major new benefit of spaceborne GPS is detailed 3D imaging of the earth's ionosphere. Today's ionospheric maps derived from upward-looking dual-frequency GPS ground data (Mannucci et al, 1998) provide little useful information on the vertical structure. Spaceborne receivers directed towards the earth's limb will provide copious horizontal cuts, tracing out the vertical electron density in great detail.

An array of spaceborne receivers will provide a richness of data enabling entirely new methods of assembling an ionospheric image. Today we depend on dual-frequency retrieval of line-of-sight total electron content (TEC). Recently we have seen Abel inversion methods used to produce electron *density* profiles from spaceborne GPS data, which can be accomplished with single-frequency data (Hajj et al, 1998). With the profusion of observations soon to be acquired from space, these piecewise methods will give way to more rigorous synthetic approaches. These will include 2D and 3D tomographic inversion, leading ultimately to direct assimilation of GPS observables into complex ionospheric models rooted in physics (Hajj et al, 2000). The latter will not only yield the most accurate and detailed images yet, but will enable reliable short term prediction of ionospheric behavior and provide insight into the onset and evolution of space storms.

4 GPS Atmospheric Occultation

The probing of planetary atmospheres by radio occultation dates to the early 1960s when Mariners 3 and 4 passed behind Mars. In this technique a radio signal from a spacecraft moving behind a planet is tracked until blockage. As the signal cuts through the atmosphere, its lengthening path delay, revealed by the observed change in phase delay or Doppler shift, can yield precise profiles of atmospheric density, pressure, temperature, moisture, and geopotential heights. Amplitude variations can expose atmospheric turbulence and wave structure.

While radio occultation has probed many planets and moons, it has as not yet found operational application to Earth, primarily because, to be of use, such measurements must be continuous, comprehensive, synoptic. We need many transmitters and receivers aloft at once, densely sampling the global atmosphere every few hours. Until the arrival of GPS and spaceborne receivers, the evident cost of such an enterprise made it impractical within Earth science programs.

In the late 1980s, a group at JPL proposed observing GPS signals from space to make atmospheric soundings by radio occultation, as shown in Fig. 3 (Yunck and Melbourne, 1990). Briefly, the observed Doppler shift in the GPS signal induced by atmospheric bending permits accurate estimation of the atmospheric refractive index. From that one can retrieve, in sequence, profiles of the atmospheric density, pressure, and temperature (or, in the lower troposphere, water vapor) with high accuracy (<1 Kelvin in temperature) and a vertical resolution of a few hundred meters. Figure 4 shows the predicted accuracy of atmospheric temperature profiles as a function of altitude, based on extensive simulation studies performed at JPL. In the lower troposphere, the uncertainty in water vapor content, particularly in the tropics, leads to a large error in the recovered temperature. In that region, since it is water vapor that is of greater consequence in weather modeling, it becomes advantageous to adopt nominal temperature lapse rates from models and instead estimate water vapor profiles.

A single satellite can recover more than 500 profiles each day, distributed around the globe; a small constellation could recover many thousands of profiles, and have a profound impact on both long-term climate studies and weather prediction. In addition, such an array would enable high-resolution 3D tomographic imaging of the ionosphere and would serve many geodetic uses as well.

In the early 1990s a group at the University Corporation for Atmospheric Research in Boulder, CO, obtained sponsorship from the U.S. National Science Foundation for the first occultation experiment called GPS/MET, which flew in

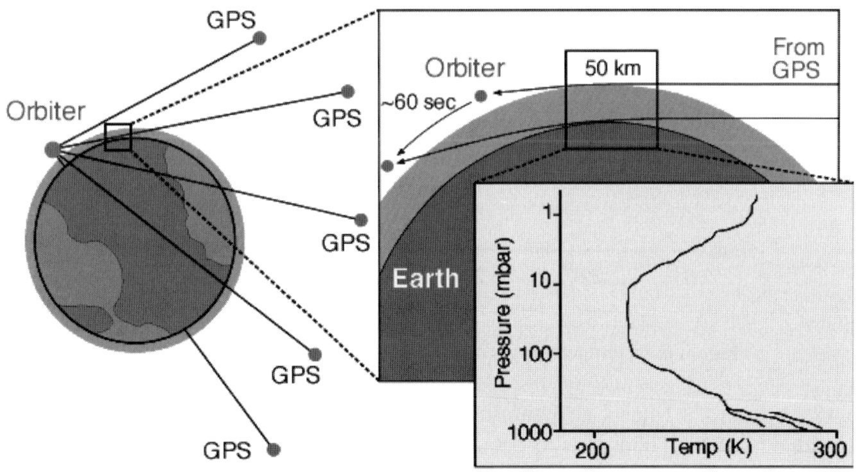

Fig. 3. Atmospheric temperature profiling by GPS occultation

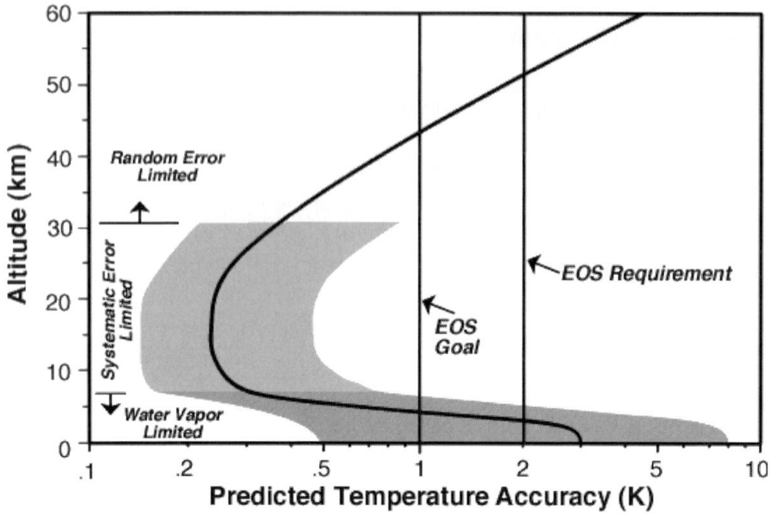

Fig. 4. Approximate occultation temperature accuracy vs altitude.

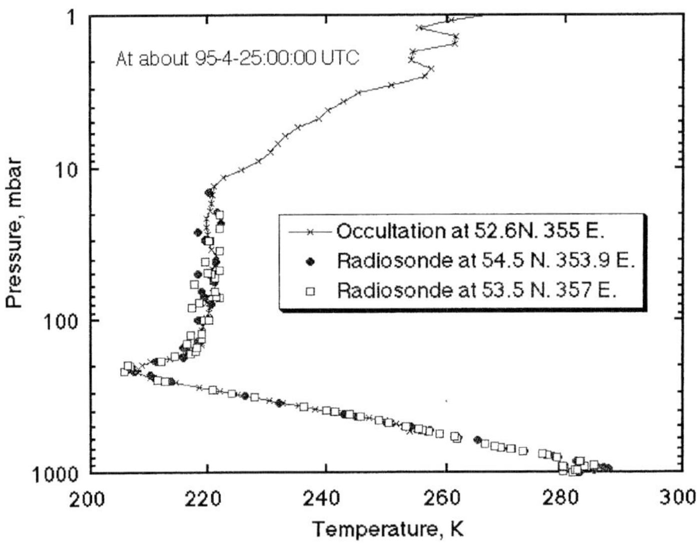

Fig. 5. GPS temperature profile compared with two radiosondes.

1995 on NASA's MicroLab I satellite carrying a modified ground receiver, first developed at JPL. Thousands of occultations were acquired and analyzed (Kursinski et al., 1997; Rocken et al., 1997). Figure 5 shows a typical temperature profile, along with nearby radiosonde measurements for comparison.

5 Bistatic GPS Altimetry

Spaceborne GPS receivers may one day perform altimetry directly by sensing GPS signals reflected off the oceans and ice sheets, a technique known as bistatic altimetry. Reflections received from low orbit are weaker than direct signals by typically 40-50 dB. A spaceborne receiver with a high gain antenna can acquire reflected pseudorange data with a precision of tens of centimeters after a few seconds of averaging. (Ocean surface roughness prevents the use of phase data.) While this is far worse than the 2 cm of conventional altimeters, a single orbiter can observe up to a dozen reflections at once. A small constellation can cover the globe in a few hours, and in a few days can collect enough data to average the height error to a few centimeters with a resolution of a few hundred kilometers, or several times better than TOPEX/Poseidon over ten days at the equator. This would make possible for the first time the detection and continuous observation of eddy-scale circulation.

Recently at JPL, analysis was performed of GPS reflections observed fortuitously by the Shuttle Radar Lab-2 in 1995. Figure 6 shows what may be the founding event of space-based bistatic GPS altimetry: the first carefully observed ocean reflection received from space (LaBrecque et al., 1998; Lowe et al, 2000). The sharply rising part of the graph provides a measure of the range; the long tail is the result of scattering about the specular point and provides information on surface roughness and other quantities.

To obtain ocean height from reflected pseudorange we must accurately model the positions of the transmitter and receiver, and the effective reflection point. The first two can be achieved to a few centimeters by techniques discussed above.

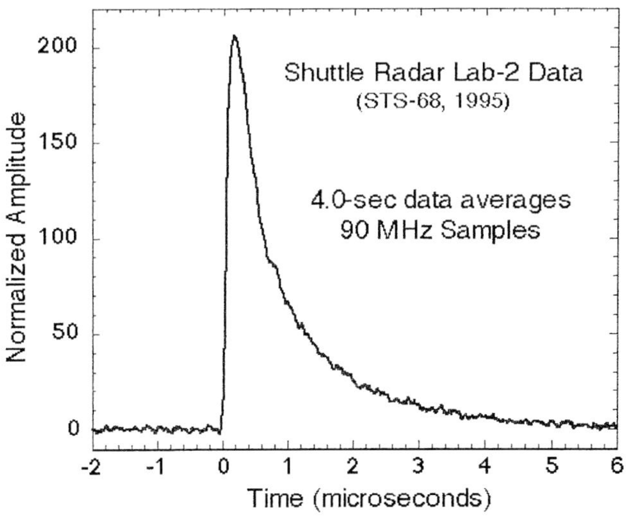

Fig. 6. Ocean-reflected GPS signal collected from space shuttle.

The reflection point can be estimated by an iterative procedure which models the nominal ocean surface through precise knowledge of the geoid and tides. Successive adjustments to the nominal reflection point are made until incidence and reflection angles agree to within a small tolerance (Wu, 1997). The chief error source, besides signal-to-noise, is the delay from tropospheric water vapor. One can use models derived from occultations and other data to calibrate this delay.

7 The Future of Spaceborne GPS Science

Soon we will have a half dozen GPS science receivers in space – CHAMP, SAC-C, GRACE (2), Jason-1, and ICESat – with others on the way. The first operational constellation, COSMIC, consisting of six small satellites for GPS occultation and geodesy, is planned for launch in 2005. While these missions will do much to advance spaceborne GPS science, they will not in themselves establish an ongoing presence of large numbers of GPS sensors in earth orbit. It is the hope of the growing GPS Earth science community that a permanent constellation of several dozen microsats will be sponsored either by government agencies or by commercial interests (eyeing a worldwide market in GPS weather products) in the near future.

There has been much speculation on how the future for this enterprise will unfold. In one view, the COSMIC model will be taken to its logical extreme: flight systems will be further miniaturized and we will see dedicated constellations of dozens or even hundreds of tiny free-flyers, each with a mass of a few kilograms and costing just a few hundred thousand dollars each to produce in volume.

There is much to be said for this view. Consider that the next generation of occultation receiver—palm-sized, low mass, consuming a few watts—will provide:

- Real time onboard position, velocity, and timing
- Real time onboard attitude and attitude rate determination
- A gigabyte of onboard data storage
- All onboard spacecraft computation and control
- All uplink extraction and command interpretation
- All tracking data for centimeter-level orbit determination
- Acquisition of occultations and ocean reflections

Equipped with a tiny cell-phone chip, each orbiter will simply dial a local number and send its data through one of the orbiting telecom systems directly to a computer which, within minutes, will generate and distribute finished science products. No special ground systems are needed either for tracking or data downlink, and operations crews would be non-existent. Very little is needed to convert these components into a finished spacecraft: enclosure, power system, magnetic torque rods for attitude control, and (optionally) micropropulsion for crude orbit maintenance. These sensorcraft could be launched in large numbers at little cost as secondary payloads, into orbits designed to decay within five to seven years.

Another view holds that these same virtues will make the GPS sensor irresistible as an add-on to constellations already planned. The miniature receiver/ processor can assume the functions of several discrete spacecraft subsystems, lowering mass, power, and cost in a cascade of economies, while providing a valuable new science dimension and a potentially lucrative data stream. The cost of adding GPS science to an otherwise unrelated constellation would be small.

Whichever direction the future takes there can be little doubt that it will be a rewarding one for those engaged in understanding the behavior of our fragile and complex Earth system.

Acknowledgement. The work described here was carried out by the Jet Propulsion Laboratory, California Institute of Technology, under contract with the National Aeronautics and Space Administration.

References

Hajj GA and Romans LJ (1998) Ionospheric electron density profiles obtained with the Global Positioning System: Results from the GPS/MET experiment, *Radio Sci*, 33:175-190.

Hajj GA, Lee LC, Pi X, Romans LJ, Schreiner WS, Straus PR and Wang C (2000) COSMIC Ionospheric Sensing and Space Weather, to appear in *Terrestrial, Atmospheric and Oceanic Sciences*, March.

Kursinski ER, Hajj GA, Schofield JT, Linfield RP, Hardy KR (1997) Observing Earth's atmosphere with radio occultation measurements using the Global Positioning System, *J Geophys Res*, 102 (D19), 23,429-23,465, Oct 20.

LaBrecque J,Lowe S,Young L, Caro E, Wu S, Romans LJ (1998) Recent Advances in the study of GPS Earth Surface Reflections from Orbiting Receivers, UNAVCO Community Meeting, Dec.

Lambeck K (1988), *Geophysical Geodesy*, Clarendon Press, Oxford.

Lemoine F, Pavlis N, Kenyon S, Rapp R, Pavlis E, and Chao B (1998) New high-resolution model developed for Earth's gravitational field, *EOS, Trans Am Geophys Union 79,* 113-118.

Lowe ST, LaBrecque JL, Zuffada C, Romans LJ, Young LE and Hajj GA (2000) First spaceborne observation of an earth-reflected GPS signal, submitted to *J Geophys Res* Jan.

Mannucci AJ, Wilson, Yuan, Ho, Lindqwister and Runge (1998) A global mapping technique for GPS-derived ionospheric total electron content measurements, *Radio Sci* 33(3), 565-583, 1998.

NASA (1991), *Solid Earth Science in the 1990s*; Volume 2—Panel Reports: NASA Tech Mem 4256, v 2, Washington, DC, 296 pp.

Rocken C, Anthes R, Exner M, Hunt D, Sokolovskiy S, Ware R, Gorbunov M, Schreiner W, Feng D, Herman B, Kuo YH and Zou X (1997) Analysis and validation of GPS/MET data in the neutral atmosphere. *J Geophys Res* 102 (D25), 29,849-29,866.

Schutz BE, Tapley BD, Abusali PAM and Rim HJ (1994) Dynamic orbit determination using GPS measurements from Topex/Poseidon, *Geophys Res Lett*, *21* (19), 2179-21-82.

Wu SC, Meehan TK and Young LE (1997) The potential use of GPS signals as ocean altimetry observables, ION National Technical Meeting, Santa Monica, CA, Jan.

Yunck TP and Melbourne WG (1990) Geoscience from GPS tracking by Earth satellites, IAG General Meeting, Edinburg, 4-11 Aug 1989, reprinted in *Global Positioning System: An Overview*, Y Bock & N Leppard (eds), Springer-Verlag, 1990.

Yunck TP, Wu SC, Wu JT and Thornton CL (1990) Precise tracking of remote sensing satellites with the global positioning system, *IEEE Trans Geosci & Remote Sensing, 28*, pp 108-116.

Yunck TP (1996) Orbit Determination, in *Global Positioning System—Theory and Applications*, B Parkinson and J Spilker (eds), AIAA, Washington, DC, pp 559-592.

The CHAMP Orbit Comparison Campaign

Henno Boomkamp

European Space Operations Centre, Robert-Bosch Strasse 5, D-64293 Darmstadt

Summary. As an initiative within the IGS LEO Pilot Project a CHAMP Orbit Comparison Campaign was started in October 2001. The main objectives of this campaign were to assess the current precision levels of CHAMP POD, and to assist the IGS LEO centers in developing their POD methods, by providing external reference orbits. This paper summarizes the status of the campaign at the time of the CHAMP Science Meeting. The most recent results are maintained at the campaign web page http://nng.esoc.esa.de/gps/campaign.html.

Key words: Orbit differences, SLR analysis, Error estimation

1 Campaign description

For taking part in the Campaign, associate analysis centers were asked to provide CHAMP orbit ephemerides for the period from May 20 to May 30, 2001. By the time of the CHAMP Science Meeting a total of twelve centers had contributed orbit solutions, which illustrates the wide interest in LEO POD based on GPS. Table 1 summarizes the contributed orbit solutions. As can be seen from the Table there was no clear preference for dynamic approaches or kinematic solutions, but with the arrival of new high degree gravity field models involving CHAMP data it can be expected that dynamic solutions will gradually gain preference.

At ESOC two complementary analysis methods are applied to all contributions, namely pair-wise orbit comparisons between all available solutions, and computation of SLR residuals to each orbit. The comparisons provide continuous orbit quality information throughout the campaign period, but only show the relative precision of a compared solution pair. On the other hand, the SLR residuals provide an indication of the absolute orbit error, but due to the rather sparse tracking it only covers a small percentage of the orbit arcs.

2 Orbit comparison results

Table 2 provides the pair-wise RMS of orbit differences in its bottom triangle. These RMS values were computed for the norm of the difference vectors in 2000 sample points per day, giving 22000 samples over the 11 day campaign period. Figure 1 provides an impression of the distribution of the orbit differences. It shows all 22000 sample points per orbit pair, sorted by size of the difference vector rather than shown in chronological order. The top triangle of Table 2 provides the order of the curves at the median point (11000) in Figure 1, where the curve for the pair marked '1' is the bottom curve and the curve '66' the top curve.

The RMS of differences between the orbit solution and the true CHAMP orbit can be seen as a measure for the orbit error, denoted as RMS_A and RMS_B for two solutions A and B respectively. If the two solutions are independent, their two orbit error signals are fully uncorrelated. The pair-wise value RMS_{AB} which represents the difference between the two solutions A and B then follows from:

$$RMS_{AB} = \sqrt{RMS_A^2 + RMS_B^2} \qquad (1)$$

The pair-wise orbit error (1) should be the value that is observed in the form of the orbit difference RMS from Table 2. It is possible that the same systematic error is present in two or more orbit solutions, for instance a bias in the GPS data. Similarly, two orbit solutions can be insufficiently uncorrelated if two centers use the same POD software. In both cases the pair-wise differences from Table 2 will be smaller than what they would be according to (1). The observed RMS of orbit differences provides a lower bound to the actual pair-wise orbit error.

Center	Kinematic	Reduced Dynamics	Dynamic	Mixed method	GPS orbits / clocks	Data differencing	Pseudo range	Phase	Accelerometer data	Comments
CSR			X		IGS	2	X	X	X	Based on TEG-4 gravity field
TUM		X			CODE	0		X		Boot-strap solution may follow
GFZ			X		GFZ	0	X	X		Based on EIGEN gravity field
GRGS		X			IGS	0	X	X		Clocks set to a priori GRGS solution
NCL				X	IGS	0	X	X		Same POD software as JPL
DEOS			X		IGS	3		X		Based on TEG-4 gravity field
JPL				X	IGS	0	X	X		2-step method, dynamic + stochastic
AIUB		X			CODE	0	X	X		Using kinematic orbit as observations
CNES		X			IGS	0	X	X		Ambiguities adjusted using code
ESA	X				ESOC	0	X	X		Time-differenced phase observables
UNB	X				IGS	0	X	X		Time-smoothed phase observables
UCAR	X			X	IGS	0	X	X		Time-differenced phase observables

Table 1. Solutions contributed to the CHAMP Orbit Comparison Campaign

	CSR	TUM	GFZ	GRGS	NCL	DEOS	JPL	AIUB	CNES	ESA	UNB	UCAR
CSR	*	2	5	7	1	8	14	25	28	37	50	58
TUM	11.82	*	10	3	4	12	16	27	22	33	47	57
GFZ	15.24	16.36	*	15	9	17	20	34	30	41	51	62
GRGS	15.67	13.56	18.64	*	6	18	19	45	23	39	46	59
NCL	9.14	12.94	15.77	15.21	*	11	13	24	26	36	48	61
DEOS	13.64	15.71	18.12	18.40	15.57	*	21	38	35	42	52	56
JPL	18.23	18.39	20.33	20.53	17.77	21.20	*	31	29	40	49	60
AIUB	24.53	25.48	27.54	44.57	24.80	29.35	27.61	*	32	44	55	65
CNES	27.18	26.41	28.82	26.85	26.92	29.73	29.35	29.10	*	43	53	64
ESA	30.66	30.33	32.26	31.80	30.71	33.26	31.34	35.91	36.53	*	54	63
UNB	61.60	60.52	61.33	60.54	60.93	61.48	61.54	79.52	63.72	63.91	*	66
UCAR	152.36	151.6	152.13	151.97	151.97	151.24	152.8	154.5	153.08	152.63	162.44	*

Table 2. Bottom triangle : RMS of orbit different vectors per solution pair (cm). Top triangle : index of the orbit comparison curves of Figure 1

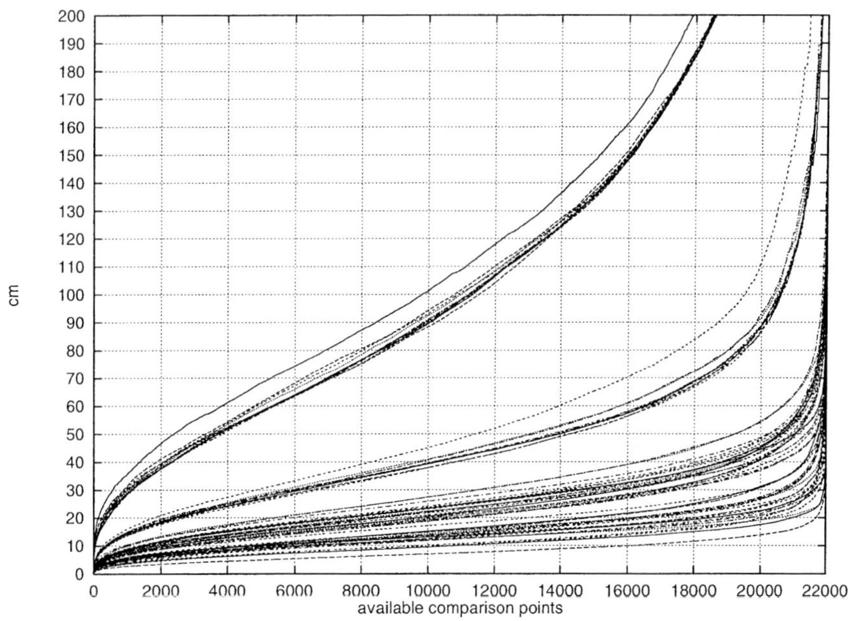

Fig. 1. Norms of the difference vectors sorted by size. The order of the curves at the median point 11000 is indicated in the top triangle of Table 2 ('1' corresponds to bottom curve)

3 SLR analysis

SLR data can provide an upper bound to the orbit error, but the SLR observations for CHAMP are so sparse that each pass can only form a momentary snapshot of the actual orbit error. Nonetheless, more than 1700 usable SLR observations were identified for the 11-day campaign period and were used to compute SLR residuals for each orbit solution separately. The SLR processing at ESOC consisted of the following:

- The satellite positions at the times of the SLR observations are obtained from the input orbits, without estimating any parameters
- The station coordinates that are used are the ITRF97 set
- No station-dependent weights are applied but some data was rejected, for any of the following reasons:
  - Two stations (1873 and 7237) were excluded due to apparent station biases
  - Individual bad measurements were rejected based on a 1.5 sigma window rejection, using the orbit that showed the lowest overall RMS, which was CSR (the window level corresponds to 21 cm 2-way range).

The set of accepted SLR observations consisted of 1710 points. The main characteristics of the SLR processing are given in Table 3. The comparison results and SLR results were obtained by processing all orbits in exactly the same way, which makes it reasonable to try and correlate the RMS of SLR residuals with the RMS of orbit error in some way. However, the latter is only available in pair-wise form from equation (1) while the SLR residuals are available in the form of separate values per orbit. If once again it is assumed that the orbits A and B in a comparison pair have independent error signals with respect to the true CHAMP orbit, the

Center	Nr of points	Mean value	Sigma	RMS
CSR	1710	0.02	4.44	4.44
TUM	1710	2.55	4.84	5.47
GFZ	1710	0.82	7.17	7.21
GRGS	1710	4.09	6.20	7.43
NCL	1710	-0.61	7.41	7.43
DEOS	1710	-1.93	8.78	8.99
JPL	1710	2.39	10.08	10.36
AIUB	1710	2.69	13.39	13.65
CNES	1710	0.47	13.67	13.67
ESA	1710	2.75	16.86	17.07
UNB	1384	0.75	27.58	27.58
UCAR	1710	6.85	72.15	72.45

Table 3. Statistics for 1-way range SLR residuals (cm).

	CSR	TUM	GFZ	GRGS	NCL	DEOS	JPL	AIUB	CNES	ESA	UNB	UCAR
CSR	*	1.678	1.800	1.810	*1.056*	*1.360*	1.617	1.709	1.891	1.738	2.205	2.099
TUM	7.05	*	1.808	1.470	1.403	1.493	1.570	1.733	1.794	1.692	2.152	2.087
GFZ	8.47	9.05	*	1.800	1.523	1.572	1.611	1.784	1.865	1.741	2.151	2.089
GRGS	8.66	9.23	10.35	*	1.448	1.578	1.610	*2.868*	1.726	1.708	2.120	2.087
NCL	8.66	9.23	10.35	10.51	*	*1.335*	*1.394*	1.596	1.730	1.650	2.133	2.087
DEOS	10.03	10.52	11.52	11.66	11.66	*	1.546	1.796	1.817	1.724	2.119	2.072
JPL	11.27	11.72	12.62	12.75	12.75	13.72	*	1.611	1.711	1.570	2.089	2.088
AIUB	14.35	14.71	15.44	15.54	15.54	16.34	17.14	*	1.506	1.643	2.584	2.096
CNES	14.37	14.72	15.45	15.56	15.56	16.36	17.15	19.32	*	1.607	2.070	2.076
ESA	17.64	17.93	18.53	18.62	18.62	19.29	19.97	21.86	21.87	*	1.970	2.051
UNB	27.94	28.12	28.51	28.56	28.56	29.01	29.46	30.77	30.78	32.44	*	2.095
UCAR	72.59	72.66	72.81	72.83	72.83	73.01	73.19	73.72	73.73	74.43	77.52	*

Table 4. Bottom triangle: pair-wise RMS of SLR residuals (cm) from relation (1). Top triangle: ratio D between RMS of orbit differences and pair-wise RMS of SLR residuals.

relation (1) can also be used to convert the single orbit RMS of SLR residuals into a pair-wise value. These pair-wise SLR residuals are presented in Table 4, as well as the ratio D between the pair-wise orbit error signal from Table 2 and the pair-wise RMS of SLR residuals.

Careful inspection of the ratios D suggests some dependencies between solution pairs, namely in cases where for a single center the comparison with one of the other 11 solutions shows a striking difference from comparisons with the other 10. For instance, the pairs CSR+DEOS are known to be both dynamic solutions based on TEG4, and the results for pair NCL+JPL suggest correlations induced by use of the same POD software. For the pair AIUB+GRGS an incidental large orbit difference seems to influence the comparison results.

Excluding 5 outliers (indicated in Table 4 with *italics*) , as well as the solutions of UCAR and UNB, which show very large SLR residuals, the mean value for D over the 40 remaining comparison pairs is 1.52 at a standard deviation of 0.18.

Elementary statistical analysis [1] shows that the mean value of the ratios D for pair-wise results will be identical to the ratio d for single orbit results, which is the ratio between the single-solution orbit error and the single-solution RMS of SLR residuals. The same applies to the standard deviation of pair-wise and single-orbit results. This standard deviation is small enough to consider the value of d to be a constant, which correlates the SLR residuals to the absolute orbit error with the following empirical relation:

$$RMS(orbit) = (1.52 \pm 0.18) RMS(slr) \tag{2}$$

Using this result, the SLR residuals from Table 3 lead to the absolute orbit error estimates for each solution in the CHAMP Orbit Campaign as presented in Table 5.

Center	RMS 1-way SLR	Estimated orbit error
CSR	4.44	6.75
TUM	5.47	8.31
GFZ	7.21	10.96
GRGS	7.43	11.29
NCL	7.43	11.29
DEOS	8.99	13.66
JPL	10.36	15.75
AIUB	13.65	20.75
CNES	13.67	20.78
ESA	17.07	25.95
UNB	27.58	41.92
UCAR	72.45	110.12

Table 5. Estimates of absolute orbit error for the contributed solutions (cm) based on the empirical relation (2).

4 Conclusions

The CHAMP Orbit Comparison Campaign has been successful in providing an assessment of the status of CHAMP POD, and has also demonstrated the wide interest that exists in this topic.

Combining orbit comparison information with SLR tracking data has lead to estimates of the absolute orbit precision for each contributed solution.

In the few months since the release of the CHAMP flight receiver data, the orbit precision levels for CHAMP have reached levels close to 10 cm RMS. The POD experience gained with CHAMP, in combination with new CHAMP-based gravity field models will form a good starting point for the GRACE mission.

Acknowledgements

The CHAMP Orbit campaign would not have been possible without the contributions from the 12 participating centers.

References

[1] The analysis can be found on the CHAMP campaign webpages at http://nng.esoc.esa.de/gps/campaign.html

CHAMP Orbit Determination with GPS Phase-Connected Precise Point Positioning

Sunil B. Bisnath and Richard B. Langley

Geodetic Research Laboratory, Department of Geodesy and Geomatics; Engineering, University of New Brunswick, Fredericton, N.B., Canada

Summary. A completely geometric approach for precise orbit determination (POD) of low earth orbiters (LEOs) is presented, which does not rely on dynamic models, but only data from a GPS receiver onboard a LEO and the International GPS Service (IGS) GPS orbit and clock products. Initial processing of CHAMP receiver data indicated measurement anomalies requiring additional pre-processing. The exising of outliers becomes a significant concern in the processing, given that this results in significant gaps in datasets. Intermediate processing results indicate that orbit comparison precision approaching 30 cm r.m.s. in each position component is attainable. However, these results are severely impacted by the near ubiquitous data gaps in the preprocessed measurements. However, initial analysis of recent data indicates quality datasets capable of providing near-decimetre-level precision.

Key words: orbit determination, GPS, point positioning

1 Introduction

The University of New Brunswick (UNB) has been performing GPS-based LEO orbit determination research since 1998. The investigations have centred on GPS-only solution strategies – referred to as *geometric* orbit determination (Bisnath and Langley 1999). The goals of the research were to answer the question: Can GPS alone be used to determine the precise orbit of a LEO, and if so, how well? With the removal of Selective Availability (SA), the objective evolved into POD with a single spaceborne GPS receiver. The CHAMP science satellite (GFZ 2001) and the geodetic-grade BlackJack receiver onboard represent an excellent opportunity to test these algorithms.

2 Methodology

To make complete, efficient use of the available GPS data, the UNB processing strategy simultaneously utilizes the *pseudorange* measurements to compute LEO *position* and the *carrier-phase* measurements to compute LEO *position change* in a kinematic, sequential, least-squares filter/smoother – a state space approach. The primary inputs are: dual-frequency pseudorange and carrier-phase measurements from the LEO receiver; precise GPS constellation ephemerides; and precise constellation satellite clock offsets from GPS Time. The ephemerides and clock data are *readily available* from the IGS. The GPS-determined positions can be interpolated to produce a LEO state throughout the orbit. This processing flow is illustrated in Figure 1.

The parameterisation of the filter models is given in Bisnath and Langley (2001). A flowchart of the least-squares filter is shown in Figure 2. As can be seen, this filter is a variant of the code-smoothed-phase filter. Therefore a minimum number of continuous phase measurements are required for filter operation without re-initialisation.

The technique then encompasses two philosophies: neither a reference receiver nor a reference receiver network is explicitly necessary, since that information is indirectly provided by the IGS data products – hence single receiver positioning; and no assumed dynamic models are required, since the time-differenced, carrier-phase observations precisely measure motion – hence platform independence.

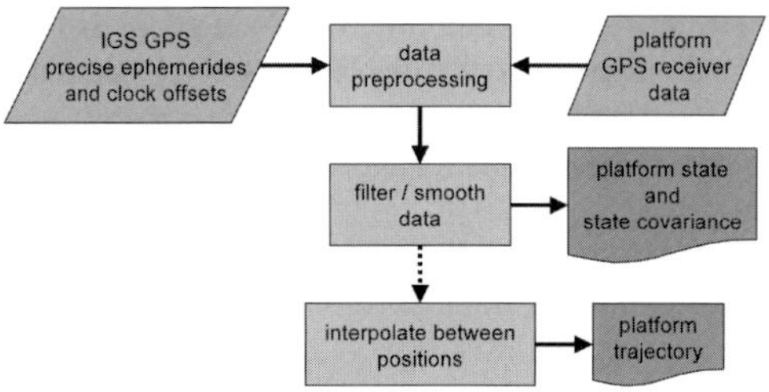

Fig. 1. Processing flow of the geometric orbit determination strategy

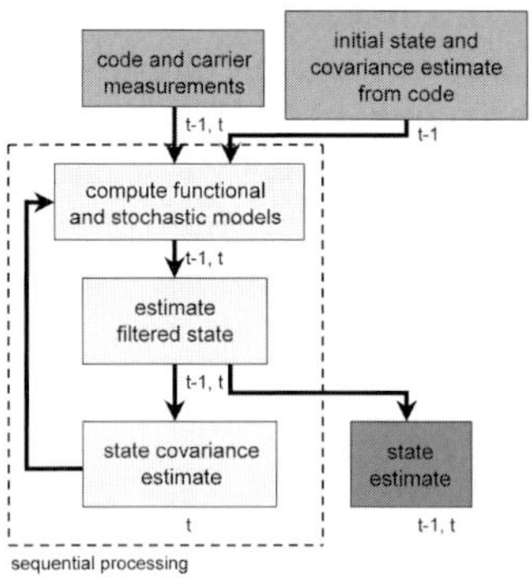

Fig. 2. Combination of pseudorange and carrier-phase observations in the kinematic, sequential least-squares filter

3 Preprocessing

It was found that the geometric CHAMP orbit solutions are very sensitive to data editing performed in the preprocessor. Data editing consists of applying a signal-to-noise filter and a rate-of-change of widelane-phase minus narrowlane-pseudorange linear combination filter. The former removes low strength signals at the measurement input stage, while the latter eliminates measurements that deviate from the norm before the initial estimation process. This preprocessing strategy therefore cleans the data in the measurement domain, without need for post-estimation residual analysis or need for a reference CHAMP orbit to constrain the definition of typical measurement behaviour.

Figure 3 illustrates an example of data editing. The GPS satellite is tracked to almost -15°, causing the signal-to-noise values to approach zero BlackJack units, and the rates-of-change of the linear combinations to deviate significantly from zero. The receiver satellite-tracking algorithm is responsible for this situation, producing large numbers of low elevation angle satellites tracked in the CHAMP anti-velocity direction. Measurements accumulated from these weak signals are deleted, resulting in intermittent poor data availability as can be seen in Figure 4.

4 Results

CHAMP data provided by the CHAMP Information System and Data Center (ISDC 2001) was processed for the period of the IGS LEO Orbit Comparison

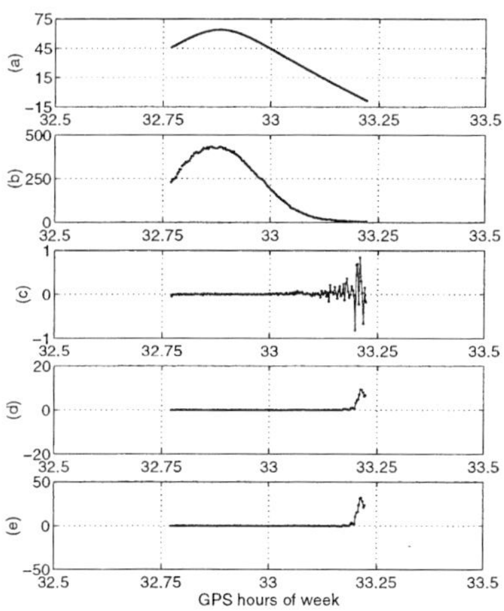

Fig. 3. Poor behaviour of measurements at low signal-to-noise values. **a**) Elevation angle (degrees); **b**) L1 phase signal-to-noise (GPS receiver units); **c**) L1 P-code - L2 P-code rate-of-change (m/s); **d**) L1 phase - L2 phase rate-of-change (m/s); **e**) widelane phase - narrow-lane pseudorange rate-of-change (m/s). GPS PRN08, day of year 148, 2001

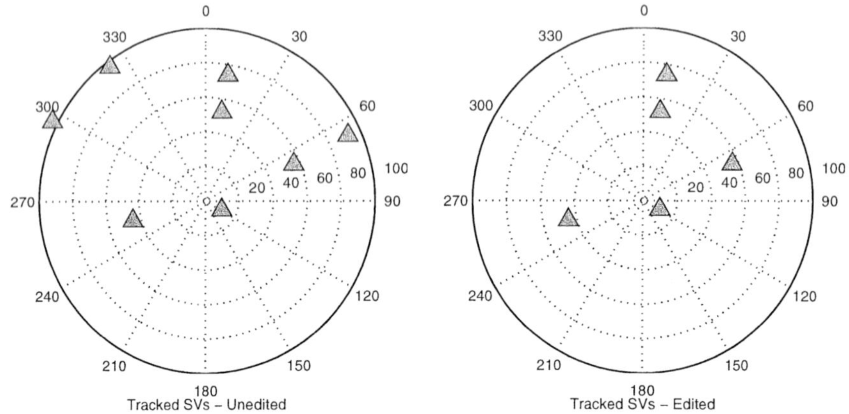

Fig. 4. Example of GPS satellite sky distribution before and after data editing. Day of year 148, 2001, 30.2389 hours. Azimuth of 180° ≈ velocity direction. Before editing: # SVs=8 and PDOP=1.7. After editing: # SVs=5 and PDOP=4.0

Campaign (ESOC 2001): day of year 140 to 150, 2001. The resulting UNB solutions were compared against Jet Propulsion Laboratory (JPL) solutions using a comparison interpolator developed by ESOC (ESOC 2001). Each 24 hour CHAMP data arc required less than 30 minutes for processing with UNB unoptimised computer code on an 85 MHz Sun server.

Figure 5 shows the position component differences for day 143 – a typical day. The radial, along-track, cross-track, and norm r.m.s. (in cm) are 36, 25, 24, and 50, respectively. The forward filter post-fit residual r.m.s. for the ionosphere-free pseudorange is 105 cm and for the time-difference ionosphere-free phase is 9 cm. The spikes and much of the noise in the positions are due to data gaps and remaining poor-quality measurements. This is also evident from the relatively large post-fit residuals.

Figure 6 illustrates the daily differences between UNB and JPL 24 hour arcs. The position difference r.m.s. ranges from: 30 cm to 57 cm in the radial direction; 21 cm to 39 cm along-track; 22 cm to 42 cm cross-track; and 43 cm to 76 cm for the norm. Notice the larger error difference in the radial component. The CHAMP radial component represents the nominal "up" component in the topocentric sense, and of course suffers from the GPS geometry predicament of having no transmitters below the receiver. The poor overall repeatability is again caused by the lack of continuous, quality GPS measurements. An indication of near-optimum solution comparison can be observed during periods of good continuous data, *e.g.*, between hours 2 and 3 of day of year 150. For this short arc, the r.m.s. differences between UNB and JPL are as small as: 13 cm, 10 cm, 7 cm; in the radial, along-track, and cross-track directions, respectively, and 18 cm in the norm.

At the end of 2001, the CHAMP tracking algorithm was altered to prevent tracking of GPS satellites below the CHAMP antenna horizon. Such a situation

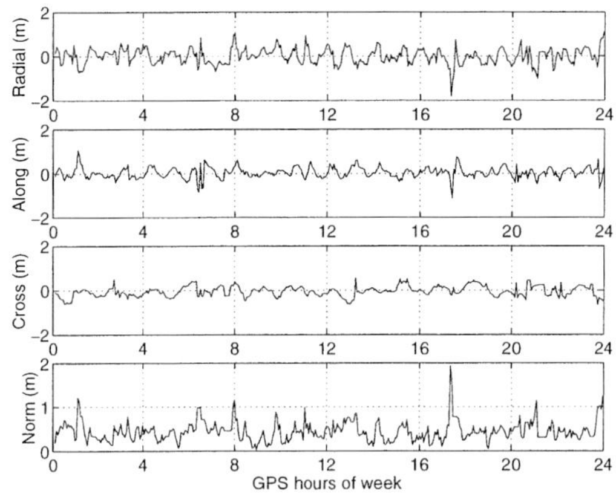

Fig. 5. Position component differences between UNB and JPL for day of year 143, 2001. r.m.s. (cm): radial=36; along-track=25; cross-track=24; norm=50. Filter post-fit residuals: ionosphere-free pseudorange=105 cm; time-difference ionosphere-free phase=9 cm

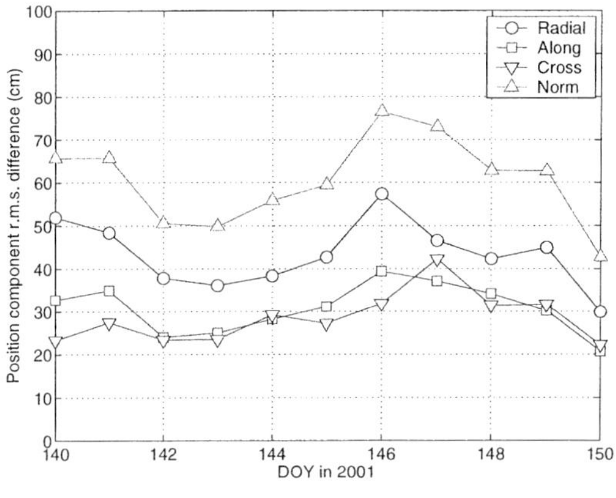

Fig. 6. Daily position component r.m.s. differences between UNB and JPL. Mean r.m.s. (cm): radial=43; along-track=30; cross-track=28; norm=60

would maximise the potential of the geometric solution strategy. From early analysis, this has resulted in a significant increase in the number of quality measurements, and tremendous reductions in position component r.m.s. differences with respect to dynamically determined orbits, with differences approaching the decimetre-level.

5 Conclusions and Further Research

An *a posteriori* LEO orbit determination strategy based solely on GPS measurements has been devised, which is simple and efficient. The processing strategy requires significant editing of the CHAMP GPS receiver measurements due to the receiver tracking algorithm. Intermediate results indicate 24 hour arc r.m.s. differences with respect to dynamical orbits of approximately 40 cm in the radial component and 30 cm in each of the along-track and cross-track components. Short arc r.m.s. differences as small as 10 cm in each component are possible. CHAMP single-receiver, GPS-only POD is attainable, but *geometrically strong*, *continuous* GPS measurements are required. Since only tens of minutes of processing time are required to process one day's worth of data, a real-time capability is possible.

A number of additional functional and stochastic modelling improvements will be made to improve the presented orbit solutions, and initial processing of recent CHAMP receiver data indicates quality datasets capable of providing near-decimetre-level precision with the UNB processing strategy.

Acknowledgements

This research was conducted with financial support from the Natural Sciences and Engineering Research Council of Canada and the GEOIDE Network of Centres of Excellence (Project ENV#14). The authors would like to thank the IGS for having produced and made available the precise GPS satellite ephemerides and clock offset files.

References

Bisnath SB, Langley RB (1999) Precise *A Posteriori* Geometric Tracking of Low Earth Orbiters with GPS. Canadian Aeronautics and Space J 45: 245-252

Bisnath SB, Langley RB (2001) Precise Orbit Determination of Low Earth Orbiters with GPS Point Positioning. In: Proceedings of the Institute of Navigation 2001 National Technical Meeting, The Institute of Navigation, Alexandria, pp 725-733

ESOC (2001) IGS LEO Orbit Comparison Campaign 2001. <http://nng.esoc.esa.de/gps /campaign.html>. Accessed October 2001

GFZ (2001) CHAMP Home Page. <http://op.gfz-potsdam.de/champ/index_CHAMP.html>. Accessed January 2001

ISDC (2001) CHAMP Information System and Data Center. <http://isdc.gfz-potsdam.de /champ/>. Accessed October 2001

Kinematic and Dynamic Determination of Trajectories for Low Earth Satellites Using GPS

Heike Bock, Urs Hugentobler, Gerhard Beutler

Astronomical Institute, University of Bern, Sidlerstr. 5, CH-3012 Bern, Switzerland *bock@aiub.unibe.ch*

Summary. With an increasing number of satellites carrying reliable GPS receivers orbit determination using the high precision and uninterrupted GPS tracking technique gains increasing attention. Different from other approaches for generating kinematic satellite trajectories based on zero- or double-differenced GPS observations the procedure developed at our institute uses zero-differences with differences from epoch to epoch for the phase observable. The approach is very efficient due to the elimination of the phase ambiguities. In addition it includes a sophisticated data cleaning procedure. It may also be applied to objects moving on other than Earth-orbiting trajectories.

Within the LEO Pilot Project of the IGS (International GPS Service) a CHAMP orbit test campaign has been established covering eleven days from May 20 to 30, 2001 (day 140-150, 2001). A dozen institutions are contributing to the project and compare their kinematic and dynamic orbit results within the campaign.

Key words: High Rate Clocks, Low Earth Orbiters, Kinematic Point Positioning, Data Screening, Orbit Determination

1 Kinematic Point Positioning

The kinematic point positioning procedure is based on zero-differenced GPS observations. GPS satellite orbits and clock corrections are introduced as fixed. To get precise clock corrections for the GPS satellites at a rate of 30 seconds the orbits, station coordinates, and tropospheric zenith delays are fixed on the IGS solutions of the CODE (Center for Orbit Determination in Europe) analysis center. Clock correction differences are estimated for each epoch based on the epoch-differenced phase observations from the IGS ground network and used to interpolate the precise 5-minute clock corrections computed for the IGS.

Different from other approaches for generating kinematic satellite trajectories based on zero- or double-differenced GPS observations – e.g. Švehla et al. [4] – our procedure computes the kinematic positions from code pseudoranges and position differences from epoch-differenced phase observations. In a subsequent step the two time series are combined to precise positions. An advantage of this approach being similar to [1] is its efficiency because of the epoch-wise processing with the phase ambiguities being eliminated. The normal equation matrix for the combination has a banddiagonal structure which can be inverted by a fast and simple algorithm. The neglection of correlations between the epochs, however, reduces the accuracy of the combined positions. In addition, jumps may occur in the kinematic orbit at epochs where the positions cannot be connected by phase differences because of few tracked satellites or data problems.

The computation of position differences from phase as well as the screening procedure relies on an a priori orbit. A reduced-dynamic orbit or a filtered trajectory based on code observations only is sufficient for this purpose.

2 Data Quality and Screening

Important for the processing of GPS data from LEOs is the data quality and screening. Figure 1 shows the distribution of the observations from day 339/2001 (December 5, 2001) as a function of the zenith distances for the two satellites CHAMP and SAC-C. On that day CHAMP was tracking eight GPS satellites during 75% of the day while SAC-C was tracking only up to seven satellites. During about 5% of the day the CHAMP receiver was tracking six or less satellites. These numbers are typical for the time period from 206/2001 to 065/2002. Before day 206/2001, during the IGS CHAMP orbit test campaign, the receiver was tracking satellites below the local horizon reducing the number of usable satellite observations by about 10%. Since day 065/2002 CHAMP is tracking up to ten satellites.

The screening algorithm to find unusable observations relies on an a priori orbit which may be derived from code observations only. Outliers are rejected by a majority voting algorithm [3]. The kinematic positions and position

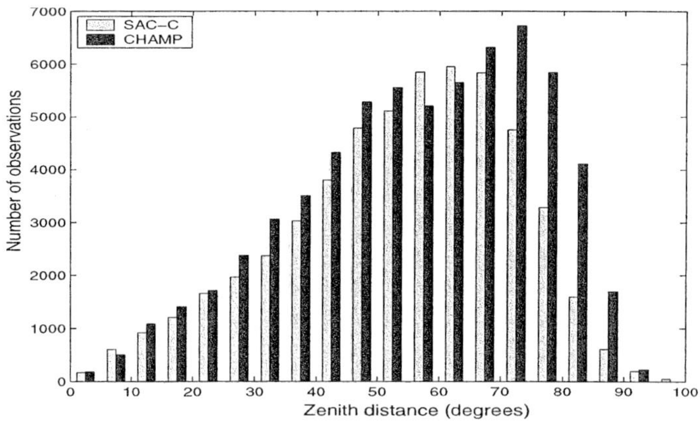

Fig. 1. Distribution of the observations for day 339/2001 vs. zenith distances.

differences are then computed using an iterative procedure. For the time period covered by the CHAMP orbit comparison campaign the screening algorithm removed about 5% of the observations of the satellite.

3 Kinematic and Dynamic Orbits

Figure 2 shows the differences between our kinematic orbit and the dynamic orbit computed by R. König at GFZ, Potsdam, for day 141, 2001. The RMS difference between the two orbits is of the order of 30 cm. The kinematic orbit shows jumps at epochs which are not connected by phase observations. The magnitude of the jumps is related to the noise of the code observations.

As a next step a dynamical orbit is fitted to the precise kinematic positions. Our procedure allows to model the non-gravitational forces (solutions A-E in Table 1) and estimate their scales factors (D, for air drag), or to introduce accelerometer observations with estimating the bias (F-I) and scale parameters of the instrument, to estimate offsets between the different arc pieces not connected by the phase to account for jumps (B, D, G, I), and to setup stochastic pulses at predefined epochs in order to account for satellite manoeuvers (C-E, H, I).

In Table 1 the results from different orbit parameterizations are listed for day 142 and 143 (May 22 and 23, 2001) together with the RMS values of the fit. Estimating the offsets between the different arc pieces (A→B, C→D, F→G, H→I) as well as setting up stochastic pulses at the manoeuver epochs (A→C, B→D, F→H, G→I) reduces the RMS values. For solution E the stochastic pulses are set up every 10 minutes which results in a reduced-dynamic orbit.

Alternatively to use the precise kinematic positions for a dynamic orbit fit, the positions determined from code and position differences from phase

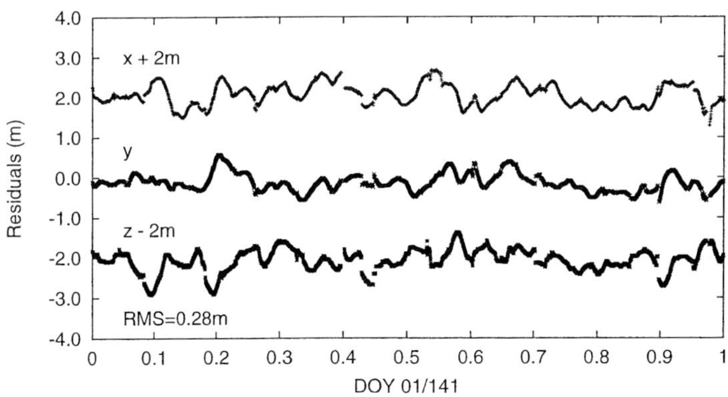

Fig. 2. Differences between GFZ dynamic and AIUB kinematic CHAMP orbit in Earth-fixed frame for day 141/2001.

Table 1. Comparison of different dynamic orbit parameterizations, RMS(m).

Solution	A	B	C	D	E	F	G	H	I
Non-grav.	mod	mod	mod	mod	mod	acc+bias	acc+bias	acc+bias	acc+bias
Empirical acc.	9	9	9	8	6	6	6	6	6
Arc offsets	-	yes	-	yes	-	-	yes	-	yes
Stoch. pulses	-	-	man	man	10min	-	-	man	man
Air drag	-	-	-	yes	-	-	-	-	-
RMS for day									
142	1.30	1.05	0.51	0.42	0.04	1.33	1.08	0.53	0.42
143	1.61	1.47	0.59	0.50	0.03	1.92	1.81	0.59	0.50

may be introduced separately with the weight of the respective observation types as pseudo-observations. Figure 3 shows the difference of a reduced-dynamic orbit (constant and once-per-rev empirical accelerations, stochastic pulses every 10 minutes) based on positions and position differences and a reduced-dynamic orbit generated by M. Rothacher and D. Švehla at Technical University in Munich. The RMS difference is between 12 and 17 cm for the eleven days of the CHAMP orbit comparison campaign.

4 Summary

The described approach allows an efficient determination of kinematic positions of LEOs and may be used for receivers moving on arbitrary trajectories, too. The currently achieved accuracy is of a few decimeters, mainly given by missing phase information introducing jumps, bad observation data which passed the screening procedure, and neglection of correlations between

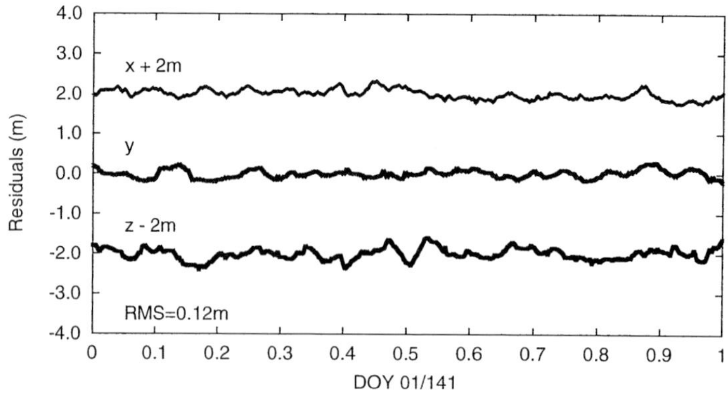

Fig. 3. Differences between reduced-dynamic CHAMP orbits from TUM and AIUB in Earth-fixed frame for day 141/2001.

epochs. The implemented dynamic orbit determination algorithm allows to deal with the jumps in the kinematic orbit, to account for manoeuvers, and to probe the range between the dynamic and reduced-dynamic regimes.

References

1. Bisnath S, Langley R (2001) High-Precision Platform Positioning with a Single GPS Receiver, ION GPS 2001, Salt Lake City, Utah, USA, September 11-14, 2001
2. Bock H, Hugentobler U, Springer T, Beutler G (2002) Efficient Precise Orbit Determination of LEO satellites using GPS, Advanced Space Research 30/2: 295-300
3. Bock H, Beutler G, Hugentobler U (2001) Kinematic Orbit Determination for low Earth Orbiters, IAG Scientific Assembly, Budapest, Hungary, September 2-7, 2001
4. Švehla D, Rothacher M (2001) Kinematic Orbit Determination of LEOs Based on Zero or Double-Difference Algorithms Using Simulated and Real SST GPS Data, IAG Scientific Assembly, Budapest, Hungary, September 2-7, 2001

CHAMP Double-Difference Kinematic POD with Ambiguity Resolution

Dražen Švehla[1] and Markus Rothacher[2]

[1] Institut für Astronomische and Physikalische Geodäsie (IAPG), Technische Universität München, Arcisstrasse 21, D-80333 München, Germany *svehla@bv.tum.de*

[2] Forschungseinrichtung Satellitengeodäsie (FESG), Technische Universität München, Arcisstrasse 21, D-80333 München, Germany *rothacher@bv.tum.de*

Summary. Purely kinematic precise orbit determination (POD) was performed for the CHAMP satellite over 11 days in the framework of the CHAMP Orbit Comparison Campaign. For the same period, reduced-dynamic orbits were computed and both types of orbits were independently validated by SLR. The tests with SLR showed that the accuracy of CHAMP reduced-dynamic POD is at the level of 4-5 cm, whereas kinematic orbits have a quality of 5-6 cm. Neither in the kinematic nor in the dynamic orbits a significant SLR range bias could be detected. In addition, we will present results concerning the ambiguity resolution techniques used.

Key words: CHAMP, LEO, kinematic orbit, dynamic orbit, POD

1 Introduction

The CHAMP satellite with its very low altitude orbit opened a new era in precise orbit determination with GPS. Low altitude orbits pose high requirements on the orbit modeling due to the highly dynamic environment: on one side one has to cope with increased effects from the gravity field and on the other side there is a big influence of the non-conservative forces like air-drag and solar-radiation pressure. From that point of view, LEO orbits are very well-suited for various Earth sensing activities like, e.g., gravity field and magnetic field recovery and atmosphere sounding but as we come closer and closer to the Earth, the gravity field and non-conservative forces, especially air-drag, become more and more difficult to model or estimate. Kinematic orbit determination is therefore an interesting alternative to the reduced-dynamic approach, because it is fully independent of the forces acting on the satellite and our orbit modeling capabilities. The accuracy of kinematic positions only depends on the accuracy of the carrier-phase measurements and the number and constellation of the GPS satellites tracked. Instead of using the enormous amount of LEO GPS measurements, one might then use the kinematic positions, with corresponding variance/covariance information, as pseudo-observations to determine gravity field coefficients and other dynamical parameters.

In Sections 2 we will give an overview of our reduced-dynamic POD approach and then focus in Section 3 on purely kinematic CHAMP orbits. Results of ambiguity resolution over the period of 11 days will be presented in Section 3 and both types of orbits will be compared with SLR measurements and with the orbits of other POD centers.

2 Reduced-Dynamic POD

The concept of using dynamical models together with measurements stemming from the LEO spaceborne GPS receiver in precise orbit determination has been very well tested and documented in various papers. Some recent CHAMP results can be found, e.g., in (Rim et al. 2001), (König et al. 2001). By measuring the total effect of all non-conservative forces (e.g. air-drag and solar-radiation pressure) acting on the surface of the LEO with accelerometers, a separation of the gravitational force from the non-conservative forces is possible and gravity field recovery can be performed. That was for the first time demonstrated using the data of the CHAMP satellite and the new gravity model EIGEN1S was derived, see (Reigber et al. 2002).

Let us now have a closer look at our approach for reduced-dynamic POD. We use carrier-phase GPS measurements only and we may compute orbits on either the zero- or double-difference level. One of the advantages of using double-differences is, that we get rid of all epoch-wise GPS and LEO clock parameters. The drawback is, that measurements from the IGS ground network have to be included. The main advantage of the double-difference approach is, however, that ambiguity resolution can be done and in this paper we will consider only the double-difference approach. More information about the zero- and double-difference approach may be found in (Švehla and Rothacher, 2001). As mathematical model we use a normal least-squares adjustment with variational equations without making use of any filtering. The basic idea of our dynamic POD can be summarized in the following two step procedure: 1) using kinematic positions as pseudo-observations all dynamic parameters are solved for and an a priori dynamic orbit is computed. The partial derivatives of the satellite position with respect to all orbit parameters are computed in this step by a simultaneous integration of the equation of motion and the variational equations; 2) using the carrier-phase measurements the a priori orbit is improved by estimating orbital parameters and pseudo-stochastic pulses. Alternatively, the second step could be replaced by using highly accurate kinematic positions as pseudo-observations for the estimation of all dynamical parameters.

After the least-squares adjustment the orbits are defined by 6 initial conditions (Keplerian elements), 9 solar radiation pressure parameters, 1 scaling factor for the air-drag and pseudo-stochastic pulses in along-track, cross-track

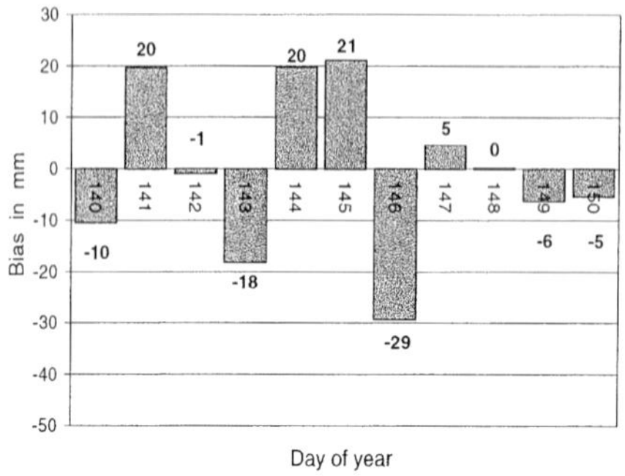

Fig. 1. Daily SLR range bias for dynamic orbit, mean SLR bias= -1.1 mm ± 44 mm over all SLR residuals.

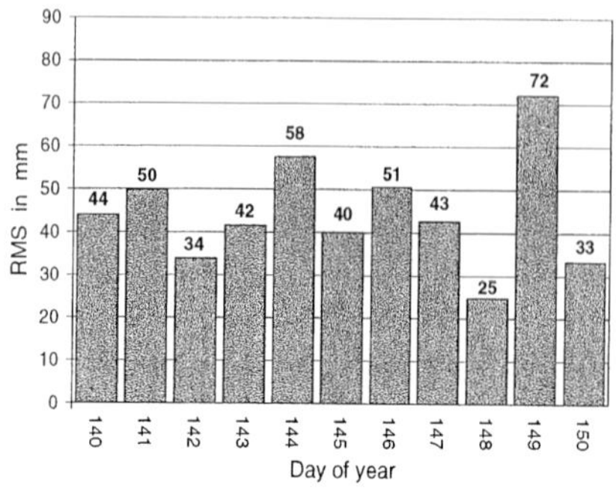

Fig. 2. Daily RMS of SLR residuals for dynamic orbit, RMS=44.3 mm over all SLR residuals.

and radial component. The pseudo-stochastic pulses are small changes in velocity which allow a better fit of the orbit to the measurements and are set up every 9 minutes in our case. The nice feature of these pseudo-stochastic pulses is, that partials can be computed as a linear combination of the partials with respect to the osculating elements. More frequent setting up of stochastic pulses may be considered equivalent to modeling the air-drag with more parameters. Air-drag densities are modeled using MSISE-90, see (Hedin 1991).

The dynamic orbits determined in arcs of 24 hours were validated by SLR measurements over the period of 11 days and the RMS of the SLR residuals is presented in Fig. 2. The SLR residuals were computed as the difference between the SLR measurements (corrected for the tropospheric delay with the Marini-Murray model) minus the distance between the SLR station and the GPS-derived orbit position. Altogether 2007 SLR residuals were obtained in this way using 69 daily station files from 17 SLR stations.

The mean RMS of the reduced-dynamic orbit over 11 days is 44.5 mm and Figure 1 clearly shows that there is no significant SLR range bias in the dynamic orbit. Station-dependent SLR biases were not estimated in this analysis, although some SLR stations might have significant range biases.

3 Kinematic POD

Kinematic POD procedures were developed for the zero- and double-difference processing of carrier-phase measurements. There was no improvement in orbit accuracy by including code measurements. Since ambiguity resolution can only be done with double-differences we prefer the double-difference POD approach. An alternative zero-difference approach based on forming differences between phase observations of consecutive epochs and avoiding ambiguity parameters can be found in (Bock et al. 2001). Our kinematic POD approach is also based on a least-squares adjustment without making use of filtering. The use of Kalman filtering in long-distance kinematic GPS is discussed in (Colombo et al. 1998), (Colombo et al. 2000) and (Bisnaith and Langley 2001). By forming double-differences between the IGS ground stations and the CHAMP satellite, epoch-wise GPS and LEO clocks are eliminated and CHAMP kinematic positions and ambiguities are the remaining parameters. In order to avoid an impact of the ionosphere on ambiguity resolution, wide lane ambiguities were estimated by making use of the Melbourne-Wübbena linear combination of code and phase observations. Figure 3(a) shows the percentage of resolved wide-lane ambiguities over 11 days and Figure 3(b) the percentage of resolved wide-lane ambiguities for each receiver type. One can immediately see that some of the IGS receivers are better performing than others. This is due the fact that not all of them provide P-code measurements on both, the L1 and L2 frequency. According to (Ray 2002) there are three main classes of GPS receivers within the IGS network, namely: 1) cross-correlators that observe C1 and P2'=C1 + (P2-P1) (ROGUE SNR-x, AOA ICS-4000Z, TRIMBLE 4000, and TRIMBLE 4700); 2) Y-codeless, non-cross-correlators that observe P1 and P2 (ASHTECH Z-XII3, AOA SNR-12 ACT, and AOA BENCHMARK ACT); 3) C1, Y-codeless, non-cross-correlators that apparently function like other modern Y-codeless receivers but report C1 (instead of P1) and P2 (TRIMBLE 5700, LEICA CRS1000, and LEICA SR9600). For those receivers that do not provide P-code measurements on both frequency the use of differential code biases (P1-C1) for the GPS satellites, available from (Schaer 2002), considerably improves wide-lane ambiguity resolution.

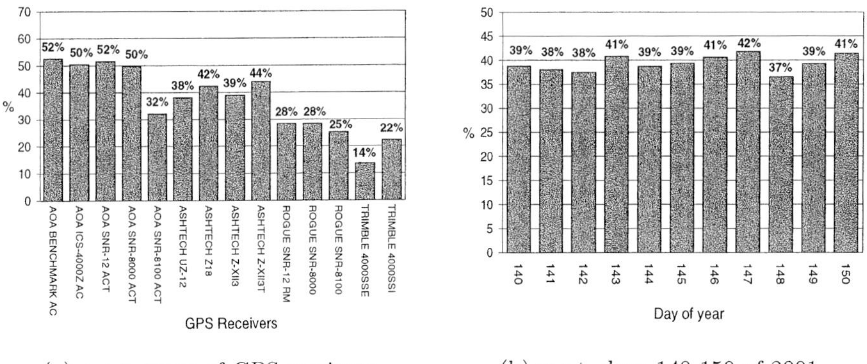

(a) w.r.t. types of GPS receivers (b) w.r.t. days 140-150 of 2001.

Fig. 3. Percentage of resolved Melbourne-Wübbena ambiguities over 11 days.

Subsequently, by using the ionosphere-free linear combination of the carrier-phase measurements and the resolved wide-lane ambiguities, an iterative resolution of the narrow-lane ambiguities (bootstrapping) can be performed. Two main methods were developed to perform this narrow-lane ambiguity resolution. In the kinematic bootstrapping epoch-wise coordinates are pre-eliminated in order to reduce the size of the normal-equation matrix. The first solution is a float solution where the ambiguities are real numbers. Then the best estimated ambiguities are set to integer numbers, the normal equation system is updated and re-inverted and the whole procedure is repeated. More about this type of bootstrapping and the criteria applied for ambiguity fixing can be found in (Švehla and Rothacher, 2001). The same procedure can also be used when estimating dynamic orbit parameters and we then speak of dynamical bootstrapping. Baseline-wise ambiguity resolution could in principle be applied for kinematic as well as for dynamic orbits, but highly accurate a priori orbits have to be available in that case. The orbits are then fixed in the baseline by baseline ambiguity resolution. The drawback of this method is, that the criteria to fix the ambiguities have to be very restrictive in order to assure that ambiguities are correctly resolved. In principle, baseline-wise ambiguity resolution can be performed iteratively: after the first baseline-wise ambiguity resolution step a new orbit is computed making use of the fixed ambiguities and a new iteration of the baseline-wise ambiguity resolution is performed with the updated orbits. Our experience with baseline-wise ambiguity resolution showed that highly accurate dynamical orbit models are a prerequisite for this method. More details about Melbourne-Wübbena wide-laning with narrow-lane bootstrapping may be found in (Švehla and Rothacher, 2001). Figure 4 shows the percentage of resolved narrow-lane ambiguities and Figure 5 the accuracy of the kinematic orbits with fixed ambiguities over 11 days (days 140-150 in 2001) obtained from SLR validation. The RMS of all SLR residuals for the kinematic orbits with fixed ambiguities over 11 days is 56.4 mm, a considerable improvement compared to the RMS of 117 mm for kinematic orbits with float ambiguities.

The SLR residuals for kinematic orbits were computed in the same way as for reduced-dynamic orbits, the only difference was, that the kinematic orbits are given with a sampling of 30s and, thus, an interpolation procedure is required in order to obtain positions at the epochs of the SLR normal points. A first order polynomial was used to interpolate kinematic positions within the sampling interval along an a priori dynamic orbit. We noticed that the validation of kinematic orbits with SLR is more difficult and the necessary interpolation may easily increase the RMS by 1-2 cm.

The impact of ambiguity resolution on the kinematic orbits can be seen in Figure 6 (day 148/2001). Figure 10(a) shows the comparison of our kinematic orbits with the dynamic orbits computed at GeoForschungsZentrum (GFZ) Potsdam and Figure 10(b) the comparison with the dynamic orbits computed at Center for Space Research (CSR), Texas (day 148/2001). The differences

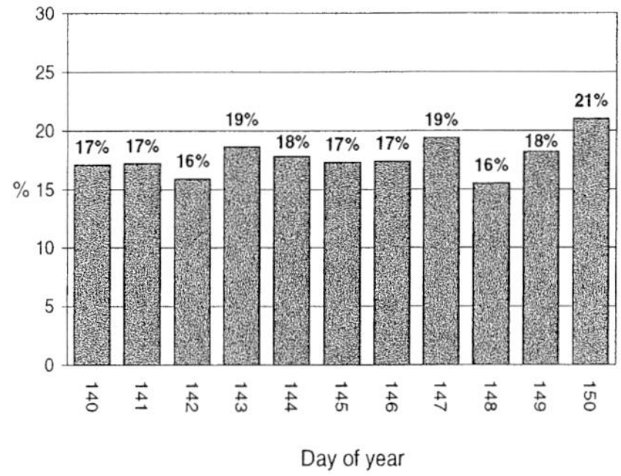

Fig. 4. Percentage of resolved narrow-lane ambiguities over 11 days.

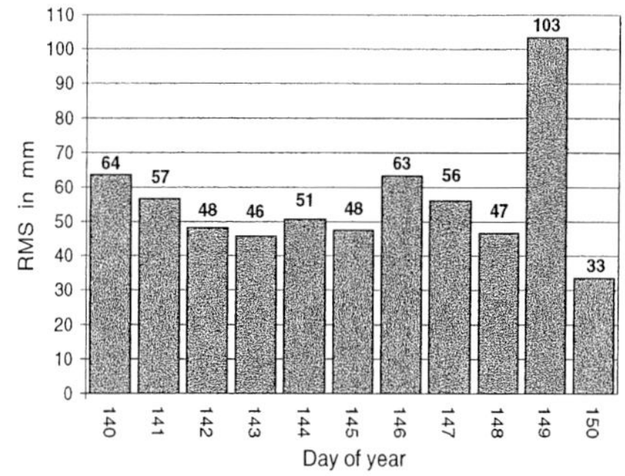

Fig. 5. RMS of SLR residuals for kinematic orbit with fixed ambiguities (using dynamic bootstrapping), RMS=56.4. mm over all SLR residuals.

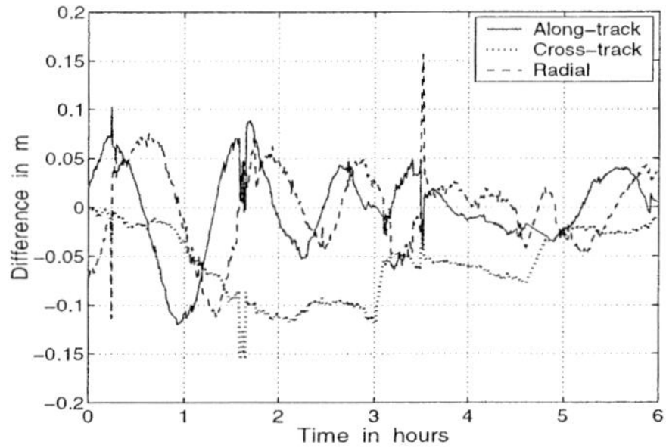

Fig. 6. Differences between kinematic orbit with float and fixed ambiguities (148/2001).

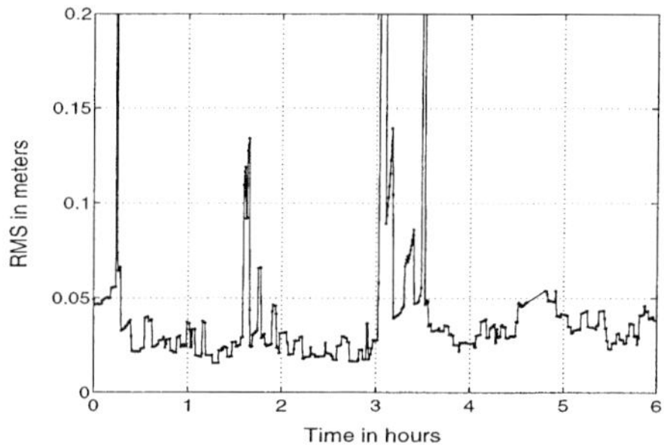

Fig. 7. Formal RMS of kinematic positions, day 148/2001.

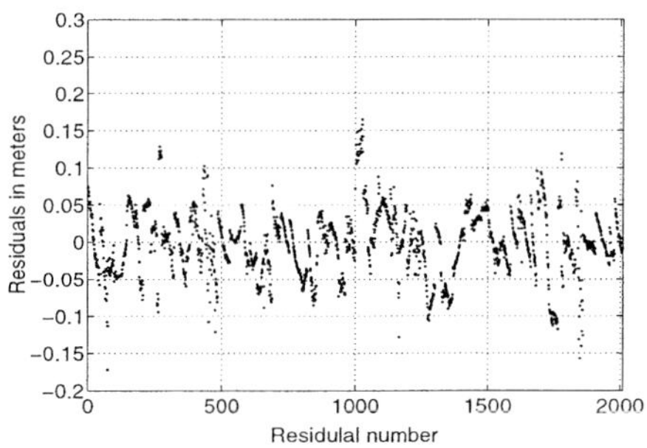

Fig. 8. SLR residuals for dynamic orbit (RMS=44.3 mm).

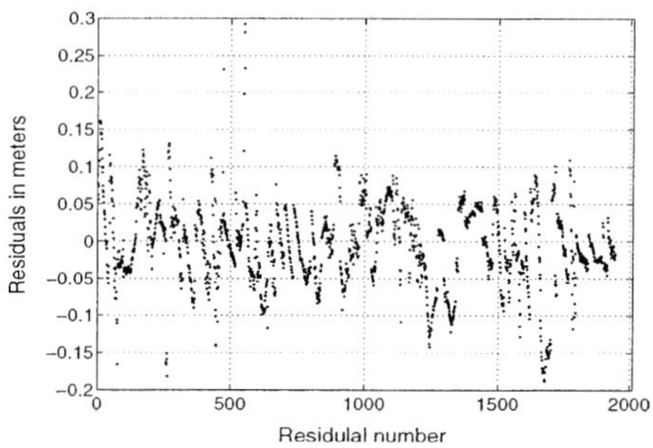

Fig. 9. SLR residuals for kinematic orbit (RMS=56.4 mm).

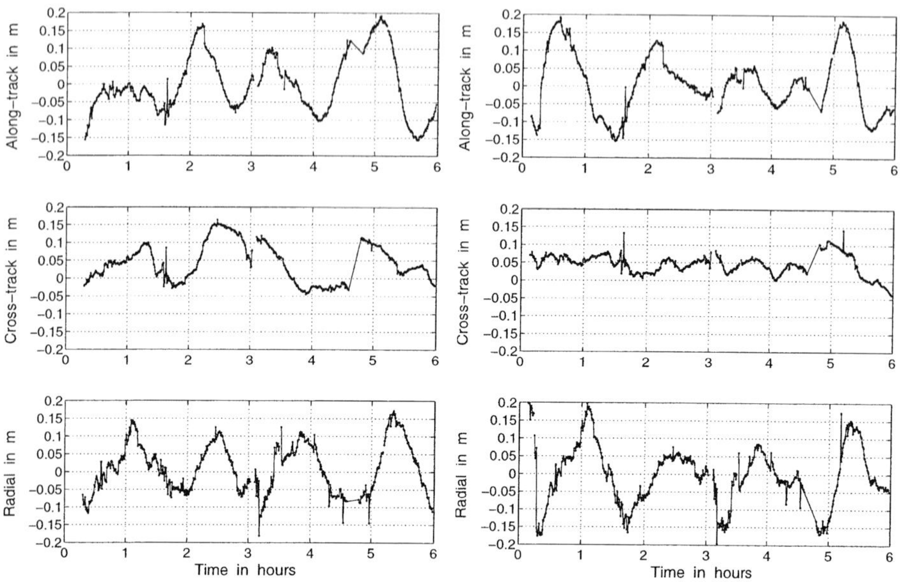

(a) Comparison with dynamic GFZ orbit (b) Comparison with dynamic CSR orbit

Fig. 10. Comparison of our kinematic orbits with GFZ and CSR dynamic orbits (days 148/2001).

clearly show interesting once per revolution terms, especially in the radial and along-track directions. One has to keep in mind that kinematic orbits do not depend on any force model or gravity field. Kinematic orbits can therefore be used as a validation tool for dynamic orbit modeling. Looking at Figure 7 with the formal a posteriori RMS of the kinematic positions we may conclude that each kinematic position has a different level of accuracy and that kinematic orbit positions should always be used together with the corresponding variance-covariance information.

4 Conclusion

Purely kinematic precise orbit determination (POD) for the CHAMP satellite was performed over 11 days in the framework of the CHAMP Orbit Comparison Campaign. In order to avoid the effects of ionosphere, ambiguity resolution was performed using Melbourne-Wübbena wide-laning with narrow-lane bootstrapping. For the same period (days 140-150 in 2001), reduced-dynamic orbits were computed and both types of orbits were validated with SLR. The accuracy of the dynamic orbits is at a level of 4-5 cm (44.3 mm), whereas the kinematic ambiguity-fixed orbits have a quality of about 5-6 cm (56.4 mm). Kinematic orbits of such a quality are extremely valuable to assess dynamical orbit models and parameterizations. No significant SLR bias could be detected in the two orbit types. It is important to note that kinematic orbit positions should always be used together with the corresponding variance-covariance information.

Acknowledgement. We are grateful to GeoForschungsZentrum Potsdam for providing GPS measurements from CHAMP satellite and German Academic Exchange Service (DAAD) for financial support through DAAD Scholarship.

References

Bisnath SB, Langley RB (2001) High-Precision Platform Positioning with a Single GPS Receiver. Proceedings ION GPS-2001, in print, Salt Lake City, Utah.

Bock H, Hugentobler U, Beutler G (2001) Kinematic Orbit determination for Low Earth Orbiters (LEOs). Proceedings in print. IAG 2001 Scientific Assembly, Budapest, Hungary.

Colombo OL (1998) Long-Range Kinematic GPS. In Teunissen PJG and Kleusberg A (Eds) GPS for Geodesy Springer Verlag, Lecture Notes in Earth Sciences.

Colombo OL, Hernandez-Pajares M, Juan JM, Sanz J (2000) Ionospheric Tomography Helps Resolve GPS Ambiguities On The Fly At distances Of Hundreds Of Kilometers During Increased Geomagnetic Activity, Colombo OL, Hernandez-Pajares M, Juan JM, Sanz J. Proceedings of the IEEE PLANS 2000, San Diego, California.

Hedin AE (1991) Extension of the MSIS Thermosphere Model into the Middle and Lower Atmosphere, J Geophys Res, 96, 1159.

König R, Baustert G, Neumayer KH, Meixner H, Schweiger V, Zhu S, Reigber C (2001): Precise orbit determination for CHAMP. Paper presented at EGS XXVI General Assembly, Nice, France.

Rim H, Kang Z, Nagel P, Yoon S, Bettadpur S, Schutz B, Tapley B (2001) CHAMP Precision Orbit Determination, AAS/AIAA 2001 AAS-01-334.

Ray J (2002) C1/P1 biases for Leica and Trimble 5700 receivers. IGSMAIL-3737. http://igscb.jpl.nasa.gov/mail/igsmail/2002/msg00075.html.

Reigber Ch, Balmino G, Schwintzer P, Biancale R, Bode A, Lemoine JM, Koenig R, Loyer S, Neumayer H, Marty JC, Barthelmes F, Perosanz F, Zhu SY (2002) A high quality global gravity field model from CHAMP GPS tracking data and Accelerometry (EIGEN-1S), Geophys Research Letter, 29, No 14, 10.1029/2002GL015064, 37-1 - 37-4, 2002.

Schaer S (2002) Personal comunication. Global Ionosphere Maps Produced by CODE: Differential P1-C1 Code Biases (DCBs). ftp://ftp.unibe.ch/aiub/, http://www.aiub.unibe.ch/ionosphere.html.

Švehla D, Rothacher M (2001) Kinematic Orbit Determination of LEOs Based on Zero or Double-difference Algorithms Using Simulated and Real SST GPS Data. Proceedings in print. IAG 2001 Scientific Assembly, Budapest, Hungary.

Approaches to CHAMP Precise Orbit Determination

Karl Hans Neumayer, Rolf König, Christoph Reigber, and Sheng Yuan Zhu

GeoForschungsZentrum Potsdam (GFZ), Division 1, Kinematics and Dynamics of the Earth.
neumayer@gfz-potsdam.de

Summary. With a tailored gravity field model, the dynamic restitution of the CHAMP orbit can be carried out quite accurately. The solution can be based on the onboard accelerometer data or on dynamical models with additional parameters. Beside the dynamic mode, also a geometric/kinematic satellite position determination is implemented in the GFZ POD software EPOS-OC. Results from the dynamic, from the geometric/kinematic, and from the mixed approach are given.

Key words: CHAMP, EIGEN gravity field, orbit determination, kinematic mode, dynamic mode, mixed approach

1 Introduction

Currently under investigation is a method that has already been successfully tried shortly after CHAMP's lift-off when our knowledge of the gravity field at CHAMP's altitude was poor and we were yet uncertain about the quality of the data the on-board accelerometer delivered. After its brief advent at the begin of the CHAMP program the method disappeared somewhat into oblivion; it is presently being resurrected and we try to find out if we still have benefits from it at this stage where the gravity field we use is fairly good.

The main idea basically is to combine the advantages of the geometric and the dynamic orbit adjustment technique and to avoid the respective disadvantages at the same time. Before we proceed towards the details, we give a brief description what is meant by geometric and dynamic orbit determination.

2 The geometric and the dynamic technique for orbit determination

The geometric or kinematic method of establishing the position of a spacecraft from GPS data is, roughly spoken, the way a commercial hand-held GPS receiver is used by a common person. At every measurement epoch, the receiver has to pick up the signals of at least four different GPS satellites in order to solve for its three spatial coordinates and the receiver clock error. The four unknowns - i.e. position and clock error - are computed from a system of non-linear equations; a procedure which, from the geometrical viewpoint, amounts to finding the common intersection point of at least four spheres.

A distinguishing feature of the kinematic mode is the perfect a-priori uncorrelatedness of the measurement epochs. No hypothesis is made whatsoever with respect to the dynamics of the vehicle the receiver is mounted on. From the conceptual viewpoint it is as if at every observation epoch we had a different GPS receiver. In order to illustrate this, let us for the moment assume that, instead of the one receiver that moves in space, occupying at given epochs t_0, t_1, t_2, ... the sequece of positions p_0, p_1, p_2, ... we had a sequence of fixed receivers just at these positions p_0, p_1, p_2, ... in space, and, for every index m, at the epoch t_m, the m-th receiver, being at p_m, logged a measurement. If we merged together this sequence of measurements to one file and fed this file to the geometric/kinematic method, the algorithm could not distinguish this from the original setting of the one moving receiver.

In the framework of dynamic GPS orbit adjustment, the GPS measurements are conceptually treated just like other measurements (Laser, PRARE, Doppler etc.) and used to adjust parameters on the right hand side of a differential equation describing the orbital motion. From the purely procedural viewpoint it is not necessary to have observations of at least four different satellites at every epoch, whereas in the kinematic mode, if there are data available of only three or less GPS satellites, it is not possible to compute the receiver position and one has therefore to discard the whole measurement epoch.

3 Combining geometric and dynamic orbit estimation

Both the kinematic (geometric) and the dynamic orbit adjustment method have their advantages and their disadvantages. As the geometric method does not make use of any kind of dynamic model, errors produced by a faulty gravity field are simply not there. The same is true for the earth tides, the surface forces etc.

On the other hand, there may be GPS measurement epochs that are flagged as being O.K. but where the geometric method produces extremely faulty positions, e.g. if we have a bad satellite geometry. These epochs that produce poor results cannot be found by relying on geometric point positioning alone.

The main disadvantage of dynamic orbit adjustment clearly is faulty dynamic modelling. The main advantage it has in comparison to the kinematic point positioning is a certain coherence between GPS observation epochs. From a given state (i.e. a given position/velocity pair) it is possible to infer

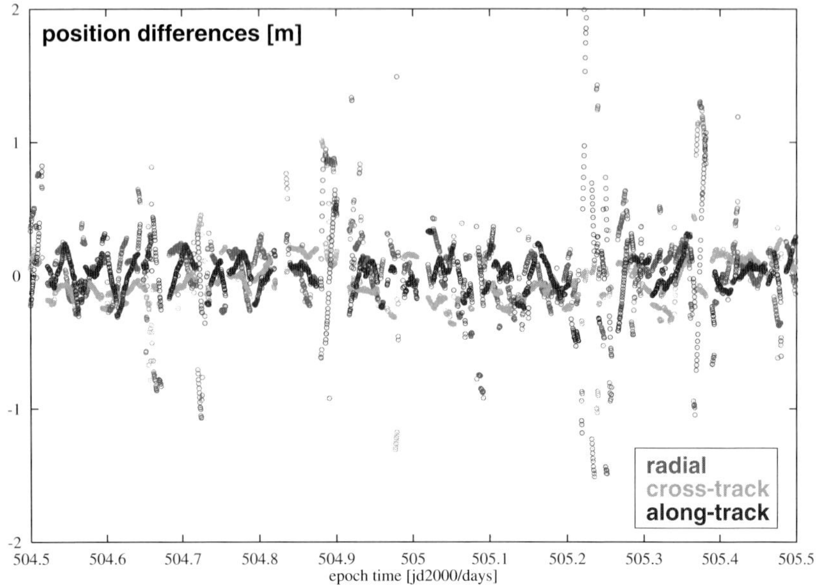

Fig. 2. Dynamically vs. geometrically restituted orbit.

data we see that the difference stays well below the sub-meter level. Thus, if a difference of e.g. 10 m appears, we can be sure to have an outlier of the kinematic orbit.

One approach of kinematic orbit points editing is to formalize the above-mentioned procedure in a straightforward manner; for an illustration see Fig. 3: First, the somewhat ragged-looking time series of coordinate differences is smoothed using a simple low-pass filter. Secondly, we prescribe around this smoothed difference a small neighbourhood, in our example of 40 cm, and those difference values that surpass this threshold are discarded, and the rest is kept.

This way of doing it demands explicit manipulation of the kinematic orbit points that come out of the geometric adjustment procedure. A far more elegant method is to feed into the kinematic adjustment part right from the beginning only those GPS measurements that produce well-behaved kinematic orbit points. But that at first sight seems to be impossible, because in order to assess the quality of kinematic orbit points and to edit them it is necessary to compute them first. We can however make use of the fact that our orbit adjustment program is capable of producing edited data for dynamic runs.

During the adjustment process that uses any conceivable kind of measurements the program does iterative outlier removal. In the case of GPS data, the dynamic adjustment process has the tendency to remove exactly those

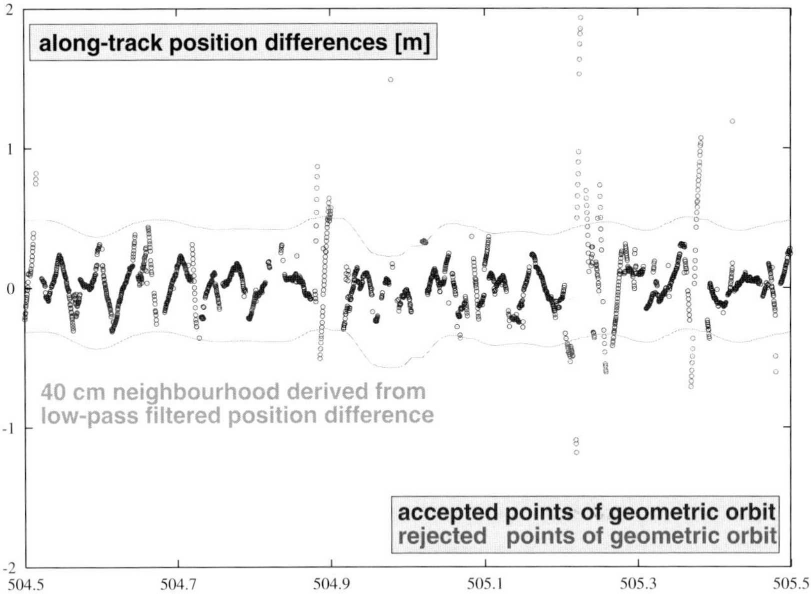

Fig. 3. Dynamically vs. geometrically restituted orbit: position differences for outlier removal.

GPS measurement epochs that produce also a poor match of kinematic and dynamic orbit states, as can be seen in Fig. 4. This is not very amazing as the epochs of bad geometry have also the tendency to poorly fit an underlying orbit with adjusted parameters.

Now these GPS data edited by the dynamic adjustment part are written onto a file, and we have only to feed this edited file into the kinematic adjustment run instead of the original data, and we are likely not to get ill-determined trajectory points.

6 Conclusions and outlook

First tentative results indicate that the presented method - at our present level of accuracy of the gravity field - gives no further visible improvement over the classical dynamical adjustment technique. With `EIGEN-CH1_S2SL4` we already have a gravity field of so high quality that additional stabilization of the adjustment process by using geometrically derived orbit points is no longer needed. The advantages of the mixed method - as already has been mentioned in the introduction - rather emerge at the begin of a mission where we still have many unknowns and uncertainties. It is clear therefore that we expect to benefit from our investigations at the begin of the coming GRACE mission.

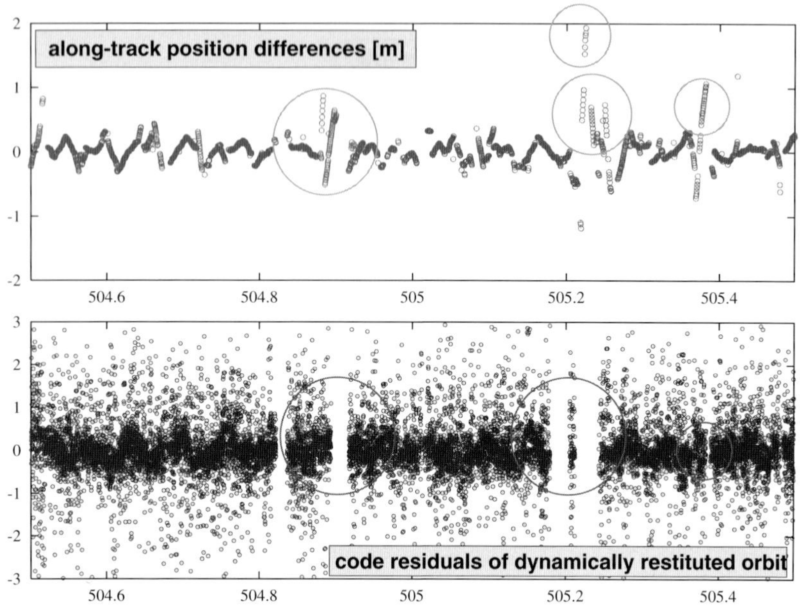

Fig. 4. The dynamically restituted orbit as a tool to find epochs of bad geometry.

Acknowledgement. This study was carried out under the grant of the German Federal Ministry of Education and Research (BMBF) No. 03F0333A.

References

1. Neumayer KH, Zhu SY, Reigber Ch, König R, Meixner H (2000) CHAMP Orbit Recovery - Algorithmic Approaches and Preliminary Results, EOS Transactions, American Geophysical Union, Vol. 81, No. 48, p. F309.

STAR Accelerometer Contribution to Dynamic Orbit and Gravity Field Model Adjustment.

Sylvain Loyer[1], Sean Bruinsma[2], Damien Tamagnan[2], Jean-Michel Lemoine[2], Felix Perosanz[2], Richard Biancale[2]

[1]Noveltis, Parc Technologique du Canal, 2 av. de l'Europe, 31520 Ramonville Saint Agne, France
[2]CNES/GRGS, 18 av. Edouard Belin, 31401 Toulouse Cedex 4, France.

Summary. We investigate the contribution of the STAR accelerometer on board the CHAMP satellite to the dynamic orbit restitution and gravity field recovery. The quality of the dynamic orbits is evaluated in term of tracking data residuals (GPS and Laser data) and by comparisons to precise reduced-dynamic orbits. The use of STAR data for dynamic orbits allow a gain of 20 centimeters in term of Laser residuals and around 50 cm in terms of 3D comparisons to reduced-dynamic orbits. The gravity fields obtained with only 1.5 months of CHAMP observations are significantly improved by the use of the STAR accelerometer data.

Key words: CHAMP satellite, STAR accelerometer, Precise Orbit Determination, Gravity field

1 Introduction

The STAR accelerometer is located at the center of mass of the CHAMP satellite and measures the linear and angular non-gravitational accelerations each second with a low noise level (around $5.10^{-9} ms^{-2}/\sqrt{Hz}$ in radial direction and $1.10^{-9} ms^{-2}/\sqrt{Hz}$ in the tangential and normal directions). However, an in-flight calibration procedure is required. STAR data are a key component (together with GPS data) for the gravity field determination experiment of the CHAMP mission. The principles of the experiment is given in (Schwintzer et al. 2000) and the detailled evaluation of the accelerometer itself is described in (Perosanz et al. 2002).

The purpose of this paper is to investigate the contribution of the STAR accelerometer to the mission in terms of orbit restitution as well as gravity field determination.

We have computed gravity field models based on different kinds of dynamical CHAMP orbits. The entire computation covers a period of 1.5 month between end of September and the beginning of November 2000. The computations based on different dynamical modelling use the same GPS tracking data to constrain the orbits. The non-gravitational accelerations are computed according to the following three different methods:

- The non-gravitational accelerations are computed using classical models and empirical terms. STAR data are not used (**Models solution**).
- The non-gravitational accelerations are provided by the STAR data
- The non-gravitational accelerations are taken from the STAR data except for the radial component, which is replaced by empirical terms (**STAR_R0 solution**). The radial acceleration provided by the STAR accelerometer contains non-expected signals (spikes, extreme sensitivity to temperature and large bias variations...) that could affect the orbits restitution and/or gravity field evaluation. When we plot the acceleration obtained by the models and the accelerations provided by STAR we clearly highlight the problem on the radial axis (see an example in Fig. 1.). This last solution was computed in order to evaluate the impact of the radial signal on the dynamical orbit restitution as well as the gravity field solution.

Employing the three methods, we compute the orbits and the gravity field and we evaluate and compare the results.

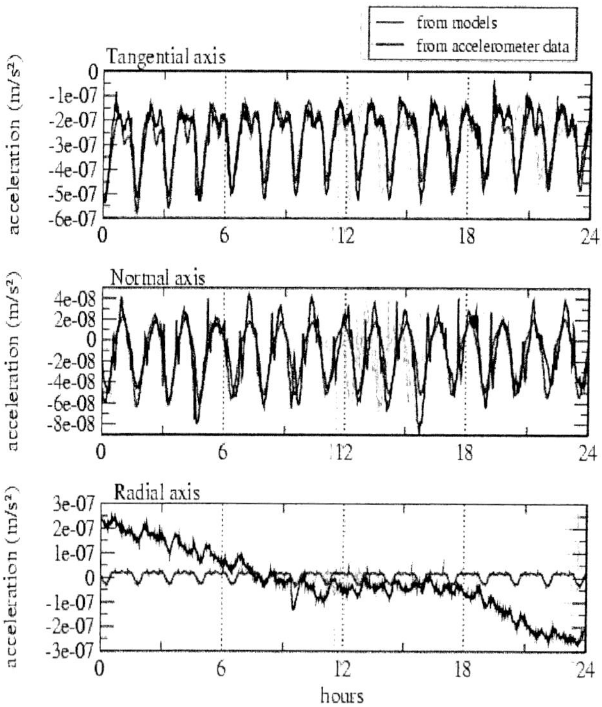

Fig. 1. Example of surface accelerations over 1 day measured by STAR and compared to the models. The anomaly of the STAR signal on the radial axis is clearly visible

2 Orbit computation and gravity field modelling

The experiments of orbit computation and gravity field modelling were done with 1.5 month of data, from 24 September through 7 November 2000. A total of 28 arcs, with lengths of 1 to 1.5 days, could be computed. To quantify the contribution of STAR data to orbit accuracy, these 28 arcs have been computed according to three methods including accelerometer data or models to account for the non-gravitational forces.

- The "Models" method uses both gravitational and non-gravitational models (drag, solar and terrestrial radiation pressure). In order to eliminate model errors, we estimate 4 drag scale factors per day and an empirical one-per-revolution term in the along track direction.

Experiment	Models	STAR	STAR_R0
Data used	-GPS Phase ($\sigma_{\text{a priori}}$ = 1 cm) -GPS Range($\sigma_{\text{a priori}}$= 100 cm) -SLR (Satelite Laser Ranging data) down weighted		
Gravitational Forces	-Gravity field: GRIM5_CHAMP_INITIAL (including 10 days of CHAMP data) -Oceanic tides -Terrestrial tides -Atmospheric pressure potential		
Non-gravitational forces	-No STAR data -4 empirical coefficients per day in factor of drag forces -an empirical force at the once per revolution period in the tangential direction (cosine and sine terms)	-STAR data -temperature correction on radial axis -an empirical force at the once per revolution period in the radial direction (cosine and sine terms)	-STAR data for T and N directions -an empirical force at the once per revolution period in the radial direction (cosine and sine terms)
STAR calibration parameters	none	-free bias on R, T and N directions -Scale factors fixed to 1.0, 0.82 and 0.9 for R, T, and N directions	-free bias on R,T and N directions -Scale factors fixed to 0.82 and 0.9 for T and N directions

Table 1. The three methods (Models, STAR, STAR_R0) tested in the experiment (R, T and N mean Radial, Tangential and Normal directions)

- The "STAR" and "STAR_R0" methods use pre-processed accelerometer data to compute non-gravitational forces. The gravitational forces are predicted with models and are the same as in the "Models" method. In the "STAR" solution, we apply a temperature correction to the radial linear acceleration. The detailled temperature correction procedure is given in (Perret et al. 2001).
- In the "STAR_R0" method the radial measurements of the accelerometer are fixed to zero.

The 28 orbits are computed employing a least-squares adjustment of the different data (GPS phase, GPS range, STAR measurements, attitude quaternions and down weighted SLR). After orbit adjustment a normal equation for the gravity field coefficients, empirical biases and initial state vector parameters is computed per arc and per method (see Table 1).

3 Dynamic orbits results

The RMS of the residuals of the different data types (SLR, GPS Range, GPS phase) are plotted below (on Fig. 2.) for the 28 orbits computed with the three methods (The SLR data do not participate in the adjustment but are used as an external criterion to assess the quality of the orbit fit).

The STAR data significantly improve the accuracy of the orbit determination. Using STAR, the average of the RMS-of-fit decreases by 50% compared to the RMS of the "Models" method. A second criterion of the contribution of STAR data to orbit accuracy is to compare the dynamic orbit with a reference reduced dynamic orbit solution. These reference orbits are computed using both gravitational and non-gravitational models. These models are corrected by estimating hundreds of 3-dimensional stochastic forces per day. The RMS of the GPS phase residuals is better than 1.5 cm and that of the SLR residuals is about 10 cm for the reference orbits.We have computed the position difference relative to the reference orbit for each arc and for the different methods as a 3D RMS. These individual 3D RMS' are shown in Figure 3 and the statistics are listed in Table 2 for the 3 directions (radial, cross-track and along-track). We obtain a mean of 78 cm for the Models

Experiment	Models	STAR	STAR_R0
Radial	0 +/- 20 cm	1+/- 11 cm	2 +/- 11 cm
Cross-track	4 +/- 20 cm	-4 +/- 17 cm	-6 +/- 17 cm
Along-track	-1 +/- 72 cm	0 +/- 23 cm	1 +/- 23 cm
RMS 3D	78 cm	32 cm	32 cm

Table 2. 3D RMS and mean of differences relative to the reference solution in the radial, cross-track, and along track directions

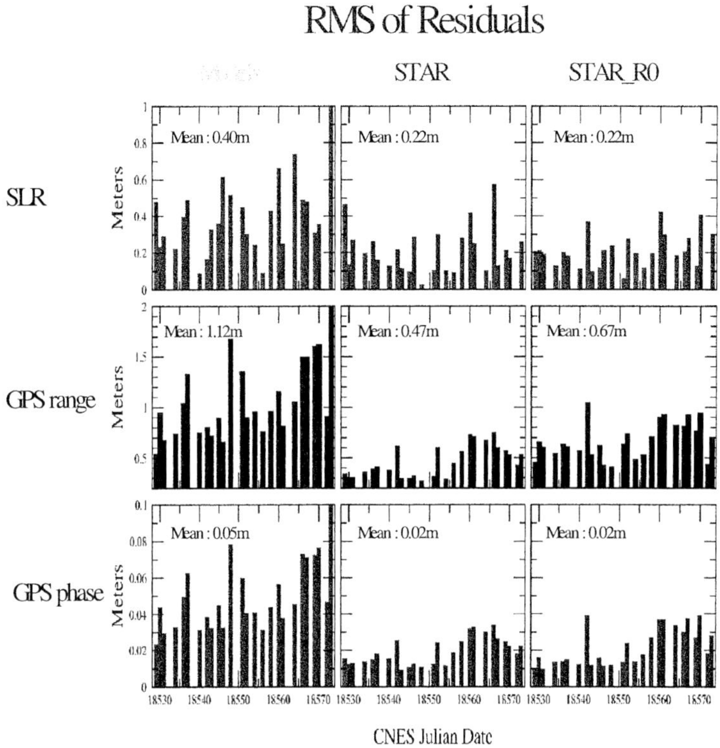

Fig. 2. RMS of measurement residuals (from top to bottom : SLR, GPS range and GPS Phase) for the 3 methods tested (from left to right : Models, STAR, and STAR_R0).

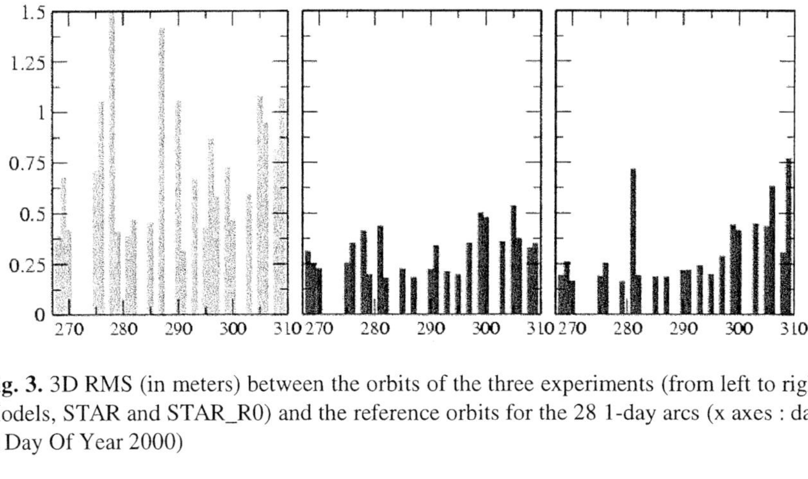

Fig. 3. 3D RMS (in meters) between the orbits of the three experiments (from left to right : Models, STAR and STAR_R0) and the reference orbits for the 28 1-day arcs (x axes : dates in Day Of Year 2000)

and 32 cm for the STAR and STAR_R0 methods. STAR data improve the orbit fit results by around 60%. Systematic biases for STAR solutions are mainly due to the scale factors that are fixed to a priori values in our computation.

4 Gravity field results

For the three methods the normal equations are subsequently accumulated and inverted. We solve for accelerometer calibration parameters, empirical terms and for the spherical harmonic coefficients of the gravity field up to degree 120 and constrained according to the Kaula rule. The obtained solutions are all CHAMP-only solutions (We do not add normal equations from other satellites). The Kaula weight applied to the "Models" solution is 4 times bigger than the Kaula weight needed for the "STAR" and "STAR_R0" solutions. We have also computed the differences between the derived geoids and the Mean Sea Surface from altimetry (CLS Mean Sea Surface) minus the oceanic circulation POCM92 and we display the height difference for the three solutions. The rms of differences between CHAMP geoids and MSS-POCM (computed from 5°x 5° grids) is 82 cm for "Models" solution, 74 cm for "STAR" solution and 76 cm for "STAR_R0" solution.

The gravity field results are presented in Figure 4 below in term of geoid height per spherical harmonic degree.

Due to the small amount of data used (because of the short time interval of the

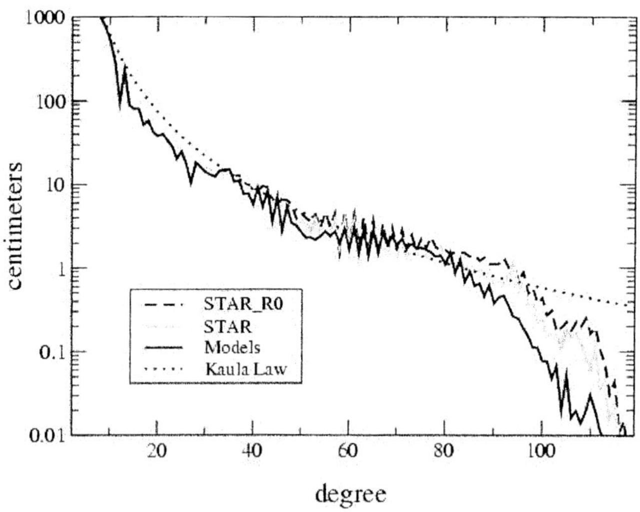

Fig. 4. Geoid height per spherical harmonic degree for the three solutions (Models, STAR and STAR_R0). The loss of power at high-degree for the "models" solution is clearly visible.

computations), some orbit track patterns were detected on the maps (not shown). Nevertheless, the comparisons are sufficient to conclude on the impact of STAR: the solutions including STAR data contain more power at all degrees (starting at degree 30 and particularly beyond degree 80). They allow a gain between 5 and 10 centimeters relatively to MSS-POCM92.

5 Conclusions

The results presented here confirm that the quality of the current non-gravitational models is not sufficient to compute accurate dynamical orbits of the CHAMP satellite. Consequently, the gravity field solution is not accurate.

The accelerometer data significantly improve the dynamical orbit restitution as well as the quality of the CHAMP-derived geoids.

Using or not-using radial data does not significantly change our results. As expected prior to the mission, the future use of both STAR and GPS data should greatly increase our knowledge of the gravity field of the Earth as well as our knowledge of non-gravitational forces acting on the satellite.

References

Perosanz F, Biancale R, Loyer S, Lemoine JM, Perret A, Touboul P, Foulon B, Pradels G, Grunwald L, Fayard T, Vales N, Sarrailh N, Duvivier E (2002) On board evaluation of the STAR accelerometer, this issue.

Perret A, Biancale R, Camus AL, Lemoine JM, Fayard T, Loyer S, Perosanz F, Sarrailh M (Mai 2001) STAR Commisioning Phase Calibration / Validation activities by CNES, Final report, Volume 1 & 2.

Schwintzer P, Kang Z, Perosanz F (2000), Accelerometry aboard CHAMP, in R Rummel, H Drewes, W Bosch, H Hornik (eds), IAG section II symposium, Munich 1998 Oct 5-9, Springer, pp 197-200.

Impact of different data combinations on the CHAMP orbit determination

Sheng Yuan Zhu, Karl Hans Neumayer, Franz-Heinrich Massmann, Chuang Shi and Christoph Reigber

GeoForschungsZentrum Potsdam (GFZ), Division 1, Kinematics and Dynamics of the Earth

Summary. For the orbit determination of the CHAMP satellite three data sets are of major importance: GPS, satellite laser ranging and the accelerometer observations. The first two can be used independently for the orbit restitution, while the third one can only be used in combination with another data type(s). Different combinations of these data types (such as GPS only; SLR only; GPS plus accelerometer; GPS plus SLR, GPS plus SLR plus accelerometer; etc.) are tested for the POD. The aim is to investigate the usefulness and contribution of each data type and to study the advantages and weakness of various data combination procedures.

Key words: Orbit, gravity, GPS, accelerometer

1 Introduction

The design of the EPOS-OC software at the GFZ allows the CHAMP orbit restitution by adopting different processing procedures: pure geometric; reduced dynamic; and full dynamic (see Neumayer et. al. 2000).

On board the CHAMP satellite there are a number of instruments. They provide measurements which are useful for orbit determination: GPS, laser and accelerometer (including attitude and thrust data). The first two can be used independently for orbit restitution, and they form a kind of 'space-collocation' of the two techniques. GPS data is the most important one. In order to use CHAMP GPS data, one needs the orbits of GPS satellites (and their clocks, if non-differenced GPS data are used). One can first solve the GPS orbits and clocks by using the GPS data from ground receivers; then fixing them using on-board GPS data alone to restitute the CHAMP orbit. This method is called 'two-step' procedure. When using CHAMP GPS data together with the data from ground stations in order to simultaneously determine the orbits of CHAMP and GPS satellites, it is the 'one-step' method.

In this study test results were given to demonstrate the contribution and importance of each data type and all possible combinations of them for the CHAMP orbit estimation.

2 Experiment results: average of 10-day interval

The experiments have generally been performed for the time interval from 2001-May-21 to May-30. The solved for parameters are: once per revolution empirical forces (each 90 min.), scaling factor Cd for drag (each 6 hour), state vector. If accelerometer data are used, the scales of accelerometer data are solved instead of Cd. Thruster factors are also estimated, when this information is considered. In most cases the SLR data are not used in the solutions, but the residuals are taken as an indicator of the orbit quality.

The results of adopting different data type combinations and using various processing methods are listed in table 1 to 3. Table 1 presents the results of 10 one-day arc solutions. First solution uses only the GPS data. In the second solution, in addition to phase and code measurements, the attitude information is considered. The arc length of the third solution is increased to 2-day. The forth solution is computed by using SLR data alone, due to the small amount of tracking data in this solution Cd is solved once per day, empirical force is estimated once per every two days.

All results given here are computed with the new CHAMP gravity model EIGEN-1s, see Reigber et. al. (2002). CHAMP data within the experiment time interval have not been used for the computation of this model. As we know, the most important factors for improving the quality of CHAMP orbits are gravity model and the number of once-per-rev. parameters being solved. A lot of tests were performed, in which different gravity models were used and different amounts of parameters were solved. Fig.1 symbolically summarizes these tests. The conclusions are: 1) the orbit accuracy (judged by laser rms fit) generally increases with the amount of empirical force parameters, at first the improvement is significant, but up to a certain amount, no perceptible effect can be seen (in some tests, the quality even gets worse); 2) if the amount of unknowns is the same, the CHAMP gravity model always gives better orbits. The more unknowns are solved the less significant the superiority of the good gravity model is. The reason is that more gravity errors are absorbed by larger amounts of once-per-rev parameters.

Once the gravity model and the amount of unknowns are fixed, combinations of different data types are tested in the solutions. The orbit quality improvement of

Data used	Arc length	SLR rms	GPS residuals Code/Phase
GPS alone	10×1day	6.8	78.8/1.20
GPS+attitude	10×1day	6.6	78.7/1.14
GPS+attitude	5×2day	6.5	75.4/1.52
SLR only	1×10day	14.2	--

Table 1. The orbit accuracy resulting from different POD methods Unit: cm

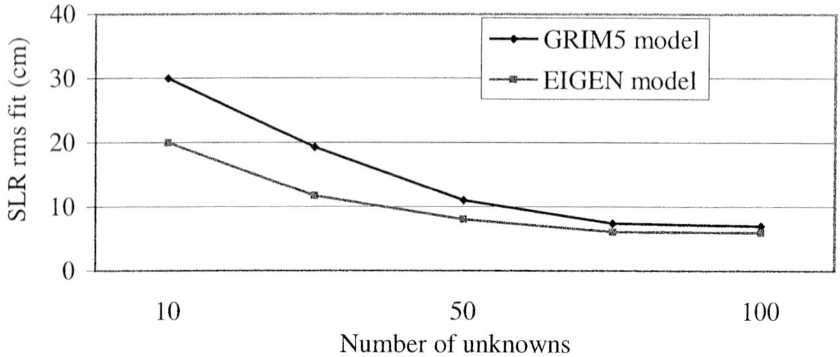

Fig. 1. Champ orbit accuracy by different number of unknowns and different gravity models.

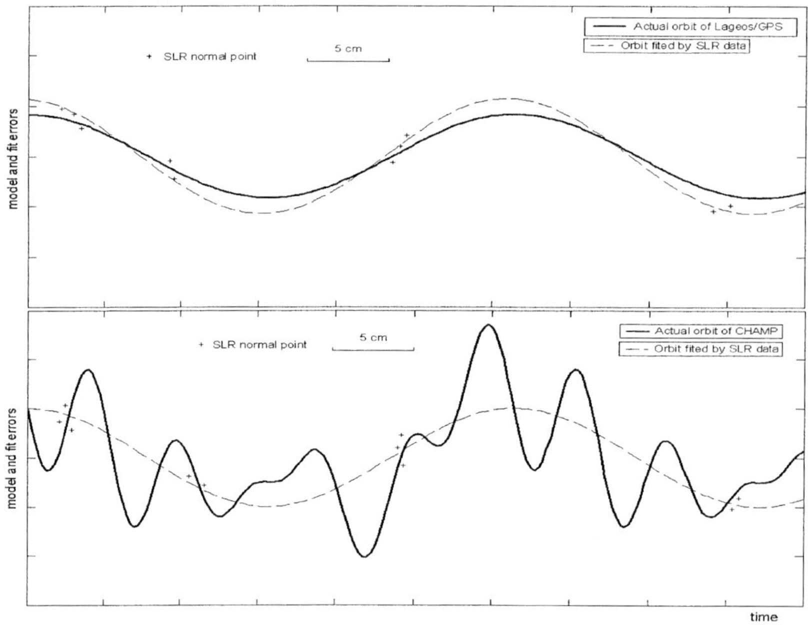

Fig. 2. SLR data used to fit high-altitude (Lageos/GPS) and low-altitude (CHAMP) orbits

the second solution in table 1 demonstrates the contribution of the attitude data. Usually, a solution based on a shorter arc gives smaller observation residuals. But it dose not necessarily give better orbits, as can be seen in the third solution.

As we know SLR data can be used independently for orbit restitution. The fourth row of table 1 presents such kind of solution. Compared with the first solution, this 10-day arc SLR orbit has 40, 30 and 134 cm error in the radial, cross- and along-track directions. By increasing the number of solved for parameters the SLR residuals reduce rapidly, but the actual orbits are getting worse. The reason is explained in Fig. 2. The upper curve stands for the effect of gravity error on GPS or LAGEOS (high-altitude) satellite, the lower for CHAMP (low-altitude). The gravity model error causes larger fluctuations (short period terms) in the orbit of low-altitude satellite. While its effect on high-altitude satellite is comparatively smooth. Although the SLR tracking data is not very abundant, the upper curve can be modeled quite well by solving a small number of parameters. This is why by using SLR data alone, one can get 10 cm (1-d) level accuracy (compared with IGS) for GPS35, (see Zhu et. al. 1997). But the number of parameters needed for modeling the low altitude satellite should be very large. The available data amount is far from enough. No matter what kind of parameterization one uses, the fitted curve has always a large deviation from the 'true' one. This is why high precession, (and especially) continuous GPS data is crucial for LEO projects.

3 Experiment results: some details

Some detailed studies of the contribution of various data type combinations for the orbit restitution are performed through a two-day arc experiment. In the first solution GPS data, accelerometer, attitude, and thrust data are all used. The scale factors of the thrust and accelerometer data are estimated instead of Cd coefficients. The amount of laser data points is 339. The orbits of all other solutions are compared with this one. The last column of table 2 shows the comparison results.

The second solution is obtained, when the SLR data are given the same weight as phase observations. SLR residual should be significantly smaller, since they are now used in the orbit determination. Even the rms of GPS observations is reduced

Data used	SLR rms	GPS residuals Code/Phase	Compared with 1 R / N / T
c+p+att.+thr.+acc (all)	5.6	70.5 / 1.65	--
all+slr	2.1	70.2 / 1.64	1.3 / 0.8 / 2.5
all-acc	4.9	70.9 / 1.53	3.3 / 3.4 / 3.1
code+phase	5.8	70.2 / 1.72	3.3 / 4.4 / 2.7
code only	11.8	64.8 / --	8.9 / 15.4 / 20.7
code+acc+att+thr	10.2	64.7 / --	9.3 / 15.1 / 19.9

Table 2. The orbit accuracy resulting from various data combinations, (2d arc d146-147) Unit: cm

slightly. This is a positive signal, especially when considering that the relation of the SLR observation numbers to the phase measurements are 339 to 29000, or 1:74.4. The 3-d orbit rms difference is 3cm. Since the 3-d precession of the CHAMP orbits is reaching the 10 cm level, this 3cm variation is not negligible.

The third solution is similar to the first one, but the accelerometer data has been excluded. The laser rms is getting smaller. This means that the use of the accelerometer data does not bring improvement for the POD over the two days period of investigation, the reason is the modeling of accelerometer data in the radial direction (see Touboul and Foulon, 2001). The fourth solution only uses GPS data. Contributions of the attitude and thrust data can be clearly seen by comparing with the third solution.

Only the code data is adopted in the fifth solution. The quality of the CHAMP orbit thus computed is better than one would have expected. The phase measurement precession is almost 100 times better than that of code. But in the two-step method, GPS orbits and clocks are fixed, which contain also uncertainty. They affects the (o-c) of code measurements in the same way as affects the phase measurements. This uncertainty is much larger than the precession of phase, but much less than that of the code. It means the actual accuracy of the phase data is degraded more significantly than code.

Adding accelerometer, attitude and thrust data to the code measurements one gets the last solution. It results in better orbits than without using these data. It seems that although the modeling of accelerometer still needs to be improved, these data still provides some useful information.

For DOY 141, CHAMP GPS data and those from 40 ground stations are simultaneously processed, the orbits of GPS and CHAMP were restituted in the same solution. Compared with the two-step solution, the accuracy (as judged by laser fits) of CHAMP orbit is improved slightly, the (CHAMP) GPS phase residuals are reduced (but those of code are increased), as can be seen from table 3. As far as the GPS orbits are concerned, in the 2-step solution the 3-d rms difference compared with IGS orbit is about 14cm, that of 1-step solution is 12cm. Due to its low altitude LEO is more sensitive to the geocenter. Combined processing method benefits from the advantages of each satellite type, thus strengthens the solution. One-step method may also help to improve the consistency of the reference frame. Further experiments are needed to fully explore the usefulness of this method.

Method	GPS_SST data	Code/phase rms	SLR residual
1-step	15441	91.1/0.82	5.4
2-step	15134	74.2/1.56	6.0

Table 3. One-step example (compared with 2-step method) Unit: cm

4 Summarizing Conclusion

1. GPS data has crucial importance for LEO orbits.
2. SLR provides an independent check of the quality of CHAMP orbit. Once the orbit position (3d) accuracy reaches 1 dm level, the SLR data should better be included in the orbit solution.
3. Accelerometer data should provide very useful information. Some hardware problems have been detected. It is expected that after the improved model being provided the contribution of this data type will be much larger (see Touboul and Foulon, 2001).
4. Thrust data, and especially the attitude data, are useful for the improvement of the quality of the orbit.
5. Combined use of various data types in a reasonable way generally gives the best orbits.
6. Although one can estimate a lot of once-per-rev (or stochastic) parameters to reduce the gravity model error to a certain extend, the improvement of gravity is the most important factor for the accuracy of the CHAMP orbit.
7. Short arc length is helpful for reducing the residuals of (O-C), but not necessary for the good orbit, especially when the gravity and other dynamic models are good to certain extend (like CHAMP EIGEN gravity model). For gravity recovery, longer arcs might be more useful.
8. One-step method may provide more consistent and accurate CHAMP orbit, although CPU time and other related efforts increase also significantly.

Acknowledgement. This study was carried out under the grant of the German Federal Ministry of Education and Research (BMBF) No. 03F0333A.

References

Neumayer KH, Zhu SY, Reigber Ch, Koenig R, Meixner H (2000) CHAMP Orbit Recovery - Algorithmic Approaches and Results. EOS, Transactions. American Geophysical Union, Vol 81, No 48, p F309.

Reigber Ch, Balmino G, Schwintzer P, Biancale R, Bode A, Lemoine JM, Koenig R, Loyer S, Neumayer H, Marty JC, Barthelmes F, Perosanz F, Zhu SY (2002) A high-quality global gravity field model from CHAMP GPS tracking data and accelerometry (EIGEN-1s), Geophys Research Letter, 29, No. 14, 10.1029/2002GL015064, 37-1 - 37-4, 2002

Touboul P and Foulon B (2001) Acceleromrtre STAR Analuse du comportement de l'electrode X3' ONERA tech. Report. 1/05559 DMPH.

Zhu SY, Reigber Ch, Kang Z (1997) Apropos laser tracking to GPS satellite. Journal of Geodesy, vol 71, p 423-431

CHAMP Rapid Science Orbit Determination - Status and Future Prospects

Grzegorz Michalak[1], Gerald Baustert[2], Rolf König[1], Christoph Reigber[1]

[1] GeoForschungsZentrum Potsdam (GFZ), Division 1, Kinematics and Dynamics of the Earth
[2] TeleOrbit Linden, GFZ Contractor

Summary. The CHAMP Rapid Science Orbit (RSO) is produced to support the atmospheric and ionospheric occultation processors as well as the magnetic/electric field analysis system of the CHAMP Science Data System. The RSO is computed using dynamical models in a two step approach. First the orbits and clocks of the GPS satellites are estimated, then the CHAMP orbit is determined by fixing the estimated GPS orbits and clocks. For quality control external GPS orbits and CHAMP laser observations are used. The GPS orbits are delivered with 14 h latency and with an accuracy of about 10 cm in 1-D position, while the CHAMP orbits are delivered with 16 h latency and an accuracy of 5 cm in 1-D position. For the future it is planned to decrease the RSO latency for use in a numerical weather prediction environment.

Key words: CHAMP, GPS, rapid precise orbit determination

1 Introduction

The GeoForschungsZentrum Potsdam (GFZ) delivers orbit ephemerides for the GPS satellites and the CHAMP satellite on a daily basis. These precise orbits are called RSOs (Rapid Science Orbits). The GPS-RSO (the RSO for the GPS satellites) is produced since the launch of the CHAMP satellite, July 15, 2000, while the production of the CHAMP-RSO commenced on March 5, 2001, when the radio occultation experiment started. Precise RSOs are needed for the evaluation of radio occultations [1] and to support magnetic and electric field studies [2], [3]. The RSOs, being routine products of the CHAMP Science Data System installed by GFZ, are calculated in a two step approach: in the first step GPS orbits and clocks are determined, in the second step CHAMP orbits are calculated keeping the GPS orbits and clocks fixed.

2 Rapid Science Orbits: Two Step Approach

2.1 GPS-RSO

The 5 minute spaced ephemerides and clocks of the GPS satellites covering 1 day are calculated dynamically using undifferenced GPS code and phase data from the GFZ/JPL low latency ground station network, which was installed prior to the CHAMP launch to support the radio occultation experiment, and from hourly and daily sites of the International GPS Service (IGS). To ensure short computation times and high accuracies of the orbits, an optimization procedure for the GPS ground network data is applied [4] that selects approximately 50 stations with optimal geographic distribution. The orbits are integrated numerically by means of a constant step size Cowell integrator. The force model for the GPS-RSO consists of the following constituents:

- Earth gravity model GRIM5-C1 12×12
- Perturbations from Sun, Moon and planets
- Solar radiation pressure
- Earth tides
- Empirical forces

In the daily runs the following parameters are estimated: the initial states for all GPS satellites, 2 solar radiation pressure coefficients for each satellite, empirical coefficients (4 per satellite revolution), satellite clock corrections, station coordinates (a-priori values are constrained to the ITRF 2000 reference frame [6]), tropospheric corrections (6 for each station), station clock corrections, and ambiguities.

The quality of the GPS orbits is monitored by comparing the GPS-RSO with the IGS Ultra Rapid Orbits (igu) [7] and later on by comparison with the IGS Rapid Orbits (igr). The result of the quality control can be seen in Fig. 1.

In the upper panel the post-fit code and phase RMS values are presented. It can be seen that the mean value for the code RMS is approximately 85 cm and for the phase about 0.55 cm. The lower panel in Fig. 1 contains the RMS values of the differences between the GPS-RSO and the igr orbits after applying a global Helmert transformation. The result of this comparison indicates an agreement with the igr orbits on a level of 10 cm in 1-D position. High RMS values visible as peaks in Fig. 1 are obtained when a single satellite differs significantly from the igr solution. In most cases this satellite is marked as less precise in the header of the igr SP3 file.

2.2 CHAMP-RSO

For each day 2 CHAMP orbits of 14-hour arc length each, with 30 s spacing of the ephemerides are dynamically calculated using undifferenced GPS Satellite-To-Satellite Tracking (SST) code and phase measurements of the

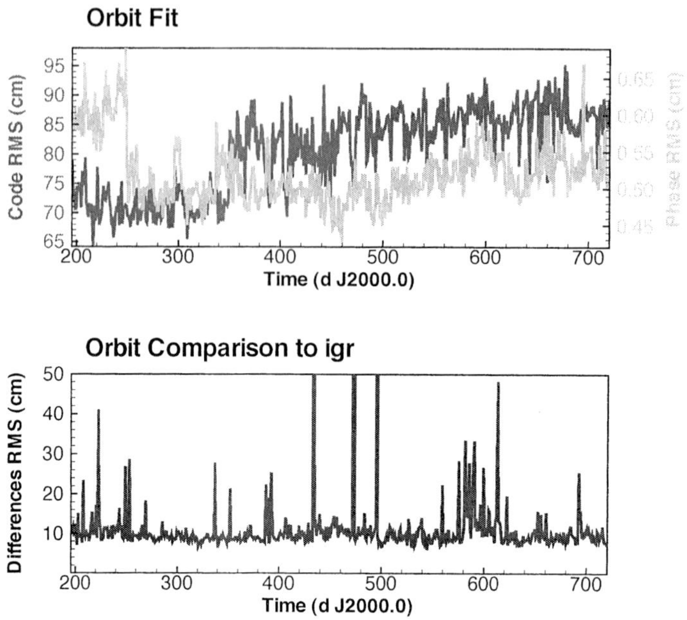

Fig. 1. The accuracy of the GPS-RSO. In the upper panel the post-fit code and phase RMS is shown, in the lower one the RMS values of differences in 1-D position between GPS-RSOs and the IGS Rapid Orbits (igr).

on-board receiver. The first orbit starts at 22 o'clock of the day before and ends at 12 o'clock of the day. The second one starts at 10 o'clock and ends at 24 o'clock. The two orbits overlap between 10:00–12:00 and 22:00–24:00. The CHAMP orbits are determined keeping the GPS orbits and clocks fixed (see Sect. 2.1). Since the fixed GPS orbits and clocks have 300 s spacing, an interpolation is made to obtain 30 s resolution for the CHAMP-RSO. In the CHAMP orbit adjustment the force model consists of:

- Earth gravity model GRIM5-C1 70×70, updated with 2 months of CHAMP data
- Perturbations from Sun, Moon and planets
- Albedo and radiation of the Earth
- Atmospheric drag
- Solar radiation pressure
- Earth and ocean tides
- Empirical forces

For the CHAMP-RSO no use is made of accelerometer data. The unmodelled nongravitational surface forces are represented by the estimated empirical forces. For each orbit the following parameters are estimated: the CHAMP initial state, 1 solar radiation pressure coefficient, 3 atmospheric drag coefficients, 72 empirical force parameters (every 1.5 hours, once and

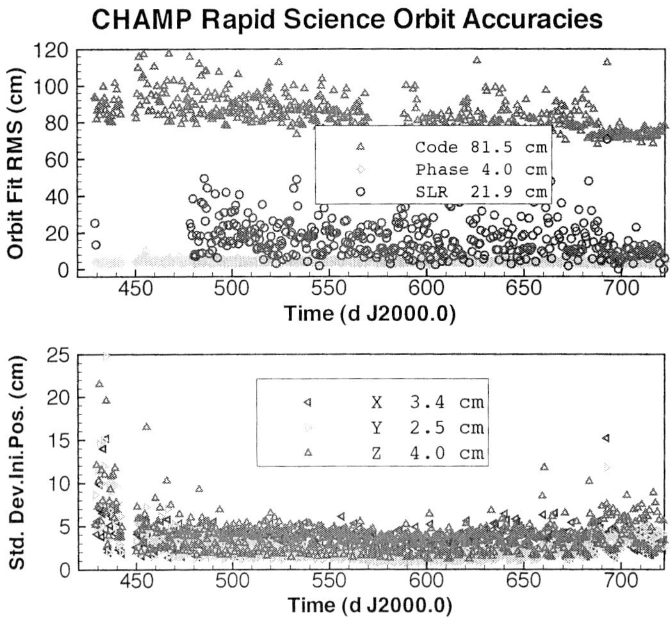

Fig. 2. Accuracies of the CHAMP-RSO. A significant improvement of SLR residuals is visible after day 682 when a new orbit parametrization was introduced: the mean SLR RMS value dropped to 12 cm.

twice per revolution in normal and transversal direction), CHAMP clock corrections, and ambiguities.

For the internal quality check of the CHAMP-RSO code and phase RMS values as well as standard deviations of the initial elements are used. The quality of the orbits is also independently monitored by the RMS values of down-weighted SLR observations. In Fig. 2 the RMS values of code, phase and the SLR observations (upper panel) and also the standard deviations of the initial positions (X, Y and Z components) are presented. The global value for code RMS amounts to 81 cm, for the phase RMS to 4 cm and for the SLR RMS to 22 cm, which can be viewed as 3-D position accuracy estimate. After day 682 (November 13, 2001) we introduced a denser parametrization of the orbit by empirical coefficients as given above. As shown in Fig. 1, this resulted in improved code, phase and SLR RMS values of 74 cm, 3 cm and 12 cm, respectively. The denser parametrization however produces higher standard deviations of the initial position (see the lower panel in Fig. 1).

Due to border effects at the beginning and at the end of dynamically restituted orbits, it is recommended for users to cut the 10:00–24:00 orbit to the period 10:50:00–22:49:30 and the 22:00–12:00 orbit to the period 22:50:00–10:49:30. The border effects are already significantly reduced by a recent implementation aiming at a better modeling of the forces (gravity field, empirical forces). After these improvements the SLR RMS is again significantly

Table 1. GPS and CHAMP RSO product specification. For a description of the CHORB format see http://op.gfz-potsdam.de/champ/more/docs_CHAMP.html

Orbit	Spacing	Arc length	Content	Latency	Volume	Format
GPS-RSO	300 s	1 d	pos., vel. & clocks	14 h	1 MB	SP3
CHAMP-RSO	30 s	14 h	pos.& vel.	16 h	0.4 MB	CHORB

reduced. Moreover, on March 5, 2002, the CHAMP GPS receiver was commanded to track simultaneously up to 10 GPS satellites in contrast to a maximum of 8 before. Global SLR RMS values of about 7 cm are now achievd after these improvements.

2.3 GPS and CHAMP RSO Products

GPS and CHAMP RSOs are delivered on a daily basis to the CHAMP ISDC (Information System and Data Center [5]). The GPS-RSO latency is about 14 h, which means that the products are available approximately at 14:00 UT at the ISDC. The CHAMP-RSO products are delivered to ISDC with a 16 hours latency and are therefore available at the ISDC at 16:00 UT. The characteristics of the GPS-RSO products, named CH-OG-3-RSO+CTS-GPS_yyyy_doy_hh in the ISDC, and the CHAMP-RSO products, named CH-OG-3-RSO+CTS-CHA_yyyy_doy_hh, are given in Table 1. The string CTS denotes Conventional Terrestrial System which is based on the ITRF 2000 reference frame as given above. The acronyms yyyy denote the year, doy the day of the year, hh the hour of the day. Products with the assignment CH-OG-3 are for public access. In the ISDC also products beginning with iCH-OG-3 for restricted access are stored. They containe the same products in the TDS (True of Date System) reference system.

3 Future Prospects

The current way of supplying the RSO can be regarded as a first attempt towards a more frequent product delivery. High frequency and short delay of RSO provision is especially important for numerical weather prediction systems for which the radio occultations should be available three hours after the last measurement. Therefore it is planned to demonstrate near-real time RSO production every three hours. The present RSO accuracies in 1-D position, 10 cm for GPS and 4-6 cm for CHAMP is above the pre-mission specifications. Thus, for this application the quality control threshold could be relaxed if needed to meet the latency requirements.

4 Summary and Conclusions

GPS and CHAMP Rapid Science Orbits calculated by the GeoForschungsZentrum Potsdam are delivered to the CHAMP Information System and Data Center on a daily basis with delays of 14 h and 16 h, respectively. Comparison of the GPS-RSO with IGS igr orbits indicate a 10 cm 1-D position accuracy. CHAMP-RSO accuracy is currently sizing at a level of 5 cm infered independently from SLR measurements. The accuracy requirements of the radio occultation experiment, 10 cm for GPS and 25 cm for CHAMP, are thus met. More frequent RSO deliveries for use in GPS meteorology are under development and will be demonstrated in the near future.

Acknowledgement. This study was carried out under the grant of the German Federal Ministry of Education and Research (BMBF) No. 03F0333A.

References

1. Wickert J, Reigber Ch, Beyerle G, König R, Marquardt C, Schmidt T, Grunwaldt L, Galas R, Meehan TK, Melbourne WG, Hocke K (2001) Atmosphere Sounding by GPS Radio Occultation: First Results from CHAMP. Geophysical Research Letters, **28**, No 17, 3262–3266
2. Lühr H (2001) The magnetic survey mission CHAMP: Early results. American Geophysical Union, Spring Meeting 2001, abstract # GP31C-01
3. Cooke DL, Roth C, Huang C et al (2001) Early validation results from the Digital Ion Drift Meter, DIDM in CHAMP. American Geophysical Union, Spring Meeting 2001, abstract #SM32A-22
4. Schwieger V, Baustert G, König R (2001) Optimization of the GPS Observation Plan for CHAMP Rapid Orbit Determination. Geophysical Research Abstracts, EGS 26th General Assembly, Nice, France
5. Ritschel B, Palm H, Schneider M (2001) CHAMP ISDC internet site http://isdc.gfz-potsdam.de/champ
6. ITRF 2000 reference frame, http://lareg.ensg.ign.fr/ITRF/ITRF2000
7. IGS Data and Products, http://igscb.jpl.nasa.gov/components/prods_cb.html

station network. Pre-launch investigations based on simulations for CHAMP and real-world data from the GPS/MET satellite had shown that accurate orbit predictions could be derived from such GPS navigation solution data. The major benefits of this data type are the almost continuous coverage of the orbit at a sufficient level of accuracy and the routinely availability of the data. For CHAMP the GPS navigation solution consists of state vectors (position and velocity) at 10 seconds intervals in the WGS reference frame computed onboard by the BlackJack receiver from GPS signals. The coverage of the CHAMP orbit from this data type is more than 99 % of the elapsed mission time. With the initial software version of the BlackJack receiver the accuracy of the position information was at the level of 40 m. After a software upgrade (February 23rd, 2001) the accuracy improved to a value of about 7 - 8 m. Precise SLR data is incorporated to further strengthen the estimation of orbital parameters used for the generation of the predicted orbit.

3.2 Method and Standards

In a first step the orbit sampled by the data (observed part) is recovered in a dynamic orbit determination method based on numerical integration of the perturbated motion of the satellite in combination with an iterative least-squares adjustment procedure implemented in the EPOS-OC software

Table 1. Standards of the Predicted Science Orbit

Data	GPS navigation solution, spacing 10 s SLR normal points Auxiliary data: polar motion, solar/geomagnetic activity
Models	Gravitational field of the Earth (CHAMP tailored GRIM5-C1, n=m=90) Ocean tides (GRIM5-C1) Tides of the solid Earth (Wahr + Corrections) Gravitational field of the Sun, Moon (point masses, DE200/LE200) Atmospheric drag (CIRA86, macro-model of the satellite) Solar radiation pressure (macro-model of the satellite) Earth Albedo (Knocke CSR, macro-model of the satellite)
Parameters	6 initial elements at epoch (position & velocity) before Nov. 28th, 2001 (old paramtrization): 1 global scaling factor for atmospheric drag after Nov. 28th, 2001 (new parametrization): 6-hourly polygonal scaling factor for atmospheric drag
Weighting	GPS-POS (20.00 m old param., 9.00 m new param.) GPS-VEL (10.00 cm/s old param., 10.00 cm/s new param.) SLR (1.50 m old param., 8.00 m new param.)

system of GFZ. Once the adjustment process has converged, based on the estimates of the set of parameters and the set of models chosen in the previous step, the orbit is extrapolated (predicted part) by the numerical integration procedure beyond the epoch of the most actual available observational data such that the predicted orbit covers a 4-day period. Table 1 lists the standards for the generation of the predicted orbit. The initial (old) parametrization (before November, 28th 2001) originates from the experiences gathered at GFZ during the operations of the ERS-1/2 and the GFZ-1 mission based on SLR data only. This rather limited set of free parameters is restricted with respect to the potential accuracy of orbit determination from the given SLR data. However, it has proven to be of benefit for the prediction of the orbit where focus is laid on the overall future trend of the motion of the satellites. Internal investigations for CHAMP during the mission had shown that the quality of the observed as well as the predicted part of the orbit could be improved by an extended parametrization including 6-hourly scaling factors for drag instead of one global scaling factor (see Sect. 4). It should be noted, that the consistent and reliable estimation of such parameters is only possible due to the quasi-continuous coverage of the orbit with GPS navigation solution, however.

3.3 Products

Table [2] summarizes the products generated by the prediction processor derived from the predicted part of the orbit. The basic products for SLR tracking of CHAMP are the ILRS standardized [4] so-called tuned Inter Range Vectors (IRV) in connection with the coefficients of the so-called drag functions (DRAG). Twoline elements are basically used for the scheduling of passes at the data dump stations in Ny Alesund and Neustrelitz. The so-called Smithonian Astronomical Observatory elements (SAO) are still used at some ILRS stations for scheduling purposes. IRVs, DRAG, TLE and SAO generated by the prediction processor are also referred to as orbit prediction products (OPP). The PreDicted Science Orbit (PDO) is the original dynamically integrated and differentially corrected orbit (observed and predicted part). The data files are transferred to the relevant data center (see Table 2) via ftp. IRVs and DRAG (and for some ILRS stations SAO) files are disseminated via ILRS mail exploders to the ILRS ground stations.

4 Accuracy of Orbit Predictions

The accuracy of the orbit predictions has been investigated by means of the orbital fit of the observational data and orbit comparisons to the more precise so-called Rapid Science Orbit (see [2] for details on the RSO). In the study for a set of 64 orbit predictions in the period September 3rd to September

Table 2. Products of the Prediction Processor

Product	Recipient
PreDicted science Orbit (PDO)	ISDC, public access
tuned Inter Range Vectors (IRV)	ILRS data centers & stations
DRAG function coefficients (DRAG)	ILRS data centers & stations
Twoline Elements (TLE)	Telemetry stations
Smithonian Astronomical Observatory elements (SAO)	some ILRS stations

30th, 2001 results were computed seperately for the observed part and the predicted part of the orbits.

4.1 Orbital fit

Fig. 1 depicts the orbital fit of the GPS navigation solution (only position) and the SLR normal points. The significant drop-off of the rms values for GPS navigation solution data from about 40 m to aprroximately 7 - 8 m due to the software upgrade is clearly visible. The rms values of the SLR data is of the order of 2 - 4 m. In principle, since SLR data is independent from the GPS navigation solution, the SLR residuals indicate the outer accuracy of the observed part of the orbit. However, one should note that despite the down-weighting of SLR data (see Table 1), the SLR data still contribute to the orbit determination. The values given in Table 1 were found empirically and represent an appropriate mutual weighting of precise laser range measurements relative to the less accurate, but outnumbering amount of data from the GPS navigation solution.

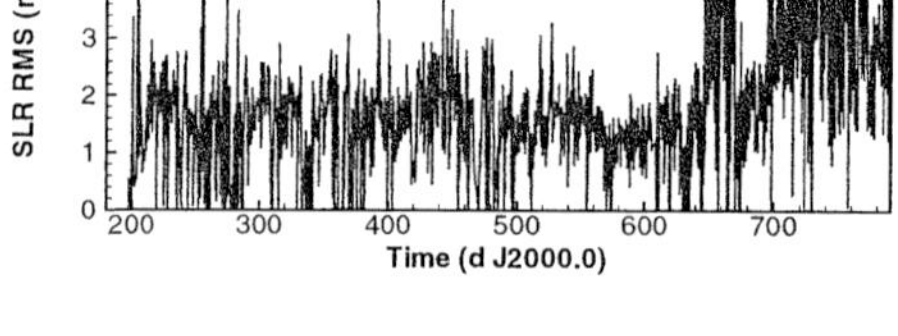

Fig. 1. Orbital Fit GPS Navigation Solution, SLR Data

4.2 Orbit Comparison - Observed Part

Table 3 summarizes the rms-values of the orbit comparisons of the observed part of the PDO with respect to the RSO in the components of the orbital system, i.e. in radial, along- and cross-track directions. The values given correspond to the extended parametrization (6-hourly scaling factors for drag); in brackets the rms-values of the old parametrization (one global scaling factor for drag) are displayed. The improvement in along-track direction due to the modified parameter set is significant, whereas in radial direction only a moderate and in cross-track direction no improvement is visible.

Table 3. Orbit Comparison PDO - RSO (observed part)

rms radial [m]	rms cross [m]	rms along [m]
1.4 (1.9)	1.8 (1.8)	3.6 (17.1)

4.3 Orbit Comparison - Predicted Part

For the evaluation of the quality of the predicted part the resulting deviations in radial, along- and cross-track directions between the PDO (predicted part) and the RSO are investigated as time series. Table 4 depicts the means of the absolute values of the development of the deviations for all 64 PDOs at 12, 24 and 36 hours after the final observation incorporated in the individual orbit. Again, in brackets results for the old parametrization are given. Typical 1/rev oscillations included in the time series were removed beforehand by fitting the individual time series to cubic, quadratic functions or linear functions (see equation 1) in a least-squares adjustment combined with a check of significance of the estimated polynomial parameters.

$$dev_i(t) = a + bt + ct^2 + dt^3 \qquad (1)$$

where $dev_i(t)$ is the time series of the deviation in radial, cross-, along-track direction. a, b, c and d are the polynomial coefficients to be determined. Based on the coefficients valid for the individual PDO, values at $t = 12, t = 24$ and $t = 36$ were computed. ¿From the resulting absolute values for each intervall the mean values of all 64 PDOs were evaluated and displayed in Table 4. As expected, due to the low orbit of CHAMP and the large impact of the variable surface forces such as drag and solar radiation pressure, the along-track component shows the largest deviations at all intervals. In radial and cross-track direction the decay of the predicted orbit is rather moderate. Again, the improvement due to the extended parametrization in along-track direction is significant, whereas in the other directions the amelioration is neglectable. From these results it is seen that in general the extended parametrization

Table 4. Orbit Comparison PDO - RSO (predicted part)

interval [h]	radial [m]	cross [m]	along [m]
12	5 (8)	1 (1)	184 (388)
24	12 (13)	3 (3)	770 (1130)
36	20 (20)	4 (4)	1860 (2290)

also improves the quality of the predicted part of the orbit in the along-track component. As of November 28th, 2001 the modified parametrization was adopted in the prediction processor. The mean value at 12 hrs intervall in along-track direction indicates, that the overall set of the 64 orbit predictions generated does not fullfill SLR requirement already after 12 hrs. The detailed investigation of the behaviour of the predicts reveal variable performance for individual orbits, however from the set of orbits given, about 18 orbits show deviations in along-track at 12 hrs smaller than or equal to the specified SLR requirement. The rest shows larger deviations up to 50 ms; 3 orbits reaching exceptional values up to 100 ms. This variable behaviour can be explained by unpredicted (and hardly predictable) solar events causing large short-time variations in drag and solar radiation pressure. This result indicates that in addition to the improvement of the modified parametrization updating at periods smaller 12 hrs is still necessary. As a consequence from this study the generation frequency was set to 3 predicts per day these days.

5 Conclusions

Accurate orbit predictions based on the GPS navigation solution from the on-board receiver and on additional SLR data are provided on a regular basis by an automated processor. Results from orbits comparisons with the RSO as a precise reference orbit indicate an orbit accuracy at the level of a few meters for the observed part of the PDO. With the implementation of an extended parametrization the accuracy of the observed as well as the predicted part in along-track direction has improved significantly. The decay of the predicted part in radial and cross-track direction is neglectable for time intervals up to 12 hrs; in along-track direction larger deviations already at the 12 hrs interval have to be compensated through updating of the predictions on a sub-12-hrs basis (status: 3 predicts/day). Depending on the actual decrease of CHAMP's altitude and the development of the solar and geomagnetic activity a further increased update rate might become necessary.

Acknowledgement. This study was carried out under the grant of the German Federal Ministry of Education and Research (BMBF) No. 03F0333A.

References

1. Schmidt R, Baustert G, König R, Meixner H, Neumayer K-H, Reigber Ch (2000) The Development of Automated Processing of Orbit Predictions for CHAMP. In: Proceedings of the International Workshop on Laser Ranging, November 2000, Matera, Italy
2. Michalak G, Baustert G, König R, Reigber Ch (2002) CHAMP Rapid Science Orbit determination - Status and Future Prospects. This issue.
3. Ritschel B, Palm H, Schneider M, (2001) CHAMP ISDC internet site http://isdc.gfz-potsdam.de/champ
4. The International Laser Ranging Service internet site http://ilrs.gsfc.nasa.gov/

Thermospheric events in CHAMP Precise Orbit Determination

Rolf König and Karl Hans Neumayer

GeoForschungsZentrum Potsdam, Telegrafenberg, D-14473 Potsdam, Division Kinematics and Dynamics of the Earth.
koenigr@gfz-potsdam.de

Summary. The accelerometer on-board the CHAMP satellite measures thermospheric events like large solar mass eruptions reaching Earth. The accelerometer data allow to replace the models for the non-conservative forces in Precise Orbit Determination (POD). The paper will discuss the impacts on POD and show some of the results.

Key words: CHAMP, solar wind, thermospheric event, accelerometry

1 Introduction

The along-track part of the acceleration measured on-board CHAMP (for a mission description see Reigber et al.(1999)) is most intimately connected with the atmospheric density at the position of the satellite. ¿From elementary gas dynamics it is known (see e.g. King-Hele (1964)) that the along-track component of the air drag F_D of a symmetric sphere that moves with a three-dimensional velocity $\dot{\xi}$ with respect to a surrounding gaseous medium is proportional to

$$F_D \sim \rho(\xi)\ \|\dot{\xi}\|_2^2, \qquad (1)$$

here ξ is the position of the sphere and $\rho = \rho(\xi)$ is the gas density; thus the latter quantity, in principle, is given by dividing the along-track acceleration by the square of the velocity.

In reality things are a little more complicated. The air drag is not the only surface force acting on the satellite body, the satellite has no spherical symmetry and makes small pitch and yaw movements up to 2 degrees away from nominal zero, and we have completely neglected the effects of a rotating earth and the dynamics of the atmosphere. Nevertheless, the along-track acceleration and the gas density remain highly correlated even if we take all these effects into account.

It is therefore not too far-fetched to assume that spatially well-localized changes in the density of the air are visible in the along-track acceleration. A quite active producer of atmospheric density fluctuations is the sun.

2 The thermospheric event of September 25th, 2001

The solar wind consists of matter that - under normal circumstances - travels at a velocity of some 400 km/s. Around September 22nd, 2001, there was a solar event (see e.g. Jansen (2002)), an ejection of charged particles, and such an event has two effects on the solar wind: the particle density increases, and the mean velocity of the charged particles can double and reach up to 800 km/s. Now given a velocity of 400 to 800 km/s of such a particle cloud, it is easy to verify that it needs between 2 to 4 days to reach the immediate environment of the earth. In our case, the aftermath of the solar event reached Earth's magnetic field some three days after the solar eruption. Its effects became visible during the last four hours of September 25th.

The magnetic field of the earth catches the charged particles and traps them like a liquid in a bottle. The only places of this magnetic trap that allow venting when saturation occurs are the polar regions of the earth. The interaction of the particles with the upper atmosphere are accompanied by a well-known optical phenomenon, the Aurorae Borealis and Australis. But there is more.

In Fig. 1, we see the comparison of directly measured accelerometer data (the ragged curve) against adjusted model surface forces (the smooth curve). This force model formally consists of the sum of air drag, solar radiation pressure and earth albedo, but the largest effect at these low altitudes is of course caused by the drag of the residual atmosphere.

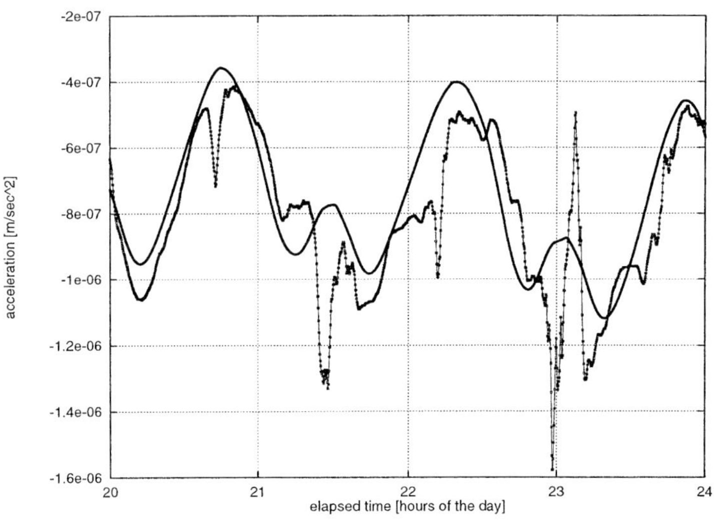

Fig. 1. Disturbed air drag measured by the CHAMP accelerometer.

In the measured accelerometer data, four distinct features stick out quite at first sight: At 20:45, at 21:30, at 22:15 and at 23:00 we see sharp downward peaks in the accelerometer measurements. The last peak is even followed by an immediate upward peak.

Downward peaks in measured accelerometer data always mean sudden and well-localized decelerations. As most of the surface force budget comes from the air drag, we suspect that the satellite has passed through regions with sharply increased atmospheric densities. Now what has happened?

It can be shown that the four anomalies occurr exactly at overflights of the magnetic north and south polar regions of the earth. Here are the very places where the above-mentioned venting of the saturated magnetic bottle takes place, and the impact of the charged particles heats up the atmosphere and lets it well up like boiling water in a cauldron. This upwelling creates zones of high gas density, and this causes the extraordinary accelerometer readings.

3 Orbit adjustment with modelled surface forces

In the daily rapid science orbit processing within the CHAMP science data system installed by the GeoForschungsZentrum Potsdam, using only the three modelled physical surface forces does not give enough parameters to get a satisfactory product. To enhance the quality, along-track and cross-track empirical coefficients once and twice per revolution are estimated; i.e. these additional forces are trigonometric polynoms with expansion coefficients that change every time that CHAMP completes one orbit around Earth.

This parametrization of the dynamics goes hand in hand with an inflation of the number of adjustable variables. To fly around the earth one time CHAMP needs 93 minutes, so it completes some 16 orbits per day. As the empirical terms are once and twice periodic for every new orbit of CHAMP, this figure has to be multiplied by 2; as empirical coefficients in two spatial directions, namely cross-track and along-track, are estimated, the number is doubled, and a last doubling is necessary as coefficients have to be adjusted for both cosine and sine terms seperately. Thus the set of parameters includes 128 additional variables per day for the empirical part of the perturbation forces alone.

For the corresponding plot of modelled and measured surface forces during the last 4 hours of September 25th see Fig. 2.

The ragged curve again is the measured acceleration, the smoother curve that now exhibits steps, due to the revolution-wise changing empirical coefficients, stands for the modelled surface forces augmented by the empirical accelerations. We see that at the period where there are atmospheric events there also is a lot of action in the modelled surface forces, but yet the latter curve still does not faithfully represent the true measured accelerations as

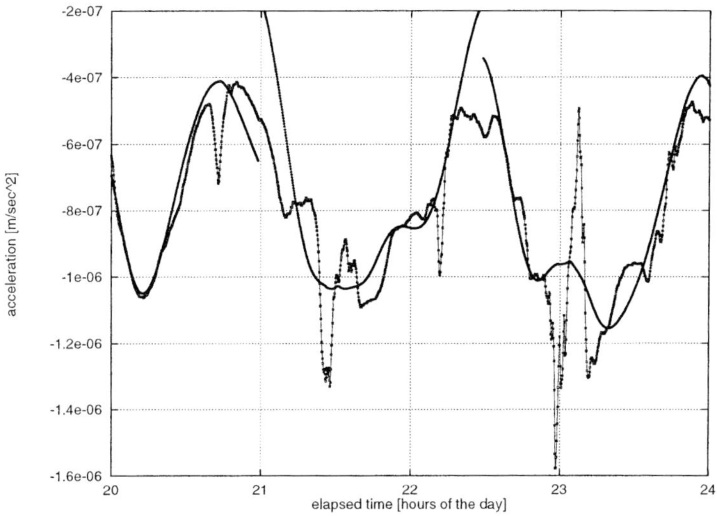

Fig. 2. Modelled non-conservative accelerations, revolution dependent empirical parameters estimated.

they are. As we have 1 per rev and 2 per rev empirical coefficients, the temporal size of a typical feature that an approach with modelled surface forces is capable to resolve is half a revolution of CHAMP, i.e. approximately 45 minutes. Anything that is shorter cannot be felt by this method.

4 Orbit adjustment with measured surface forces

Instead of modelling the surface forces acting on the CHAMP spacecraft body we can use the accelerometer data and adjust accelerometer calibration parameters. The parameters determining the surface forces are now scaling factors and biases for the accelerometer output.

For the orbit of September 25th, global scaling factors were estimated for the along-track and the cross-track channel as well as polygonal biases for all three spatial directions: radial, along-track and cross-track. The epochs where the polygon corners are located in time have a separation of approximately one orbit period of CHAMP, i.e. 93 minutes.

All estimated biases were held completely free.

In Fig. 3, we see the original accelerometer data for the along-track channel plotted over the same acceleration endowed with calibration parameters. In the upper half of the same plot, there is a diagram of the estimated bias parameter polygon. It is remarkable how stable the along-track bias comes out, especially if we take into account that no a-priori sigmas for the polygon corners were prescribed.

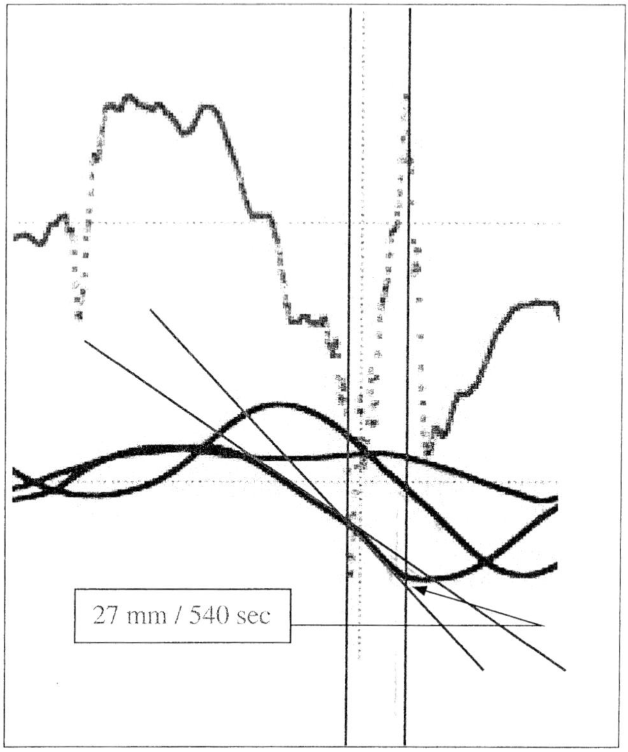

Fig. 5. Net amount of orbit change due to thermo burst: 0.05 mm/s.

It is vital for such an assessment to have something like an independent external reference, i.e. SLR (Satellite Laser Ranging). What we would need is an SLR pass approximately three minutes after the thermo burst. Now, apparent from September 25th, two other events in 2001 were checked, but no SLR data were available at the critical epochs. Unfortunately, the kind of event presented here, due to its nature, has the tendency to occur in the vincinity of the North and the South pole of the earth, i.e. at lattitudes where we have no Laser stations to monitor them.

Acknowledgement. This study was carried out under the grant of the German Federal Ministry of Education and Research (BMBF) No. 03F0333A.

References

1. Reigber Ch, Schwintzer P, Lühr H (1999) The CHAMP geopotential mission. Boll Geof Theor Appl **40**, 285-289

2. King-Hele D (1964) Theory of satellite orbits in an atmosphere. Butterworths, London
3. Jansen F (2002) Projekt WeltraumWetterWarte. Internet, http://www.physik.uni-greifswald.de/~sterne/Sternwarte/html/alt.html

New Global Gravity Field Models from Selected CHAMP Data Sets

Christoph Reigber [1], Georges Balmino [2], Peter Schwintzer [1], Richard Biancale [2], Albert Bode [1], Jean-Michel Lemoine [2], Rolf König [1], Sylvain Loyer [2], Hans Neumayer [1], Jean-Charles Marty [2], Franz Barthelmes [1], Felix Perosanz [2], Sheng Yuan Zhu [1]

[1]GeoForschungsZentrum Potsdam (GFZ), Division Kinematics & Dynamics of the Earth, Potsdam, Germany
[2]Groupe de Recherche de Géodésie Spatiale (GRGS), Toulouse, France

Summary. Several sets of data were carefully selected out of the first 12 months of CHAMP GPS and STAR micro-accelerometer observations to derive a new model of the Earth's static gravity field. These data have been dynamically processed as 63 orbital arcs, each of one to one and a half day duration. The approach consists of two steps: the GPS satellite orbits and clocks are first determined, then the CHAMP orbital arcs are adjusted using undifferenced phase and pseudo-range SST observables, and the residuals of those are used for retrieving the gravity harmonic coefficients. Micro-accelerometer, attitude and manoeuver information are taken into account to determine the surface accelerations as well as the residual thrusts acting on the spacecraft. Temporal variations of the gravity field are introduced in the observation equations for further usage in subsequent models, but are not here solved for. Several solutions have been derived, with or without additional information coming from the previous GRIM 5 satellite model, or from other laser satellite observations over a time period encompassing the CHAMP first year mission, with regularization (based on Kaula's rule) and with different weighting factors. The solution which is presented is complete to degree and order 120, although a lack of power in the coefficients power spectrum is obvious above degree 40 approximately, as may be expected from a satellite solution, due to the orbit decreasing sensitivity to gravity at the present altitude. The quality of the field is assessed mainly through orbit determinations of several geodetic satellites, of CHAMP itself (especially with the residuals of the laser range measurements - not included in the gravity modelling process), and through comparisons with the geoidal surface derived from a mean altimetric sea surface - corrected for the ocean circulation, at a comparable resolution. A significant improvement in the long to medium wavelength harmonics of the model can already be seen as compared to previous satellite solutions.

Key words: Gravity field, global model, satellite-to-satellite tracking, micro-accelerometry

1 Introduction

The determination of gravity field models from Earth's satellites has traditionally been based on observations of their trajectories by different techniques from the ground, with uneven coverage of the orbits and inhomogeneous quality. Therefore the solutions of this inverse celestial mechanics problem in which the model parameters are estimated from the orbit perturbations, have had to make use of many satellites in different geometries (inclinations) but have been rather weak. Besides, the resolution of all models has been limited by the quasi-exponential decay of the orbital perturbations due to the relatively high mean altitude of the observable satellites resulting in too weak signals – apart from resonance occurrence. Also, the modeling uncertainties in the surface forces (due to air drag and solar radiation pressures) acting on most spacecraft, have been a large error source. In terms of spherical coefficients of a representation of the gravitational potential, the number of terms which may be considered well determined (decorrelated from each other and with uncertainties below the 1% level) is limited to a few hundreds (say maximum degree and order around 20 – not speaking of time variations). Of course current satellite models, for instance GRIM5-S1 (Biancale et al., 2000), are computed to higher degree and order by applying some mathematical regularization to the solution. The highest resolution models, such as EGM 96 (Lemoine et al., 1998) with an improvement in the overall quality of representation of the field, have been obtained by combination of the satellite data with surface observations: gravity measurements (with disputable confidence, depending on the area) and satellite altimetry derived geoid heights over the oceans (which require an ocean circulation model).

However the need for a higher resolution and more reliable satellite only solution remains, especially for studying the absolute ocean circulation where a precise geoid is used as reference surface together with the sea surface itself as mapped by satellite altimetry. The CHAMP mission (Reigber et al., 1999) was designed for this purpose, and fulfills for the first time most important criteria for a gravity mapping mission as reported above: (i) low mean altitude (450 km at launch , 300 km towards the end of the mission) and quasi-polar-circular orbit, (ii) continuous coverage by the Global Positioning System by which the satellite-to-satellite tracking principle (SST) in the high-low mode is fully realized, (iii) direct measurements of the surface forces by means of an on-board micro-accelerometer, STAR (Touboul et al., 1999).

CHAMP was launched on July 15, 2000 and has since provided a lot of data, all of which have not yet been processed for gravity field modeling because of unexpected complexities in the STAR sensor calibration (Perosanz et al., *these proceedings*), also due to the heavy processing work related to the GPS data usage. The results reported here are based on about three months of carefully selected data; they validate the preprocessing procedures, the processing and accelerometer calibration algorithms ; they demonstrate the gravity field modeling precision achievable with CHAMP and provide CHAMP-based Earth gravity field models as a first reference.

2 CHAMP gravity field data and processing strategy

For the first CHAMP-based models, 88 days of the mission data were processed in 63 arcs: 50 arcs of one and a half day and 13 arcs of one day length. The period covers July 30-Aug. 10 and Sept. 24-Dec.31, 2000 (with some gaps up to several days in between).

The tracking data are the GPS-Blackjack SST observations (with a maximum of 7 satellites per epoch), desampled from 10 sec. to 30 sec., representing 1,135,000 code and phase measurements. Laser ranging observations were not included in the present work for they provide an external and excellent test data set for assessing the quality of the solutions in orbit determination. The surface force accelerations are taken at 10 sec. interval (normal points from 1 sec. values provided by the STAR three-axes accelerometer). The spacecraft orientation is taken from the star camera quaternions every 10 sec. (desampled from the one second original data file). The thruster firing events (epoch and duration for each thruster pair) are also considered and introduced in the orbit dynamical model.

A two step approach has been adopted: (i) the GPS orbits and 30 second clock corrections are first computed for each arc (with 15 min overlaps for adjacent arcs); (ii) the CHAMP orbit is determined and the normal equation system (in the least squares context) is generated for each arc, with the GPS orbits and clocks fixed.

The parameterization of the normal equation systems is the following:

- Internal and instrumental parameters: one CHAMP state vector per arc; CHAMP clock offsets (1 per 30 sec-epoch), GPS-SST ambiguities (about 675 per 1.5 day arc); accelerometer biases and scale factors per axis, plus temperature and temperature gradient dependence parameters in the radial acceleration bias; attitude thruster mismatch linear accelerations.
- External (physical model) parameters: 10738 static gravitational geopotential harmonics, complete to degree-order 91, plus zonals and resonant terms (seven bands) up to degree 119 and order 111 (from the sensitivity analysis done with the CHAMP orbit for the considered time period); ocean tide potential harmonic coefficients solved for the diurnal and semi-diurnal tides (long-period tides have been fixed); the central term in the geopotential model (GM) and spherical harmonics' temporal variations have not been solved for with this limited data set (but partial derivatives have been computed for future solutions).

The weighting of the data for this first solution corresponds to 200 cm and 9 cm white noise for the GPS code and phase measurements respectively, which may be thought as being pessimistic in the absolute sense. In any event, these numbers are globally adjusted when performing the combination with other satellite information since a variance factor for each normal system is then determined.

3 First CHAMP and other CHAMP-based combined solutions

The accumulated normal equation system for the 63 selected arcs is solved with Kaula's law stabilization equations for geopotential harmonics above degree 30 (only). The resulting model has been named **EIGEN-CH1** (for **E**uropean **I**mproved **G**ravity model of the **E**arth by **N**ew techniques – **CHAMP** solution **1**).

From the quality tests performed on this first solution, which are reported below, also and naturally from a scientific perspective and to quantify the gain coming from the new mission, it was desirable to combine the CHAMP equations with those coming from other satellites.

Firstly, recent data sets on the laser satellites LAGEOS 1 and 2, STARLETTE and STELLA, covering the whole year 2000, have been processed and the resulting normal equations combined with those of CHAMP : the solution, as it appears in the next section in the evaluation step, is named **EIGEN-CH1-SL4**.

Second, the normal equation base of the GRIM5-S1 model (Biancale et al., *ibid*) was complemented by new equations from the LAGEOS 1,2 and STARLETTE satellites (i.e. laser data non included in the model before, and which extend the data set over several years in the past – thus providing strong constraints on the zonal coefficients): this new pre-CHAMP model was named GRIM5-S2. Subsequently another solution was derived from the CHAMP equations augmented by the GRIM5-S2 normal equations plus those from the four laser satellites-year 2000 data quoted above: this solution is named **EIGEN-CH1-S2SL4**. It was derived by solving also for GM, time variations of the zonals of degree 2, 3 and 4, long period ocean tides, and the harmonics of degree two and order one were kept separate for the GRIM5 and CHAMP observation periods; regularization was also applied as before.

Figure 1 shows the signal amplitudes per degree of the harmonic coefficients of the three models, in terms of geoid heights; for comparison are also shown the signals of the EGM 96 and GRIM5-S1 models. Compared to the combined solution EGM96 which has assimilated all available surface data, the satellite solutions logically show a lack of power above degree 30-35; however, the striking fact is that all CHAMP-based models, although in a very preliminary stage due to the limited amount of used data, exhibit a significantly higher power which demonstrates the gain coming from the low altitude orbit and from the continuous tracking coverage of CHAMP.

4 Evaluation of the solutions

The three new solutions can be evaluated from the statistical viewpoint, by their performance in orbit determination and by the precision of their associated geoid, some of these tests being meaningful only in the significant harmonic bandwidth of the models.

The geoid formal error amplitudes per harmonic degree are shown in figure 2,

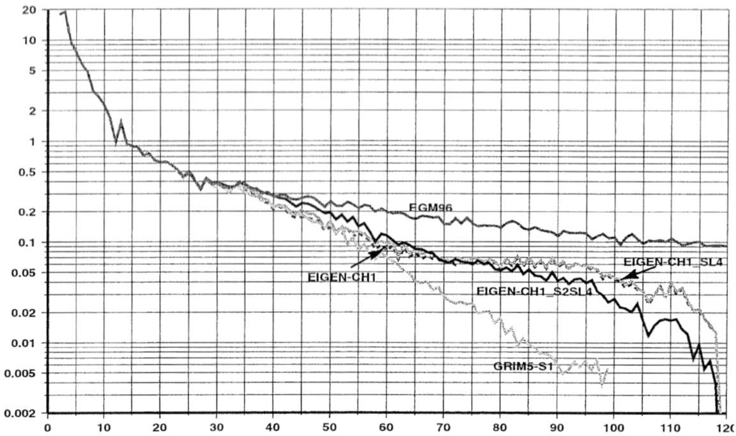

Fig. 1. Signal amplitudes per degree in terms of geoid heights [m] for 5 different gravity field models (power spectra)

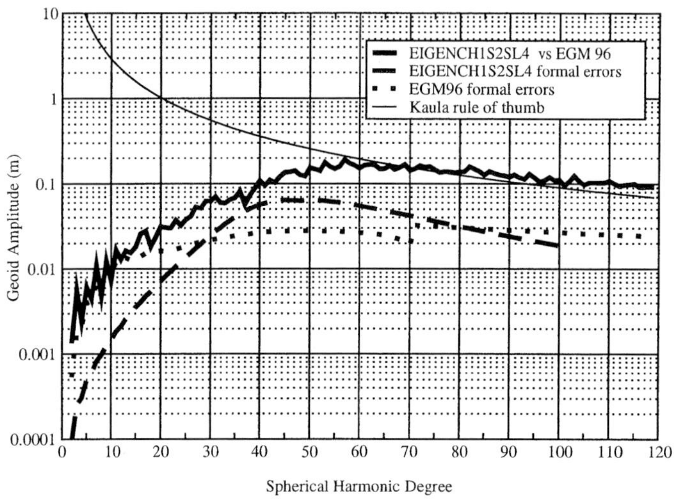

Fig. 2. Kaula's signal degree amplitude and difference and error degree amplitudes in terms of geoid heights [m]

for EIGEN-CH1-S2SL4 and EGM96, and compared with the difference between the two models and with the geoid signal itself (Kaula's rule of thumb).

The model differences and the EGM96 errors agree fairly well up to about degree 15, which is a confirmation that the longest wavelength part of the geopotential is equally well modelled; the EIGEN error curve, which shape (similar to other models' such as, here, EGM96) reflects that regularization is taking over from degree ~40, would indicate that the formal errors are somewhat underestimated and require some tuning. However figure 3 shows that the cumulated geoid differences to EGM 96 are systematically smaller than the formally predicted ones above degree ~25 for all EIGEN models, and systematically larger below degree 25 which seems to contradict the previous statement, except if one thinks that there might be larger longwavelength errors in EGM96 than expected – which ought to be clarified when more CHAMP data are processed and new models produced.

The precise orbit determination tests have been conducted for the CHAMP satellite itself and for a selection of others of geodetic interest (see table 1). The independent laser ranging observations on CHAMP are a way to externally assess the performance of the models on this low orbit. From these results, it is obvious that: (i) the CHAMP data have been well assimilated in the EIGEN models; (ii) the new models are not capable of fitting well the orbit data of other low altitude satellites (not LAGEOS 1-2), which is an indication that other mismodeling effects exist for these satellites, which have been absorbed in the model equations (i.e. the resulting model parameters are able to represent the disturbing effects in a taylored way); (iii) the CHAMP orbit precision is to-day at the decimeter level (from the laser residuals), but this could change (improve) rapidly with processing more CHAMP data.

While orbit determination tests are capable of detecting coefficients of specific orders which are not so well determined with respect to the high sensitivity (i.e. higher perturbations) of some satellites to such coefficients, comparisons with surface data – which are done in the spatial domain, can assess the quality of a model homogeneously up to a certain maximum degree (i.e. down to a given resolution). From the power spectra behaviour, which was previously discussed, it seems adequate with the present preliminary models to carry such comparisons up to degree

Satellite	Gravity Field Model			
	GRIM5-C1	GRIM5-S1	EIGEN-CH1S2SL4	EIGEN-CH1
CHAMP GPS phase	10.4 cm	19.8 cm	1.2 cm	1.2 cm
independent SLR (4 x 1.5 d)	62 cm	95 cm	11 cm	11 cm
GFZ-1 (5 x 6d SLR)	30 cm	20 cm	21 cm	208 cm
STARLETTE (5x5d SLR)	2.9 cm	3.0 cm	3.0 cm	15 cm
STELLA (5x7d SLR)	2.5 cm	2.5 cm	2.9 cm	29 cm
LAGEOS-1 (3x10d SLR)	2.0 cm	2.0 cm	2.0 cm	2.0 cm
LAGEOS-2 (3x10d SLR)	1.8 cm	1.8 cm	1.7 cm	1.9 cm
TOPEX (3x5d) SLR	-	2.5 cm	2.5 cm	-
DOP	-	0.53 cm/s	0.53 cm/s	-
X-Over	-	6.6 cm	6.7 cm	-

Table 1. Orbital fit results on selected satellite arcs. Values are RMS residuals for the considered data type (SLR: satellite laser range, DOP: Doppler-DORIS, X-Over: altimetric cross-over) and over N arcs of D days each (noted N x D d). None of the data used here were included in the gravity field models.

* EGM 96 S is the satellite solution associated to EGM 96 (Lemoine et al., ibid)

Gravity model	EIGEN-CH1	EIGEN-CH1 -SL4	EIGEN-CH1 -S2SL4	GRIM5-S1	EGM 96S*
rms (5°x5°)	0.29 cm	0.30 cm	0.27 cm	0.50 cm	0.45 cm

Table 2. Geoid differences between the EIGEN models and the altimetry derived ocean geoid. Comparison with the GRIM5-S1 and EGM 96S geoids are also given.

order 36 (maximum resolution of ~5 degrees). Moreover, we have decided at this early stage to show the tests with the ocean geoid surface only because of known large variations in the surface gravity anomaly errors (in the available worldwide data sets) and because geoid heights over continental areas, derived from GPS positioning and leveling, require to account for terrain effects before making such comparisons. Quality assessment tests were therefore done over 5°x5° blocks by comparing the ocean geoid derived from the mean sea surface (MSS) and from an ocean circulation model; we used respectively a MSS internal solution computed at GFZ from ERS and TOPEX/Poseidon altimetry, and the POCM model.

Table 2 summarizes the results, which demonstrate a significant improvement over previous satellite solutions in the 500 - 20,000 km half wavelength range for all three CHAMP-based models.

5 Conclusion

Nearly continuous GPS satellite-to-satellite tracking and STAR accelerometry data on the low altitude, highly inclined satellite CHAMP allow a significant improvement in long wavelength global gravity field recovery even with a limited three months data set. The geoid models corresponding to the three EIGEN solutions fit the altimetry derived ocean geoid to 30 cm or better over 5°x5° blocks. The precise orbit determination of other low altitude geodetic satellites still requires a combination with the tracking data (normal equations) of such satellites. The high quality of the long wavelength part in the CHAMP gravity solutions should allow the recovery of temporal variations at longest wavelength.

Note: The EIGEN-CH1-S2SL4 solution is the one with greatest interest. For sake of simplicity it has been renamed **EIGEN-1S** and its spherical harmonic coefficients can be downloaded via Internet from: op.gfz-potsdam.de/CHAMP/results.

Acknowledgements

The GfZ part of this study was carried out under the grant of the German Federal Ministry of Education and Research (BMBF) No. 03F0333A. On the French side the research was supported by the Centre National d'Etudes Spatiales (CNES).

References

Biancale R, Balmino G, Lemoine, J-M, Marty J-C, Moynot B, Barlier F, Exertier P, Laurain O, Gegout P, Schwintzer P, Reigber Ch, Bode A, König R, Massmann F-H, Raimondo J-C, Schmidt R, Zhu SY (2000) A New global Earth's Gravity Field Model from Satellite Orbit Perturbations: GRIM5-S1. Geophys Res Letters *27*:3611-3614.

Lemoine FG, Kenyon SC, Factor JK, Trimmer RG, Pavlis NK, Chinn DS, Cox CM, Klosko SM, Luthcke SB, Torrence MH, Wang, YM, Williamson RG, Pavlis EC, Rapp RH, Olson TR (1998) The development of the joint NASA GSFC and the National Imagery and Mapping Agency (NIMA) geopotential model EGM96. NASA Technical Paper NASA/TP-1998-206861, Goddard Space Flight Center, Greenbelt.

Perosanz F, Biancale R, Loyer S, Lemoine JM, Perret A, Touboul P, Foulon B, Pradels G, Grunwald L, Fayard T, Vales N, Sarrailh M, Duvivier E (2002) On Board Evaluation of the STAR Accelerometer. Proceedings of the first CHAMP Science Meeting Potsdam, January.

Reigber Ch, Schwintzer P, Lühr H (1999) The CHAMP geopotential mission. Boll Geof Teor Appl, 40:285-289.

Touboul P, Foulon B, Willemenot E (1999) Electrostatic space accelerometers for present & future missions. Acta Astronautica *45*, 10:605-617.

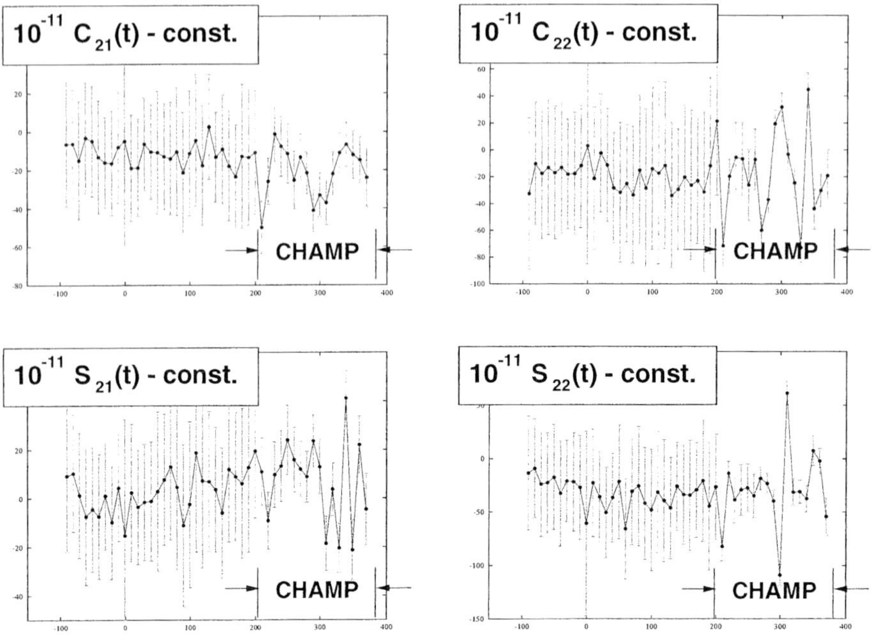

Fig. 2. The Stokes coefficients of degree two.

3 Physical causes of seasonal variations of zonal Stokes coefficients

Seasonal variations of Stokes coefficients must be attributed to similar mass changes caused by variations of ocean dynamics and the global hydrological cycle, because the coefficients derived from the CHAMP mission data are corrected for the influence of atmospheric mass redistributions.

The influence of ocean dynamics is estimated from the ocean model OCMT (Thomas et al 2001; Wünsch et al. 2001). The influence of the global hydrological cycle on the Stokes coefficients depends on water mass redistribution between the oceans and continental water storage. This process is discussed in two ways.

From TOPEX/Poseidon altimeter data the annual period $\Delta h_{oc}(t)$ of global sea level changes was estimated (Jochmann et al. 2001). Using $\Delta h_{oc}(t)$ the annual variations of zonal Stokes coefficients are obtained according to

$$\Delta \bar{C}_{n0} = \frac{4\pi\rho a^2\,(1+k'_n)}{M\,(2n+1)}\;a_{n0}\;\Delta h_{oc}(t) \tag{1}$$

where ρ is the density of sea water and a_{n0} is a zonal coefficient of the ocean function (see e.g. Lambeck 1980). k'_n is the load Love number.

The variation of the continental water storage was estimated using an inverse solution. Its influence on the Stokes coefficients is obtained if in (1) $\Delta h_{oc}(t)$ is replaced by

$$\Delta h_{co} = -\frac{S_{oc}}{S_{co}} \Delta h_{oc}(t + \Delta t) \tag{2}$$

where S_{oc} and S_{co} are the areas of the oceans and the continents. The phase delay $\Delta t = -150°$ was obtained applying the inverse solution to the estimation of excitation functions of polar motion (see Jochmann et al. 2001). This approach showed that for evaluating the influence of the hydrologic cycle on polar motion the assumption of an equally distributed variation of continental water storage is sufficient. This is also valid for estimating the impact of this process on low degree Stokes coefficients.

In a second approach the global sea level induced seasonal variation of Stokes coefficients was combined with a continental water storage variation derived from soil moisture models according to Huang et al. (1996); the latter monthly gridded data for 1968 - 1978 were expanded into spherical harmonics. Both estimations yield nearly the same seasonal variations of Stokes coefficients.

Seasonal variations of the Stokes coefficients derived from geophysical processes agree well with those obtained from satellite missions for the coefficient $\bar{C}_{20}$ only. For higher degree zonal coefficients the temporal variations are very much larger than seasonal variations derived from geophysical mod-

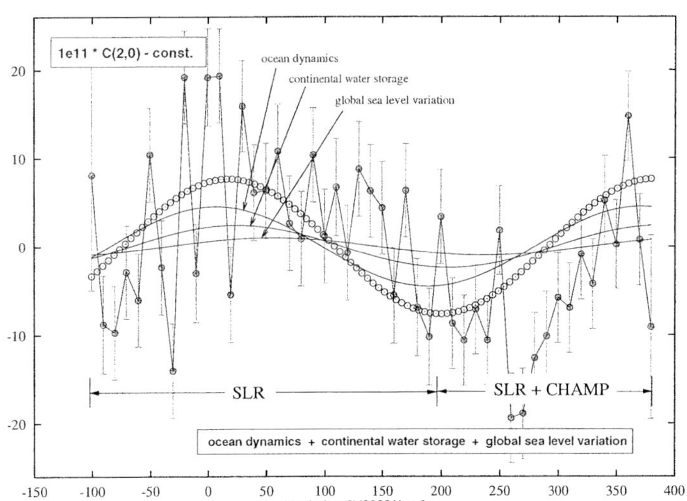

Fig. 3. The time series for $\bar{C}_{20}$

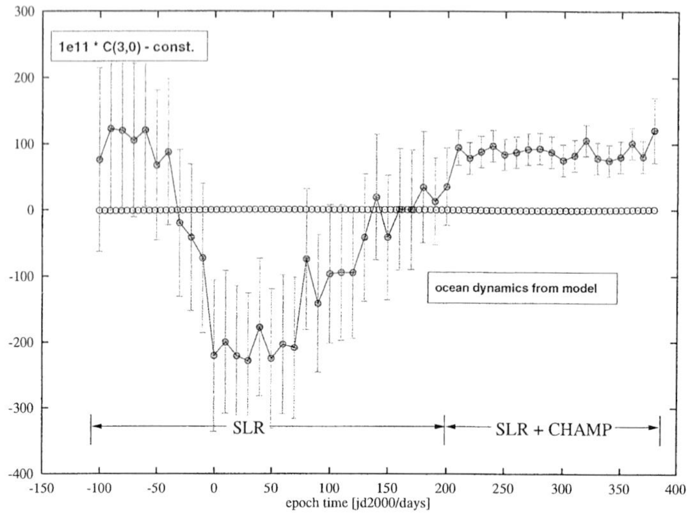

Fig. 4. The time series for $\bar{C}_{30}$

els (see Fig 4). Table 1 and Fig.3 show the results for $\bar{C}_{20}$. They are given in the notation $\Delta\bar{C}_{20} = a_1 \cos\omega t + a_2 \sin\omega t = a \cos(\omega t - \varphi)$.

It is seen that ocean dynamics and the global hydrological cycle can sufficiently explain the amplitude of the seasonal variation of $\bar{C}_{20}$. Since the discussed values of $\bar{C}_{20}$ were estimated for different time intervals, phase differences must be expected. Another cause of the phase differences could be the estimated influence of ocean dynamics, because different ocean models yield different phase angles (Jochmann et al. 2001).

Table 1. Annual variation of $\bar{C}_{20} \times 10^{11}$ estimated from ocean dynamics and the global hydrological cycle

Geoph. process	a_1	a_2	a	φ
(1) ocean dyn.	4.51	0.47	4.54	6.0°
(2) global sea level	0.60	0.80	1.00	53.1°
(3) water stor. inverse	2.22	0.95	2.41	23.1°
(4) water stor. Huang et al.	2.20	1.06	2.44	25.7°
(1)+(2)+(3)	7.33	2.22	7.65	16.8°
(1)+(2)+(4)	7.31	2.33	7.67	17.7°
CHAMP	4.45	7.71	8.90	60.0°

4 Conclusions

Stokes coefficients of the Earth's gravity field have been estimated at 10 day intervals from more than a year of SLR observations together with some 90 days of CHAMP data. For the coefficient $\bar{C}_{20}$ the time series can be sufficiently explained by a theoretical curve provided by an ocean model. The introduction of CHAMP data into the normal equations tends to stabilize the time series estimation process in general. However, more than a year of CHAMP data has to be processed in order to assess the true impact of GPS-CHAMP SST data combined with accelerometry on the solvability of time-dependent gravity field coefficients.

Acknowledgement. This study was carried out under the grant from the German Federal Ministry of Education and Research (BMBF) No. 03F0333A. We are indebted to our colleagues of GRGS (Toulouse) for providing us with normal equations derived from Lageos 1/2, Stella and Starlette data.

References

1. Cheng M, Tapley BD (1999) Seasonal variations in low degree zonal harmonics of the Earth's gravity field from satellite laser ranging observations. Journal of Geophysical Research **104** No B2, 2667-2681
2. Huang J, Van den Dool HM & Georgakakos KP (1996) Analysis of model-calculated soil moisture over the United States (1931-1993) and applications to long range temperature forecasts. J Climate, **9**, 1350-1362
3. Jochmann H, Reigber Ch, Greiner-Mai H, Wünsch J & Barthelmes F (2001) Geophysikalische Prozesse, Erdrotation und zeitlich variables Schwerefeld. Scientific Technical Report STR01/11 GeoForschungsZentrum Potsdam
4. Lambeck K (1980) The earth's variable rotation. Cambridge Univ Press, Cambridge
5. Thomas M, Sündermann J & Maier-Reimer E (2001) Consideration of ocean tides in an OGCM and impacts on subseasonal to decadal polar motion excitation. Geophys Res Lett, **28**, 2457-2460
6. Wünsch J, Thomas M & Gruber Th (2001) Simulation of oceanic bottom pressure for gravity space missions. Geophys J Int, **147**, 428-434

CHAMP Gravity Field Recovery with the Energy Balance Approach: First Results

Christian Gerlach[1], Nico Sneeuw[2], Pieter Visser[3], and Dražen Švehla[1]

[1] Institut für Astronomische und Physikalische Geodäsie, Technische Universität München, *gerlach@bv.tum.de, svehla@bv.tum.de*
[2] Department of Geomatics Engineering, University of Calgary, *sneeuw@ucalgary.ca*
[3] Delft Institute for Earth Oriented Space Research, Delft University of Technology, *Pieter.Visser@lr.tudelft.nl*

Summary. Using the principle of energy conservation has been considered for gravity field determination from satellite observations since the early satellite era, see e.g. O'Keefe (1957), Bjerhammar (1968), Reigber (1969) or Ilk (1983). CHAMP is the first satellite to which the energy balance approach can be usefully applied, now that near-continuous orbit tracking by GPS is available, aided by accelerometry. Simulation studies show the feasibility of the approach. One concern is the sensitivity to velocity errors. As a next step CHAMP's Rapid Science Orbits (RSO) are used. Their error level is sufficiently low to demonstrate the feasibility of the energy balance approach with real data. The accelerometer data (ACC), used for modeling the non-conservative forces, give rise to further concern.

Key words: gravity field, energy integral, accelerometer bias

1 Energy Integral

The concept is based on the total energy E of a satellite, which is assumed to be a constant of motion. According to Landau and Lifschitz (1976) it reads for a unit mass

$$E = \frac{1}{2}\,\dot{\boldsymbol{r}}^2 - \frac{1}{2}\,(\boldsymbol{\Omega} \times \boldsymbol{r})^2 + U, \qquad (1)$$

where $\boldsymbol{\Omega}$ is the rotation vector of the earth and $\boldsymbol{r}$ and $\dot{\boldsymbol{r}}$ are position and velocity of the satellite in an earth fixed frame. Equation (1) is also referred to as energy integral. For the potential energy U the sign convention of physics is employed (opposite to geodetic convention). In order to derive from (1) the gravitational potential of the earth, U must be corrected for the tidal potential of third bodies.

In reality, however, E is not constant in time, due to non-conservative forces like air drag or solar radiation pressure. In the case of CHAMP these forces are measured by accelerometry and the respective loss of energy can be determined by integration of the measured accelerations $\boldsymbol{a}$ along the orbit. Equation (1) can thus be rearranged to

$$V = \frac{1}{2}\dot{\boldsymbol{r}}^2 - \frac{1}{2}(\boldsymbol{\Omega} \times \boldsymbol{r})^2 - V_t - \int_{\boldsymbol{r}} \boldsymbol{a} \cdot \mathrm{d}\boldsymbol{r} - E, \tag{2}$$

where V is the gravitational potential of the earth (now with the geodetic sign convention) and V_t is the tidal potential. V can be determined up to the unknown constant E. Subtraction of a normal field gives the disturbing potential T as time series along the orbit. In the sequel T is used as pseudo observation, with $\boldsymbol{r}$,$\dot{\boldsymbol{r}}$ and $\boldsymbol{a}$ being measured.

2 Analysis Method

A semi-analytical approach, proposed by Sneeuw (2000), was used for gravity field analysis. Under the assumption of constant orbit height and inclination, potential coefficients of different spherical harmonic (SH) order m are uncorrelated and the normal equation matrix shows a block-diagonal structure. Therefore the inversion of the matrix can be performed very efficiently. Errors due to the applied approximations can be minimized by iteration.

The potential, up to SH-degree L, along the orbit can be expressed as

$$V(u, \Lambda) = \sum_{m=-L}^{L} \sum_{k=-L}^{L} A_{mk}\, e^{i(ku+m\Lambda)} \tag{3}$$

$$A_{mk} = \sum_{l=\max(|m|,|k|)}^{L} \frac{GM}{R} \left(\frac{R}{r}\right)^{l+1} F_{lmk}(I)\, K_{lm}, \tag{4}$$

where u is the argument of latitude, Λ the longitude of the ascending node and the lumped coefficients A_{mk} are computed from inclination function F_{lmk} and the potential coefficients K_{lm}. From equation (3) the A_{mk} can be determined by a 2D Fourier-transformation. For this purpose the disturbing potential along the orbit is generated in a regular grid on the surface of a torus. This is the proper space domain for a 2D Fourier-series (see Sneeuw, 2001).

3 Simulation Studies and Error Budget

In order to test the feasibility of the energy approach, simulation studies were carried out. From an *a priori* gravity field (EGM96s was used) an orbit was generated by numerical integration. Simulated positions and velocities along the orbit were then used to compute the disturbing potential according to the equations given in section 1. In order to decrease the approximation error induced by the analytic assumptions (see section 2) the disturbing potential was reduced to a constant orbit height using the radial gradient $\partial T/\partial r$. The time series of pseudo observations T was then used for estimation of potential coefficients. In principle it should be possible to recover the *a priori* field

caused by a bias in the ACC data. It can be determined at cross-over points of the satellite tracks. The derived potential values at those points (reduced to constant orbit altitude) are expected to be equal. For biased accelerations the energy integral does not hold and the values at identical points differ by

$$\Delta T = \sum \Delta \boldsymbol{r}'_e \cdot (\boldsymbol{R}^s_e \cdot \boldsymbol{b}), \tag{6}$$

where $\boldsymbol{b}$ is the bias. Equation (6) is the mathematical model for the estimation of $\boldsymbol{b}$ from the potential differences at cross-over points. In our computations we have only estimated the along-track component of the bias. A mean value of $-3.4 \cdot 10^{-6}$m s^{-2} was derived for it.

The slope s of the drift is given by $s = \Delta T/\Delta t$, where ΔT is the potential difference at the cross-over points and Δt is the time span between the corresponding epochs. The slope (see figure 2, right) reveals variations in the drift of the order of ± 20 m^2 s^{-2}. This corresponds to a temporal variation of the bias. For the used period the bias differs between -3.54 m s^{-2} and -3.43 m s^{-2}. A polynomial of 9th order was used to model the bias. The variation of the slope indicates periodic variations in the bias.

4.3 Results

After estimation of the along-track bias, the energy integral was recomputed (a scale factor of 0.8 was used for all components) and a gravity field determined from the derived disturbing potential. No correlations between the pseudo observations were modeled in the estimation process. The field was compared to the GRIM5-C1 potential model. The coefficients $C_{2,1}$ and $S_{2,1}$ could not be estimated to a sufficient accuracy, which might be due to different reference frame definitions in the two fields. Both coefficients were disregarded in the following comparison. Figure 3 shows the difference in geoid height between the two fields at orbit height. The latitude-RMS shows

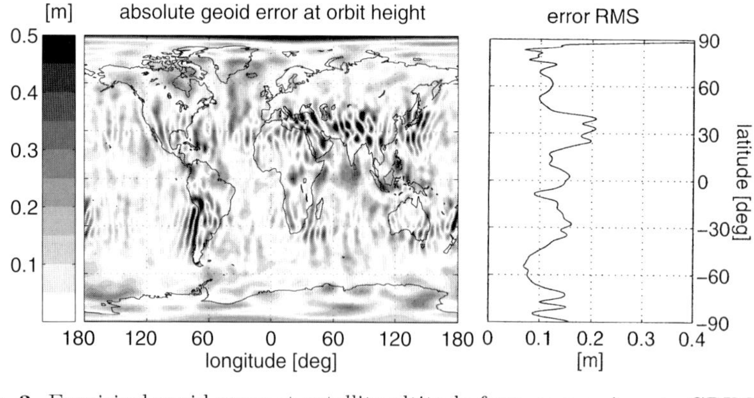

Fig. 3. Empirical geoid error at satellite altitude from comparison to GRIM5-C1.

values of 10–20 cm, which corresponds to the accuracy one can expect from RSO data (see section 3). The structure of the empirical error shows correlations to orbit tracks. This again indicates that the errors are to a large extent caused by orbit errors. It is worthwhile noticing, that in case of poor orbit quality a similar error structure is a well known phenomenon in geoids based on satellite altimetry.

5 Summary

The energy balance approach, proposed since the early days of satellite geodesy, could be applied for the first time with real data. A gravity field has been computed from 11 days of CHAMP RSO and ACC data. Comparison to GRIM5-C1 shows an accuracy of 10–20 cm in geoid height at satellite altitude. This corresponds to the expected accuracy level obtained from an error simulation. The energy integral is highly sensitive to velocity errors. This shows that a very precise orbit is necessary to recover the gravity field to a reasonable quality.

Furthermore it has been shown, that the semi-analytical torus approach can be successfully applied in gravity field recovery and that the energy balance approach is useful for accelerometer calibration.

Acknowledgement. We are grateful to GeoForschungsZentrum Potsdam for providing ACC and RSO data.

References

Bjerhammar A (1968): On the energy integral for satellites. Res. Note 29, U.S. Army Eng. Topogr. Lab.

Ilk KH (1983): Formulierung von Energieaustauschbeziehungen zur Ausmessung des Gravitationsfeldes. In: Schneider M (ed): Die Arbeiten des Sonderforschungsbereichs 78 Satellitengeodäsie der Technischen Universität München im Jahre 1982. Astronomisch-Geodätische Arbeiten der Bayerischen Kommission für die Internationale Erdmessung, no 43

Landau LD, Lifschitz EM (1976): Lehrbuch der Theoretischen Physik, Band I. Akademie-Verlag, Berlin

O'Keefe JA (1957): An application of Jacobi's integral to the motion of an earth satellite. Astronomical Journal vol 62, no 1252, pp 265-266

Reigber C (1969): Zur Bestimmung des Gravitationsfeldes der Erde aus Satellitenbeobachtungen. Deutsche Geodätische Kommission, Reihe C, no 137

Sneeuw N (2000): A semi-analytical approach to gravity field analysis from satellite observations. Deutsche Geodätische Kommission, Reihe C, no 527

Sneeuw N (2001): Dynamical satellite geodesy on the torus: Block-diagonality from a semi-analytical approach. In: Sideris, MG (ed): Gravity, Geoid and Geodynamics 2000. IAG Symposia, vol 123, Springer-Verlag, Berlin, New-York

Preliminary Analysis of CHAMP State Vector and Accelerometer Data for the Recovery of the Gravity Potential

Eva Howe and Carl Christian Tscherning

Department of Geophysics. Juliane Maries Vej 30, DK-2100 Copenhagen Ø, Denmark
eva@gfy.ku.dk, cct@gfy.ku.dk

Summary. From energy considerations the relationship in inertial space between the gravitational potential and the state vector of the satellite can be found. The state vector from the rapid science orbit of CHAMP has been used after the velocities have been converted to the inertial system.

The satellite loses kinetic energy due to external forces. We consider the tidal potential from the sun and the moon, energy dissipation and the explicit time variation of the gravitational potential in inertial space.

Data from two 4 day periods have been analysed. The GRS80 earth normal potential and a constant were subtracted in order to obtain potential differences. Using Bruns' theorem we calculated the height anomaly at the satellite position. The height anomalies were compared with values calculated from EGM96. The differences show a piecewise bias and trend. When correcting for a bias and trend in a one-day period the differences had a standard deviation of 0.14 m. Otherwise the differences were 0.5 and 1.4 m for the two periods, respectively. This show that the accelerometers have biases which may be identified and eliminated by comparing with a high degree and order gravity potential reference field.

Key words: Energy principle, height anomalies, accelerometer biases

1 Introduction

We will here use energy considerations to find the relationship in inertial space between the state vector of CHAMP and the gravitational potential. The CHAMP orbit is based on GPS and the gravity field model GRIM5-C1 using the reduced dynamics method to determine the position and the velocity (the state vector).

3 accelerometers are also onboard CHAMP. We have used the pre-processed accelerometer data in order to calculate the energy dissipation. In the pre-processing the accelerations have been re-sampled from 1 Hz to 0.1 Hz. However there are scale errors, biases and possibly drifts in the accelerometer, which we have to estimate. A comparison with external data has here been used to estimate the bias correction. No scale and drift errors were estimated.

The rapid science orbit state vector data (x, y, z, v_x, v_y, v_z) and the pre-processed accelerometer data (a_x, a_y, a_z) for two 4 day periods have been analysed. The two randomly chosen periods are May 18-21 2001 and August 18-21 2001.

2 Data analysis

The state vector velocities were converted to inertial velocities and the kinetic energy was calculated $T=\frac{1}{2}*v^2$. From this we have subtracted the potential of the sun V_{sun} and the moon V_{moon}, the rotation potential $\omega*(x*v_y - y*v_x)$ (Jekeli 1999), the frictional energy F, a constant E_0 and the GRS80 earth gravity normal potential U without centrifugal term. The height anomaly ζ was calculated by dividing the potential difference by normal gravity, g. The height anomaly can be expressed as

$$\zeta = (\tfrac{1}{2}*v^2 - V_{sun} - V_{moon} - \omega*(x*v_y - y*v_x) - F - E_0 - U)/g$$

The frictional energy F has been calculated in different ways, or put equal to zero. We calculated F as $\int |v| * a_y dt$ or as $\int \bar{v} \bullet \bar{a} dt$ where a_y is the along track acceleration and a is the acceleration vector. The first procedure for the calculation of the frictional energy requires alignment of v with the satellite. The quaternions which express the alignment had unfortunately large errors in the two 4 day periods.

The use of the frictional energy without any corrections to the accelerations does not give better results than setting it equal to zero. We therefore had to assume that the accelerometer had biases, scale factor errors and possibly drifts. The bias has been determined as a tilt of the height anomalies. The total bias correction is determined to be $-0.364*10^{-5}$ m/s^2 and $-0.352*10^{-5}$ m/s^2 for the two periods but it seems like a drift is present also. The scale factor is determined to be between 0.7 and 0.8. We have initially only subtracted a bias in the along track acceleration and a trend (given by the CHAMP data centre) in the radial acceleration.

The height anomaly was compared to values computed from EGM96 (Lemoine et al 1998), see Table 1. Note that applying the corrections for friction does not give a better fit to EGM96. If the full inner product is used the results become very erroneous.

	Dif. May		
	With F=0	With $F = \int \lvert v\rvert * a_y dt$	With $F = \int \bar{v} \bullet \bar{a} dt$
Mean	0.23	0.23	0.14
Std. dev.	0.30	0.49	26.26

Table 1. Comparisons between EGM96 and CHAMP height anomalies at satellite altitude

	Dif. August		
	With F=0	With $F = \int \lvert v \rvert * a_y dt$	With $F = \int \bar{v} \bullet \bar{a} dt$
Mean	0.40	0.40	0.31
Std. Dev.	1.37	1.39	26.01

Table 1. (cont.)

The height anomalies for the two periods are shown on Figure 1, where we have not corrected for energy dissipation. It can be seen that both datasets gives a good picture of the geoid at satellite altitude.

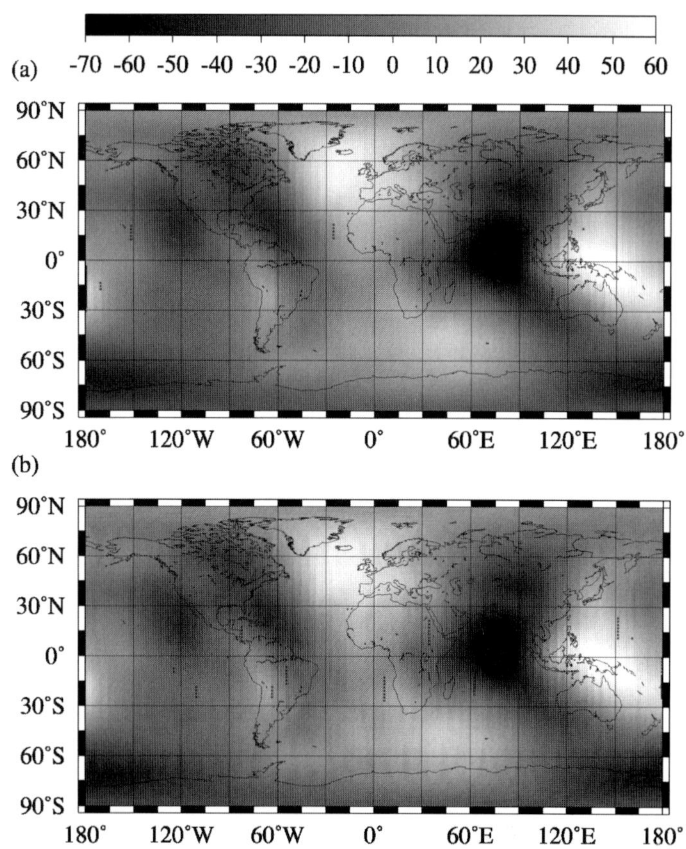

Fig. 1 a) is the height anomalies at satellite altitude for May 18-21 2001. b) is the height anomalies at satellite altitude for August 18-21 2001.

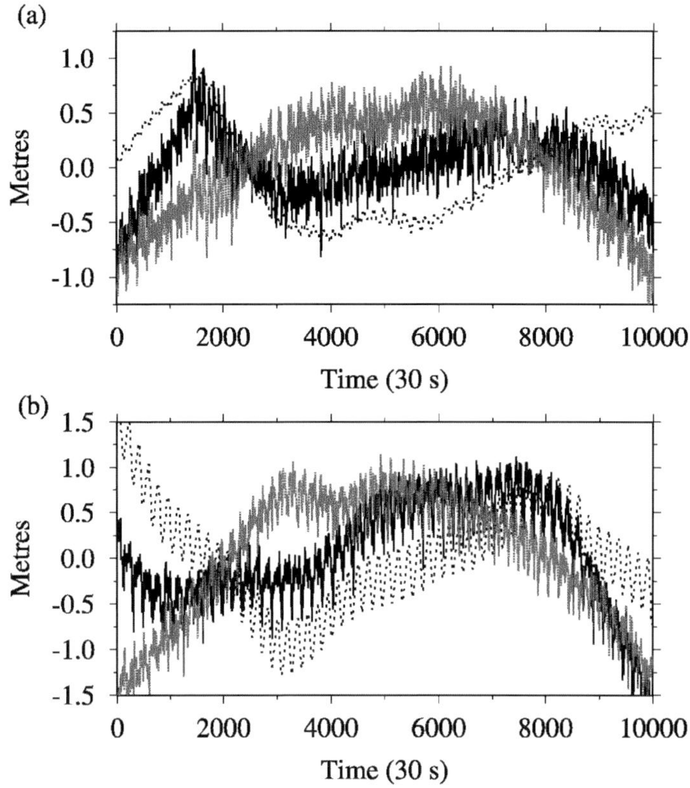

Fig. 2 a): The differences between CHAMP height anomalies from May 18-21 2001 and values calculated from EGM96 (black). The friction energy divided by gravity is plotted as dots. The grey curve is the difference between CHAMP height anomalies where the corrected acceleration is considered and values calculated from EGM96. b): As in a) for August 18-21 2001.

The differences between CHAMP height anomalies and values calculated from EGM96 are shown in Figure 2 together with $F = (\int |v| \bullet a_y dt) / g$. The difference and the correction seem both to be piecewise linear.

There is a large difference between the two periods. We found a simple explanation for this by analysing more closely the accelerometer data. The acceleration is supposed to be numerically largest when the satellite perigee is over the Equator. But this depends on atmosphere density, which changes from day to night. If the satellite is at its perigee at night, this compensates for the closeness to the Earth. This is what happens in the May period. The opposite happens in August, which leads to an amplification of the accelerations. The satellite is at perigee during the day. It shows that the day-night difference in atmosphere density is important. The bias and scale errors have a larger influence when the signal amplitude is larger.

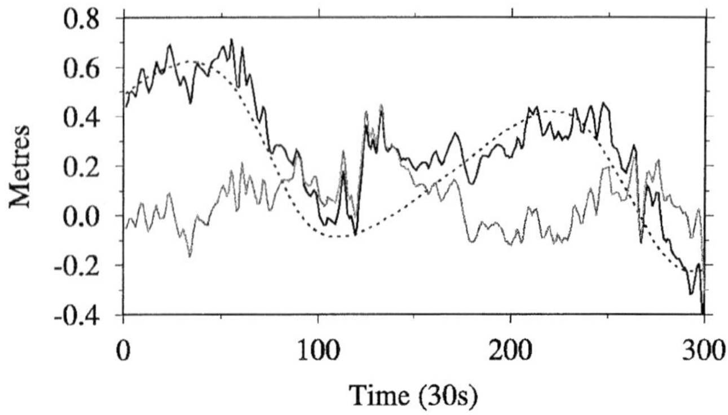

Fig. 3 The differences between CHAMP height anomalies and values calculated from EGM96 for approximately August 18 2001 (black). Only the first 300 measurements are plotted. The frictional energy divided by gravity is plotted as dots. The grey curve is the difference between CHAMP height anomalies where the corrected acceleration is considered and values calculated from EGM96. We see an agreement of 0.14 m with EGM96 for the time period. The spike of 0.4 m corresponds to the difference calculated in a point over a mountain range in Antarctica.

We tried a further adjustment of the accelerations using a best fit to EGM96 for reduced periods when the differences had a linear behaviour. This gave an improved agreement of 0.14 m with EGM96 for the time period, see Figure 3.

3 Conclusion

We have seen the importance of the relationship between atmosphere density and the accelerations, which can be used to correct the accelerations. Furthermore the residuals have a piecewise linear trend originating from the frictional energy loss, which may be parameterised and determined by least-squares adjustment like when adjusting a gravity network. This may be an alternative procedure for correcting the observed accelerations. Another possibility is to use density values from an atmospheric model and disregard the very noisy acceleration and attitude data.

The first CHAMP gravity field model, EIGEN-1S gives a new possibility to correct the accelerations. We will use this to determine better estimates of the scale factor, the biases and the drifts.

Acknowledgement

Thanks to the CHAMP data centre for providing the data. Thanks to L. Stenseng for debugging the software. This paper is a contribution to the project SAGRADA sponsored by the Natural Science Council of Denmark.

References

Jekeli C (1999) The determination of gravitational potential differences from satellite-to-satellite tracking. Celestial Mechanics and Dynamical Astronomy, Vol 75, pp 85-101

Lemoine FG, Kenyon SC, Factor JK, Trimmer RG, Pavlis NK, Chinn DS, Cox CM, Klosko SM, Luthcke SB, Torrence MH, Wang YM, Williamson RG, Pavlis EC, Rapp RH, and Olson TR (1998) The Development of the Joint NASA GSFC and the National Imagery and Mapping Agency (NIMA) Geopotential Model EGM96. NASA/TP-1998-206861, Goddard Space Flight Center, Greenbelt, MD

Wessel P and Smith WHF (1998) New improved version of Generic Mapping Tools released. EOS Trans. American Geophysical Union, Vol 79 (47), pp 579

CHAMP Precise Orbit Determination and Gravity Field Recovery

Philip Moore, James F. Turner and Zhang Qiang

Department of Geomatics, Newcastle University, UK. *philip.moore@ncl.ac.uk*

Summary. In this chapter we outline the methodology in use at Newcastle University to recover the static and time varying gravity field from the CHAMP mission. Details are presented of the reduced dynamic process adopted within the JPL software *GIPSY-OASIS II* to recover precise ephemerides from the GPS code and phase data with orbit overlaps and tracking residuals used to quantify the orbit. The GPS orbital positioning is subsequently used to recover accelerometer scaling factors and offsets and thruster accelerations alongside the gravity field harmonics in the *FAUST* software. Estimations of these parameters are presented, as well as results from a study of the modelling of the thruster accelerations. For the latter, the use of constant accelerations across a single step in the multi-step integrator is investigated as a tool for modelling thruster misalignment and mismatch. Finally, preliminary results from an enhancement of the gravity field are presented.

Key words: gravity field, reduced dynamic orbits, accelerometer, thrusters

1 GPS Reduced Dynamic Tracking of CHAMP with *GIPSY-OASIS II*

Precise Cartesian positioning of the CHAMP satellite has been determined from the GPS code and phase data utililising a reduced dynamic procedure within the JPL software *GIPSY-OASIS II* (Bertiger et al, 1994; Kuang et al, 2001). This is a two-stage process whereby the orbit is first converged in a completely dynamic procedure based on integration of the equations of motion with parameterisation of the force model described below. In the second stage stochastic empirical accelerations are introduced to incorporate a kinematic element into the orbit determination process. This reduced dynamic iteration holds fixed the orbital parameters estimated in the first stage. In order to weight the kinematic/dynamic elements, the process noise sigmas controlling the stochastic process are determined by a tuning process in which the orbit overlap between successive arcs are minimised. The minimisation was performed over a 10day period 31 May to 9 June, 2001 (i.e. days 151-160 of 2001 inclusive). Orbital arcs for a particular day commenced 21hr on the preceeding day and terminated at 3hr on the following day. The 30hr orbital arcs thus permitted examination of the positioning differences over the 6hr overlap between successive arcs. Stage 1 dynamic orbits were generated with the following strategy:

- GPS code and phase data decimated to 30sec intervals
- Precise GPS orbits and clocks from JPL
- 10 deg elevation cut-off
- Space craft (S/C) modelled by 6 planar surfaces plus cylinder for boom
- EGM-96 gravity field (Lemoine et al, 1998)
- Nominal CHAMP attitude model with S/C axes aligned to the radial, crosstrack and alongtrack directions
- Estimated parameters: Initial position/velocity, drag scale factor, once-per-revolution alongtrack and crosstrack empirical accelerations, CHAMP clock correction and CHAMP-GPS phase biases

After the dynamic orbit determination overlap agreement was typically near 4m. Stochastic accelerations were subsequentlty added into the dynamic solution with the degree of adherence to the dynamic orbit controlled by the process noise sigmas. Figure 1 depicts the behaviour of the rms overlap differences in the radial, crosstrack, and alongtrack directions as the process noise sigmas range from zero (dynamic orbit) to 200.D-9 ms^{-2} (kinematic orbit). The selected values of the process noise sigmas were 10.D-9 ms^{-2}, 30.D-9 ms^{-2} and 60.D-9 ms^{-2} in the radial, crosstrack and alongtrack directions respectively. Reduced dynamic orbits were subsequently generated with the tuned process noise sigmas for days 140 to 200, 2001 (i.e. 20 May to 19 July, 2001). Figure 3 and Figure 4 show that the reduced dynamic GPS pseudorange and phase residuals are typically 50cm and 1.6cm respectively. A plot of the overlaps for the 60day period is given as Figure 2 where it is seen that the reduced dynamic procedure has considerably improved the agreement to the 4-6cm level. Of course, overlap comparisons are a measure of the internal consistency of the procedure and should not be used to quantify the true accuracy of the orbits. An independent measure of orbital accuracy is provided by satellite laser ranging (SLR). Utilising 1908 SLR measurements over days 140-150 the rms residual was 14.3cm. Although geographically limited the rms value supports accuracy of the order of $15/\sqrt{3}$cm in a Cartesian coordinate.

2 Thruster misalignment and mismatch: An equivalent model for a multistep integrator

Attitude control of the CHAMP space craft is maintained by firing thrusters when the angular displacement exceeds a preset limit. With thrusters operating as a couple in paired combinations the ideal result is a rotation about the centre of mass. Any misalignment or mismatch of the thruster pair will result in a linear acceleration which must be considered in precise orbit compuations. As most orbital software employ a multistep numerical orbit integrator a change of step length to accommodate a thruster event can be problematic, through not insurmountable. To enable the construction and validation

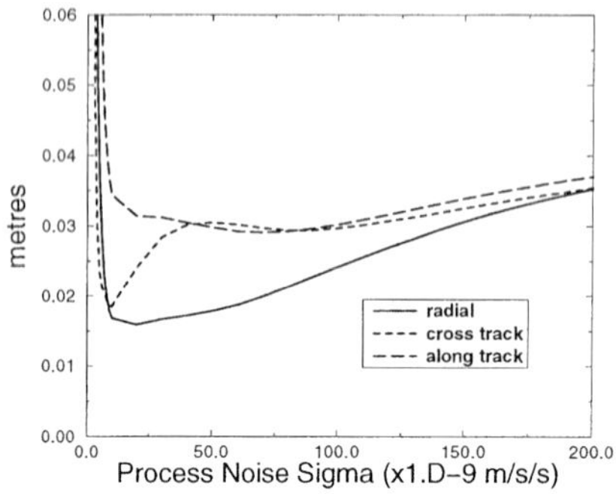

Fig. 1. Behaviour of the orbit overlap error with respect to the kinematic weighting

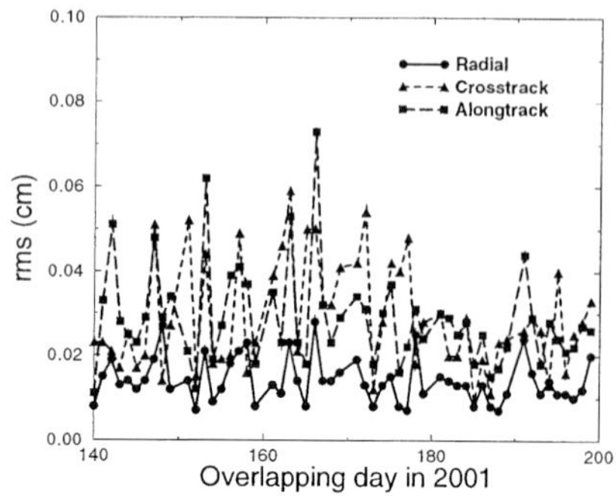

Fig. 2. Orbit overlap differences of GPS based orbits for 60 days in 2001

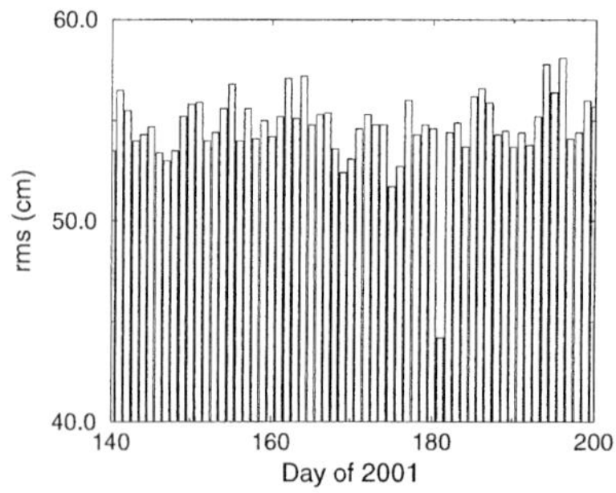

Fig. 3. Reduced dynamic GPS code residuals

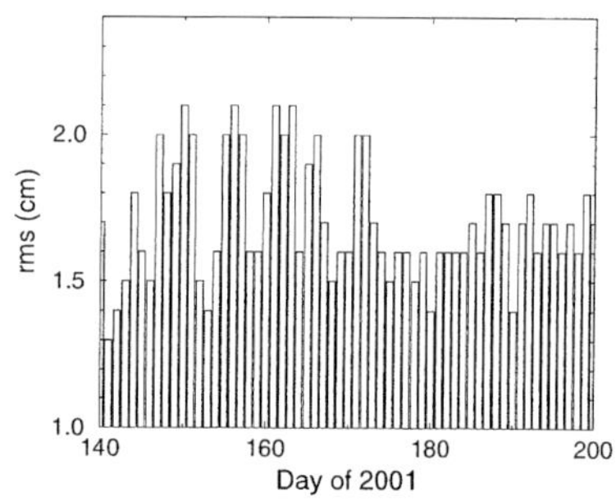

Fig. 4. Reduced dynamic GPS phase residuals

of an equivalent thrust model for use within *FAUST* the standard 8th order multistep Gauss-Jackson integrator has been complemented by an adjustable single step 8th order Runge-Kutta method. The latter scheme modified the step length close to a thruster event with the nodes corresponding to the start/stop times. In Figure 5, t_0 and t_3 denote the start and end time of the integration step and t_1 and t_2 the start and stop time of the thruster event. Two scenarios are illustrated namely a thruster event totally within the step interval and a thruster event that straddles the start or stop time of an integration step. If $\underline{c}$ is the unknown magnitude of the thruster mismatch/misalignment then the magnitude of the required acceleration spanning the step interval, $t_3 - t_0$, is $\underline{d} = \underline{c}(t_2 - t_1)/(t_3 - t_0)$. $\underline{d}$ was recovered as part of the orbit determination process for each of the 12 thruster pairs. To illustrate the error incurred on using the above simplification and the multistep integrator, ephemerides were computed using the Runge-Kutta in-

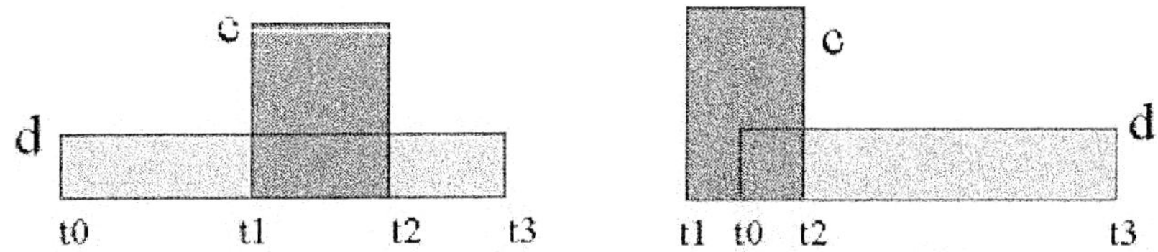

(a) Thrust event inside step interval (b) Thrust event straddling step node

Fig. 5. Multistep thrust model

tegrator with step adjustment for the thruster events. These were then used as tracking data with solution for accelerometer scale factors and biases and thruster magnitudes. RMS agreement in the orbital solutions was 6.1cm radially, 5.3cm crosstrack and 8.2cm alongtrack. Although such orbital differences are dependent on the magnitude of the thruster mismatch/misalignment, it is clear that further studies are necessary to quantify the effect on gravity field recovery which may confirm the need for improved schemes associated with thruster events.

3 Gravity Field Enhancement

Cartesian coordinates in the ECF (Earth centred fixed) frame at 60s intervals were derived from the reduced dynamic procedure coded in *GIPSY-OASIS II*. These coordinates were considered as independent tracking data of CHAMP; that is, each component x_{ECF}, y_{ECF} and z_{ECF} at time, t, was assumed to be a separate measure of the orbital positioning giving a total of 4320 data points per 24hr arc. These measurements were implemented within *FAUST*, Newcastle's dynamic orbit determination software. Linear accelerations from the CHAMP accelerometers, attitude quaternions and thruster events were used in place of non-conservative force models. A preliminary enhancement of the EGM96 gravity field to degree and order 70 has been generated with the 60 days of CHAMP data. CHAMP orbits were estimated from *FAUST* using the x_{ECF}, y_{ECF} and z_{ECF} as tracking data in 60 1-day arcs. The level of fit to the tracking data and to all satellite laser ranging data is summarised in Table 1. It is to be noted that the high rms values reflect errors in the force modelling, particularly the gravity field, rather than inaccuracies in the tracking data. For example, the SLR rms residual at 137cm is an order of magnitude larger than the 14.3cm fit to the reduced dynamic positioning and is indicative of deficiencies in the gravity field for CHAMP. The fully dynamic procedure described above permitted recovery of the gravity field normal equations along with the following solved-for parameters:

– Initial position and velocity (daily)

- Accelerometer biases and scale factors (daily)
- 3-D thruster misalignment and mismatch for each thruster pair (global solution over the 60 days)

The gravity field normal matrix, N_{CHAMP}, with right-hand side vector, $\underline{b}_{CHAMP}$, was constrained by the EGM96 covariance matrix, C_{EGM}, i.e.

$$(w_t N_{CHAMP} + C_{EGM}^{-1})\underline{x} = w_t \underline{b}_{CHAMP}$$

where the weight w_t was chosen to produce a tailored field that was appropriate for both CHAMP and high satellites such as LAGEOS1 and LAGEOS2, and to a lesser extent TOPEX/Poseidon. In comparison with EGM96, the CHAMP field, EGM96-C, gives marked improvements in the quality of dynamic orbit determinations of CHAMP with *GIPSY-OASIS II* (see Table 2; and both the x_{ECF}, y_{ECF} and z_{ECF} residuals and SLR residuals in *FAUST* (Table 1). Further validation has been performed through orbit comparisons with LAGEOS1, LAGEOS2, TOPEX/Poseidon, Stella and Starlette (Table 3).

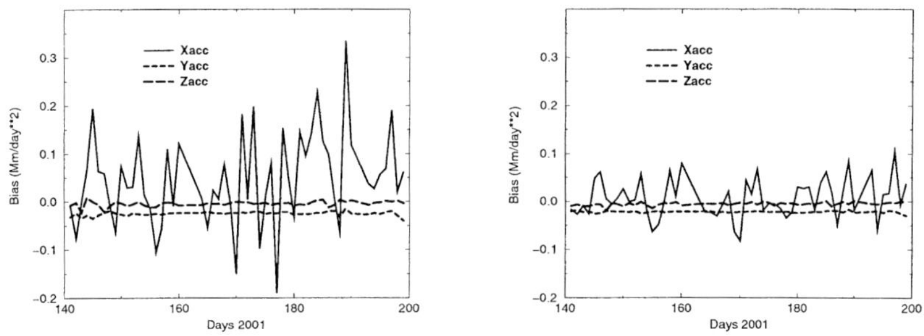

(a) EGM96 (b) EGM96-C

Fig. 6. Estimated accelerometer biases

Table 1. *FAUST* rms residuals (cm) of CHAMP tracking data from dynamic orbits for days 140 to 200, 2001

	x_{ECF}	y_{ECF}	z_{ECF}	SLR
EGM96	133	167	141	137
EGM96-C	48	53	62	57

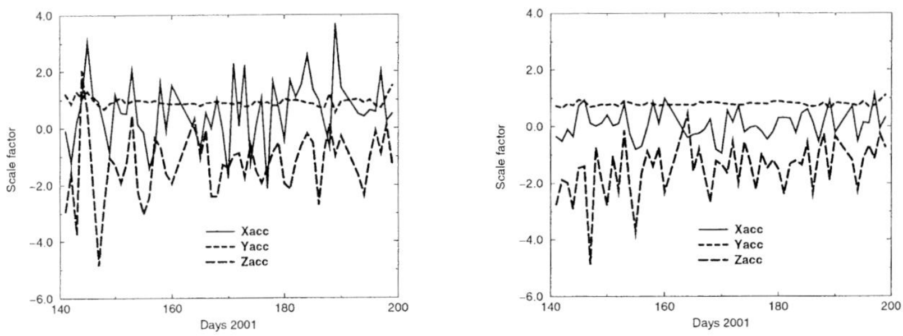

(a) EGM96 (b) EGM96-C

Fig. 7. Estimated accelerometer scale factors

Table 2. *GIPSY-OASIS* rms residuals (cm) of CHAMP GPS tracking data from dynamic orbits

	GPS rms (cm)	code data	GPS rms (cm)	phase data
EGM96	328.9	11253	25.6	12916
EGM96-C	239.0	12808	10.8	12918

Table 3. Test of the CHAMP based gravity field with other geodetic satellites; SLR tracking data (cm).

	inclination deg.	altitude km	EGM96 rms (cm)	EGM96-C rms (cm)	field size
LAGEOS1	109.8	5900	3.20	3.20	20 x 20
LAGEOS2	52.6	6900	2.83	2.83	
TOPEX	66.5	1330	5.75	5.81	70 x 70
STELLA	98.7	800	6.13	8.02	
STARLETTE	49.8	800	4.53	5.60	

4 Discussion and conclusions

GPS positioning of CHAMP has been derived and validated using *GIPSY-OASIS II* by a reduced dynamic approach with the dynamic/kinematic weighting parameters estimated by a tuning process which minimises orbit overlap differences. The process yields a 60 day phase residual of 1.7cm rms with a total overlap difference of 4.4cm. SLR residuals of 14.3cm rms for a 11 day period confirm that the reduced dynamic Cartesian positioning is accuracy to about $15/\sqrt{3}$cm in each component. The reduced dynamic Cartesian positioning and accelerometer data is subsequently used to tailor the EGM96 gravity field for the CHAMP orbit. Bias and scale factors for the 3 axes of the accelerometer are plotted as Figure 6 and Figure 7, for both EGM96 and the enhanced field. The figures show that crosstrack, (Z_{Acc}), and alongtrack, (Y_{Acc}), biases are consistent but that the radial (X_{Acc}) bias is variable due to the well-documented electrode problem in this direction, although the enhanced field has reduced the variance. The scale factor for the alongtrack acceleration is the most consistent with a value of 0.80 ± 0.08. The crosstrack bias remains variable even after gravity field enhancement and probably reflects residual gravity field signatures rather than deficiencies in the accelerometer. Orbital computations with the enhanced gravity field using high geodetic satellites (LAGEOS1, LAGEOS2) and to a lesser extent TOPEX/Poseidon show that the long-wavelength component of the recovered field is useful globally with the implication that long-wavelength temporal variations will be recovered by this procedure. That the field does not perform as well as existing models on lower satellite orbits (Stella and Starlette, for example) in Table 3 is not unexpected and an artefact of the recovery procedure. Additional normal equations specific for these lower satellites are needed to enhance the recovery.

References

1. Bertiger WI et al (1994) GPS precise tracking of TOPEX/POSEIDON: Results and implications, J Geophys Res, **99**, 24449–24464.
2. Kuang D et al (2001) Precise orbit determination for CHAMP using GPS data from BlackJack receiver, to appear in Proceedings of the ION National Technical Meeting, Long Beach, California, 2001.
3. Lemoine FG et al (1998) The development of the joint NASA GSFC and the National Imagery and Mapping Agency (NIMA) geopotential model EGM96. NASA/TP-1998-206861.

Gravitational Field Modelling From CHAMP-Ephemerides by Harmonic Splines and Fast Multipole Techniques

Oliver Glockner

Geomathematics Group, University of Kaiserslautern, PO-Box 3049, 67653 Kaiserslautern, glockner@rhrk.uni-kl.de

Summary. The recovery of the Earth's gravitational field from satellite data is an ill-posed problem and demands regularization. A projection method in terms of harmonic spline bases combined with Tikhonov regularization is presented to stabilize the solution process. Powerful numerical algorithms are needed to process the huge amount of observational data with reasonable time and storage requirements. For the solution of the occuring large and dense linear systems of equations we developed a fast multipole accelerated iterative solver. This solver performs the matrix multiplication in each iteration step in $\mathcal{O}(N)$ operations and with $\mathcal{O}(N)$ storage requirements. The applicability and efficiency of the spline model in combination with the fast multipole technique for gravitational field modelling is documented on the basis of given CHAMP ephemerides.

Key words: Harmonic Splines, Fast Multipole Method, Tikhonov Regularization, Gravitational Field

1 Introduction

The relation between the gravitational potential U and the observational function G can be formulated as Fredholm integral equation of the first kind $\Lambda U = G$, where Λ denotes a compact integral (pseudodifferential) operator. Since we only consider isotropic observations we assume high-low SST to provide the first order radial derivative of the gravitational potential whereas SGG provides the second order radial derivative at discrete positions of the satellite. While SGG "directly" measures the second order derivatives, some preliminary work has to be done in high-low SST to obtain the gradient vector from which the radial derivative can be extracted. Newton's law of inertial motion provides the connection between the position respectively the acceleration vector of the satellite and the gravitational field (cf.[4]).

The major drawback of acquiring the observations at satellite's altitude is that the downward continuation process, i.e. the derivation of the gravitational field at Earth's level from satellite measurements, is exponentially ill-posed and therefore requires regularization (see e.g., [3] or [5]). Mathematically, the operator Λ does not have a continuous inverse. Several regularization methods are well-known in mathematical literature, such as truncated singular value decomposition, Tikhonov regularization, projection methods or combinations of those (cf. e.g., [1]). The discretization of the operator equation is done by choosing suitable finite-dimensional ansatz and test spaces (here we choose harmonic spline spaces) within a reproducing kernel Hilbert space framework (cf. [3], [5]) which allows to avoid the calculation of expensive inner products (integrals) to assemble the linear system. Since the occuring linear systems of equations are badly conditioned an additional Tikhonov stabilization is introduced.

However, a dense linear system of equations has to be solved. Evidently, a direct solution of those equations is only reasonable if the number of equations is not too large since it requires $\mathcal{O}\left(N^3\right)$ operations and the essential effort of an iterative solver is $\mathcal{O}\left(N^2\right)$ for the matrix multiplication in each iteration step. When using special reproducing kernel Hilbert spaces, such as spaces where the singularity kernel is the reproducing kernel, it will be shown that it is possible to accelerate the matrix multiplication in an iterative solver to $\mathcal{O}(N)$ operations by use of the fast multipole technique (cf. [5], [6]). The fast multipole method (FMM) has shown to be a powerful tool in many physical applications involving Laplace's as well as Helmholtz's equations in two and three dimensions. The simulations (numerical experiments) show that the proposed method is capable to model the Earth's gravitational potential from satellite data.

2 Mathematical Formulation

We are looking for a harmonic function outside the Earth's surface Σ, i.e. the Earth's gravitational potential. Consider an internal (Bjerhammar) sphere Ω_R. Then the Runge-Walsh theorem (cf. [3]) tells us that to each potential U, being harmonic outside Σ there exists a potential V which is harmonic outside Ω_R and which approximates U arbitrarily well in uniform sense. Furthermore, $\mathcal{L}^2(\Omega_R)$-approximation on Ω_R means locally uniform approximation outside Ω_R. Thus, if we are able to find an $\mathcal{L}^2(\Omega_R)$-approximation V_N of V on Ω_R, we have a uniform approximation of U on and outside Σ. Furthermore, let V, restricted to Ω_R, and G, restricted to Ω_r, which is a sphere inside the satellite orbit, be elements of separable reproducing kernel Hilbert spaces $\mathcal{H}$ and $\mathcal{K}$, respectively, of square-integrable functions.

2.1 Operator Equation

We connect the wanted function $V \in \mathcal{H}$ with the given function $G \in \mathcal{K}$ via the operator $\Lambda : \mathcal{H} \to \mathcal{K}$, where

$$(\Lambda V)(x) = (K_\Lambda(\cdot, x), V)_{\mathcal{H}} = G(x)$$

with

$$K_\Lambda(\cdot,\cdot) = \sum_{n=0}^{\infty} \sum_{k=1}^{2n+1} \frac{1}{A_n^2} \sigma_n Y_{n,k}^R(\cdot) Y_{n,k}^r(\cdot).$$

$\left\{1/A_n\, Y_{n,k}^R\right\}$ and $\left\{1/A_n\, Y_{n,k}^r\right\}$ are complete orthonormal systems (spherical harmonics) in $\mathcal{H}$ and $\mathcal{K}$, respectively. The system $\left\{\sigma_n, 1/A_n\, Y_{n,k}^R, 1/A_n\, Y_{n,k}^r\right\}$

with $\sigma_n = \left(\frac{R}{r}\right)^n (n+1)$ is called the singular system of Λ. Λ is a compact operator with infinite-dimensional range, hence Λ^{-1} is not bounded on $\mathcal{K}$ (see [1]).

The function spaces $\mathcal{H} = \mathcal{H}(\Omega_R; A_n)$ and $\mathcal{K} = \mathcal{K}(\Omega_r; A_n)$ are chosen to be reproducing kernel Hilbert spaces with reproducing kernels $K_{\mathcal{H}}(\cdot,\cdot)$ and $K_{\mathcal{K}}(\cdot,\cdot)$, respectively. We further assume that the measurements constitute a set of linearly independent bounded linear functionals (radial derivatives) on $\mathcal{K}$:

$$\mathcal{L}_x^i G(x) = \left(\mathcal{L}_x^i K_{\mathcal{K}}(x,\cdot), G\right)_{\mathcal{K}} = \left(x \cdot (\nabla V)(x)\right)\big|_{x=x(t_i)}, \quad i = 1,\ldots,N.$$

The radial derivative can be extracted from the gradient vector ∇V, which can be derived from the CHAMP-ephemerides by numerical differentiation and Newton's law of inertial motion $(\nabla V)(x(t)) = \ddot{x}(t)$ (for details about numerical differentiation of the ephemerides the reader is directed to [2]).

2.2 Discretization and Regularization

For numerical calculations the operator equation has to be discretized. First, we have to specify suitable finite-dimensional subspaces $\mathcal{K}_N \subset \mathcal{K}$ (test space) and $\mathcal{H}_N \subset \mathcal{H}$ (ansatz space):

$$\mathcal{K}_N = span(\mathcal{L}_x^1 K_{\mathcal{K}}(x,\cdot),\ldots,\mathcal{L}_x^N K_{\mathcal{K}}(x,\cdot)),$$

$$\mathcal{H}_N = span(\Lambda^* \mathcal{L}_x^1 K_{\mathcal{K}}(x,\cdot),\ldots,\Lambda^* \mathcal{L}_x^N K_{\mathcal{K}}(x,\cdot)),$$

where Λ^* denotes the adjoint operator. The wanted approximation V_N is then assumed to be an element of $\mathcal{H}_N$, i.e.

$$V_N = \sum_{i=1}^{N} a_i \Lambda^* \mathcal{L}_x^i K_{\mathcal{K}}(x,\cdot) \in \mathcal{H}_N$$

with

$$\Lambda^* \mathcal{L}_x^i K_{\mathcal{K}}(x,\cdot) = \sum_{n=0}^{\infty} \sum_{k=1}^{2n+1} \frac{1}{A_n^2}(n+1)\left(\frac{R}{r}\right)^n Y_{n,k}^R(\cdot)\left(\frac{R}{|x_i|}\right)^{n+1} Y_{n,k}^r(x_i).$$

This projection method is known as the dual least-squares method (see, e.g. [1]). The unknown coefficients a_i, $i = 1,\ldots,N$ are determined by the Galerkin equations

$$(\mathbf{B} + \gamma \mathbf{I})\, a = b \tag{1}$$

with

$$B_{ij} = \left(\Lambda^* \mathcal{L}_x^i K_{\mathcal{K}}(x,\cdot), \Lambda^* \mathcal{L}_x^j K_{\mathcal{K}}(x,\cdot)\right)_{\mathcal{H}},$$

$$b_i = \left(\mathcal{L}_x^i K_{\mathcal{K}}(x,\cdot), G\right)_{\mathcal{K}} = \widetilde{v_i} = v_i + \widetilde{\varepsilon}_i.$$

If we choose the reproducing kernel Hilbert spaces to be defined by the sequence $A_n = (n+1/2)^{1/2} h^{-n/2}$, then the singularity kernel ([3], [5]) defines the reproducing kernel in $\mathcal{K}$ and by applying the Kelvin transformation $y^* = h\frac{R^2}{|y|^2}y$ we deduce

$$B_{ij} = \frac{1}{2\pi|y|}\left(\left((y^* \cdot \nabla_{y^*})^2 + 2\,(y^* \cdot \nabla_{y^*})\right)\frac{1}{|x-y^*|} + \frac{1}{|x-y^*|}\right)\Bigg|_{(x_i,x_j^*)} . \quad (2)$$

Since the entries of the matrix consist of the fundamental solution of the Laplace equation and directional derivatives of it the solution of the linear system by a fast multipole accelerated solver becomes possible. Details about the derivation of this aspect are described in [5].

2.3 The Fast Multipole Technique

In order to reduce the computational effort and the storage requirements for solving the linear system (1) we adapted the fast multipole technique to accelerate the matrix multiplication in an iterative (GMRES) solver. Theory and algorithmic of the FMM are very sophisticated and their description would go beyond the scope of this article. For details we refer, e.g. to [5], [6] and the references therein. In this subsection we briefly outline the ideas and the basic aspects of the fast multipole method (FMM). The subject of interest is the fast evaluation of a matrix multiplication $\sum_{i=1}^{N} a_i B_{ij}$, $j = 1, \ldots, N$, B_{ij} as in (2), N large. The idea of the FMM is based on the localization property of the kernel defining the entries B_{ij}, i.e. on the fact that most of the energy is contained within a small vicinity of x_j^*. The sum is split up into a near-field part (of course, defined with respect to x_j^*), which is evaluated explicitly, and a far-field part, for which we seek a "fast" approximation with "sufficient accuracy".

The algorithm is started by embedding the computational domain into an (initial) cube, followed by an adaptive and hierarchical subdivision of each cube into eight child cubes. Adaptivity means that a cube is only subdivided if it contains a minimum number of nodal points and is essential for the efficiency.

The approximation of the far-field is performed on the basis of the well-known expansion of $1/|x-y|$ in terms of inner and outer harmonics. The kernel is expanded for both variables with respect to the centers x_0 and y_0 of the cubes (instead of the origin) and the expansion is truncated at a certain (low) degree p. Conditions for the convergence of the expansion can easily be derived. An expansion for $(y \cdot \nabla_y)\,k(x,y)$ can be found indirectly by applying $(y \cdot \nabla_y)$ to the expansion of $k(x,y)$ (see [5]). Since the error of the expansion is controllable only in a small vicinity of (x_0, y_0), we have to perform several expansions with respect to different centers (x_0, y_0) to obtain a global approximation. A lot of work can be done in an a-priori step independent of the number and the location of the target points, resulting in a small set of so-called far-field coefficients containing the information from distant measurements. The hierarchical subdivision in connection with the translation and conversion theorems of spherical harmonics (see [5]) allow to calculate the far-field coefficients in level l recursively from those of the parent cube in level $l+1$ so that an explicit calculation is only needed for the finest level. Altogether we obtain an algorithm which performs a matrix multiplication in $\mathcal{O}(N)$ operations and with order N storage requirements. Details and extensive numerical experiments concerning the efficiency of the method can be found in [5].

3 Numerical Results

For the application of our method we generated a 90 days CHAMP orbit by an orbit prediction program (MOVSAT) with a sampling rate of 0.2 Hz in an artificial gravitational field generated by EGM96 up to degree and order 90, resp. 128. The first order radial derivatives of the potential have been derived by numerical differentiation and Newton's law of motion (cf. [2]). For our modelling a special uniformly distributed set of measurements has been chosen from the complete set (see [5], [2]). The optimal regularization parameter γ has been determined by the L-curve method ([5], [6]). The results of the simulations are presented in the following table and figures.

Degree	Unknowns	γ	Max. error	Mean error	Rmse
90	9152	10^{-15}	21.92 m^2/s^2	3.09 m^2/s^2	3.97 m^2/s^2
128	17370	10^{-14}	64.53 m^2/s^2	3.19 m^2/s^2	5.05 m^2/s^2

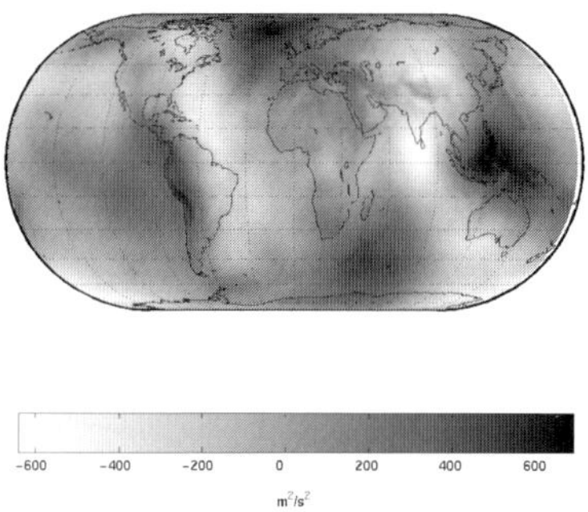

Recovered potential, degrees 3-90

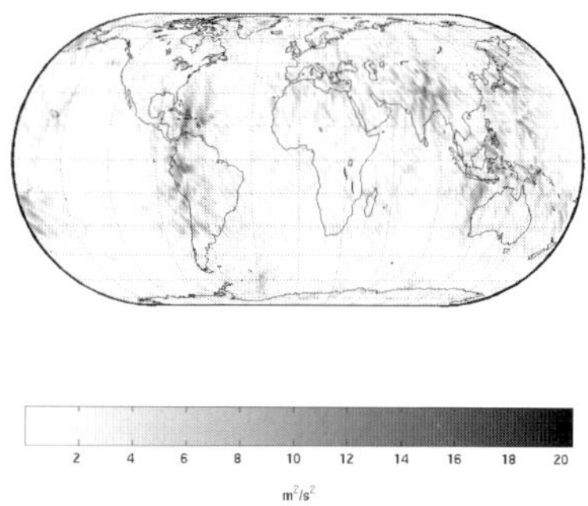

Error plot

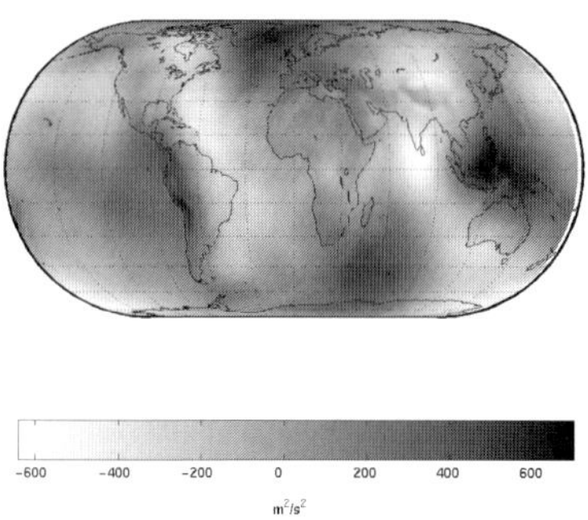

Recovered potential, degrees 3-128

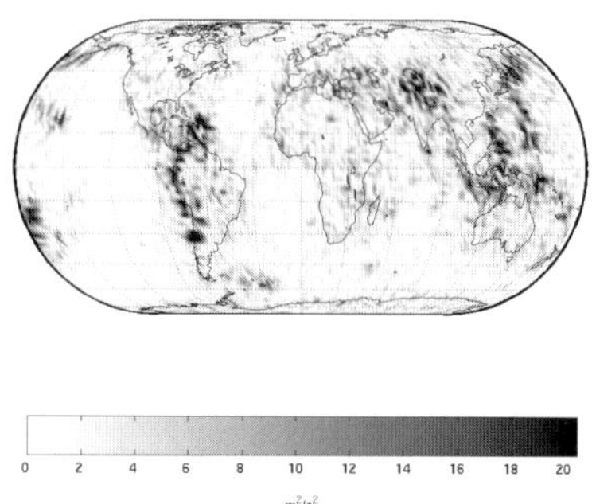

Error plot

In order to demonstrate that our method also works for GOCE observables and for local problems we present two additional experiments. A 90 days GOCE orbit has been generated with corresponding second order radial derivatives, perturbed with white noise, at the satellite positions. The results are shown below.

Problem	Unknowns	γ	Max. error	Mean error	Rmse
global	40980	10^{-10}	$19.44\ m^2/s^2$	$1.05\ m^2/s^2$	$1.5\ m^2/s^2$
local	7688	10^{-12}	$10.19\ m^2/s^2$	$1.55\ m^2/s^2$	$2.02\ m^2/s^2$

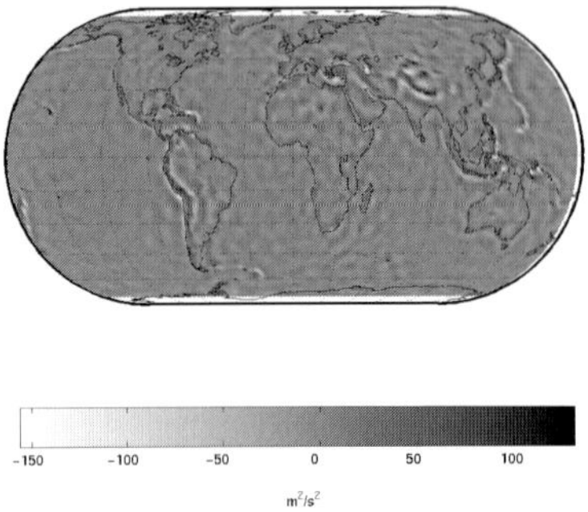

Recovered potential, degrees 36-200

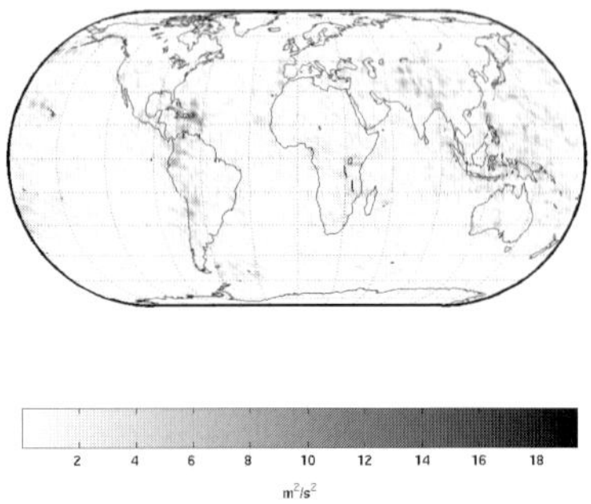

Error plot

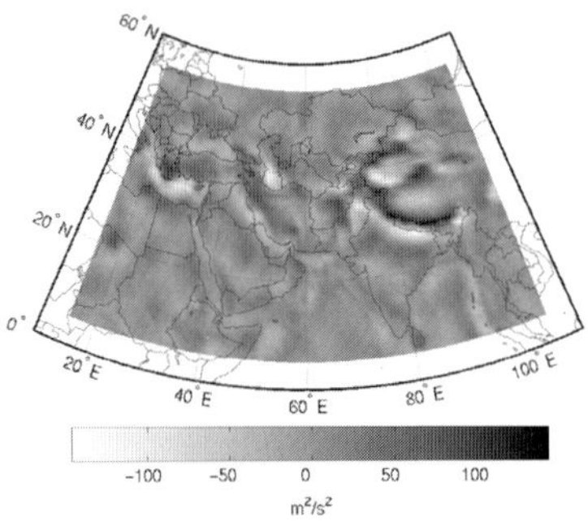

Local recovery of the potential, degrees 36-200

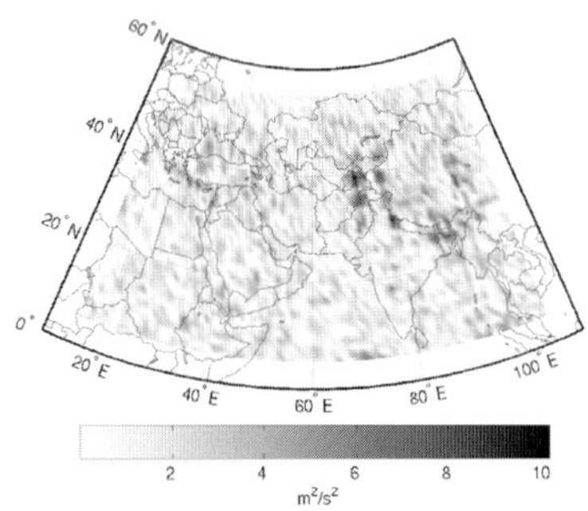

Error plot

References

1. Engl HW, Hanke M, Neubauer A (1996) Regularization of inverse problems. Kluwer, Dordrecht, Boston, London
2. Fengler M (2001) Multiscale modelling of the Earth's gravitational potential from discrete noisy CHAMP-ephemerides. Diploma Thesis, University of Kaiserslautern, Geomathematics Group
3. Freeden W (1999) Multiscale modelling of spaceborne geodata. BG Teubner, Stuttgart, Leipzig
4. Freeden W, Glockner O, Thalhammer M (1998) Multiscale gravitational field recovery from GPS-satellite-to-satellite tracking. Stud Geoph Geod 43: 229-264
5. Glockner O (2001) On numerical aspects of gravitational field modelling from SST and SGG by harmonic splines and wavelets (with application to CHAMP data). PhD-Thesis, University of Kaiserslautern, Geomathematics Group
6. Michel D (2001) On the combination of harmonic splines and fast multipole methods for CHAMP data modelling. Diploma Thesis, University of Kaiserslautern, Geomathematics Group

Evaluation of Geoid Models with GPS/Levelling Points in Sweden and Finland

Mirjam Bilker, Matti Ollikainen, Markku Poutanen

Finnish Geodetic Institute, Geodeetinrinne 2, FIN-02430 Masala, Finland, Mirjam.Bilker@fgi.fi, tel:+358-9-295550

Summary. Five different geoid models are analysed to estimate the possible error budget of new global geoids based on the data from the gravity satellites CHAMP, GRACE and GOCE. We used the global models OSU91A and EGM96, the European model EGG97, the Nordic model NKG96 and the newest model for Finland FIN2000. FIN2000 is fitted to the Finnish height system and is the most accurate of the five. A total of 245 points in Finland and Sweden was used in the analysis. For all points, GPS-determined heights were available as well as orthometric or normal heights based on spirit levelling. From these heights the geoid height was calculated and compared with corresponding geoid heights obtained from the geoid models, allowing us to estimate the geoid model error and its spatial variation.

Key words: Geoid models, GPS-levelling

1 Introduction

The satellite mission CHAMP, GRACE and GOCE will improve the knowledge of the Earth's gravity field considerably. Global geoid models will improve and regional models will get better in areas where not much gravity information has been available up to now. During the preparation for the coming information from the gravity missions, we investigated the geoid models that are at present available for Scandinavia.

Dense datasets of geodetic observations in this region are available, covering a long period. The data could either be used for independent validation of the gravity mission products or for combination with the data from the gravity missions. By comparing heights from GPS and levelling with geoid heights from existing geoid models, we get an inside in the signals that were not modelled in the geoid models at different levels.

2 Data

For the investigation we used geoid height grids of five geoid models: The global models OSU91A and EGM96, the European model EGG97, the Nordic model NKG96, and the model for Finland FIN2000.

The OSU91A model (Rapp et al. 1991) was produced by the Ohio State University from a satellite only model combined with altimeter and surface gravity observations. The geoid height grid we used has an interval of 0.125 degrees.

The EGM96 model (Lemoine et al. 1998) was a combined effort of the NASA Goddard Space Flight Centre (GSFC), the National Imagery and Mapping Agency (NIMA), and the Ohio State University (OSU). The model was made using improved satellite and surface data. The geoid height grid has an interval of 0.25 degrees.

The EGG97 model (Denker et al. 1996) is a geoid model for Europe computed by the Institut für Erdmessung at the University of Hannover. The EGM96 model was used for the long wavelength gravity field information. The gravity field information for the shorter wavelengths was obtained from land gravity data with a resolution of at least 10 km and ship data combined with altimeter data. The geoid height grid has a resolution of 1.0 by 1.5 minutes.

The NKG96 model (Forsberg et al. 1996) covers Fennoscandia and the Baltic. With the EGM96 model as a reference model, the NKG96 model was calculated using gravimetric data densely covering the area and altimeter data covering the Baltic Sea. Finally, the geoid model was corrected with a constant to fit better to the local height systems. This correction term was determined by fitting the model to GPS/levelling data in the region. The NKG96 grid has a resolution of 0.025 by 0.05 degrees. The above-mentioned models are non-tidal quasi geoids.

The FIN2000 model was made by fitting the NKG96 model with a fourth order polynomial to 159 GPS/levelling points in Finland. The model covers only Finland and can be used to convert ellipsoidal heights directly to the Finnish N60 height system.

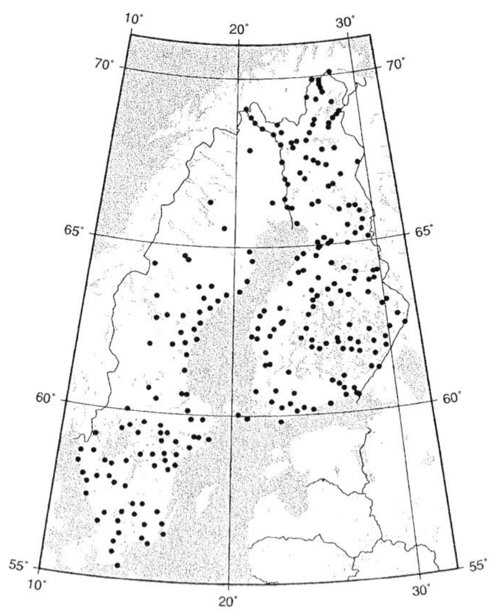

Fig. 1. Points used in study

In the evaluation, a total of 245 points was used covering Finland and big parts of Sweden. GPS heights as well as levelling heights were available for all the points. The points are shown in Figure 1.

For Finland 154 points were available. The GPS heights refer to the EUREF-FIN reference system, of which the epoch for the heights is 1997.0. The levelling heights are orthometric heights in the Finnish N60 height system, referring to epoch 1960.

For Sweden 91 points were used. The Swedish GPS heights refer to the SWEREF99 system, of which the heights refer to epoch 1999.5. The levelling heights are normal heights in the RH70 system, which refers to epoch 1970.

3 Comparison of geoid heights

To be able to compare the geoid heights and the GPS and levelling heights with each other, they should all be referring to the same kind of geoid. This is the case for the Swedish heights, but not for the Finnish heights.

The Finnish levelling heights are orthometric heights referring to the mean geoid. For the comparison with the OSU91A, EGM96, EGG97, and NKG96 geoid heights, these levelling heights should be transformed to normal heights above the non-tidal quasi geoid. The difference between geoid and quasi geoid heights is in Finland very small and will be neglected in this study. Ekman (1989) gives the transformation between heights above a normal geoid and a non-tidal geoid. The transformation is constant over an area and will therefore later on cancel out when we fit a plane through the geoid height differences.

Of bigger influence than the geoid types is the postglacial rebound in Fennoscandia. When comparing heights from different epochs, the land uplift will show up as part of the signal. The reference epochs of the levelling heights and GPS heights differ 37 years in Finland and 29.5 years in Sweden. This will result in a relative growth of the calculated geoid heights between the north and the south of up to 40 cm. To avoid this, the levelling heights are brought to the same epoch as the GPS heights using the land uplift map from Ekman (1996). The geoid itself rises as well, but at a much lower rate (see Ekman and Mäkinen 1996). The geoid rise is therefore ignored in this study.

Differences in geoid heights can now be calculated with:

$$\Delta N = h_{\mathrm{GPS}} - H_{\mathrm{levelling}} - N_{\mathrm{geoid\,model}} \tag{3.1}$$

To adjust the geoid heights to the local height systems, a 1st order polynomial was then fitted through the geoid height differences:

$$\Delta N = a_{00} + a_{01}(\lambda - \lambda_0) + a_{10}(\varphi - \varphi_0). \tag{3.2}$$

Separate polynomials were calculated for the Swedish and the Finnish points. No polynomial is fitted through the differences with the FIN2000 geoid model as it is already fitted with a 4th order polynomial to the Finnish height system.

Figure 2 shows the remaining differences between GPS/levelling heights and geoid heights from the models OSU91A, EGM96, EGG97, and NKG96 after a first order polynomial was fitted through the data. Figure 3 shows the differences

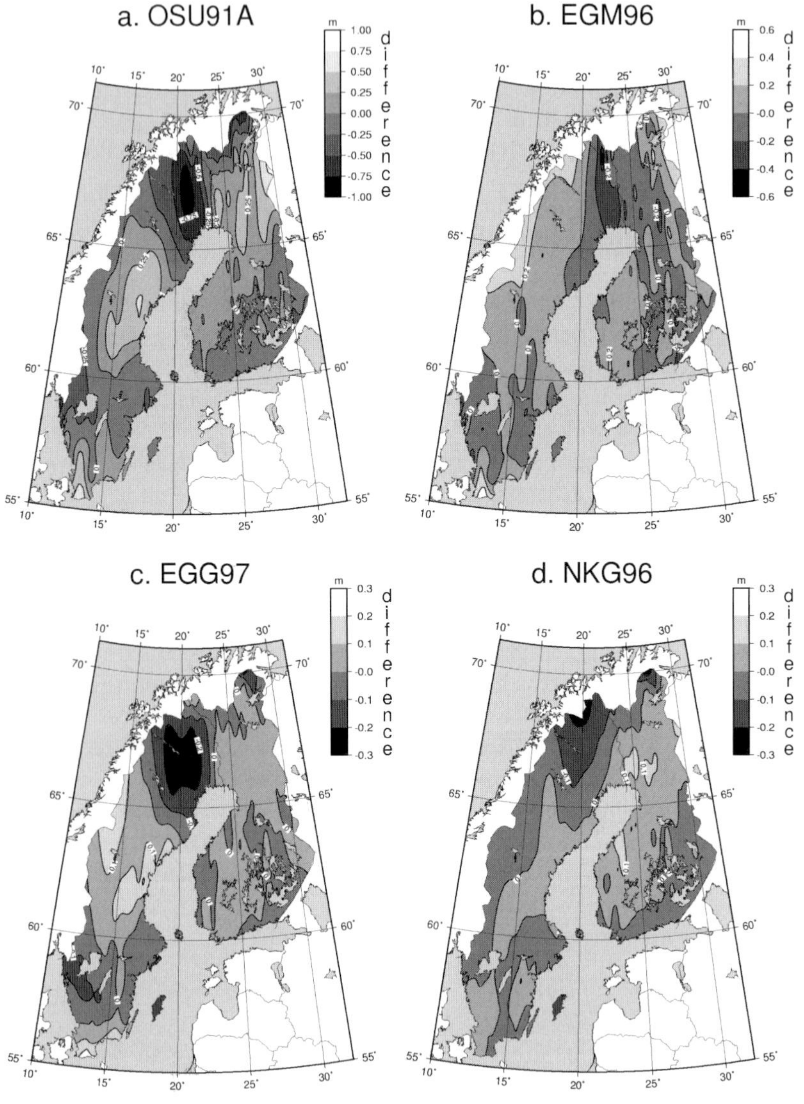

Fig. 2. . Differences in geoid heights after a 1st order polynomial fit for the geoid models: **a.** OSU91A, **b.** EGM96, **c.** EGG97, and **d.** NKG96.

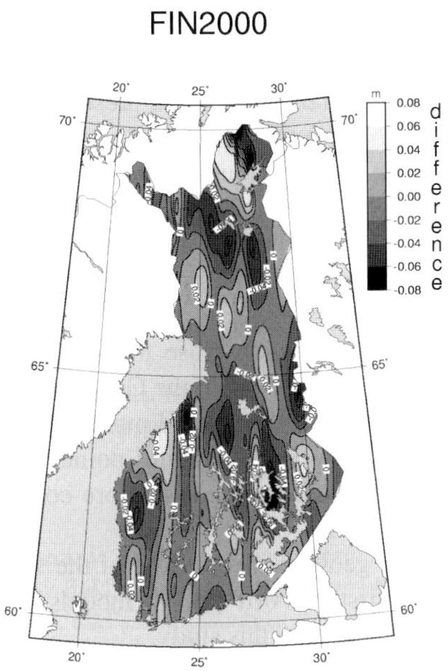

Fig. 3. Differences in geoid heights for the FIN2000 geoid model.

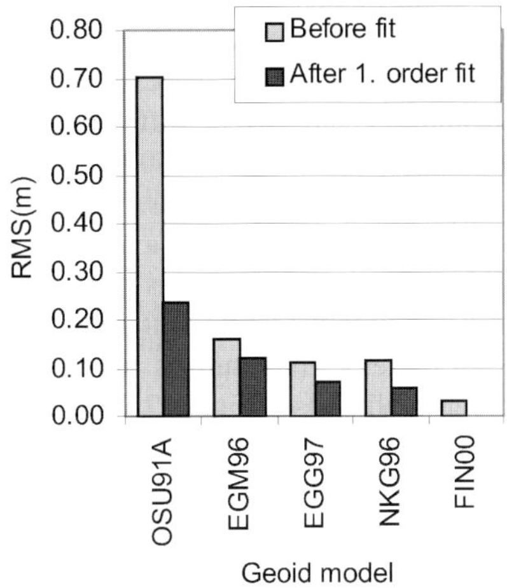

Fig. 4. RMS values of the geoid height differences before and after the first order polynomial fit.

*No 1st order polynomial fit calculated

Geoid	Minimum [m]	Maximum [m]	Average [m]	RMS [m]
OSU91A	-0.908	0.696	0.000	0.234
EGM96	-0.452	0.363	0.000	0.121
EGG97	-0.236	0.220	0.000	0.069
NKG96	-0.232	0.127	0.000	0.060
FIN2000*	-0.083	0.058	-0.010	0.030

Table 1. Statistics of the geoid height differences after 1st order polynomial fit.

between the GPS/levelling heights and the FIN2000 geoid heights. The statistical values corresponding to figures 2 and 3 are shown in figure 4 and table 1.

4 Discussion

One can see from the figures that the FIN2000 geoid model fits the best. This was expected, because the model is already fitted to the Finnish height system. The geoid height differences remaining (see figure 3) include errors in GPS, levelling

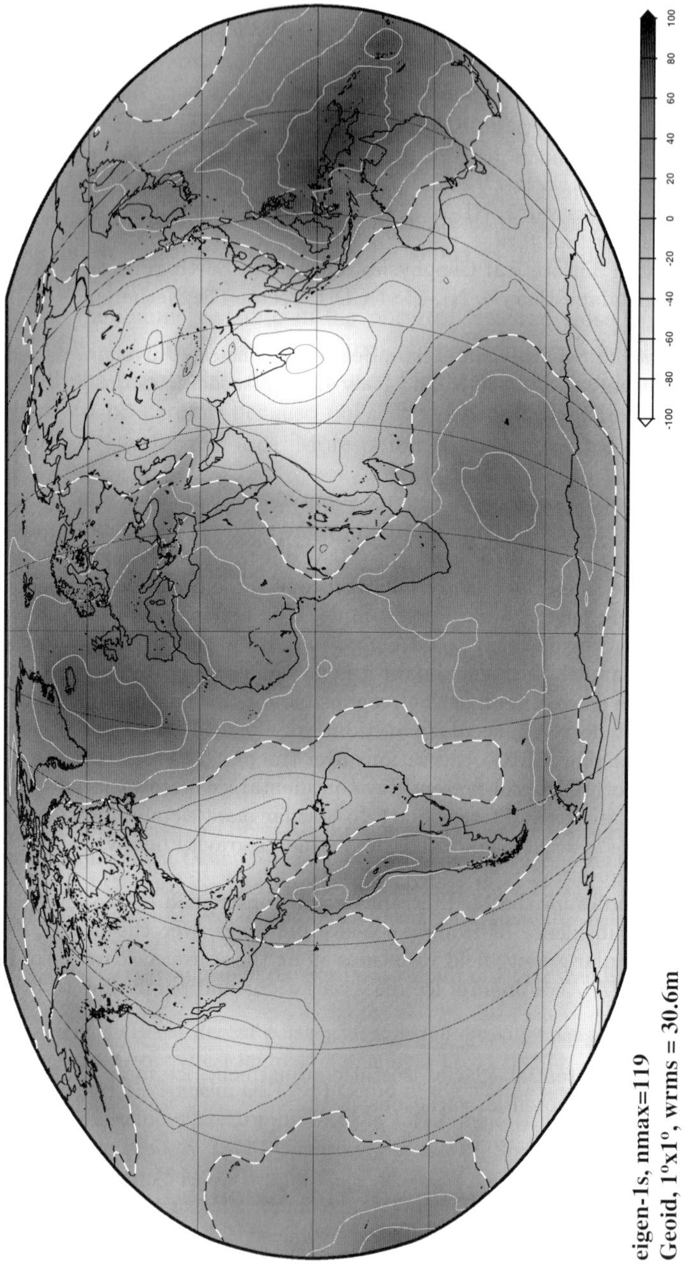

Fig. 1. Geoid model *Eigen-1s* derived from CHAMP data

into the mantle. Other geoid highs are related to anomalously hot regions, for example near Iceland and south-east of South Africa. At these latter regions, the raised, dynamically sustained, topography overcomes the *negative* geoid contributions of the lower (hotter) mantle densities in producing net *positive* geoid anomalies. Indeed, [16] and [15] showed that effects of both mantle density anomalies and their induced dynamic topographies including those of internal mantle layers, must be taken into account to reproduce the geoid signals associated with mantle convection in general and subduction of oceanic lithosphere in particular. The question whether a mantle density anomaly produces a geoid high or a geoid low even depends on the rheology of the mantle, i.e. mantle viscosity.

The most conspicuous geoid low in Fig. 1 is centered below India. Obviously, this deep geoid low might in some way be connected with the fast northward movement of the Indian plate during the Cretaceous, that resulted in the formation of the Himalayas during the Cenozoic. Another geoid low is centered about Canada. It is tempting to associate this with a deviation from isostasy resulting from deglaciation of the Pleistocene Laurentide ice sheet. Some 20,000 years the great Pleistocene ice mass complexes above Canada, the Nordic countries in Europe, Russia and Antarctica started to melt. This process ended about 6,000 years ago, with the diminished Antarctic and Greenland ice sheets as 'left overs', leaving geoid lows above the formerly glaciated areas. Today, the Earth has not completely rebounded to the isostatic condition, so that the geoid low above Canada might be associated with the mass deficit in this region resulting from the termination of the last Ice Age.

However, glacial isostatic adjustment (GIA) simulations produce geoid lows above Canada that are about a factor of two smaller than the geoid low of about 50 m at Hudson Bay that can be deduced from Fig. 1. The observed geoid low of about 50 m at Hudson Bay seems to be a robust feature, which is also visible with about the same value in the *EGS96* model ([10]). In older geoid maps, e.g. in *GEM* of [11], this geoid low is even higher, about 85 m, so even further away from what GIA models predict. One can vary the parameters of such models to some extent, e.g. mantle viscosity which is the least constrained parameter, so that simulations and observations of the geoid deep agree, but then the simulations lead to large mismatches with respect other observables like relative sea-level variations derived from raised beach lines around Hudson Bay.

Another look at Fig. 1 shows that the geoid low above Canada might be connected to some extent with other geoid lows, e.g. with the one at the western Atlantic near the east US coast. Indeed, many authors have pointed out (e.g. [12], [14], [19] and [9]) that the most likely explanation for the discrepancy between simulated and observed geoid low is a contribution from mantle convection.

So, although it often might seem obvious that a particular geoid anomaly is related to a certain geodynamical process, this example of the geoid low above Canada shows that one has to be careful in making these kind of associations. A similar problem of uncertainties concerning the causes of observed geoid features arises in studies on secular variations of the low-degree geoid harmonics, as is discussed in the next section.

2 Secular Low-Degree Geopotential Variations

It remains to be seen in how far CHAMP and GRACE will be able to deduce *secular* changes in the geoid. Twenty-five years of satellite laser ranging (SLR) to dedicated satellites as Lageos have taught us that many short-term effects can perturb the outcome of secular solutions (e.g. El Niño). Even nowadays only the degree two term has been well established. Tables 1 and 2 give an overview of observed secular changes in even (Table 1) and uneven (Table 2) zonal harmonics. Note that even for degree 4 the results between the various solutions differ strongly; even the sign of the secular change has not been established unambiguously (see, e.g., [4]).

On the interpretative site, all is not clear either. It is generally thought that two geodynamical processes are responsible for these secular variations: GIA and present-day continental ice sheet melt or decay, but in how far

Table 1. Even degree zonal secular drift ($\times 10^{-11}$ per year)

Author	$\dot{J}_2$	$\dot{J}_4$	$\dot{J}_6$
Yoder et al., 1983 [22] Lageos-1	-3		
Rubincam, 1984 [17] Lageos-1	-2.6±0.6		
Cheng et al., 1989 [2] Starlette	-2.5±0.3	0.3±0.6	
Gegout and Cazenave, 1993 [8] Lageos-1	-2.8±0.4		
Eanes, 1995 [7] Lageos-1	-2.4±0.2		
Nerem and Klosko, 1995 [13] Lageos-1,2;Ajisai;Starlette	-2.8±0.3	0.2±1.5	
Cazenave et al., 1995 [1] Lageos-1,2	-3.0±0.5	-0.8±1.5	
Cheng et al., 1997 [3] Lageos-1,2;Starlette;Ajisai;Stella;Etalon-1,2;BE-C	-2.7±0.4	-1.4±1.0	0.3±0.7
Devoti et al., 2001 [4] Lageos-1,2;Starlette;Stella	-2.9±0.2	0.6±0.5	0.3±0.3

Table 2. Odd degree zonal secular drift ($\times 10^{-11}$ per year)

Author	$\dot{J}_3$	$\dot{J}_5$	$\dot{J}^{(a)}_{odd}$	$\dot{J}^{(b)}_{odd}$
Nerem and Klosko, 1995 [13] Lageos-1,2;Ajisai;Starlette			1.6±0.4	
Cheng et al., 1997 [3] Lageos-1,2;Starlette;Ajisai;Stella;Etalon-1,2;BE-C	-1.3±0.5	2.1±0.6	0.5	0.6
Devoti et al., 2001 [4] Starlette				0.5±0.2

$^{(a)}$ using the constraint $J_{odd} = J_3 + 0.837 J_5$ from [13]
$^{(b)}$ using the constraint $J_{odd} = J_3 + 1.04 J_5 - 0.53 J_7 + \ldots$ from [4]

tectonic and mantle convection processes might contribute to the secular changes of Tables 1 and 2 is an open question.

Given all these observational and interpretative uncertainties, it still makes sense to try to deduce information from comparing forward models of GIA and present-day ice mass variations with observational data of the secular changes in the low geopotential harmonics (e.g., [21]). In Fig. 2 the secular changes $\dot{J}_n$ of degree n derived from the observations of [3] and [4] and forward models for GIA and present-day ice mass changes are depicted as a function of the lower mantle viscosity. Only the GIA models are dependent on this parameter; the observations and the present-day ice mass variations are not dependent on mantle viscosity, but for the effects of the present-day ice mass variations on the geoid variations the elasticity of the Earth has been taken into account.

An interesting pattern is emerging from the panels of Fig. 2. In some cases the solid lines, delineating the effects of GIA derived from GIA relaxation models (e.g., [6]) with the Pleistocene ICE-3G ice melt history model of [20], cross one or both of the observations. The same is the case when a value of about 5×10^{14} kg per year contemporary ice melt of the Antarctic ice sheet is added (the dashed lines). However, the majority of the solutions straddle the observations in all cases. Note also that although the $\dot{J}_4$ observations of [3] and [4] are not in agreement with each other (even having opposite signs!), the solutions with and without present-day Antarctic ice sheet decay straddle *both* observations. This remains true if present-day ice sheet decay of the Greenland ice sheet of about 1.5×10^{14} kg per year is added.

Apparently, present-day ice sheet decay of Antarctica (and perhaps of Greenland) is necessary to reconcile the geodynamic simulations with the SLR-derived observations in Fig. 2, but with a value that is less than 5×10^{14} kg per year. Formal inversions by [18] show that the same lower mantle viscosity solution of about 10^{22} Pa s is found for all panels in Fig. 2 if the Antarctic ice sheet decays presently by a value of $2.50 \pm 0.80 \times 10^{14}$ kg per year.

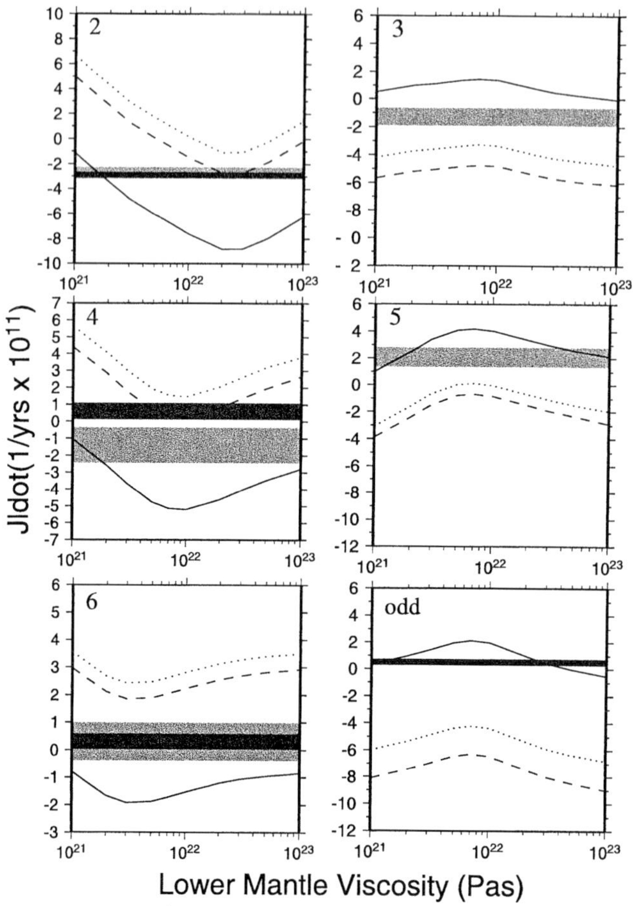

Fig. 2. $\dot{J}_n$ as function of lower mantle viscosity, ranging from 10^{21} to 10^{23} Pas, with harmonic degree n indicated at the top left corner of each panel. Upper mantle viscosity is fixed at 5×10^{20} Pas; elastic lithosphere thickness at 120 km. The horizontal bars in the panels are SLR observed values: light-grey ones are derived from [3]; dark-grey ones from [4]. The width of the bars denotes the standard deviation of the observation. Solid curves correspond to Pleistocene deglaciation only; dashed ones include an ice-mass loss of 5.04×10^{14} kg per year of the Antarctic ice sheet; dotted ones include an ice-mass loss of 1.44×10^{14} kg per year of the Greenland ice sheet.

A contribution from the Greenland ice sheet cannot be discerned from these data. The reason for this is quite simple: the center of the Pleistocene Laurentide ice sheet complex above Canada, which plays a dominant role in the GIA models, was geographically quite close to Greenland. A change in the mantle viscosity which causes a change in the $\dot{J}_n$ contributions from GIA, can therefore have the same effect on the total values of $\dot{J}_n$ as a contemporary mass variation of the Greenland ice sheet. Low-degree values n smaller than 6 are simply not discriminative enough to break this trade-off effect. If it would turn out to be possible to discern secular variations of the J_n with CHAMP or GRACE for higher values of n, then such a discrimination between GIA and present-day ice mass variations of the Greenland ice sheet might be possible.

3 High Degree Geopotential Harmonics

In the former two sections it has been shown that the low-degree geoid harmonics and their secular variations are rather due to combinations of geophysical processes than to a singular one. This leads to the problem of how we could discriminate between these various geodynamical contributions to the total geoid or secular geopotential variations. In the former section such a possible discrimination has been discussed for the low-degree $\dot{J}_n$'s under the assumption that only GIA and present-day continental ice mass variations are contributing to the observed values. Here an example of 'pattern

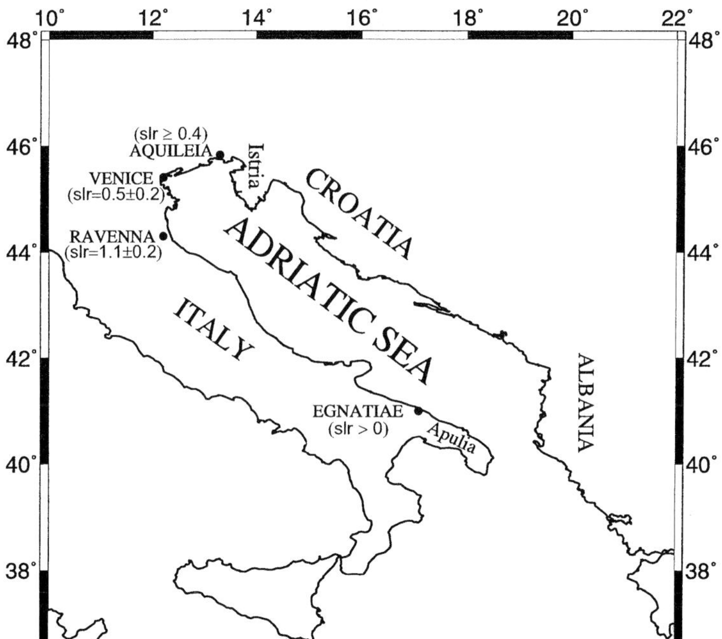

Fig. 3. Late Holocene sea-level changes in mm/yr for the Adriatic coast of Italy as derived from sea-level records of ancient Roman buildings.

recognition' for high harmonics is given to illustrate that this might be an important tool in helping to discriminate between the various contributions.

Fig. 3 shows the sea-level rise for the past few thousand years as determined from ancient artifacts along the Italian side of the Adriatic Sea. Going from Aquileia in the north to Apulia in the south, the sea-level changes have the shape of an inverted parabola. [5] have shown that such an inverted parabola shape cannot be reproduced by either GIA or a tectonic process alone. Convergence of the African and Eurasian plates combined with Tyrrhenian subduction beneath the Calabrian Arc results in a pattern of sea-level rise that is high in the northern part and low in the southern part of the Adriatic coast of Italy, GIA models produce an opposite pattern: low sea rise in the north and high in the south. The combined effect of the two processes is such that the inverted parabola shape of Fig. 3 is quite well reproduced (see [5]). Although the resulting geoid variations are too small to be detected by CHAMP, this indicates that 'pattern matching' might be an important tool in discriminating the various contributions to the observed geoid signals.

Acknowledgements: B. Vermeersen would like to thank Ch. Reigber and D. Wolf for their invitation, and F. Barthelmes and P. Schwintzer for providing the grey scale version of the *Eigen-1s* model of Fig. 1. This figure has been produced by the GMT package.

References

1. Cazenave AA, Gegout P, Ferhat G, Biancale R (1995) Temporal variations of the gravity field from Lageos 1 and Lageos 2 observations. In: Rapp RH, Cazenave AA, Nerem RS (eds) Global Gravity Field and Its Temporal Variations. IAG Symp. **116**, Boulder
2. Cheng MK, Eanes RJ, Shum CK, Schutz BE, Tapley BD (1989) Temporal variations in low-degree zonal harmonics from Starlette orbit analysis. Geophys Res Lett, **16**, 393-396
3. Cheng MK, Shum CK, Tapley BD (1997) Determination of long-term changes in the Earth's gravity field from laser ranging observations. J Geophys Res, **102**, 22,377–2,2390
4. Devoti R, Luceri V, Sciarretta C, Bianco G, Di Donato G, Vermeersen LLA, Sabadini R (2001) The SLR secular gravity variations and their impact on the inference of mantle rheology and lithospheric thickness. Geophys Res Lett, **28**, 855–858
5. Di Donato G, Negredo AM, Sabadini R, Vermeersen LLA (1999) Multiple processes causing sea level rise in the central Mediterranean. Geophys Res Lett, **26**, 1769–1772
6. Di Donato G, Vermeersen LLA, Sabadini R (2000) Sea-level changes, geoid and gravity anomalies due to Pleistocene deglaciation by means of multilayered, analytical Earth models. Tectonophys., **320**, 409–418

7. Eanes RJ (1995): A study of temporal variations in Earth's gravitational field using Lageos-1 laser range observations. CSR Report **95–8**, University of Texas Austin
8. Gegout P, Cazenave AA (1993): Temporal variations of the Earth's gravity field for 1985-1989 from Lageos. Geophys J Int, **114**, 347–359
9. Kaufmann G (2000): Ice-ocean mass balance during the Late Pleistocene glacial cycles in view of CHAMP and GRACE satellite missions. Geophys J Int, **143**, 142–156
10. Lemoine FG, Kenyon SC, Factor JK, Trimmer RG, Pavlis NK, Chinn DS, Cox CM, Klosko SM, Luthcke SB, Torrence MH, Wang YM, Williamson RG, Pavlis EC, Rapp RH, Olson TR (1998) The development of the joint NASA GSFC and the National Imagery and Mapping Agency (NIMA) Geopotential Model EGM96. NASA/TP Report **206861**, GSFC Greenbelt Maryland
11. Lerch FS, Klosko SM, Laubscher RE, Wagner CA (1979) Gravity model improvement using GEOS 3 (GEM 9 and GEM 10). J Geophys Res, **84**, 3897–3916
12. Mitrovica JX, Peltier WR (1989) Pleistocene deglaciation and the global gravity field. J Geophys Res, **94**, 13,651–13,671
13. Nerem RS, Klosko SM (1995) Secular variations of the zonal harmonics and polar motion as geophysical constraints. In: Rapp RH, Cazenave AA, Nerem RS (eds) Global Gravity Field and Its Temporal Variations. IAG Symp. **116**, Boulder
14. Peltier WR, Forte AM, Mitrovica JX, Dziewonski AM (1992) Earth's gravitational field: Seismic tomography resolves the enigma of the Laurentian anomaly. Geophys Res Lett, **19**, 1555–1558
15. Ricard Y, Fleitout L, Froidevaux C (1984) Geoid heights and lithospheric stresses for a dynamic Earth. Ann Geophys, **2**, 267–286
16. Richards MA, Hager BH (1984) Geoid anomalies in a dynamic earth. J Geophys Res, **89**, 5987–6002
17. Rubincam DP (1984) Postglacial Rebound observed by LAGEOS and the effective viscosity of the lower mantle. J Geophys Res, **89**, 1077–1087
18. Sabadini R, Di Donato G, Vermeersen LLA, Devoti R, Luceri V, Bianco G (2002) Ice mass loss in Antarctica and stiff lower mantle viscosity based on a geophysical interpretation of the long wavelength time dependent gravity field and true polar wander. Geophys Res Lett, *submitted*
19. Simons M, Hager BH (1997) Localization of the gravity field and the signature of glacial rebound. Nature, **390**, 500–504
20. Tushingham AM, Peltier WR (1991) ICE-3G: A new global model of late Pleistocene deglaciation based upon geophysical predications of postglacial relative sea level change. J Geophys Res, **96**, 4497–4523
21. Vermeersen LLA, Sabadini R, Devoti R, Luceri V, Rutigliano P, Sciarretta C, Bianco G (1998) Mantle viscosity inferences from joint inversions of Pleistocene deglaciation-induced changes in geopotential with a new SLR analysis and polar wander. Geophys Res Lett, **25**, 4261–4264
22. Yoder CF, Williams JG, Dickey JO, Schutz BE, Eanes RJ, Tapley BD (1983) Secular variation of Earth's gravitational harmonic J_2 coefficient from Lageos and nontidal acceleration of Earth rotation. Nature, **303**, 757–762

CHAMP, Mass Displacements, and the Earth's Rotation

Richard S. Gross

Jet Propulsion Laboratory, California Institute of Technology
Mail Stop 238-332, 4800 Oak Grove Drive, Pasadena, CA 91109-8099, USA

Summary. The angular momentum of the atmosphere and oceans changes as both the distribution of mass within the atmosphere and oceans changes and as the direction and speed of the winds and currents change. Since, in the absence of external torques, the angular momentum of the solid Earth-atmosphere-ocean system is conserved, the changing atmospheric and oceanic angular momentum will cause the solid Earth's angular momentum to change, or, in other words, will cause the Earth's rotation to change. The changing distribution of mass within the atmosphere and oceans also causes the Earth's gravitational field to change, an effect being measured by the CHAMP satellite mission. By measuring changes in the second-degree spherical harmonic coefficients of the Earth's gravitational field, which are related to the elements of the Earth's inertia tensor, CHAMP is, in effect, directly measuring changes in the Earth's rotation caused by mass redistribution. Thus, independent measurements of the Earth's rotation can, in principle, be used to validate the time-varying second-degree coefficients of the Earth's gravitational field measured by CHAMP after the effects of winds and currents on the Earth's rotation are removed. In addition, and perhaps surprisingly, the effects of sufficiently large earthquakes on the trace of the Earth's inertia tensor must also be removed from length of day measurements before comparing them to observed changes in the second-degree zonal gravitational field coefficient.

Key words: Gravitational field, Earth rotation, earthquakes

1 Introduction

The Earth is a dynamic system — it has a fluid, mobile atmosphere and oceans, a continually changing global distribution of ice, snow, and ground water, a fluid core that is undergoing some type of hydromagnetic motion, a mantle both thermally convecting and rebounding from the glacial loading of the last ice age, and mobile tectonic plates. These dynamic processes change the mass distribution of the Earth (on vastly disparate time scales) and hence change its gravitational field.

Any dynamic Earth process that rearranges the Earth's mass will, in general, change the Earth's inertia tensor, and hence change the Earth's rotation (Munk and MacDonald 1960). As discussed below, in terms of a spherical harmonic expansion of the Earth's gravitational potential, changes in the degree two terms are related to changes in the Earth's inertia tensor (Chao and Gross 1987) and hence to changes in the Earth's rotation. Thus, many (if not all) of the dynamic Earth processes that cause temporal variations of the Earth's gravitational field also cause variations of the Earth's rotation.

Since mass displacements change both the Earth's rotation and gravitational field, independent observations of the Earth's rotation can be used to assess measurements of the changing second-degree coefficients of the gravitational field (Chen et al. 2000) such as those being taken by the CHAMP satellite mission. Because changes in the angular momentum carried by the atmospheric winds and oceanic currents affect the Earth's rotation, but not the gravitational field, the effects of these and other motion terms must be removed from the Earth rotation observations before they can be used to assess the CHAMP measurements. In addition, it is shown here that the effects on the trace of the Earth's inertia tensor of sufficiently large earthquakes must also be removed from length of day observations before they are used to assess CHAMP measurements of changes in the second-degree zonal gravitational field coefficient.

2 Earth Rotation and Gravitational Field

Gross (2001) has recently derived from first principles the relationship between changes in the second-degree gravitational field coefficients and changes in: (1) the x- and y-components, $\chi_x(t)$ and $\chi_y(t)$ respectively, of the polar motion excitation functions, and (2) the length-of-day $\Delta\Lambda(t)$:

$$\chi_x(t) = \frac{1.61}{\Omega\,(I_{zz} - I_{xx})}\left[\Delta h_x(t) - \frac{\sqrt{5/3}}{1.44}\,\Omega M a^2\,\Delta C_{21}(t)\right] \quad (1)$$

$$\chi_y(t) = \frac{1.61}{\Omega\,(I_{zz} - I_{xx})}\left[\Delta h_y(t) - \frac{\sqrt{5/3}}{1.44}\,\Omega M a^2\,\Delta S_{21}(t)\right] \quad (2)$$

$$\Delta\Lambda(t) = \frac{\Lambda_o}{C_m \Omega}\left[\Delta h_z(t) + \frac{0.756}{3}\,\Omega\,\Delta \mathrm{Tr}(\mathbf{I}) - \frac{0.756}{3}\sqrt{20}\,\Omega M a^2\,\Delta C_{20}(t)\right] \quad (3)$$

where Λ_o is the nominal length-of-day of 86400 seconds, $\mathbf{I}$ is the inertia tensor of the Earth with trace $\mathrm{Tr}(\mathbf{I})$ and elements I_{xx} and I_{zz}, C_m is the greatest principal moment of inertia of the Earth's crust and mantle, Ω is the mean rotation rate of the Earth, M is the mass of the Earth, a is the radius of the Earth, the factors of 0.756 and 1.44 account for the yielding of the crust and mantle to imposed surface loads and the factor of 1.61 additionally accounts for the decoupling of the core and mantle. Changes in the x-, y-, and z-components of the angular momentum due to motion relative to the rotating, body-fixed terrestrial reference frame, such as that due to atmospheric winds and oceanic currents, are denoted by $\Delta h_x(t)$, $\Delta h_y(t)$ and $\Delta h_z(t)$, respectively.

From Eqs. (1) and (2) it is seen that the polar motion excitation functions $\chi_x(t)$ and $\chi_y(t)$ are linearly proportional to both: (1) changes in the relative angular momentum $\Delta h_x(t)$ and $\Delta h_y(t)$, and (2) changes in the normalized second-degree gravitational field coefficients $\Delta C_{21}(t)$ and $\Delta S_{21}(t)$. From Eq. (3), it is seen that length-of-day changes are proportional to: (1) changes in the z-component $\Delta h_z(t)$ of the relative angular momentum, (2) changes in the fully normalized second-degree zonal gravitational field coefficient $\Delta C_{20}(t)$, and (3) changes in the trace of the inertia tensor. Thus, it is straightforward to use observed polar motion excitation functions to assess CHAMP measurements of $\Delta C_{21}(t)$ and $\Delta S_{21}(t)$ once the effects of the relative angular momentum $\Delta h_x(t)$ and $\Delta h_y(t)$ have been removed from the polar motion excitation functions. Models of the atmospheric and oceanic general circulation can be used to compute the angular momentum of the winds and currents in order to remove these effects from the observed polar motion excitation functions.

The situation is a bit more complicated for length-of-day because of its dependence on changes in the trace of the Earth's inertia tensor. Length-of-day measurements, from which motion effects have been removed, can be used to assess the fidelity of CHAMP measurements of $\Delta C_{20}(t)$ only if the mass displacements that are changing the length-of-day and $\Delta C_{20}(t)$ do not change the trace of the inertia tensor. As discussed by Rochester and Smylie (1974), the trace of the inertia tensor is invariant under a wide class of deformation processes. They show that it is invariant under any process whose deformation potential has no spherical harmonic degree l=0 term. Thus, it is invariant under tidal deformations as well as under deformations due to surface loading as long as the loading is not accompanied by a net change in mass. However, the trace of the inertia tensor is not invariant under processes whose deformation potential includes a degree l=0 term or that have radial body forces such as earthquakes.

3 Earthquakes

Since the trace of the inertia tensor does not vanish for mass displacements caused by earthquakes (Chao and Gross 1987), care must be taken when using length-of-day observations to assess CHAMP measurements of $\Delta C_{20}(t)$. If the change in the trace of the Earth's inertia tensor due to earthquakes is large enough, it must be modeled and removed from the length of day observations before comparing them to measurements of $\Delta C_{20}(t)$. Chao and Gross (1987) modeled the cosesimic effect of earthquakes on the length of day itself, finding that a great earthquake such as the 1964 Alaskan event should have changed the length of day by 6.8 microseconds (μs). They also studied the effect on the length of day of the 2146 largest earthquakes that occurred during 1977–1985, finding a cumulative change in the length of day due to these earthquakes of –0.9 μs. They also found that the greatest change in the length of day during 1977–1985, amounting to 0.33 μs, was caused by the 1977 Sumba event which had a moment magnitude M_w of 8.3.

In order to assess the relative importance of the $\Delta\mathrm{Tr}(\mathbf{I})$ term to earthquake-induced changes in the length of the day, the coseismic change in the trace of the Earth's inertia tensor due to earthquakes is modeled here following the procedure

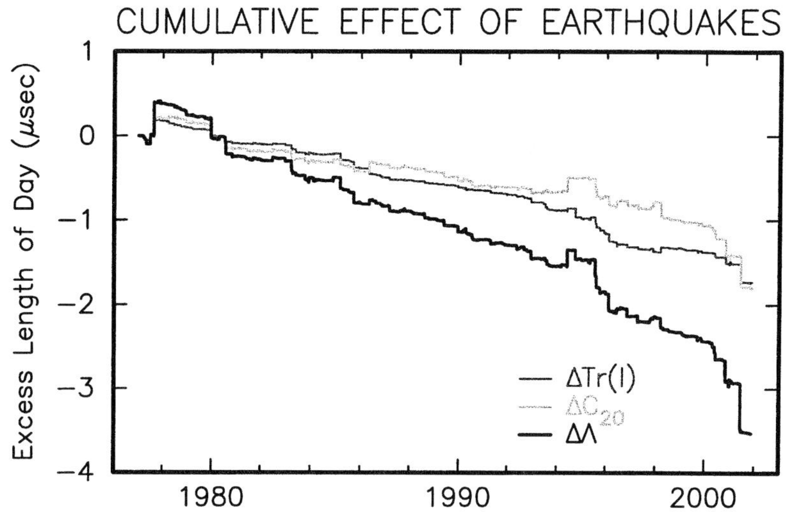

Fig. 1. Cumulative earthquake-induced change in: (1) the trace of the Earth's inertia tensor $\Delta\mathrm{Tr}(\mathbf{I})$ (middle thin black curve), (2) the second-degree zonal coefficient of the Earth's gravitational potential $\Delta C_{20}(t)$ (top gray curve), and (3) the length of day $\Delta\Lambda(t)$ (bottom thick black curve) due to the largest 18553 earthquakes that occurred during 1977–2001

of Chao and Gross (1987) as updated by Gross and Chao (2001). Fig. 1 shows the modeled change in the trace of the inertia tensor (middle thin black curve), in equivalent length of day units, for the 18553 earthquakes spanning January 1977 through November 2001 currently contained in the Harvard centroid-moment tensor catalog. The Harvard centroid-moment tensor catalog (Dziewonski et al. 2001 and references therein) contains the source properties of all earthquakes having moment magnitude M_w greater than about 5.5 that have occurred since 1977. Also shown in Fig. 1 is the contribution of the $\Delta C_{20}(t)$ term to earthquake-induced changes in the length of day (top gray curve), as well as the total change in the length of day (bottom thick black curve), being the sum of the $\Delta\mathrm{Tr}(\mathbf{I})$ and $\Delta C_{20}(t)$ terms. Since the earthquake is modeled as occurring instantaneously, there is no relative angular momentum associated with it (Chao and Gross 1987). During this longer time interval, the cumulative earthquake-induced change in the length of day is seen to be about –4 μs, with the largest change, amounting to –0.5 μs, being due to the 2001 Peruvian event which had a moment magnitude M_w of 8.4.

From Fig. 1 it is seen that the contribution of the $\Delta\mathrm{Tr}(\mathbf{I})$ term to earthquake-induced changes in the length of day is comparable to the contribution of the $\Delta C_{20}(t)$ term. However, the size of these effects is rather small, each of their cumulative effects during 1977–2001 being only about –2 μs. Thus, although in principle earthquake-induced changes in the trace of the Earth's inertia tensor should be removed from length of day observations before comparing them to measured changes in the second-degree zonal gravitational field coefficient, in practice this is not likely to be necessary since the size of the effect, amounting to at most a few microseconds, is much smaller than the current length of day measurement uncertainty of about 20 μs. It would be necessary to remove the effect of earthquakes on the trace of the Earth's inertia tensor from length of day observations only if a truly great earthquake, such as the 1960 Chilean event, were to occur during the lifetime of the CHAMP mission.

4 Summary

Changes of both the Earth's rotation and gravitational field arising from mass displacements caused by the same underlying dynamic process must, of course, be consistent with each other. Thus, independent observations of the Earth's rotation can be used to assess the fidelity of measurements of changes in the second-degree gravitational field coefficients. Of course, motion effects such as those due to atmospheric winds and oceanic currents must first be removed from the Earth rotation observations since they do not affect the Earth's gravitational field. In addition, the length of day is also affected by processes that change the trace of the Earth's inertia tensor, notably earthquakes. So, at least in principle, the effect of earthquake-induced changes in the trace of the Earth's inertia tensor on the length of day must also be removed before comparing length of day observations to measured changes in the second-degree zonal gravitational field coefficient. The importance of this earthquake effect was evaluated by modeling the coseismic change in the trace of the inertia tensor due to the 18553 largest earthquakes that occurred during January 1977 through November 2001. It was shown that the contribution of earthquake-induced changes in the trace of the Earth's inertia tensor to length of day changes during this time period is much smaller than the length of day measurement uncertainty. Thus, this effect can probably be neglected when comparing length of day observations to measured changes in the second-degree zonal gravitational field coefficient. It need only be considered if a truly great earthquake, such as the 1960 Chilean event, were to occur.

Acknowledgements

The work described in this paper was performed at the Jet Propulsion Laboratory, California Institute of Technology, under contract with the National Aeronautics and Space Administration. Support for this work was provided by the Solid Earth and Natural Hazards Program of NASA's Office of Earth Science.

References

Chao BF, Gross RS (1987) Changes in the Earth's rotation and low-degree gravitational field induced by earthquakes. Geophys J R astr Soc 91:569–596

Chen JL, Wilson CR, Eanes RJ, Tapley BD (2000) A new assessment of long-wavelength gravitational variations. J Geophys Res 105:16271–16277

Dziewonski AM, Ekström G, Maternovskaya NN (2001) Centroid — moment tensor solutions for July–September 2000. Phys Earth Planet Inter 124:9–23

Gross RS (2001) Gravity, oceanic angular momentum, and the Earth's rotation. In: Sideris MG (ed) Gravity, Geoid, and Geodynamics 2000. Springer-Verlag, New York, pp 153–158

Gross RS, Chao BF (2001) The gravitational signature of earthquakes. In: Sideris MG (ed) Gravity, Geoid, and Geodynamics 2000. Springer-Verlag, New York, pp 205–210

Munk WH, MacDonald GJF (1960) The Rotation of the Earth: A Geophysical Discussion. Cambridge University Press, New York

Rochester MG, Smylie DE (1974) On changes in the trace of the Earth's inertia tensor. J Geophys Res 79:4948–4951

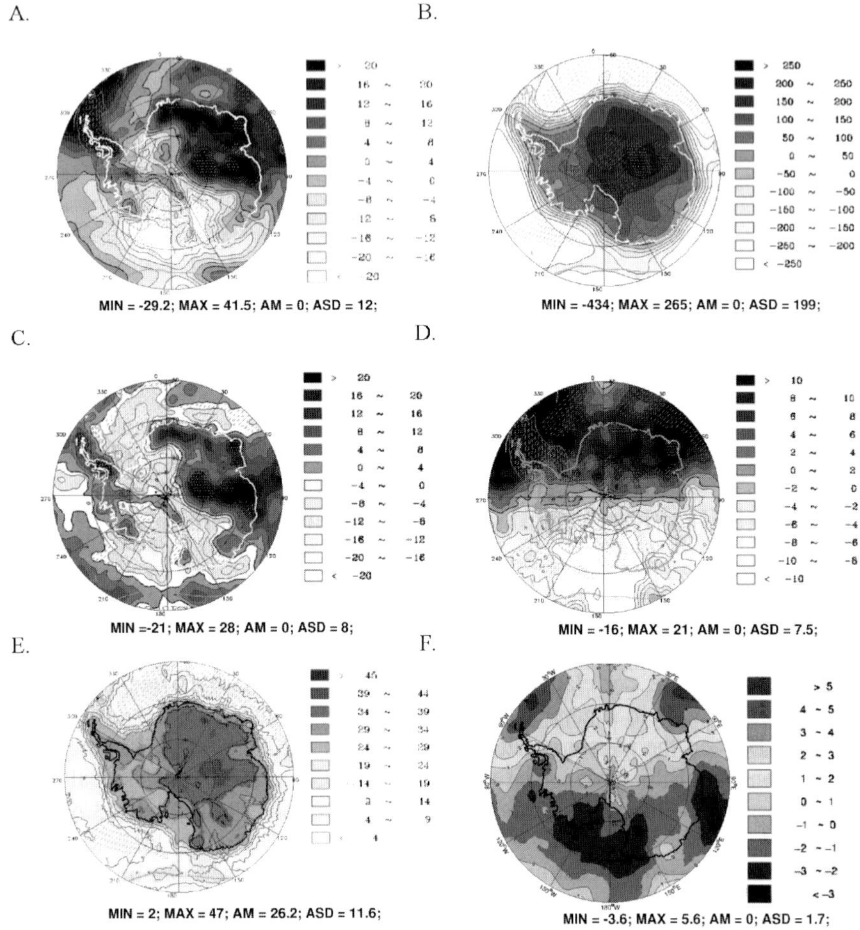

Fig. 2. Crustal and subcrustal modeling from the spectral correlation analysis of A) degree and order 360 EGM96 free-air anomalies with B) the terrain gravity effects. The C) terrain-correlated and D) terrain-decorrelated free-air anomalies were interpreted, respectively, for E) crustal thickness variations and F) the relief in km of the core-mantle boundary. Additional map attributes include the amplitude range (AR) of (MIN, MAX)-values, the amplitude unit (AU), altitude (Z), and grid interval (GI).

For example, Figure 2.F shows the relief of the core-mantle boundary (CMB) as modeled from the terrain-decorrelated anomalies that are positively correlated with EGM96 free-air gravity estimates up to degree 4. The CMB was estimated by least squares inversion using Gauss-Legendre quadrature integration with a 4-gm/cm^3 contrast in the density of the core relative to the mantle across the seismically determined mean core radius of 2,900 km. To first order, these CMB estimates may reflect strongly subducting lithospheric flow in the Pacific. However, the estimates also can constrain core flow, topographic core-mantle coupling,

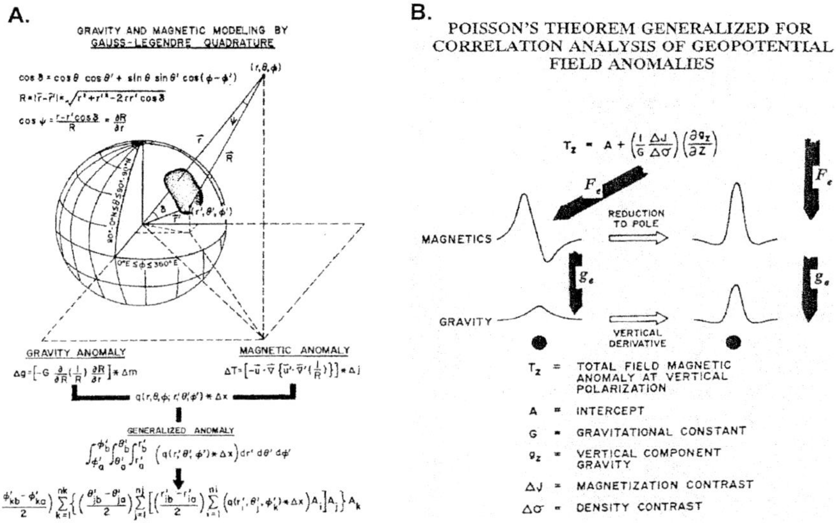

Fig. 3. A) Spherical coordinate gravity and magnetic anomaly modeling of a geologic body with arbitrary shape and physical properties by Gauss-Legendre quadrature integration [4]. B) Poisson's relation generalized for correlation analysis of geopotential field anomalies [5]. Anomaly signals are for a point source (large dot) subjected to the Earth's magnetic ($\boldsymbol{F_e}$) and gravity ($\boldsymbol{g_e}$) fields.

core-mantle thermal coupling and torques that may lengthen the day, and the non-dipolar properties of the core field. We expect to improve these CMB estimates significantly using the enhanced terrain-decorrelated free-air anomalies from CHAMP and new terrain data for the Antarctic [2].

We also plan to study the remaining higher degree (> 4) terrain-decorrelated anomalies for constraints on mantle mass flow patterns and thermal properties. A critical problem in the Antarctic, for example, is to establish the subsurface properties for a presumed large mantle plume related to Cenozoic volcanism in Marie Byrd Land and the West Antarctic Rift System. This plume has been largely inferred from limited outcrops of bimodal alkaic volcanic rocks in Marie Byrd Land, the Ross Sea and Transantarctic Mountains [1]. The CHAMP gravity and magnetic observations provide a truly unique opportunity to establish the thermal and mass dynamics in the mantle and crust for the presumed Marie Bryd Land Plume. These results in turn can expand our understanding of the region's Cenozoic tectonism and its relationships to extensional plate movements between the West Antarctic microplates and East Antarctica that occurred during the Mesozoic fragmentation of Gondwana.

3 Satellite Gravity and Magnetic Anomaly Correlations

With the CHAMP mission, we have for the first time co-registered complementary satellite magnetic and gravity observations to identify and study regionally correlative lithospheric mass and magnetization variations via Poisson's relationship (Figure 3.B). The anomaly correlations provide important physical constraints on regional petrological, structural, and thermal variations of the Antarctic lithosphere. They also help minimize interpretational limitations due to anomaly superposition effects and source ambiguity that is inherent to the analysis of potential fields [5; 7].

Figure 4 gives an example where we evaluate anomaly correlations in the initial gravity and magnetic observations from CHAMP. Following Poisson's relationship that is summarized in Figure 3.B, we compare the first vertical derivative (FVD) free-air gravity anomalies in Figure 4.A with the differentially-reduced-to-pole (DRTP) magnetic anomalies of Figure 4.B at 450-km altitude. The FVD anomalies were evaluated from an equivalent point gravity pole model obtained from Figure 1.A by least squares inversion [3]. Similarly, the DRTP anomalies were obtained from a least squares equivalent point magnetic dipole model of the CHAMP scalar total magnetic field anomalies [9].

The FVD gravity anomalies were normalized [6] in magnitude to the standard deviation of the DRTP magnetic anomalies. The normalization facilitates visual correlation analysis because the same plotting parameters (e.g., contour range and interval, color scale, etc.) can be used to represent both data sets. Multiplying the amplitudes in Figure 4.A by the normalization factor (NF) recovers the original

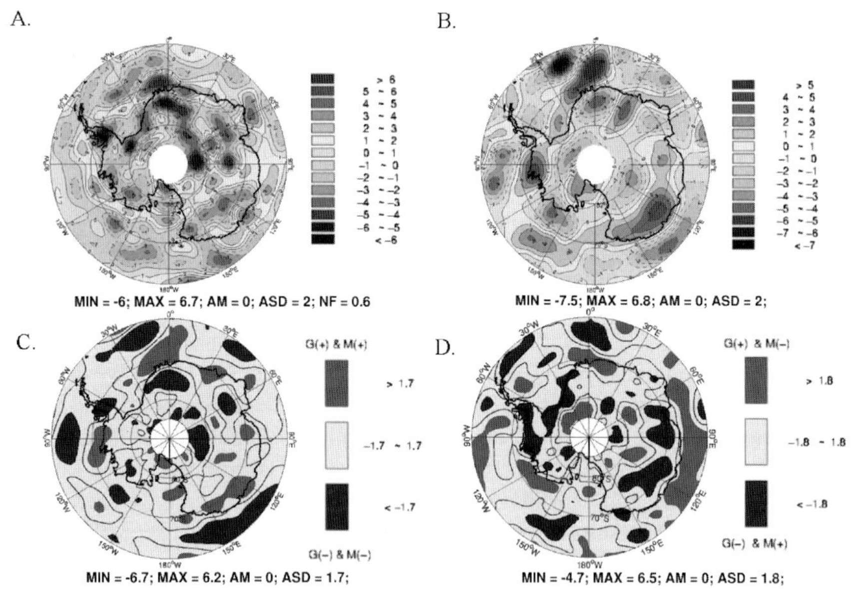

Fig. 4. Spectral correlation analysis of A) normalized first vertical derivative gravity anomalies (μGals/m) and B) differentially reduced-to-pole magnetic anomalies (nT) from CHAMP at 450-km altitude. Anomaly correlations are quantified by the unitless C) summed and D) differenced local favorability indices.

FVD anomaly amplitudes. Inspection of Figures 4.A and 4.B reveals numerous anomaly associations, both positive and negative, to constrain the regional density and magnetization properties of the lithosphere. The stronger positive and negative associations are quantified by the local favorability indices [6] in Figures 4.C and 4.D, respectively.

Local favorability indices are derived from the sums and differences of correlation filtered data that have been standardized by removing the data mean and dividing through by the standard deviation of the data. For example, summing the standardized results from inversely transforming all positively correlated FVD and DRTP wavenumber components yields the summed local favorability indices (SLFI) in Figure 4.C. These positive and negative SLFI are plotted to emphasize the stronger peak-to-peak and trough-to-trough anomaly correlations, respectively. For the negatively correlated FVD and DRTP wavenumber components, on the other hand, we difference the standardized outputs for differenced local favorability indices (DLFI). In Figure 4.D, the positive and negative DLFI are plotted to emphasize the stronger FVD peak-to-DRTP trough and FVD trough-to-DRTP peak associations, respectively.

CHAMP provides a truly exceptional opportunity to investigate the lithospheric utility of regional gravity and magnetic anomaly correlations. These anomaly correlations are subject in practice to numerous complicating geological variations including sources with appreciable magnetic remanence not aligned with the present Earth's field, inhomogeneous magnetization and density properties, and significant differences in their depths of origin. However, the correlations also can reveal regional patterns of intrusive activity for new insights on Antarctic tectonic evolution and plate dynamics, as well as crustal thickness and thermal perturbations that concentrate crustal stresses producing earthquakes, volcanism, and other neotectonic effects.

Acknowledgements. We thank Peter Schwintzer of the GFZ for making available the preliminary CHAMP gravity results used in Figure 1.A. Elements of this research were supported by NASA grant NAG5-7645.

References

1. LeMasurier WE (1990) Late Cenozoic volcanism on the Antarctic plate – An overview. In: LeMasurier WE, Thompson MRA (eds) Volcanoes of the Antarctic Plate and Southern Oceans. Ant Res Ser, **48**, 1-19.
2. Lythe MB, Vaughan DG, Bedmap Consortium (2001) BEDMAP: A new ice thickness and subgracial topographic model of Antarctica. J Geophys Res, **106**, 11335-11351.
3. von Frese RRB, Hinze WJ, Braile LW, (1981) Spherical earth gravity and magnetic anomaly analysis by equivalent point source inversion. Earth Planet. Sci Lett, **53**, 69-83.
4. von Frese RRB, Hinze WJ, Braile LW, Luca AJ (1981) Spherical earth gravity and magnetic anomaly modeling by Gauss-Legendre quadrature integration. J Geophys, **49**, 234-242.

5. von Frese RRB, Hinze WJ, Braile LW (1982) Regional North American gravity and magnetic anomaly correlations. Geophys J R Astr Soc, **69**, 745-761.
6. von Frese RRB, Jones MB, Kim JW, Kim H-J (1997) Analysis of anomaly correlations. Geophysics, **62**, 342-351.
7. von Frese RRB, Jones MB, Kim JW, Li WS (1997) Spectral correlation of magnetic and gravity anomalies of Ohio. Geophysics, **62**, 365-380.
8. von Frese RRB, Tan L, Kim JW, Bentley CR (1999) Antarctic crustal modeling from the spectral correlation of free-air gravity anomalies with the terrain. J Geophys Res, **104**, 25275-25296.
9. von Frese RRB, Kim HR, Taylor PT, Kim JW (2002) CHAMP, Ørsted, and Magsat magnetic anomalies of the Antarctic lithosphere. This volume.

Assimilation of Altimeter and Geoid Data into a Global Ocean Model

Verena Seufer[1,2], Jens Schröter[1], Manfred Wenzel[1], and Wolfgang Keller[2]

[1] Alfred Wegener Institute for Polar and Marine Research, Bremerhaven, Germany
http://www.awi-bremerhaven.de

[2] Geodetic Institute, Stuttgart University, Germany
http://www.uni-stuttgart.de/gi

Summary. A dynamical global ocean model is used to determine the general circulation and its associated transports from data. As a reference a surface geostrophic velocity can be calculated from the slope of the sea surface height, which is measured by satellite altimetry. To get absolute values for the sea surface elevation relative to the equipotential surface we want to assimilate not only altimeter data but also geoid heights into our global ocean model. Until nowadays the assimilation of geoid heights has been done only in few cases, because the accuracy of the available gravity models is not sufficient. With the CHAMP satellite and future satellite missions we will get highly improved gravity models. Since these data are not available yet we use the EGM96 geoid models for our recent experiments. It is planned to esitmate corrections not only for the ocean model but also for the geoid models. We want to present the technique which will be used to assimilate the geoid data into the global ocean model.

Key words: Ocean model, satellite altimetry, geoid

1 Introduction

A major problem in physical oceanography is the determination of the general circulation and its associated transports from data. A geostrophic circulation can be derived from measured temperature and salinity only relative to some unknown reference velocity field. For this reference the surface geostrophic velocity can be calculated from the slope of the sea surface elevation (SSE). SSE is the deviation of the sea surface height (SSH) as measured by satellite altimetry from the equipotential surface which is equivalent to the geoid.

Therefore it is necessary to assimilate the measured SSH referenced to the geoid to constrain ocean models. For this it seems to be reasonable to employ a geoid model that is independent of altimeter data, i.e. we must use a so called satellite-only geoid which is estimated without utilizing altimeter data. Otherwise we encounter the problem that the altimetric surface and the marine geoid are not independent ('Münchhausen problem' [4]).

Until now the assimilation of geoid heights into ocean models has been made only in few cases because of the accuracy of the available gravity models.

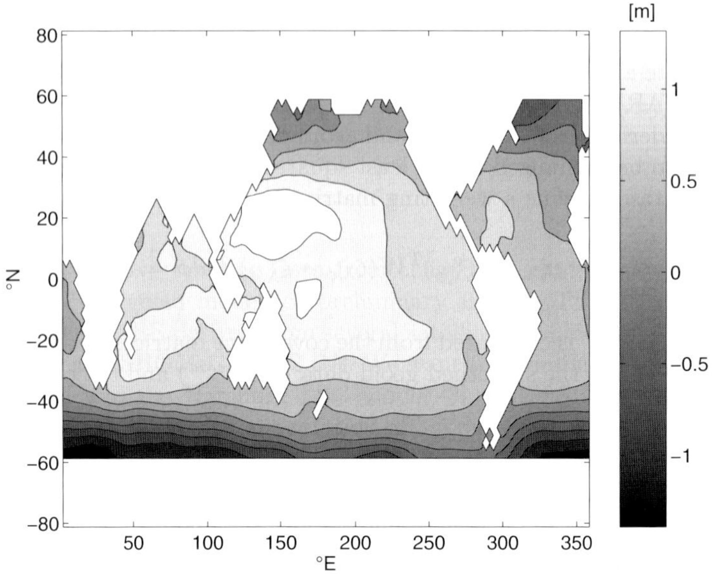

Fig. 2. Mean sea surface elevation (SSE) derived from the global ocean model LSG without assimilation of geoid data (max. 1.5896 m, min. -1.3928 m)

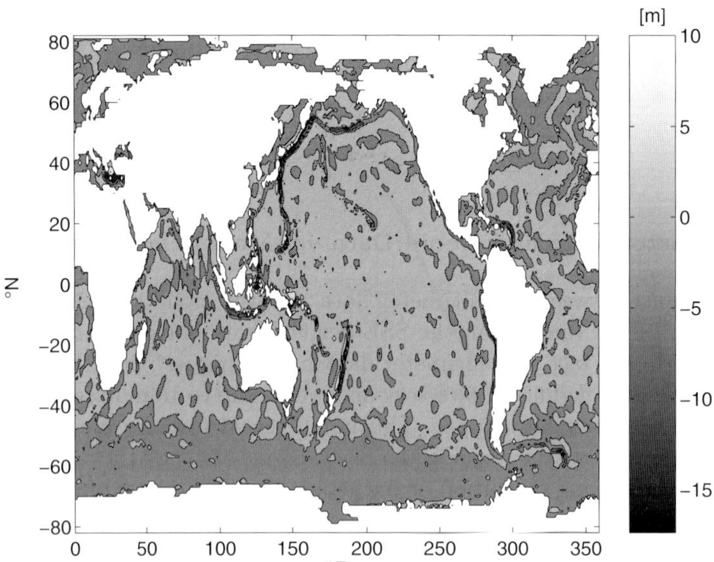

Fig. 3. Difference between altimetry and EGM96S geoid heights (max. 13.6740 m, min. -17.3807 m)

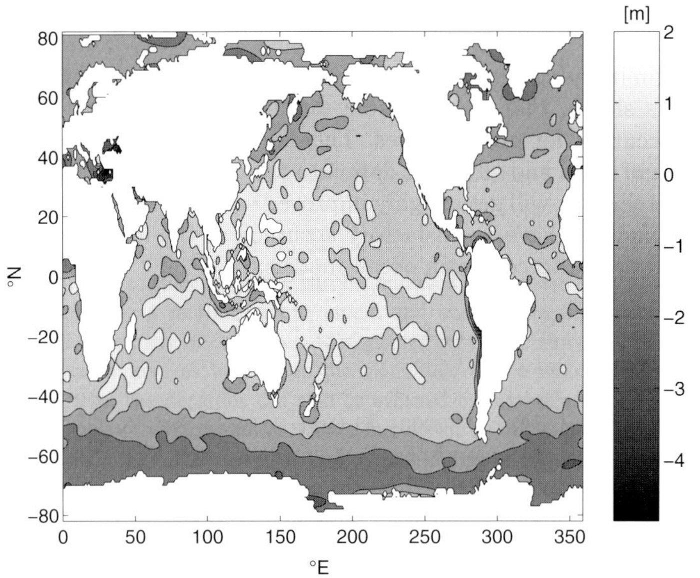

Fig. 4. Smoothed difference between altimetry and EGM96S geoid heights (max. 2.9075 m, min. -4.8630 m)

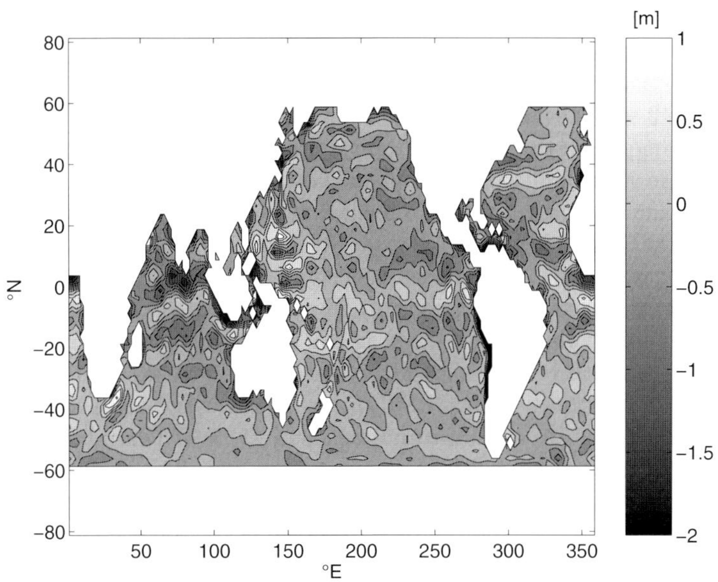

Fig. 5. Smoothed difference between the mean SSE derived from the LSG model and the mean SSE computed from altimeter and geoid heights

	Altitude (km)	Latitude (°)	Mean flux (sfu)	Uncertainty (%)
DTM database	200 - 1000	-90 ; 90	70 - 175	5 – 15
CACTUS	225 - 500	-30 ; 30	70 - 90	2 – 5
STAR	250 - 500	-87 ; 87	90 - 210	<5

Table 1. The temporal and spatial coverage and the uncertainty the DTM, CACTUS, and STAR (after 5 years of nominal operations) total atmospheric density datasets

- the correct calibration parameters are applied to the STAR data;
- an accurate non-gravitational force model is employed.

2 Density retrieval procedure

The spurious peaks in the raw STAR data are removed first, using absolute thresholds. Next, a low-pass filter with a cut-off frequency of 0.1 Hz is applied to the data, which effectively removes all undesired, high frequency signals including those due to manoeuvres. This actually smoothes out fast variations in drag also. However, the resolution of the DTM model is of the order of hundreds of kilometres, while a STAR data-sampling rate of 0.1 Hz corresponds to a resolution of approximately 75 km. The STAR data are uncalibrated, so the last step in the data pre-processing is the application of a bias and scale factor to the data. These calibration parameters could not be determined on ground for technical reasons and were estimated by means of a dynamic orbit computation procedure. They are not constant but depend on the accelerometer and data handling electronics in operation (main or redundant), which means that calibration parameters are valid between events, listed in the CHAMP housekeeping data.

STAR measures the non-gravitational accelerations acting on CHAMP. Only the tangential component of the measured acceleration is used in this study because it is most sensitive to atmospheric drag, which is proportional to density. Secondly, its calibration parameters are determined best. In order to extract the drag component, the radiation pressure components have to be removed from the measurements. The 'observed' total density, ρ^{STAR}, can be calculated by scaling the density predicted with a model, ρ^{model}, as follows:

$$\rho^{STAR} = \frac{a_{total}^{STAR} - a_{presol}^{\text{model}} - a_{albedo}^{\text{model}} - a_{IR}^{\text{model}}}{a_{drag}^{\text{model}}} \cdot \rho^{\text{model}} \tag{1}$$

In this equation, a_{total}^{STAR}, $a_{presol}^{\text{model}}$, $a_{albedo}^{\text{model}}$, a_{IR}^{model}, and a_{drag}^{model} are the (along-track) accelerations measured by STAR, the modelled solar radiation pressure acceleration, the accelerations due to the Earth's albedo and IR radiation, and the modelled drag acceleration, respectively. Therefore, a satellite macro-model, providing reflectivity coefficients (for visible- and IR wavelengths) and superficies of the satellite panels, as well as knowledge pertinent to terrestrial albedo and IR emissivity, is required in order to compute the modelled non-gravitational accelerations accurately. In this study, a 15-plate macro-model, correctly oriented in inertial space by the telemetered attitude quaternions, is employed. Terrestrial albedo- and emissivity top-of-atmosphere observations are collected by the European Centre for Medium-range Weather Forecast (ECMWF) and were made available to us in the form of 6-hourly 0.5°x0.5° grids. The Earth radiation pressure force is computed using the model described in (Knocke et al. 1988). The solar radiation pressure force, with Earth- and Moon shadowing effects, is computed using a regularising shadow function, calibrated to the CHAMP orbit (Biancale et al. 2001). In the along-track direction, the radiation pressure forces are less than one percent of the total non-gravitational force, which main contributor is atmospheric drag.

The general expression for the atmospheric drag acceleration of a satellite, a_{drag}, in which the atmospheric density ρ is in this study predicted with DTM2000 (Bruinsma and Thuillier 2000), is as follows:

$$a_{drag} = -\tfrac{1}{2} C_D \frac{A}{m} \rho v^2 \tag{2}$$

where C_D is the drag coefficient, A the satellite's effective cross-sectional area, m its mass, and v its velocity relative to the atmosphere, which is assumed to be co-rotating with the Earth. The drag coefficient gives a measure of the momentum exchange between the satellite surface and the colliding particles. It is modelled under the assumptions of hyperthermal free-molecular flow and accommodated diffuse re-emission of the neutral atmospheric particles according to (Cook 1965). The expression of the drag coefficient for a flat plate at incidence ϕ to the flow is:

$$C_D = 2\left(1 + \tfrac{2}{3}\sqrt{1 + \alpha\left(\frac{T_{sat}}{T_{atmo}} - 1\right)}\sin\phi\right), \quad \alpha = \frac{3.6u}{(1+u)^2} \tag{3}$$

where T_{sat} is the temperature of the satellite surface, T_{atmo} the kinetic temperature of the ambient gas molecules, and u the ratio of the mean mass of the incident gas atom to the mass of the satellite surface atom. The effective cross-sectional area, which is perpendicular to the satellite's speed relative to the atmosphere, can be determined accurately thanks to the attitude quaternions. The satellite mass history is monitored, and is decreasing approximately 0.2 kg per month for an initial total mass of 522 kg.

The CHAMP tracking- and accelerometer data were processed using the GINS orbit determination software, which was also used in the construction of the GRIM5 (Biancale et al. 2000) and EIGEN-1S (Balmino et al., 2002) gravity fields. GINS was employed to compute CHAMP orbits, fitted to the GPS and SLR tracking data, to obtain the co-ordinates of the STAR measurements in the fixed terrestrial reference frame as well as parameters pertinent to density modelling. The accelerations of the non-gravitational models and the DTM2000 density were computed in parallel in order to derive the 'observed' density using equation (1). The output of GINS of a density run consists of a file containing: Julian date, altitude (above GRS-80 ellipsoid), geocentric latitude and longitude, solar local time, mean- and daily solar radio flux ($F_{10.7}$), and the planetary geomagnetic index Kp.

The accuracy of the total density derived from STAR measurements and model predictions using equation (1) depends on two kinds of errors: those causing (mainly) a systematic off-set of the observations with respect to their absolute values, and secondly, those affecting their noise-level.

A source of the first kind of error is due to the uncertainty in the instrumental calibration parameters, which is presently smaller than 5%. The tangential bias and scale factor were estimated simultaneously with gravity field coefficients during the calibration activities (Perret et al. 2001), and mean values of $-3.075\ 10^{-6}$ ms^{-2} and 0.82, respectively, were determined for the period 24/9-7/11 2000. These values have been used as a-priori in the computation of the CHAMP normal equations used in the construction of EIGEN-1S. The tangential scale factors, estimated on a daily basis, simultaneously with the gravity field coefficients of EIGEN-1S are displayed in Figure 1. This figure also shows where known onboard events, causing the calibration parameters to change, took place. The bias (not shown) over the same period varied less than 3%.

The second significant source of systematic error is the drag coefficient model. Such an error also directly affects the value of the reconstituted density, as may be seen in equation (2). The uncertainty of Cook's model is probably less than 10% because of the validity of the hyperthermal flow regime assumption at the low altitude of CHAMP.

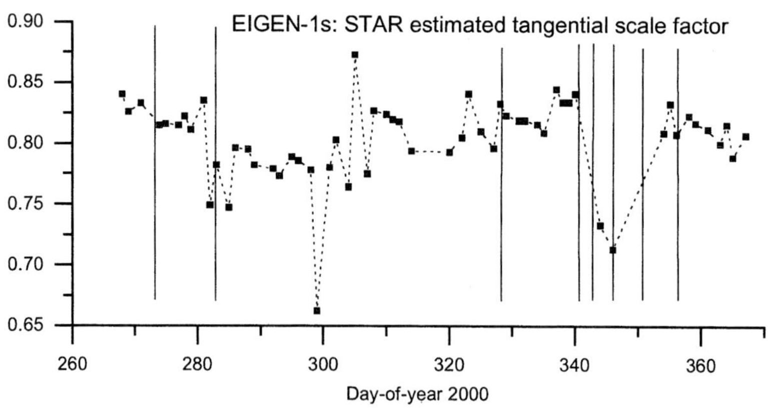

Fig. 1. The tangential scale factors of STAR estimated in the framework of EIGEN-1S. Vertical lines show where onboard events that may change the calibration took place

The minimum or instrumental noise-level of the derived density observations is determined by the accelerometer's tangential sensitivity of $3 \cdot 10^{-9}$ m/s^2/Hz$^{1/2}$ in the measurement bandwidth (10^{-4} to 10^{-1} Hz), which corresponds to 1-3%. The most important factor determining the noise in the observations is due to the unknown amplitude and direction of the neutral winds, which can be significantly different from the case of a co-rotating atmosphere. The accelerations cannot be accurately corrected for winds by employing a thermospheric wind model because of their large uncertainties. The maximum error occurs when the wind is directed along the satellite velocity vector, so along the accelerometer's tangential axis, and amounts to approximately 5% per 200 m/s. The strongest winds are encountered at high latitudes during geomagnetic storms (Kp>6), when they attain 600-1000 m/s, in which case they even become large (200 m/s) down to tropical latitudes. These amplitudes have been obtained from the Thermosphere General Circulation Model (TGCM output available at the CEDAR database), a model that is consistent with observed upper atmosphere winds.

The remaining contributions to the noise are negligible. These are for example due to attitude errors, errors in the modelled radiation pressure accelerations, and the mass history of the spacecraft. The sum of the remaining errors contributes less than 1% to the noise in the derived density.

3 First results

Applying the procedure described in section 2, total densities were derived from STAR data for the periods 1/9-30/11 2000 as well as 15/5-31/8 2001. A constant bias and scale factor of $-3.075\ 10^{-6}$ ms^{-2} and 0.82, respectively, were applied. These values are less than 3% larger than a recently determined set of calibration parameters for the period 15/5-31/12 2001. They are also within a few percent of the values shown in Figure 1. Only data for which the geomagnetic activity Kp was less than 3^+ were processed in order to limit noise due to unmodeled winds to less than 5%. These densities are compared to DTM2000 predictions based on the solar activity indicator Mg II index (Thuillier and Bruinsma 2001) instead of the usually employed $F_{10.7}$. The results are presented in the following table, where the observed-to-calculated ratios (called residual in the following) and their rms (about mean) are displayed.

Table 2 shows that the results are not homogeneous in terms of rms of the residuals and, in a lesser degree, their means. The cause of the large rms over the period 15/5-30/6 2001 is presently under investigation, but it is probably due to erroneous modelling as a function of local time. The local time over that period varies from 6-12 on the night-side and from 18-24 on the day-side of the orbit. In September 2000, the local time was sampled in that interval also. This may be a reason for the large rms of the residuals for that month. The best results are obtained for the months where the local time is sampled in the 2-8 and 14-20 hr range (10 and 11 2000, and 7 and 8 2001). Erroneous modelling of the atmospheric response

period	mean O/C	rms O/C	number of data
09/2000	0.78	0.21	22353
10/2000	0.97	0.18	29352
11/2000	0.91	0.18	7269
05/2001	0.99	0.32	43186
06/2001	1.04	0.28	41022
07/2001	0.87	0.19	64882
08/2001	0.80	0.13	26320

Table 2. The mean and rms of the observed-to-calculated ratio, employing the DTM2000 model, and the number of observations per period of one month

to solar activity probably causes the large steps in the means from one month to the next, for example from June to July 2001.

To evaluate the difference in spectral content between the STAR-derived densities and the DTM2000 densities, power spectra were computed using the periodogram method. This has been done for the months June-July 2001, employing both Mg II and $F_{10.7}$. The summed power of the periodograms computed with modelled densities is larger than that contained in the observations, 12% and 18% larger employing Mg II or $F_{10.7}$, respectively. This is consistent with the results shown in Table 2. However, when the periodograms are computed using all data (all Kp's), then the summed power is larger in the observations by approximately 11%. The amplitude of the 1/rev line is within a few percent of the observations using Mg II, but it is 12% larger using $F_{10.7}$. The fact that there is significantly more power at approximately the daily period in the observations than in the predictions, and significantly more power at approximately the 1.2 day period in the predictions than in the observations seems to point at a small error in the parameterisation of the delay applied to the solar activity indicators.

4 Conclusion and outlook

The density data retrieval procedure is presently operational. The uncertainty of the derived densities is smaller than 5% when periods with a Kp smaller than 3^+ are processed. Analysis of the few months of CHAMP/STAR density observations considered in this study shows that they can significantly contribute to the improvement of atmospheric density models. The time series will be a year in length in May 2002, which is the minimum duration required for a dataset to be used in a modelling effort. Therefore, assimilation of the data in DTM2000 is scheduled for summer 2002.

References

Barlier F, Berger C, Fallin JL, Kockarts G, Thuillier G (1978) A thermospheric model based on satellite drag data. Ann Geophys 34: 9-24

Balmino G, Reigber Ch, Schwintzer P, Biancale R, Bode A, Lemoine JM, König R, Loyer S, Neumayer H, Marty JC, Barthelmes F, Perosanz F, Zhu SY (2002) New Global Gravity Field Models from Selected CHAMP Data Sets, this issue

Biancale R, Perosanz F, Lemoine JM, Bruinsma S, Loyer S (2001) STAR accelerometer data processing and results: part I, paper AAS 01-332. AAS/AIAA Astrodynamics Specialist Conference, Québec, Canada

Biancale R, et al (2000) A new global Earth's gravity field model from satellite orbit perturbations: GRIM5-S1. Geophys Res Lett Vol 27, 22: 3611-3614

Berger C, Biancale R, Ill M, Barlier F (1998) Improvement of the empirical thermospheric model DTM: DTM94 – a comparative review of various temporal variations and prospects in space geodesy applications. J of Geodesy 72/3: 161-178

Bruinsma SL, Thuillier G (2000) A revised DTM atmospheric density model: modelling strategy and results. EGS XXV General Assembly, session G7, Nice, France

Cook GE (1965) Satellite drag coefficients. Planet Space Sci 13: 929-946

Knocke PC, Ries JC, Tapley BD (1988) Earth Radiation Pressure Effects on Satellites. Proceedings of the AIAA/AAS Astrodynamics Specialist Conference, Am Inst Of Aeronaut Astronaut, Washington DC, pp 577-587

Perret A, Biancale R, Camus AL, Lemoine JM, Fayard T, Loyer S, Perosanz F, Sarrailh M (2001) STAR commissioning phase calibration/validation activities by CNES. CNES report DSO/ED/TU/EI 65-2001, Toulouse, France

Thuillier G, Bruinsma S (2001) The Mg II index for upper atmosphere modelling. Ann Geophys 19: 219-228

Villain JP (1980) Traitement des données brutes de l'accéléromètre Cactus. Etude des perturbations de moyenne échelle de la densité thermosphérique. Ann Geophys 36: 41-49

Earth Magnetic Field

Instruments, Main Field Studies, Lithospheric Magnetisation, Induction Studies and Ionospheric Current Systems

CHAMP ME Data Processing and Open Issues

Martin Rother, Sungchan Choi, Hermann Lühr, and Wolfgang Mai

GeoForschungsZentrum Potsdam (GFZ), Division 2, Solid Earth Physics and Disaster Research

Summary. One goal of the CHAMP mission is to collect high precision magnetic field data on a gobal scale over several years. In order to maintain an absolute accuracy a calibration concept has been implemented partly in the payload design and partly in the data processing. Prime magnetic data are the vector field measurements. The basic concept for verifying the principle parameters of the Fluxgate magnetometers is to perform a scalar calibration against an Overhauser magnetometer on a daily base. Slow changes of the instrumental characteristics will thus be determined by this in-flight calibration. Perturbations taking place on time scales shorter than a day have to be corrected before in-flight calibration. The vector field measurements are transformed with the help of orbit and attitude data into a well defined geophysical coordinate system. Precise attitudes are derived from the merged readings of two star trackers looking at directions about 102° apart. After applying all corrections we claim to obtain an absolute vector accuracy of less than 2 nT and a differential of less than 0.5 nT for arc lengths of half an orbit.

Key words: CHAMP, magnetic measurement, processing, calibration, attitude

1 Introduction

The preprocessing for the magnetic field has to assemble daily data files containing time series, housekeeping data and other supplemental information including the actual calibration parameters and calculate in a first step preliminary corrected, in a second step finally corrected, averaged data sets, rotated into a usable Earth-Fixed-Local (EFL) system.

2 Data Organisation Outline

Magnetic field data processing is structured in four logical levels:

Level 1: Raw data series in engeneering units and housekeeping data (temperatures, satellite activity and flags) derivied from the telemetry data.

Level 2: Preliminary corrected time series, averaged vector and scalar data with positions, at a 1 Hz time resolution. The vector data are processed with a set of parameters updated regulary. The data are given in two frames, in the sensor system and in an EFL system (North-East-Center). In addition processed ASC star camera data are made available.

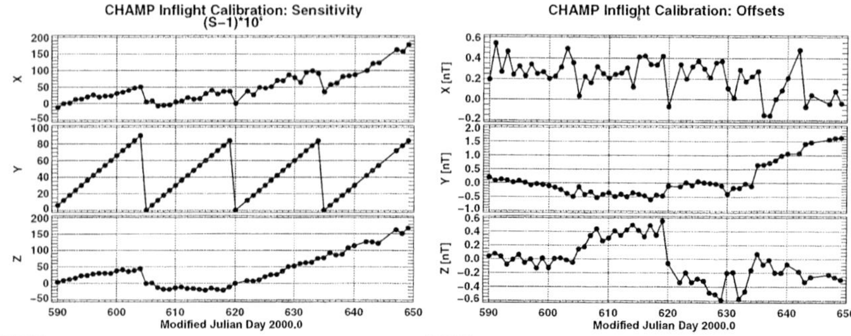

Fig. 2. Sensitivity and Offset Correction

Following corrections are made (see overview in Fig. 1):

- Time correction; there is a static correction and a small sensor internal phase time shift, which is reported in the data.
- Correction of the nine internal FGM sensor parameters: Sensitivity, Misalignment and Offset.
- Temperatur correction of the CSC sensor.
- Torquer correction, see OVM.

5 Parameters and Calibration

Following areas are subjects of inflight calibration and evaluation: the FGM sensor parameters, the measurement time adjustment, the transformation beetween sensors and the torquer correction.

5.1 Routine Inflight Calibration

Because of expected and also unexpected variablity of internal experiment parameters, most ground measured parameters have become partly obsolete. Inflight Calibration is usually performed for the period of one day, trying to fit the free parameters of the FGM magnetic vector data to the OVM scalar measurements by iterative least-square minimizing of a linearized equation.

The sensitivities are following the expected and documented trend. An initial parameter set for the preprocessing is routinely updated every 15 days. Because it's not well determined, the sensitivity of the Y-component is currently not calculated by default but is propagated with fixed slope. Special CHAMP attitude maneuvers are performed to get actual updates. As a typical example the left frame of Fig. 3 shows the residuals of the total field calculated from calibrated vector data and the corrected OVM scalar measurements.

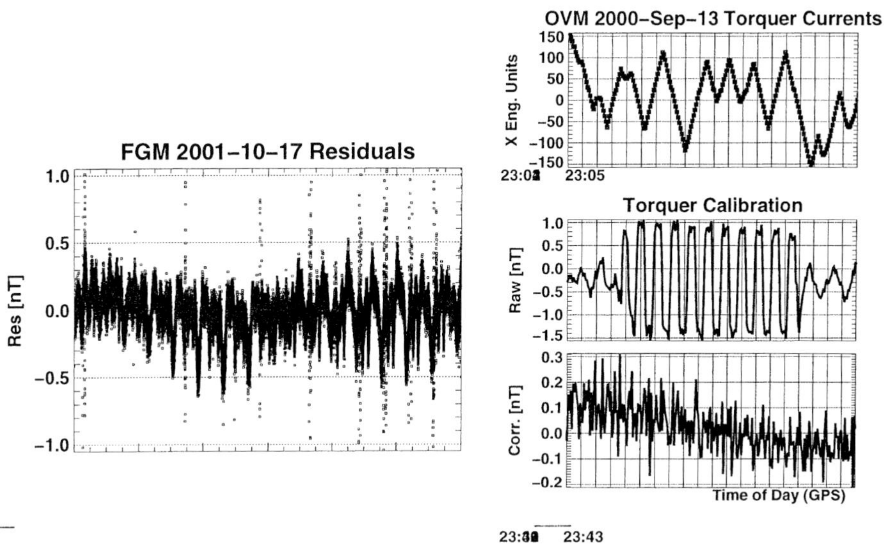

Fig. 3. Left: Residuals of a day, FGM-OVM comparison. Right: Torquer Calibration Tests.

5.2 Torquer

Since the satellite attitude control system is using mainly magnetic torquer coils, their quite huge signal has to be corrected in the magnetic field readings. The characteristic sawtooth shape of the torquer currents (see right part of Fig. 3, upper frame) could be seen often clearly in early OVM datasets, indicating an incomplete torquer correction.

During the commisioning phase it was part of the calibration attempts to run large, but simple signals (right part of Fig. 3, middle frame) on the torquer coils allowing to adjust the torquer-current conversion matrices derivied from ground calibration. The corrected signal is shown in the lower frame.

5.3 Time Adjustment

An important problem is the precise dating of the measurements, for both science and housekeeping data. We take the vector field data as the reference and determine the latency to the other relevant devices: switching of the magneto-torquer current (including the response function), scalar data from OVM and attitude data from ASC. A lot of calibration effort has been put in this task both on ground and in orbit to archive the required ± 5 ms accuracy.

5.4 Orientation of ASC Sensor Heads

The orientation of the two star tracker head with respect to the so-called *Common Reference* frame (CR) varies with time. Besides the gradual change

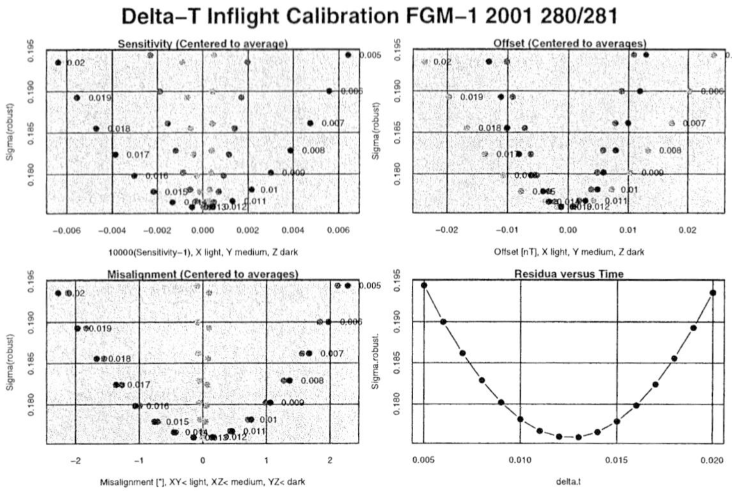

Fig. 4. Time Adjustment

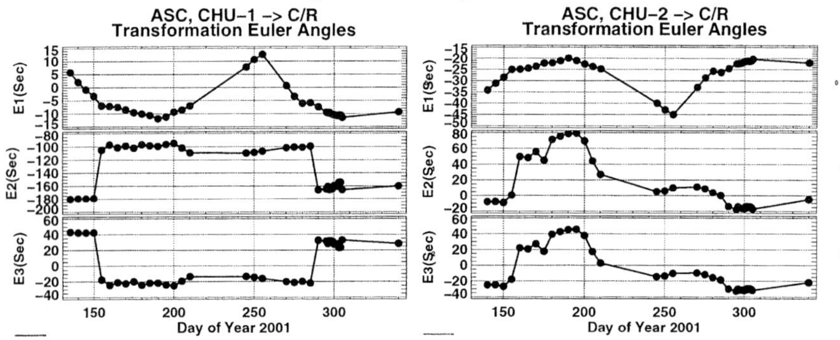

Fig. 5. Time dependent Euler angles of ASC head rotation.

(Fig. 5, top panels) related to local time there are rapid changes at certain events (Fig. 5, lower panels). The major changes are confined to the rotation about the boresight and do not influence the attitude data from dual head processing.

6 Star Camera, Attitude and Transformations

The knowledge of the spacecraft (S/C) attitude is an important parameter for the interpretation of the high resolution data of Fluxgate Magnetometers (FGM), which can be fully used only, if the attitude is known to an arc-second accuracy. For this purpose two star cameras are mounted together with the FGM sensors on an optical bench in the middle of the boom. For

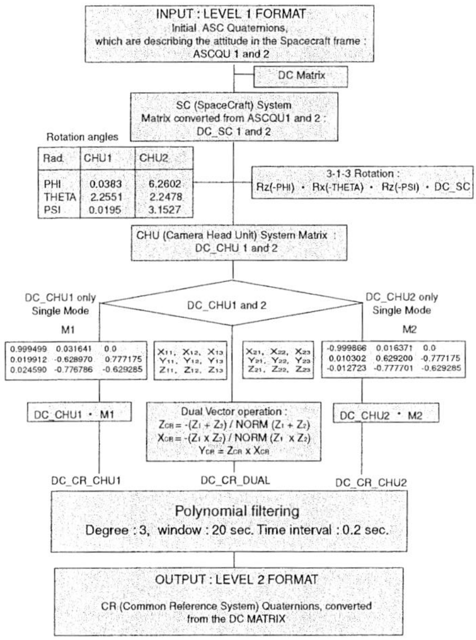

Fig. 6. ASC Preprocessing Overview

the convenience of the attitude control system on CHAMP, a quartet of quaternions are describing the attitude of the S/C frame. The quaternions can equivalently be presented as Euler angles of a 3-2-3 rotation matrix, where the angles describes the *Right Ascention*, *Declination* and *Rotation* about the boresight direction.

6.1 Star Camera Preprocessing

The first processing step (see Fig. 6) is to perform a 3-1-3 rotation from the S/C back into the camera head system (DC_CHU1 and 2) by using the direction cosine (DC) matrix. To calculate the DC matrix in the common reference (CR) frame (DC_CR_DUAL) from the associated two matrices, we perform the dual vector operation. If attitude data from only one head is available, the transformation is using matrix a M1 for DC_CHU1 and M2 for DC_CHU2 respectively, to arrive at the CR frame. For this case the required M1 and M2 can be computed as:

$$M1 = \text{DC\_CR\_DUAL} \cdot \text{DC\_CHU1}^T$$
$$M2 = \text{DC\_CR\_DUAL} \cdot \text{DC\_CHU2}^T .$$

Finally, for the level 2 format, the attitude data in CR frame are smoothed by a polynomial filtering (usally of degree 3) and resampling to the required time.

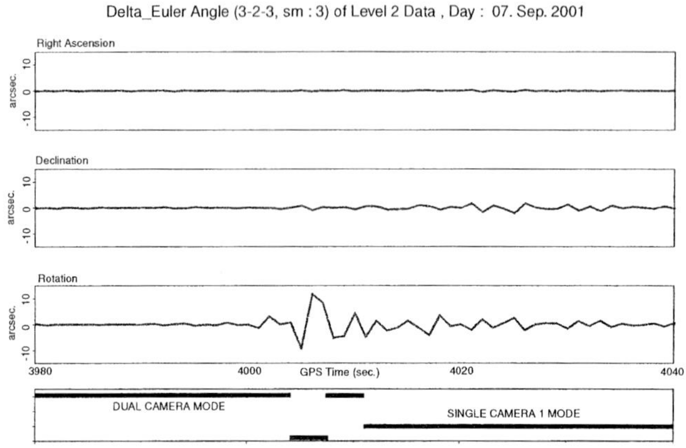

Fig. 7. Dual/Single mode change-over

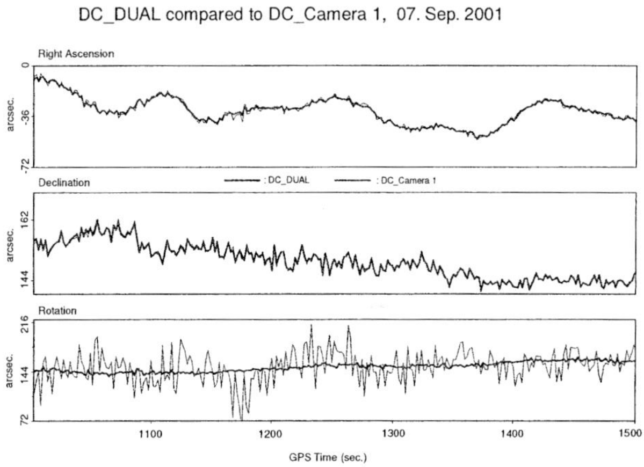

Fig. 8. Noise in Dual and Single mode

6.2 Star Camera Data Precision

The initial data precision of level 1 varies between -60 and 60 arc-seconds. After the preprocessing, typical values for precision are ± 2 arc-seconds after Dual mode data processing and ± 10 arc-seconds after single mode processing (see Fig. 8).

In the change-over area (4010 in Fig. 7)) from the dual to the single mode, an unexpected data undulation is revealed, which means that the obtained matrices M1 and M2 are not appropriate. Data inaccuracy in that area is caused by the anisotrophically distributed noise among the three angle components. As shown in Fig. 8, the noise of the *Rotation* component is significantly higher than the other components. To estimate the difference between attitudes of Dual and Single mode data, we are using

$$DC\_DUAL = \mathrm{M1}^t \cdot \mathrm{DC\_CR}$$
$$DC\_Camera\_1 = \mathrm{M1}^t \cdot \mathrm{DC\_CR\_CHU1}.$$

For the Level 3 processing we plan to determine the M1 and M2 matrices on a daily base in orther to keep the change-over discontinuities small.

Acknowledgement

The GFZ part of this study was carried out under the grant of the German Federal Ministry of Education and Research (BMBF) No. 03F0333A.

Ion Drift-Meter Status and Calibration

David L. Cooke[1], Capt. Wallace Turnbull[1], Christopher Roth[2], Alvin Morgan[3] and Robert Redus[3]

[1]Air Force Res. Lab., Space Vehicles Directorate, Hanscom AFB, MA, USA
[2]Radex Inc., Beford, MA, USA
[3]Amptek, Bedford, MA, USA

Summary. The Digital Ion Drift Meter (DIDM), developed by Amptek, Bedford, MA USA, for AFRL, was launched on July 15, 2000 aboard the Challenging Minisatellite Payload (CHAMP) - developed by the GeoForschungsZentrum, Potsdam, Germany into a 421 x 475 km polar orbit. All components and functions of DIDM are performing nominally except for two problems; the immediate loss after launch of one of the two nearly redundant sensors, and an uneven gain evolution of the micro-channel-plate used for the ion detection that has required development of an in-space calibration procedure. This procedure is now complete and we now have reasonable DIDM results for selected periods of the mission, and will soon be extending the new procedure to the entire mission. This paper will review the instrument design, the laboratory and in-space calibration procedures, and present sample results.

Key words: Ion drift-meter, electric field instrument, instrument calibration, Ionospheric dynamics

1 Instrument Overview

The DIDM measurement objective is to detect the angle of arrival of thermal plasma ions encountered by a spacecraft from which may be derived the local ion drift velocity. The DIDM drift meter, DM, principle of operation is that of a pinhole camera where the ion distribution function is imaged on a detection plane through an aperture that is very small compared to the separating distance between aperture and image plane. A Micro-Channel Plate, MCP, is converts the ions into a pulse of ~ 10^6 electrons, and a position sensitive anode locates the pulse onto the pixel array containing an image of the ion distribution. The image is reduced on-board to a set of raw moments and returned in telemetry. Ground-based software matches the moments against a tabulated set of 'expectations' to recover the desired ion parameters; drift angle, temperature and density.

DIDM also functions as a Retarding Potential Analyzer, RPA, which determines the energy of incoming ions and the ion velocity if the ion mass is presumed to be that of Oxygen. Deviations from the expected energy spectrum may be attributed to the drift velocity magnitude and the effect of the spacecraft potential. The retarding effect of the RPA is spherical and so yields the total ion energy or velocity, which may be combined with the drift angle to yield all three components of the ion velocity. The RPA has a 8 selectable 0.25 – 2 sec sweeps that all use a small number of discrete voltage steps, typically 4 to 8. This reduces both telemetry and statistical fluctuation at each step. The ground-based analysis also uses an expectation matching method that reduces processing time to much less than real time.

DIDM also carries a Planar Langmuir Probe, PLP, that monitors the spacecraft potential, ion density, and electron temperature. This is a 152 x 203 mm rectangular plate with an included 106 x 156 mm sensing area that is alternatively floated for 14 sec to track the spacecraft potential, and stepped in Voltage for 1 sec to both verify the floating potential and determine the electron temperature.

The DIDM sensors are designed such that either can be operated as a DM or RPA or both. Two sensors are used to provide simultaneous DM and RPA measurements, increased measurement rates, greater dynamic range (with different aperture size), and redundancy. The two sensors were also biased away from the forward direction by +/- 10 degrees. One reason for this bias was to move the images away from a blind spot on the position sensitive anode that results from anode element scheme. Another was to improve the angular coverage for the horizontal ion drift.

Two versions of the instrument have been constructed and flown. DIDM-I was launched on the STEP-4 satellite in 1996, but the satellite failed to deploy its solar array panels. DIDM-II is flying on the CHAMP satellite and is returning satisfactory data although some problems have been encountered. Immediately upon first power-up, it was noticed that one of the two sensors was returning extremely low count rates. The sensor passes all diagnostic tests including a test pulse injection directly to the anode preamplifiers. The failure has been attributed to melting of the sensor's thin 50% transmission nickel outer screen due to dynamic heat loading as a result of a low altitude payload faring ejection. The undamaged sensor had a 12% transmission screen and thus an improved thermal conduction path. Fortunately, we have been able to program the remaining single sensor to perform both RPA and DM functions, but the statistical quality of the data is reduced by both the split duty cycle and the lower geometric factor. This low geometric factor was selected to guarantee that one sensor would not experience dead-time effects during periods of extreme high plasma density.

Figure 1 illustrates the DIDM-I and II sensor assembly. The aperture area and important features of the back-plane design are called out in the figure. To separate the ions from photons, the MCP potential is used to turn the ions through a large angle while photons are collected on the blackened back-plane.

Ions enter DIDM through a series of three gate screens. The gate is periodically closed to allow the pulse locating electronics to relax before the next ion strike. The DIDM field of view is a 35 degree half-angle conic. After passing through the aperture, the ions enter the RPA 'can' which is biased to the RPA potential except for the aperture piece that is held to ground potential. This arrangement eliminates any retarding grid near the aperture and generates spherical equipotential contours that retard ions regardless of entry angle.

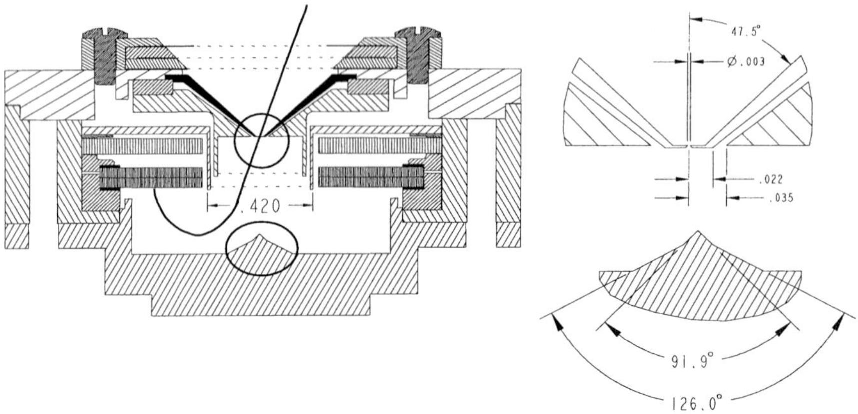

Fig. 1. DIDM cross section and detail.

After exiting the RPA ground screen, ions are rapidly accelerated and turned by the –2000 Volts MCP potential. The back-plane is held at ground. Careful electrostatic simulation was required to shape the back-plane such that ion entrance angles would map uniformly to a radial displacement on the MCP surface. The pair of chevron style MCP plates amplify the single ion strikes to cloud of about 10^6 electrons that are collected onto a three element, position sensitive 'Wedge Strip and Z' type anode. The areas of the anode elements vary according to location producing three separate pulses that uniquely locate the ion strike position by their relative amplitudes. A Field Programmable Gate Array is used to bin the strike coordinates into a pixel array of 2048 pixels with the aid of the instrument microprocessor

I DIDM can return the complete image but it also computes and returns two types of velocity moment over the image during a sample period selectable from 1/16 to 1 sec. One moment method, the Fixed Quadrature, sums counts in the 4 quadrants that evenly divide the anode image azimuthally about the central axis. The other method, the Synthetic Aperture Adaptive Quadrature, SAAQ, identifies the peak count pixel and establishes on the pixel array a perimeter or synthetic aperture about the peak. The enclosed pixels are then quartered into 4 approximately equal and predefined areas and the pixels summed. The raw moments are the orthogonal ratios of these four sums. There is an additional temperature moment formed from the ratio of counts inside the SAAQ region to the total counts. These moments are returned in telemetry and matched in real (but delayed) time by ground-based software against pre-computed simulated 'expectation' moments to identify the final geophysical quantities of drift velocity and ion temperature.

A computer program SimDIDM computes the expectation moments and RPA curves by carefully integrating assumed distributions over instrument geometric response that was determined during pre-flight calibration at AFRL. It is important to be aware that the pin-hole approach leads to images and thus moments that are dependant on the ion temperature as well as drift velocity. In general, the images should also be dependant on the spacecraft floating potential. Because CHAMP was designed to have a low floating potential, SimDIDM assumes that the spacecraft potential does not affect the drift angle, but does affect the total velocity measured by the RPA. The assumed ion distributions are Maxwellian functions of drift angle, α_y, α_z, and the Mach number, $M = v_D / \sqrt{2kT_i / m_i}$. These parameters are stepped over a 1 degree step orthonormal grid of angles and over 90 steps in M that correspond to about 125 m/sec steps in drift velocity, v_D and .03 Volt steps in T_i. At each step, the distributions are numerically integrated over the same angle grid with an analytic integration over v. DIDM's ion optics are good but not perfect, and the calibration beam is very narrow, but still has finite angular width. As a result, we have found it difficult to develop a '1 to 1 and onto' mapping between the anode and space angles that may be inverted to allow SimDIDM to integrate on the anode space and evaluate the free space distributions. Since at each point in the calibration, an anode image of the calibration beam was recorded, we have adopted the approach of always mapping the distributions onto the anode image using the calibration beam images recorded at each integration point. This limits the integrations to the calibration angular grid, however, at 1 degree steps, this has not been a severe limitation. For every M, α_y, and α_z, the integration is performed and the Fixed Quad results are tabulated as well as the RPA curve. Because of statistics, the SAAQ could have been centered on any pixel, so the integration results must be re-summed and tabulated for each of the possible 2048 pixels. The DIDM expectation tables can be quite large, ~ 1 Gbyte and take many days to compute, however, they are indexed to allow any set of moments to be matched to a best fit in a small fraction of a second on a common Unix/Linux computer.

2 DIDM Calibration

Figure 2a shows an image of the calibration beam. As the beam is scanned over the field of view, a catalog of these images is collected. The beam flux is monitored periodically and each image is re-normalized to account for drift in the beam flux. Using this catalog, simulated expectation images are computed as previously described. One such image is shown in figure 2b. Figure 2c shows an image from early in the mission. There are two ways to match expectation with observed images. The calibration method performs a least squares summation over each of the 2048 pixels until the best image fit is found. Since an observed image takes over 10 sec of telemetry to return, and a full catalog of expected images requires many 10's of GByte storage, this is not a practical method for regular operations. The 'operational' method matches the velocity and temperature moments that are both computed on-board and from the simulation. Figure 3a shows an observed image after about 9 months on orbit showing distribution that does not match any Maxwellian expectation. After much diagnostic onvestigation, it has been determined

problem was detected. Thus, although the software has been functioning, the overall calibration is just now nearing completion. At this time, only selected periods have been calibrated and processed. Two of these periods are shown in figures 5 and 6. Figure 5 show a calibration orbit maneuver from early in the mission and uses only the lab-calibration. Figure 6 is from 18 October 2001, well after the MCP gain problem began, and utilizes the in-space calibration. Prior to the in-space calibration, the drift data showed no sensible trends.

The data presented in each of figures 5 and 6 covers 1½ orbit. The top 2 panels in each figure show the azimuth (horizontal) and elevation (vertical) drift angles computed by different methods compared to the satellite attitude. Since the RPA function distorts the ion image for voltages over about 2 Volts, we can only use the DM data from about ¼ of each second. To compensate for statistical fluctuations, the data presented is a center average over 3 seconds. Over an orbit there will be many geophysical sources of plasma drift and we do not expect the drift data to always track the spacecraft attitude. The SAAQ method has been described above. The Fixed Quad moment method was also described, but in the figure we can see that there is a choice of using the ion temperature from either the SAAQ temperature moment, or the RPA derived value. These temperatures should be the same, but they are not. Whether we match the moments, or the entire images with expectations, we find that the image temperatures are unrealistically large. This effect is being investigated with the current hypothesis attributing this to a lens effect of the spacecraft potential just in front of the sensor. On STEP-4, DIDM was placed behind a flat conducting panel to avoid this effect. On CHAMP however, integration considerations made this undesirable and it was decided that with a sufficiently low spacecraft potential this would be tolerable. We expect that it will be tolerable, but it will still need to built into the expectations.

Inspection of the data in figures 5 and 6 also indicates perturbations that likely not geophysical in origin, but do seem to arise from the spacecraft interaction with the environment. Figure 6c presents the PLP offset potential, the potential difference between the PLP gold plate and the spacecraft chassis. This offset plus an additional deduced contact potential is used to correct the RPA analysis for the spacecraft potential and the velocity magnitude that is shown in 6d. It can be observed that this process works well most of the time, but there is clearly an additional effect between 05:00 and 05:20. The bottom panel of figure 6 gives additional geographic information showing that the apparent velocity perturbation at 05:00 is possibly related to DIDM and the PLP being sunlit on front side. The sunlit curve comes from a ray trace analysis of the satellite for shadowing effects. Front versus rear illumination can be deduced from the Velocity vector/Sun angle line in the figure with a zero angle denoting flight towards the sun.

4 Conclusion

The original DIDM/CHAMP ion drift measurement objective was a precision of 25 to 50 m/sec. It is clear that due to the loss of the high rate sensor and difficulties with the MCP gain, that objective will not be attained. The data so far suggests that with the in-space calibration and a moderate level of smoothing, it will be possible to attain a precision of about 1 degree or 130 m/sec. Although the required frequency for in-space calibration has not been determined, it is not expected to prevent reprocessing the entire mission's data soon.

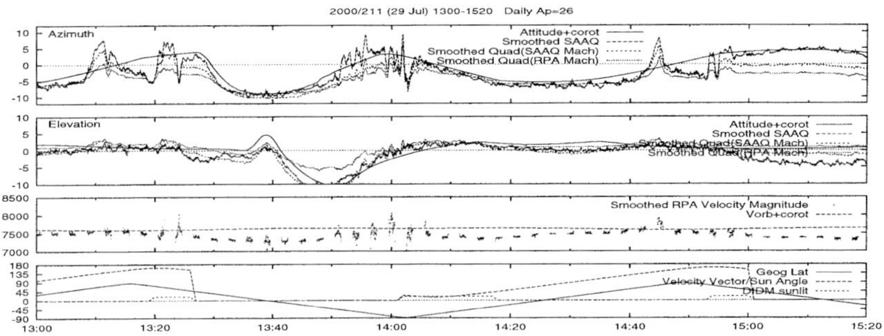

Fig. 5.

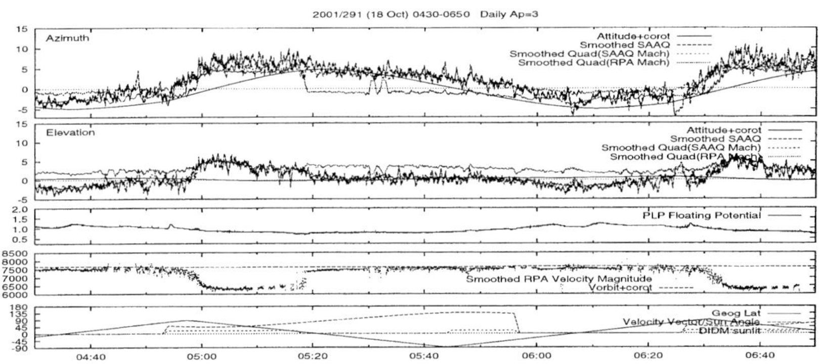

Fig. 6.

CO2 – A CHAMP Magnetic Field Model

Richard Holme[1,2], Nils Olsen[3], Martin Rother[1], and Hermann Lühr[1]

[1] GeoForschungsZentrum Potsdam, Division 2 (Physics of the Earth and Disaster Research), Germany
[2] Now at: University of Liverpool, UK *holme@liv.ac.uk*
[3] Danish Space Research Institute, Copenhagen, Denmark

Summary. We describe here a model of the magnetic field created specifically to be used with data from the CHAMP satellite, valid for the initial period of the CHAMP mission (July 2000 - December 2001). The model contains initial field, linear secular variation and external field contributions. Although CHAMP data provide the most important input to the modelling, we also use data from other satellites (Ørsted and Ørsted-2/SAC-C) and ground-based observatories, to increase the robustness of the model. The model will be improved upon in the future (in particular as we better understand the local time dependencies of the external and induced magnetic fields) but for now is the recommended standard model for use with applications of CHAMP data.

Key words: magnetic field modelling, modelling methodology

1 Model parameterization

We describe here briefly the modelling methodology (for more detail, see [4]). We represent the magnetic field as the gradient of a scalar potential $\mathbf{B} = -\nabla\Phi$ where

$$
\begin{aligned}
\Phi = a\Bigg\{ & \sum_{n=1}^{N_{MF}} \sum_{m=0}^{n} \left(g_n^m \cos m\phi + h_n^m \sin m\phi\right) \left(\frac{a}{r}\right)^{n+1} P_n^m(\cos\theta) \\
& + \sum_{n=1}^{N_{SV}} \sum_{m=0}^{n} \left(\dot{g}_n^m \cos m\phi + \dot{h}_n^m \sin m\phi\right) (t - t_0) \left(\frac{a}{r}\right)^{n+1} P_n^m(\cos\theta) \\
& + \sum_{n=1}^{2} \sum_{m=0}^{n} \left(q_n^m \cos m\phi + s_n^m \sin m\phi\right) \left(\frac{r}{a}\right)^{n} P_n^m(\cos\theta) \\
& + D_{st} \cdot \left[\left(\frac{r}{a}\right) + Q_1 \left(\frac{a}{r}\right)^2\right] \times \\
& \quad \left[\tilde{q}_1^0 P_1^0(\cos\theta) + \left(\tilde{q}_1^1 \cos\phi + \tilde{s}_1^1 \sin\phi\right) P_1^1(\cos\theta)\right] \Bigg\}.
\end{aligned}
$$

(r, θ, ϕ) are the standard Earth-centred spherical coordinates (radius, colatitude, longitude), with a reference Earth radius of $a = 6371.2$ km. P_n^m are the associated Legendre polynomials of degree n and order m. t is time in years, with the model epoch defined $t_0 = 2001.0$. $\{g_n^m, h_n^m\}$ are the internal Gauss coefficients (calculated to degree $N_{MF} = 29$), $\{\dot{g}_n^m, \dot{h}_n^m\}$ the coefficients of main field secular variation (calculated to degree $N_{SV} = 13$), and $\{q_n^m, s_n^m\}$ (calculated to degree 2) the large scale external field coefficients. The coefficients $\widetilde{q}_1^0, \widetilde{q}_1^1$ and $\widetilde{s}_1^1$ account for the variability of contributions from the magnetospheric ring current (parameterised for ease of use by the Dst index) plus their internal, induced counterpart (considered by the factor $Q_1 = 0.27$). The $n = 1, 2$, $m = 0$, terms incorporate an annual and semi-annual variation:

$$q_n^0(\tau) = q_{n,0}^0 + \left(q_{n,1c}^0 \cos\tau + q_{n,1s}^0 \sin\tau\right) + \left(q_{n,2c}^0 \cos 2\tau + q_{n,2s}^0 \sin 2\tau\right)$$

(and similar for g_n^0) for $n = 1$ and 2, where $\tau = 2\pi(t - t_0)$. In total, the model has 1121 free parameters.

It is most common to determine the model coefficients by a simple least-squares fit to the data. This approach assumes implicitly that the errors on the data are Gaussian distributed, but in practice the distribution of misfit to a model has much longer tails. To account for this, we use *Iteratively Reweighted Least Squares* with *Huber weights* ($c = 1.5$) [4]. Examination of data residuals justifies this choice *a posteriori*.

2 Data

We use CHAMP and Ørsted scalar and vector data between August 2000 and December 2001, and Ørsted-2 scalar data between January 2001 and December 2001. CHAMP data were selected preferentially to ensure that the model is most appropriate for CHAMP applications. We use the Kp index to restrict the data to quiet times, specifically requiring $Kp \leq 1+$ for the time of observation and $Kp \leq 2o$ for the previous three hour interval. CHAMP data are further restricted to extended quiet-time periods ($Kp \leq 1+$ for at least one day) to allow an independent calibration of the transformation of the vector measurements into Earth-centred coordinates. We use the Dst index to define periods when the large-scale external field is weak and stable, requiring Dst within ± 10 nT and $|d(Dst)/dt| < 3$ nT/hr. The effect of polar cap ionospheric currents is minimised by excluding data in the polar caps for which the dawn-dusk component of the interplanetary magnetic field was $|B_y| > 3$ nT. Only night-side data (between 18:00 and 06:00 Local Time) were used, to reduce contributions from ionospheric currents at middle and low latitudes. Vector data have been taken for dipole latitudes equatorward of $\pm 50°$, scalar data were used for regions poleward of $\pm 50°$ or if attitude data were not available. Sampling rate was 60 seconds; weights $w \propto \sin\theta$ are applied, to simulate an equal-area distribution. Ørsted vector data show anisotropic errors due to attitude uncertainty from calculating orientation from only one star camera [5], as do CHAMP data in a dawn-dusk orbit when one of its two cameras is sun-blinded: this is explicitly modelled [2, 1] in the inversion.

In addition to the satellite data, we also use observatory data (linear estimates of secular variation, obtained from quiet periods of the years 1998-2000) to constrain further the secular variation coefficients $\{\dot{g}_n^m, \dot{h}_n^m\}$.

3 Model

The model obtained from this analysis is named CO2 (standing for CHAMP and the two Ørsted satellites from which the bulk of the data are taken). It is available from the authors, and also from the CHAMP data centre web site `isdc.gfz-potsdam.de/champ`. The model shows broad agreement with previous models; in particular, the internal field component with $n > 14$ shows very good agreement with a dedicated model of the long-wavelength lithospheric field [3] derived from all available CHAMP scalar data.

The statistics of the fit of the model to the data are given in Table 1. The fit is within expectations in a root mean squares sense. The residuals are calculated in a camera boresight coordinate system to allow for anisotropic attitude error in Ørsted and some CHAMP data [1]. The high percentage resolution of model parameters by CHAMP data also arises because of CHAMP's lower orbit compared with Ørsted and SAC-C, leading to a much greater sensitivity to large n coefficients. Of the 1121 model parameters, 500 come from the lithospheric field, degrees n =20-29.

Table 1. Statistics of model fit to data. Resolution totals (percentage of parameters resolved by each data source) combined for all data types. F_{polar} are from above 50° dipole latitude. Vector data considered in $(B, \perp, 3)$ coordinate system, where B is the local field direction, $\perp$ the direction perpendicular to both $\mathbf{B}$ and the star camera boresight axis, and 3 the third direction perpendicular to B and $\perp$. Although most CHAMP data are unaffected by attitude uncertainty, for ease of presentation a CHAMP "virtual boresight" is defined in the radial direction

	resolution	component	N	mean (nT)	rms (nT)
CHAMP	85.5%	F_{polar}	21646	-0.68	6.99
		$F_{\text{nonpolar}} + B_B$	62765	-0.38	3.81
		$B_\perp$	31595	0.01	4.01
		B_3	31595	0.02	4.44
Ørsted	11.2%	F_{polar}	22561	-1.57	6.83
		$F_{\text{nonpolar}} + B_B$	35223	0.03	3.10
		$B_\perp$	18913	-0.64	6.66
		B_3	18913	0.11	3.51
Ørsted-2	5.4%	F_{polar}	13582	-2.47	5.72
		F_{nonpolar}	19655	0.56	3.13
Observatories	1.9%	$\dot{B}_{r,\text{obs}}$	119	-0.67	7.09
		$\dot{B}_{\theta,\text{obs}}$	119	2.85	6.68
		$\dot{B}_{\phi,\text{obs}}$	119	-0.34	5.40

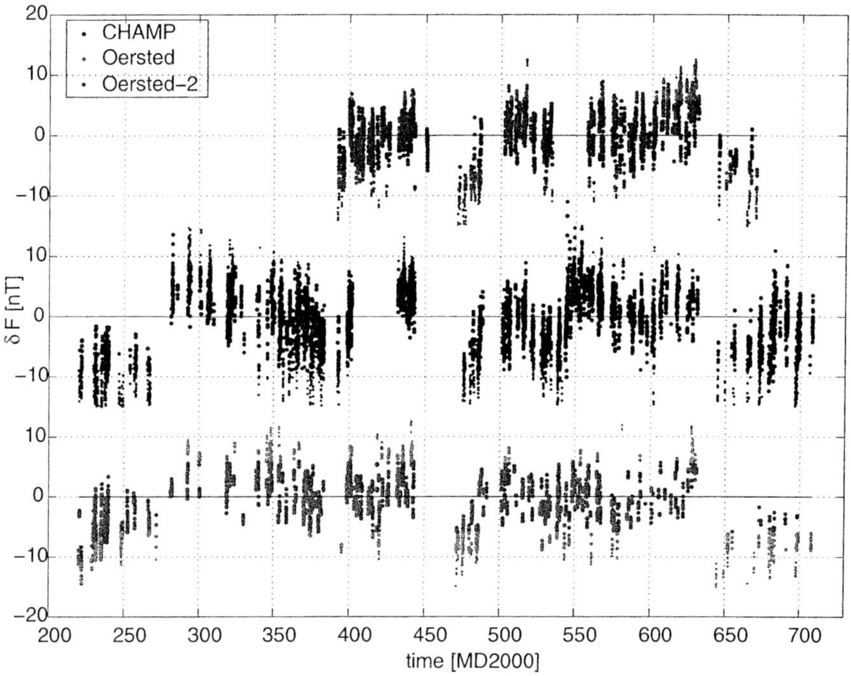

Fig. 1. Low-latitude ($< \pm 10^o$) scalar residuals as a function of time (day after midnight, 1/1/2000)

Whilst the model residuals are within specification, more careful study of their detailed structure reveals problems. In Figure 1 we show the low-latitude scalar (F) residuals as a function of time through the mission. We see that the time parameterisation of the model is clearly insufficient, as there is considerable time-dependent signal remaining in the residuals. Further, these are strongly correlated between the different satellites, and so is unlikely to result from differences in local time of the satellites orbits, but instead from inadequate representation of the axisymmetric component of the ring current.

4 Comparison with a particular data set

These problems with unmodelled external field have important implications for the application of the CO2 model and for future modelling of CHAMP data. To illustrate this, we examine one part of the CHAMP data set in detail, focusing on days 615-620, a quiet period in September 2001. In Figure 2 we show residuals of data to various field models. Top left (a) are the residuals to the CO2 model. A clear signal can be seen in the X and Z residuals. Top right (b) we plot the same residuals, but calculated in a local coordinate system

Decadal and Subdecadal Secular Variation of Main Geomagnetic Field

Ingo Wardinski[1] and Richard Holme[1,2]

[1] GeoForschungsZentrum Potsdam, Div. 2.3, Potsdam, Germany
ingo@gfz-potsdam.de

[2] Now at Department of Earth Sciences, University of Liverpool, England
holme@liv.ac.uk

Summary. We present a preliminary study of the secular variation over the period 1980 - 2000. We are motivated by the fact that this time span is bookended by high-quality satellite vector data from Magsat in 1980 and Oersted and CHAMP in 1999 - 2001. For the intervening period we consider annual means of permanent magnetic observatories, to model secular variation and the main field. We use the field model to examine the geomagnetic jerk of 1991. It is important to understand how such a discontinuous feature should be modeled in the framework of optimally smooth models. Therefore we analysed the 1970 jerk by deterministic modeling of transient behaviour of the Y component of a single observatory. The results suggest a significant contribution from external field sources to the jerk–signal. Our global model gives separation of internal and external sources, and so will allow better characterization of the origin of the jerk.

Key words: Geomagnetic jerk, Magnetic secular variation

1 Introduction

The geomagnetic field shows a rich spectrum of temporal changes from audio frequencies to thousands of years. The observed time variation is known to be the result of two primary processes: **i**) secular variation of the main field of internal origin and **ii**) variation of the external field, the sources of which are located in the ionosphere, magnetosphere and interplanetary space. In general the short time variations are of external origin, the long term variation, with time scales of decades and longer of internal origin. At intermediate time scales it is less clear; there are certainly external variations with a period of 11 years and there are fluctuations with short duration which seem to be of internal origin, the so-called *Geomagnetic jerks*.

In this study we are particularly interested in the secular variation over the period 1980-2000. This time interval is bracketed by high-quality satellite vector data (from Magsat in 1980 and Ørsted and CHAMP in 2000). Our aim is to model the magnetic field and secular variation throughout this period by a fit to all available data. The use of the two satellite epochs as "bookends" will enable a much higher resolution of secular variation modelling than has been possible to date. In particular, we hope to answer the question as to whether we can see subdecadal or shorter period variation in the internal field, which in turn will provide important constraints on processes in the core, and also on the electrical conductivity of the deep mantle.

Our approach follows that of Bloxham and Jackson [1992], who decomposed the magnetic potential into spherical harmonics, and then expanded each harmonic coefficient on a basis of cubic B-splines in time. They then sought fields that were smooth in time, in the sense of minimising a norm of the quantity

$$\int \oint_{\mathrm{CMB}} \left(\frac{\partial^2 B_r}{\partial t^2} \right)^2 d\Omega dt \ . \tag{1}$$

In other words, they sought a field that provided a desired fit to the data while minimising the square of the second time derivative of the radial field (the secular acceleration), integrated over the core-mantle boundary. We intend to follow a similar procedure (see section 3). However, this approach has one major problem. One feature of the secular variation that is of great interest is the presence of jerks, apparant discontinuities in the secular acceleration. Following Malin and Hodder [1982], they are assumed to be of internal origin. It is not clear that in the presence of such features, a uniformly smooth model of the field is appropriate.

To try to answer this question, and hence to give insight into how we should allow for the influence of geomagnetic jerks in a smooth field model, we return to individual time series of magnetic data. First, we look at a variety of measures of magnetic activity to explore the periodicity which we might expect in magnetic signals (section 2). Knowledge of such periodicities is also important when considering how best to generate models from satellite data. We then focus on a time series from one particular observatory, Fürstenfeldbruck in southern Germany. We examine in particular the records around 1969; while this is outside our main period of interest, the most well-known and strongest geomagnetic jerk appears to have occured at this time. We seek to eliminate the external field by a parameterized model. We also try to limit its influence by considering only magnetically quiet data. Using quiet time data only has its drawbacks: because the distribution of quiet days is highly uneven, simple yearly or monthly means become more difficult to apply, and therefore estimation of time derivatives of the field can no longer be done by simple first differencing. However, to locate more precisely in time any effect of the internal field, such elimination of the external field seems necessary.

2 Deterministic Model of the external field variation

In order to quantify typical time scales of the external magnetic signal we perform a spectral analysis of daily sunspots number (SSN) and daily means of magnetic indices – Aa-index from 1860 – 1999, Ap-index from 1932 – 1999 and D_{ST} from 1957 – 1999. Equation 2 gives us an estimate of the power spectral density

$$PSD(\omega) = \frac{2}{N} \left| \sum_{i=1}^{N} x'(t_i) e^{-i\omega t_i} \right|^2 \tag{2}$$

where $x'(t_i) = x(t_i) - \langle x(t_i) \rangle$.
The spectra of the SSN and magnetic indices show a very complex structure. We do not discuss this here in detail, but our spectral analysis of the magnetic indices shows at least 7 substantial periodicities, which we consider in our model (eq. 3): $\{T_i\} = \{13, 18, 27, 70, 183, 365, 4015\}$ (in days). In general the disturbances vary in a transient and irregular manner and the predominant duration of these is of the order of seconds to hours, a simple time averaging of the field cannot be expected to remove the effect of disturbance phenomena [Stewart and Whaler, 1992], because of the modulation of the solar cycle or longer cycles. We adopt an approach introduced by Gavoret et al. [1986] and developed by Stewart and Whaler [1992] to separate external and internal signals from the geomagnetic field. We model the magnetic time series as linearly correlated with the magnetic indices, sine and cosine of typical time scales and with an underlying variation represented by a cubic polynomial.

$$\begin{aligned} X(t)^{\text{model}} &= \alpha_0 + \alpha_1\,\text{AA}(t) + \alpha_2\,\text{AP}(t) + \alpha_3\,\text{D}_{\text{ST}}(t) \\ &+ \sum_i [\beta_i^s \sin(t/T_i) + \beta_i^c \cos(t/T_i)] \\ &+ \sum_{k=1}^{3} \gamma_k\, t^k \end{aligned} \tag{3}$$

$X(t)^{\text{model}}$ represents a modeled daily field value at a certain time t. The coefficients α, β and γ we assume to be time invariant. We fit this model in a least–squares sense to the data, and derive the residuals (eq. 4), which we expect to be the internal signal.

$$R(t) = X(t)^{\text{observation}} - X(t)^{\text{model}} \tag{4}$$

$R(t)$ contains a contribution from induction caused by external variation, but the amount of this contribution is small, because the transient behaviour of the induction signal is similar to that of the external variation, and therefore this is covered by the model. However, including induction and coestimating it in a deterministic model will be the topic of a future work. We apply our method first to the daily mean values of the **Y** component recorded at Fürstenfeldbruck, then to quiet daily means (Ap-index <7). As one might see the residuals for all daily means (fig. 1) shows a significant deflection around spring 1970, which we interpret as the occurrence of the jerk. In both cases the residuals of daily means (fig. 1) and the residuals of quiet daily means (fig. 2) are below 1σ (confidence level). We would expect the jerk to show up

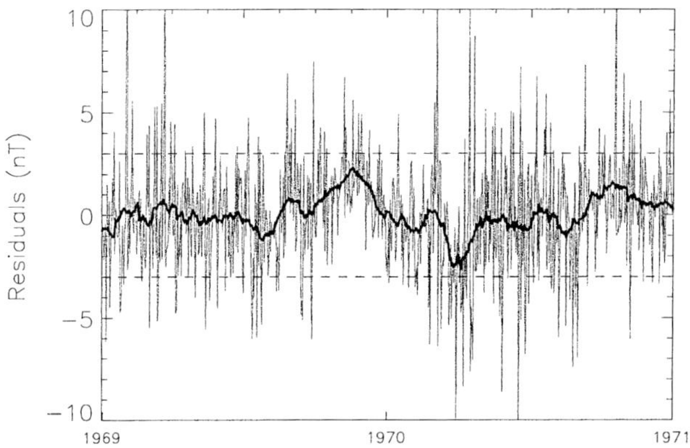

Fig. 1. The computed residuals of daily mean values from 1. Jan. 1969 to 31. Dec. 1970 (grey line). The dashed lines give the 1σ level. The thick black curve depicts a moving average over 30 days of the residuals.

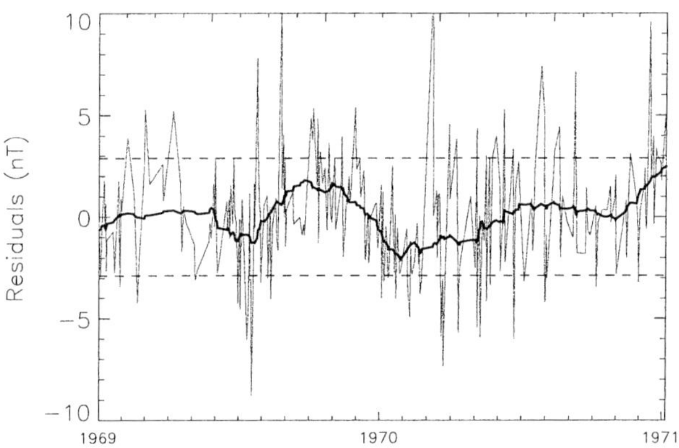

Fig. 2. The computed residuals of daily means of quiet days from 1. Jan. 1969 to 31. Dec. 1970 (grey line). The dashed lines give the 1σ level. The thick black curve depicts a moving average over 30 quiet days of the residuals.

much more strongly in the residuals of quiet daily means than in the daily mean. We therefore argue that the geomagnetic jerk of 1969/1970 is partially caused by external sources. This might explain thc fact that jerks and years with the sunspot maximum coincide.

However these kind of approach has a serious shortcoming. The disadvantage of using a single station such as one magnctic observatory is that we lose the ability to separate internal and external field sources based on the orthogonality of spherical harmonics. Thus, our time-series analysis will not produce a final answer to the question of whether jerks are of internal or external origin. However, used in parallel with global modelling (next section), we should be able to locate the precise time and sharpness of jerks much more accurately, which in turn will yield contraints on the processes which could give rise to them.

3 Time dependent field modelling

In contrast to the previous section we now focus on a more classical approach to quantify the time dependent behaviour of the Geomagnetic main field. In future we will compare the results of these two approaches.
The observations are related to a set of model parameter m (Gauss coefficients) by a model equation

$$\gamma = \mathsf{f}(\mathbf{m}) + \mathbf{e} \tag{5}$$

where γ is the data vector, $\mathsf{f}(\mathbf{m})$ is a function of $\mathbf{m}$ and $\mathbf{e}$ is the error vector. Like [Bloxham and Jackson, 1992] we seek the smoothest model for a given fit to the data, where $\mathbf{m}$ minimizes the data misfit, two model norms and additionally the 'distance' to the priori information. The two model norms $\Psi(\mathbf{m}), \Phi(\mathbf{m})$ measuring the spatial roughness and temporal roughness, respectively [Lowes, 1966; Shure et al., 1982; Bloxham and Jackson, 1992].

$$\Psi(\mathbf{m}) \sim \oint_{\mathrm{CMB}} (\nabla_h \; B_r)^2 d\Omega = \mathbf{m}^T \mathsf{S} \mathbf{m} \qquad \text{(spatial norm)} \tag{6}$$

$$\Phi(\mathbf{m}) \sim \int \oint_{\mathrm{CMB}} (\partial_t^2 B_r)^2 d\Omega dt = \mathbf{m}^T \mathsf{T} \mathbf{m} \qquad \text{(temporal norm)} \tag{7}$$

Defining $\mathbf{\Lambda} = \lambda_s \mathsf{S} + \lambda_t \mathsf{T}$ where λ_s and λ_t are damping parameters, we finally end up to minimize the objective function

$$\Upsilon(\mathbf{m}) = [\gamma - \mathsf{f}(\mathbf{m})]^T \, \mathsf{C}_e^{-1} [\gamma - \mathsf{f}(\mathbf{m})] + \mathbf{m}^T \Lambda \mathbf{m} + (\mathbf{m} - \mathbf{m}_0)^T \Gamma (\mathbf{m} - \mathbf{m}_0) \, . \tag{8}$$

C_e^{-1} is the data error covariance matrix. The last term on the right hand side of equation (8) controls the departure of our model vector $\mathbf{m}$ from priori models $\mathbf{m}_0$ at the endpoints 1980.0 and 2000.0. Γ is comparable to the previously discussed damping matrix Λ. The priori main field models are derived from MAGSAT data and Ørsted data and are of order 13. At this initial state we do not consider external coefficients.

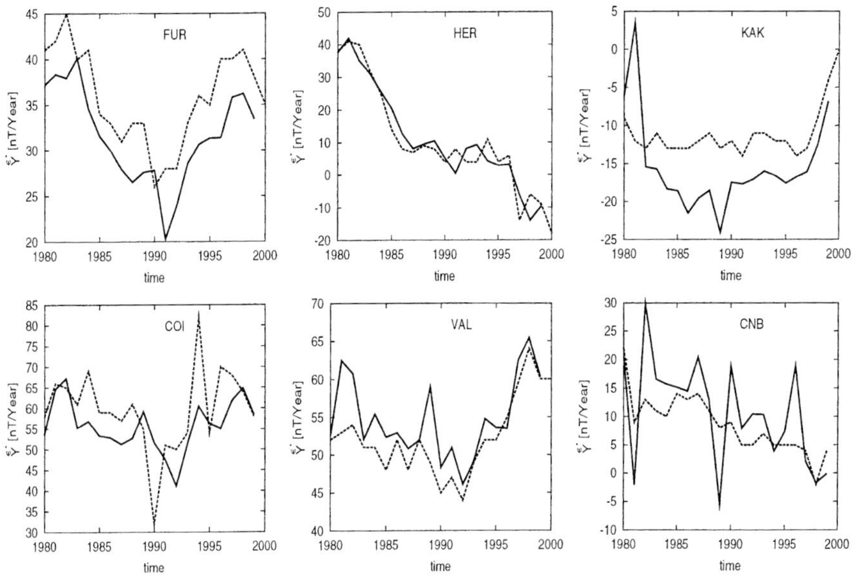

Fig.3. First differences of Y component (solid line) and modeled secular variation (dashed line) for six observatories from left to right: Fürstenfeldbruck, Hermanus, Kakioka, Coimbra, Valentia, Canberra

We apply this modelling approach to observatories annual mean first differences of the period 1980 to 2000 (first differences of observatories annual mean are the same as secular variation). The advantage of using first differences one can easily avoid contribution from the crustal field. Fortunately, the time span of our studies covers at least one jerk around 1991 [Macmillan, 1996] and maybe a second one around 2000 [Mandea et al., 2000]. So that we can study how our approach behaves in a presence of a jerk. At this initial state our model is in a strikingly good agreement with the observatory data in NGK, FUR and HER. They differ only slightly (2 – 4 nT/year) for the whole period and especially around 1991. We hope to minimize this by altering the damping and applying external corrections. The end effects at 2000 might be caused by an interfering jerk or, what is reasonable, through a lack of data.

References

Bloxham J Jackson A (1992) Time-dependent mapping of the magnetic field at the core-mantle boundary. Journal of Geophys Research, 97:19,537–19,563.

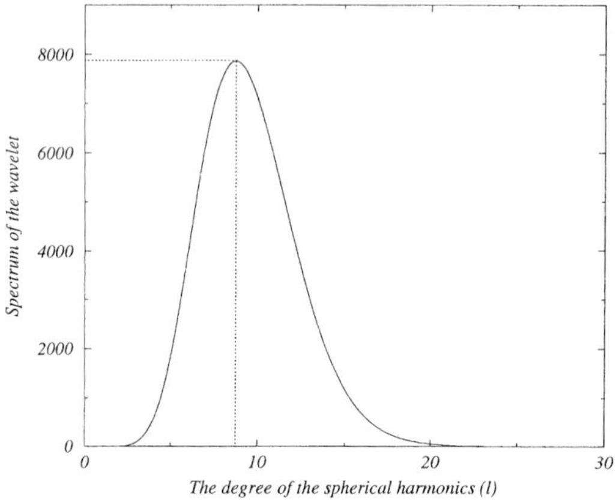

Fig. 1. Spectrum of the wavelet of scale $a = 0.625$ and degree $n = 5$.

The numbers A, B are called the framebounds. An orthonormal basis is a frame with framebounds equal to 1. The continuous collection of wavelets may be reduced to a discrete set if we sample the position-scale space. This can be intuitively explained as follows. For fixed scale a and degree n, the spectrum of the wavelet g_a covers only a part of the spaces W_ℓ. Indeed, a look at figure 1 shows that 90% of the energy is contained within a few wavelengths *around* the central peak n/a.

3 Mathematical background

Geophysical information systems have long had a need to represent sampled data on the sphere. In the following we present the expansion of scalar fields into continuous wavelets.

The generating functions. The same generating functions as for spherical harmonics are used:

$$\frac{1}{|\mathbf{X}-\mathbf{X}_0|} = \sum_{\ell=0}^{\infty} |\mathbf{X}_0|^{\ell}.p_\ell\left(\frac{\mathbf{X}}{|\mathbf{X}|}.\frac{\mathbf{X}_0}{|\mathbf{X}_0|}\right) \tag{2}$$

where p_ℓ is the Legendre Polynomial of degree ℓ, $\mathbf{X}$ the position at the surface and $\mathbf{X}_0$ the position in space of the considered "source".

The auxiliary functions.

$$H_a^m = a^m.(r_0\frac{d}{dr_0})^m\left(\frac{1}{|\mathbf{X}-\mathbf{X}_0|}\right) = \sum_{\ell=0}^{\infty}(a\ell)^m.e^{-a\ell}.p_\ell(\hat{\mathbf{X}}.\hat{\mathbf{X}}_0) \tag{3}$$

$$H_a^m = a^m.(z_0\frac{d}{dz_0})^m\left(\frac{1}{\sqrt{x^2+y^2+(z-z_0)^2}}\right) \tag{4}$$

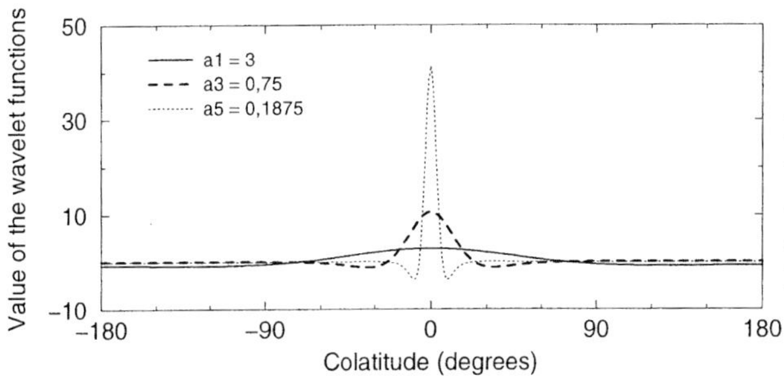

Fig. 2. The Poisson wavelets, with order 3 and different scales a, as function of θ the co-latitude on the sphere.

The basis functions.

$$W_a^m = H_a^m + \frac{2}{a}.H_a^{m+1} \tag{5}$$

Some properties of the wavelets.

– convolution on the sphere:

$$(f * g)(X) = \int_{S(1)} T_X f(Y).g(Y).dY \tag{6}$$

– scalarproduct between two wavelets:

$$< W^m_{a_1,\mathbf{r_1}}, W^m_{a_2,\mathbf{r_2}} > (\widehat{\mathbf{r_1}.\mathbf{r_2}}) = \left(\frac{a_1.a_2}{(a_1+a_2)^2}\right)^m .W^{2m}_{a_1+a_2}(\widehat{\mathbf{r_1}.\mathbf{r_2}}) \tag{7}$$

– norm:

$$\|W_a^m\| = \sqrt{W_{2a}^{2m}(0,0,z_0).4^{-m}} \tag{8}$$

– spectrum:

$$R_\ell(a) = (a\ell)^m.\exp(-a\ell).\sqrt{2\ell+1} \tag{9}$$

4 Application to synthetic data

Implementation of positioning system. The directions θ, ϕ on the sphere are defined by considering a cube, each of its six sides being divided in 4^{j-1} squares, where (j) denotes the maximum degree of sampling (figure 3).

Our positioning system is completed by defining the radius of shells. The radii of shells increase linearly, and the spheres become closer and closer to the limit given by the unit sphere (figure 4). Indeed,

$$R_j = \exp\left(-C.p^{j-1}\right) \tag{10}$$

where j is the degree of sampling the cube, $0 < p < 1$ the increasing parameter, and C a constant depending on the order of the chosen wavelet basis. In the present study the adopted values are $p = 1/2$ and C the order of the wavelet in the given basis. The latest value for C ensures that the wavelet $W_{a=1}$ is concentrated around a spherical harmonic $\ell = 1$.

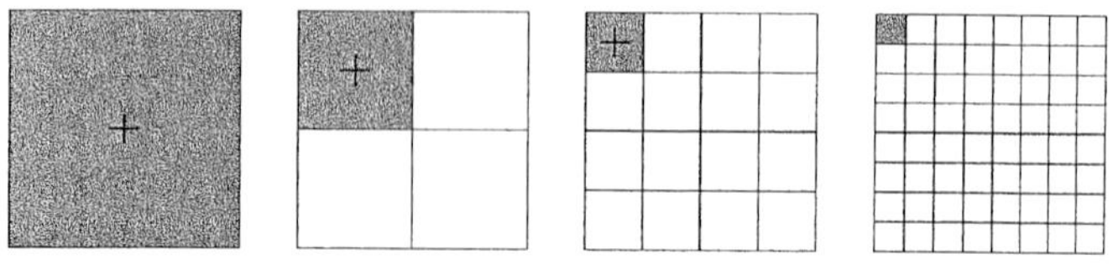

Fig. 3. Degree of sampling j of a cube.

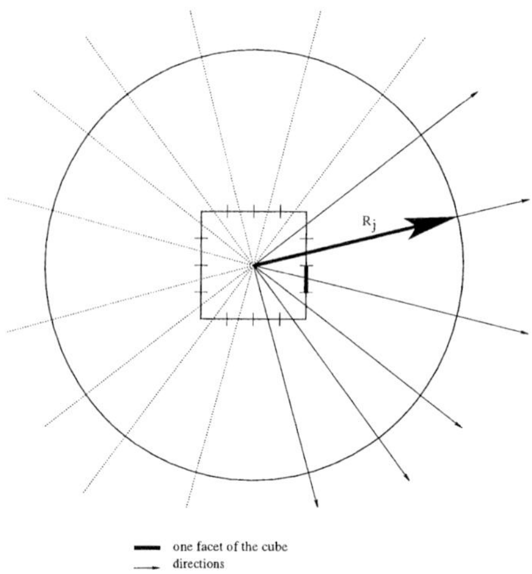

Fig. 4. Representation of a unit sphere and a cube situated in the center of the sphere for $j = 3$.

Implementation of wavelet basis. A scalar field can be represented on the sphere using:

$$s = \sum_{0}^{\infty} \alpha_n g_n . \tag{11}$$

The coefficients α_n are computed using a finit number N of functions (the value of N is obtained, for example, from the spectrum analysis). The method used is the least squares which is a statistical approach to estimate an expected value with the highest probability from the observations with random errors. According to this method, the best fitting has the property that:

$$\chi^2 = \sum_{i=1}^{N_{mes}} \left[S_i^{mes} - S^{mod}(p, \mathbf{r_i})\right]^2 \longrightarrow minimum. \tag{12}$$

$$\left[\sum_{n,n'=1}^{N} \left(\sum_{i=1}^{N_{mes}} g_n(\mathbf{r_i}).g_{n'}(\mathbf{r_i})\right)\right] \cdot \left[\sum_{n=1}^{N} \alpha_n\right] = \left[\sum_{n'=1}^{N} \left(\sum_{i=1}^{N_{mes}} S_i^{mes}.g_{n'}(\mathbf{r_i})\right)\right] \tag{13}$$

The last equation can also be written as:

$$M_{nn'}.\alpha_n = b_{n'} \tag{14}$$

where the matrix $M_{nn'}$ is the covariance matrix (Gramm matrix) and its elements are the scalarproducts between the functions g_n; each element of $b_{n'}$ contains the sum of scalarproducts between a wavelet of the basis and the scalar measurements. The $M_{nn'}$ matrix provides us with a way to measure the correlation between the functions.

Approximation of a spherical harmonic in the wavelet basis. To approximate a spherical harmonic in a wavelet basis is important, as thereafter it is possible to compare a spherical harmonic basis with a wavelet basis. All the spherical harmonics can be described by a wavelet basis. Many tests were done, for different spherical harmonics and different wavelet basis. Figure 6 shows one of them: (a) the spherical harmonic for degree $\ell = 9$, and order $m = 4$ is represented in the classical way; the same spherical harmonic is represented in our wavelet basis (b) formed by 1000 wavelets, with degree $n = 3$ and their scales varying from $a_1 = 1.0$ to $a_{1000} = 0.0625$. The two representations are similar, underling our capability to represent a spherical harmonic in a given wavelet basis.

5 Conclusions

In our attempt to produce a new representation of the magnetic field on the sphere, we choose an approach which always makes a direct relation between the spherical harmonics and wavelets. The main result is the theoritical description of the wavelets on the sphere in order to use them in field description. The first comparisons between the spherical harmonic basis and

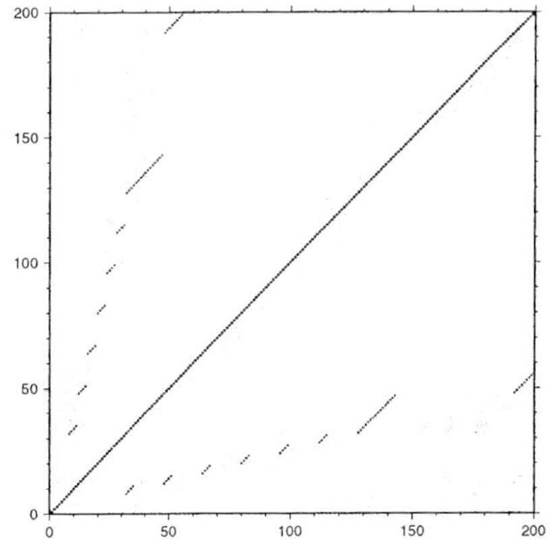

Fig. 5. Covariance matrix for $n = n' = 200$, degree $n = 3$, initial scale $a = 1$. In the ideal case covariance matrix is diagonal (the off-diagonal terms are zero because of the orthogonality properties of the functions). The correlation between functions gives off-diagonal terms in the $M_{nn'}$ matrix, which can be considered as an indicator of the departure of the actual situation from the ideal one. In fact the diagonal terms are well-defined, moreover other small segments are also present, as a result of the different interactions between the different scales of wavelets on the sphere.

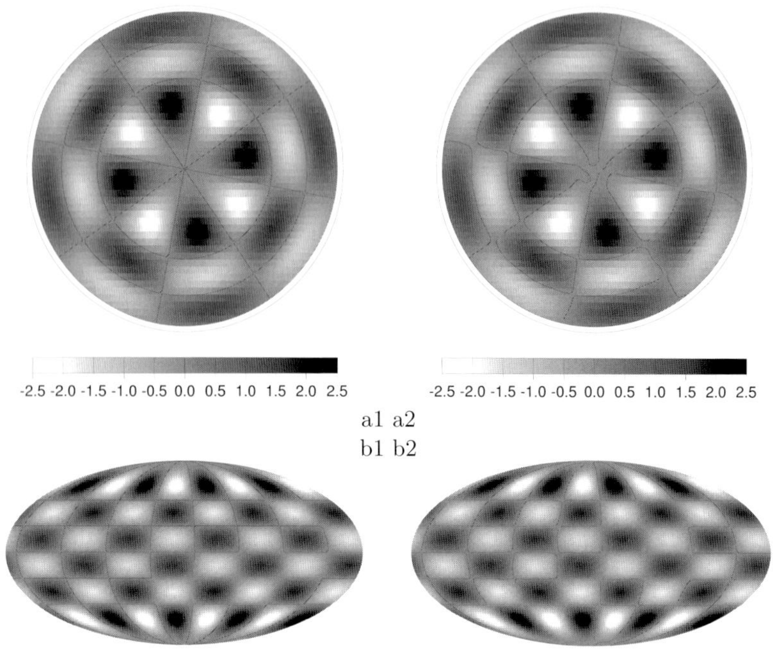

Fig. 6. Representation of the spherical harmonic [$\ell = 9, m = 4$]: **a)** directly and **b)** in the basis of the 1000 wavelets.(**1**:North Pole of the sphere, **2**:Mollweide projection.)

wavelets basis show how well the wavelets are able to reproduce the spherical harmonics. Of course, more work as to be done, as regard of the chose of the optimal wavelets basis. For example, we have to study the evolution of scales, the degree, the number of wavelets and how much the wavelets induced a distorsion on the spherical harmonics.

In the future, a primary concern is that the wavelet search and the implementation of the wavelet transformation and excision process be carried out in a computationally efficient manner. It is our goal to perform these operations in a real-time digital signal processing at extremely high sample rates, to match that of the magnetic satellites.

Improved Parameterization of External Magnetic Fields from CHAMP Measurements

Judith Schwarte[1], Hermann Lühr[1] and Richard Holme[1,2]

[1] GeoForschungsZentrum Potsdam, Section 2.3, Potsdam, Germany
judith@gfz-potsdam.de, hluehr@gfz-potsdam.de
[2] Now at University of Liverpool, Department of Earth Sciences
holme@liv.ac.uk

Summary. We describe a method for determining the influence of external field contributions to satellite scalar magnetic field data in a more reliable way. External contributions to CHAMP measurements arise from the large-scale magnetospheric current systems, e.g. the magnetopause current, ring current and tail current. In main field modelling to date, the variable 'ring current effect' has been parameterised in terms of a linear dependence on the Dst index. For the development of a better parameterisation of external fields we take advantage of the three-day CHAMP orbital repeat cycle which occurred during the early part of the mission. By subtracting two successive repeat tracks from each other the lithospheric contribution and constant model errors are eliminated while the differences of the magnetic effects from current systems are emphasised.

Our results show a clear linear relation between the degree one internal and external SHA coefficients with the corresponding difference in Dst index. However, the scaling factors differ significantly from published values, and show a clear local time dependence. Around local noon where ionospheric currents are prominent our method is not reliable.

Key words: field modelling, external magnetic field, parameterization, ring current

1 Introduction

When describing the magnetic field of the Earth in terms of spherical harmonic functions it is possible to distinguish between internal and external sources. In case of satellite data the orbit height defines the separation between the two source regions. The topic of this study is an investigation of large-scale magnetospheric current systems including the ring current.

In main field modeling a typical interval of 3 to 5 quiet days is used and the variable ring current effect has been parameterized in terms of a linear dependence on the D_{ST} index e.g. Magsat [1] or Ørsted [2]. However, while successful to an extent, this method has clear short comings. In particular, the dominant residual between field model and data often has a clear magnetospheric signature [3]. Here we made use of CHAMP satellite data towards developing a better parameterization of external fields.

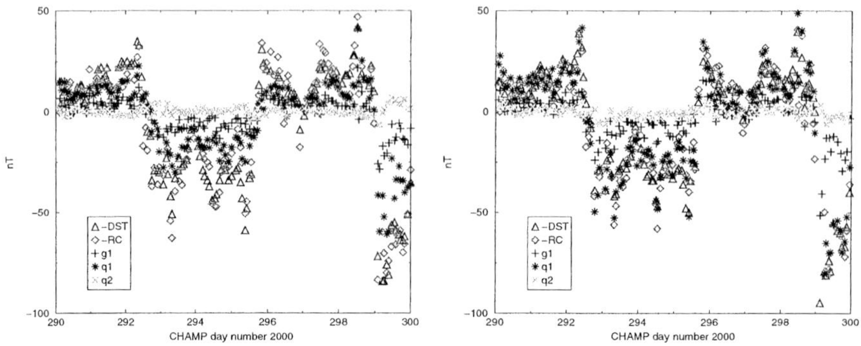

Fig. 3. Magnetic indices and coefficients for a period of 14 days in October 2000, left: dawnside, right: duskside. The day number reflects the earlier track.

4 Results

A comparison of obtained coefficients of the modeled differences of the ring current (RC in Figs. 1 and 2) with the ΔRC derived from the data (model in Figs. 1 and 2) shows a good correlation.

A problem is apparent in the dayside curve of Fig. 2. At about 90° geomagnetic colatitude the signature of the equatorial electrojet can clearly be seen. Perhaps due to the influence of the electrojet and the S_q currents the modeled curve (Fig. 2, left) does not fit as good, as the other model results.

Two possible magnetic indices representing the large-scale magnetospheric currents are the D_{ST} and RC^* index[1]. Since we are considering the differences between repeat tracks, also the differences in D_{ST} and RC^* between the epochs of CHAMP equator crossings are used (Fig. 3). The q_1^0 term follows the $-D_{ST}$ and $-RC^*$ indices rather closely. There is only a difference in amplitude. While in dusk sector the values are comparable, they are smaller in the dawn sector (further information for the December period see [4]). Also the internal term g_1^0 tracks the indices reasonably well, but at a lower level. In contrast, q_2^0 varies around zero and shows no correlation.

After having presented the qualitative relations we now will look closely into correlation properties of the quantities. We tested both magnetic indices. The results of the RC^* index are very similar to those from the D_{ST}. Since the D_{ST} index is readily available from geomagnetic data centers (e.g, http://swdcdb.kugi.kyoto-u.ac.jp/) it will be used subsequently.

[1] The 'ring current' index, RC^* is determined from observatories in the midnight sector [3].

We found that good correlations exist between D_{ST} and q_1^0 and also between D_{ST} and g_1^0 (Fig.4). All correlation results are summarized in Table 1.

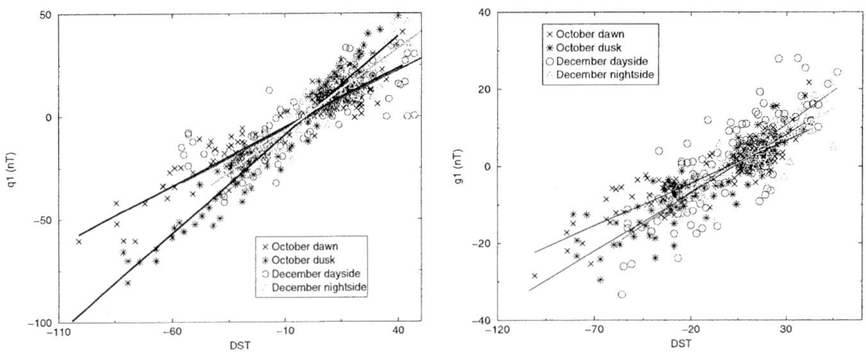

Fig. 4. Correlation between D_{ST} and q_1^0 (left) and between D_{ST} and g_1^0 (right)

Correlation Coefficient					
	D_{ST} vs RC^*	D_{ST} vs q_1^0	RC^* vs q_1^0	g_1^0 vs q_1^0	D_{ST} vs g_1^0
Oct, dawn	0.97	0.94	0.92	0.93	0.89
Oct, dusk	0.96	0.97	0.96	0.94	0.89
Dec, dayside	0.96	0.83	0.79	0.60	0.81
Dec, nightside	0.93	0.93	0.93	0.94	0.88
Slope					
	D_{ST} vs RC^*	D_{ST} vs q_1^0	RC^* vs q_1^0	g_1^0 vs q_1^0	D_{ST} vs g_1^0
Oct, dawn	0.96	0.57	0.57	2.28	0.22
Oct, dusk	0.99	0.96	0.93	2.72	0.30
Dec, dayside	1.09	0.57	0.48	0.96	0.35
Dec, nightside	1.01	0.80	0.74	2.53	0.28
Intercept					
	D_{ST} vs RC^*	D_{ST} vs q_1^0	RC^* vs q_1^0	g_1^0 vs q_1^0	D_{ST} vs g_1^0
Oct, dawn	-0.48	0.63	0.68	-0.07	0.09
Oct, dusk	-0.64	0.75	1.03	2.08	-0.82
Dec, dayside	-1.09	-0.28	0.27	-0.58	0.41
Dec, nightside	0.98	1.04	0.79	3.06	-0.51

Table 1. Numerical results of the correlation analysis

5 Conclusion and Outlook

We determined the ring current effect in scalar satellite data by directly inverting the measurements using repeat tracks. This procedure removes a lot of problems caused by biases and slow trends both in satellite data and ring current indices.

We obtained good correlations of the homogeneous external field with ring current indices and with the induced field for most local times. The derived coefficients differ markedly from [1], [3].
Poor results are obtained on the dayside where the E region currents are significant.
In a next step it is foreseen to extend this study by using CHAMP vector field data.

6 Acknowledgement

N. Olsen kindly provided the 'ring current' index, RC^* for the year 2000. This study is supported by the DFG special research program SPP 1097.

References

1. Langel RA, Sabaka TJ, Baldwin RT, Conrad JA (1996) The near-Earth magnetic field from magnetospheric and quiet-day ionospheric sources and how it is modeled. Physics of the Earth and Planetary Interiors 98:235-267
2. Olsen N, Sabaka TJ, Tøffner-Clausen L (2000) Determination of the IGRF 2000 model. Earth Planets Space 52:1175-1182
3. Olsen N (2002) A Model of the Geomagnetic Main Field and its Secular Variation for Epoch 2000 Estimated From Ørsted Data. Geophys. J. Int. 149:454-462
4. Schwarte J, Lühr H, Holme R (2001) Secular Variation from CHAMP Magnetic Field Measurements. DFG project report

Monitoring Magnetospheric Contributions using Ground-Based and Satellite Magnetic Data

Nils Olsen, Susanne Vennerstrøm, and Eigil Friis-Christensen

Danish Space Research Institute, Juliane Maries Vej 30, DK-2100 Copenhagen, Denmark, *e-mail: nio@dsri.dk*

Summary. Good models of the time-space structure of magnetic field contributions from large-scale magnetospheric current systems, like the ring-current, are necessary to obtain improved models of the core and crustal fields. In addition, the estimation of external and induced contributions enables the derivation of transfer functions from which the electrical conductivity of the mantle can be obtained.

We have presented a method that allows for a joint analysis of simultaneous magnetic data from various data sets (satellite and ground-based) by describing the temporal behavior of the spherical harmonic expansion coefficients with cubic B-splines. We demonstrate the capability of the method by applying it to data from CHAMP, Ørsted, and Ørsted-2/SAC-C as well as to observatory data.

Key words: Magnetosphere, Ring-current, Spherical Harmonic Analysis

1 Spherical Harmonic Analysis of magnetospheric sources

The magnetic field perturbation (observation minus value from a field model) $\delta\mathbf{B} = -\mathrm{grad}\, V$ is derived from a magnetic scalar potential V which is expanded in spherical harmonics:

$$V = a \sum_{n,m} \left\{ \left(g_n^m(t) \cos m\tau + h_n^m(t) \sin m\tau\right) \left(\frac{a}{r}\right)^{n+1} + \left(q_n^m(t) \cos m\tau + s_n^m(t) \sin m\tau\right) \left(\frac{r}{a}\right)^n \right\} P_n^m(\cos\theta) . \quad (1)$$

$a = 6371.2$ km is the mean radius of the Earth, θ is geomagnetic colatitude, and τ is Magnetic Local Time (MLT). $P_n^m(\cos\theta)$ are the associated Schmidt semi-normalized Legendre functions of degree n and order m, and (g_n^m, h_n^m) and (q_n^m, s_n^m) are the Gauss coefficients describing sources internal and external to the Earth, respectively. Note the dependency on magnetic local time τ rather than longitude, and the explicit dependency of the expansion coefficients on time t.

Estimation of the coefficients at (equidistant) time instances, t_i, that are given by the natural time discretisation of the data (e.g., hourly mean values for observatory data, orbit period for satellite data) is the usual approach when analysing observatory or single satellite data. However, this

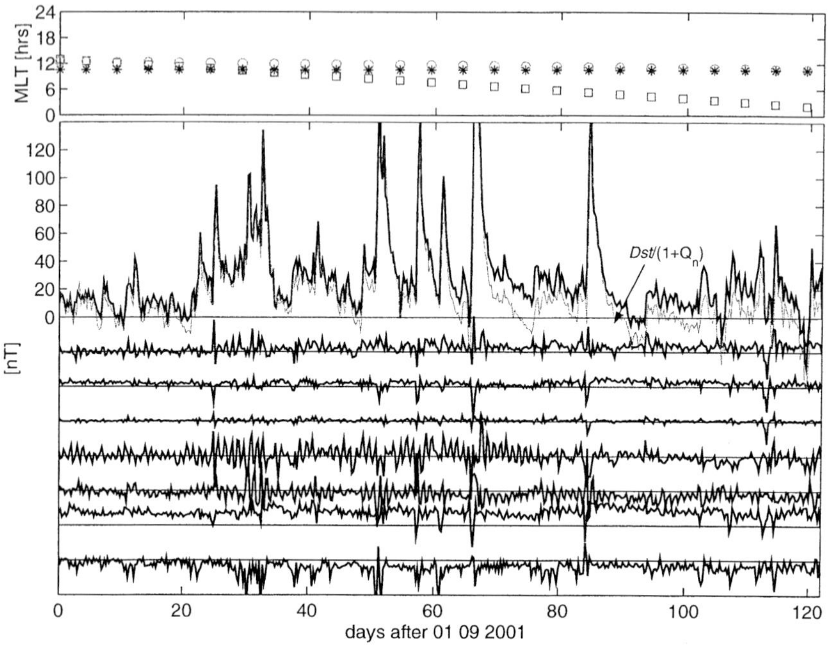

Fig. 1. Top: Local time distribution of the equator crossing of the south-going tracks of CHAMP ( ), Ørsted (○) and SAC-C (∗) for Sep 1 - Dec 31, 2001. Bottom: Time series of external expansion coefficients, estimated from satellite data. Also shown is an estimate of q_1^0 based on the *Dst*-index (thin line).

approach is not possible for the analysis of *simultaneous* data from different sources/satellites due to the lack of a common natural time basis (e.g., the equator crossings of the satellites do not occur simultaneously). To overcome this problem, we parameterize the time dependency of the Gauss coefficients by cubic *B*-splines.

As an alternative to eq. (1), internal, induced, contributions can be treated by assuming a fixed ratio Q_n between induced and inducing coefficients. Neglecting the time lag between induced and external sources (which means assuming an infinite conductor at some depth and an insulating mantle above it) results in

$$V = a \sum_{n,m} \{q_n^m(t) \cos m\tau + s_n^m(t) \sin m\tau\} \left\{ \left(\frac{r}{a}\right)^n + Q_n \left(\frac{a}{r}\right)^{n+1} \right\} P_n^m . \quad (2)$$

2 Application to satellite data

In our first example we applied the method to data from CHAMP, Ørsted and Ørsted-2/SAC-C spanning 4 months (Sep 1 – Dec 31, 2001). Only nightside data (Magnetic Local Time $\tau > 18{:}00$ or $\tau < 06{:}00$; vector data only equatorwards of $\pm 60^\circ$ dipole latitude) were used, and values from the core and crustal field part of the model described in OLSEN (2002) were subtracted from the observations. The data are equally distributed in time and weights $\propto \sin\theta$ are used to simulate an equal area distribution of the data points on the globe. If available, we used all three vector components (scalar data were used for orbits for which vector data were not available and poleward of $\pm 60^\circ$ dipole latitude), and estimated 8 Gauss coefficients ($q_n^0, n = 1-4;\ q_n^1, s_n^1,\ n = 1-2$) of the expansion coefficients of eq. (2) for each of the 739 knots (122 days, knot separation 4 hours) using *Iteratively Reweighted Least Squares* with Huber weights (cf. OLSEN (2002)). For the ratio of induced to external coefficients we used $Q_n = 0.28$, which is a typical values for field variations of a few days.

Time series of the expansion coefficients, together with the local time of the south-going tracks of the satellite orbits, are displayed in Figure 1. As expected, q_1^0 is the coefficient with the largest time-variation. It mostly reflects magnetic field variations connected with magnetic storms. Since the *Dst*-index monitors magnetospheric contributions to the horizontal component at the magnetic equator (and thus includes external *and* induced contributions), an estimate of the external coefficient q_1^0 can be obtained from $-Dst/(1 + Q_n)$. This value is shown by the thin line. The well-known offset of the ground-based estimate is obvious (the true q_1^0 is about 20 nT higher than that based on *Dst*, as found for the first time by LANGEL & ESTES (1985)). However, the difference between q_1^0 derived from satellite data and from *Dst*, respectively, is smaller during September compared to the last two months of the year. This is probably caused by base-line uncertainties of the provisional *Dst*-index that has been used in this study.

Contributions from higher harmonics are mostly limited to the main-phase of geomagnetic storms; q_3^0 decays for instance more rapidly than q_1^0. There is a pronounced daily variation of the coefficients with $m = 1$, indicating the existence of an asymmetric ring-current that varies with UT.

3 Application to observatory data

For our next example we used three years of hourly mean values (1998-2000) from observatories equatorwards of $\pm 60^\circ$ dipole latitudes. After subtracting a linear trend from the quiet-time local midnight values, we used the residuals in the geomagnetic North- and East-direction to estimate the external expansion coefficients q_n^0, $(n = 1-4)$ and q_1^1, s_1^1 for each hour. Only observatories in the night-side sector were used; those at local midnight are given maximum weight $w = 1$; the weights decrease toward dawn/dusk according to $\max(\cos\tau/2, 0)$. Due to the uneven spatial distribution of observatories, the (weighted) number of observatories varies between $\Sigma w = 13.0$ at $UT = 8$ and $\Sigma w = 48.7$ at $UT = 20$.

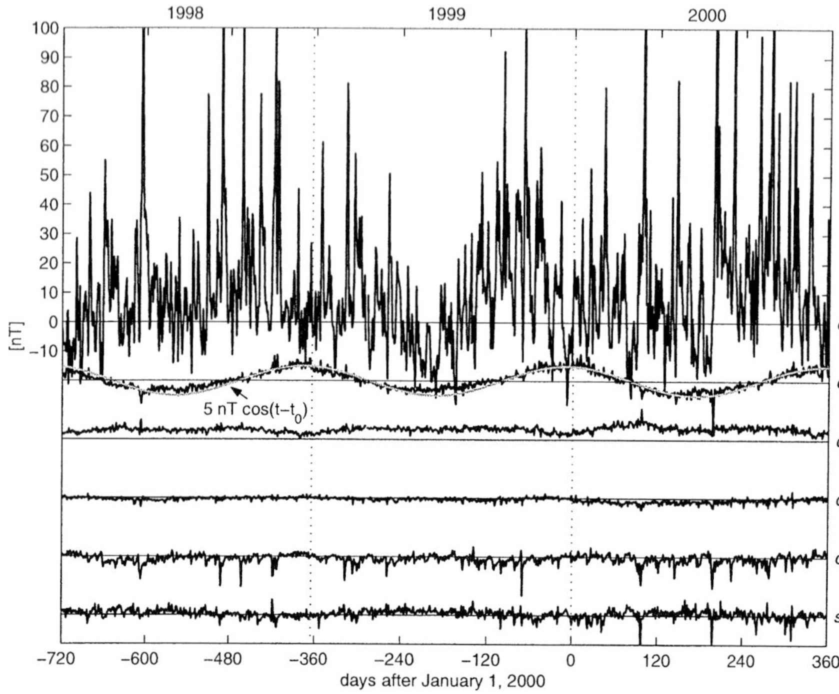

Fig. 2. Time series of external expansion coefficients, estimated from observatory data. The grey line presents 5 nT $\cos(t - t_0)$, where t is season (in radians), and t_0 corresponds to December 22.

Figure 2 shows time series of the estimated expansion coefficients. There is a very pronounced annual variation in q_2^0 of about 5 nT amplitude. This is more than twice the value that was found in satellite data (HOLME ET AL. (2002) found an amplitude of 2 nT when solving simultaneously for main-field and external field contributions using CHAMP, Ørsted and Ørsted-2 data). This discrepancy could indicate an annual variation of ionospheric contributions (which are external as seen by an observatory, but internal as seen from satellites). Note the non-zero mean of q_3^0; averaging over all three years yields a value of 3.4 nT. As in the previous example, coefficients with $m = 1$ contain a pronounced daily variation of several nT.

4 Estimation of transfer functions between induced and external terms

In our final example we demonstrate the ability of our method for estimating transfer functions between internal (induced) and external (inducing) expansion coefficients. We used one year (January – December 2001) of data from

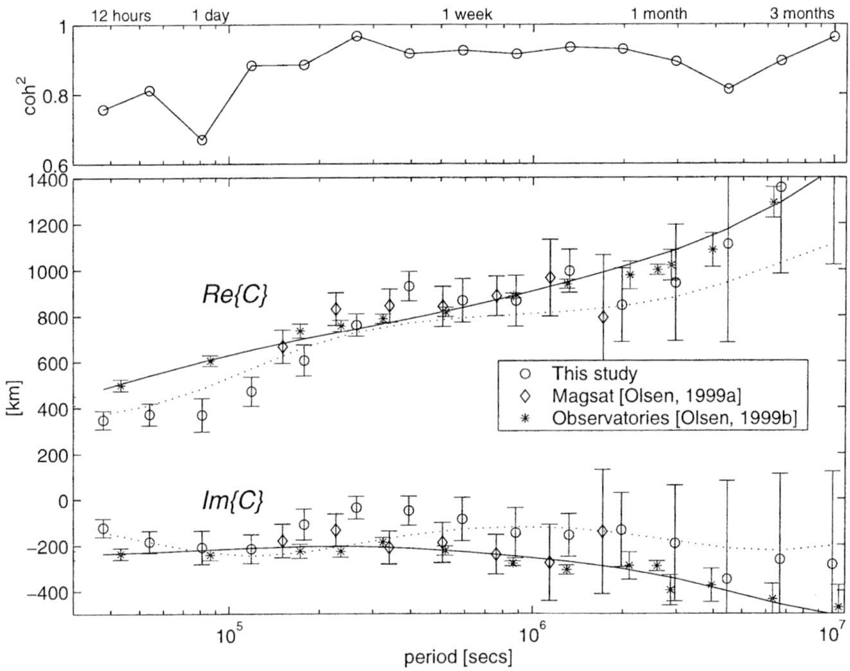

Fig. 3. Top: Squared coherency between g_1^0 and q_1^0. Bottom: C-response estimated in this paper, compared to values found from different data sources. The solid line is the theoretical response of a model fitted to the observatory values (OLSEN (1999*a*); the dashed line is that of a model fitted to the satellite values of the present study.

CHAMP, Ørsted and Ørsted-2 and solved for the 4 external expansion coefficients $q_n^0, n = 1 - 4$ and their induced counterparts g_n^0, and applied the approach of OLSEN (1999*b*) to time series of $\varepsilon = q_1^0$ and $\iota = g_1^0$, respectively: Each time series is split up into L segments of length T, and the Fourier components corresponding to this period are estimated. 15 values of T between 10 hours and 115 days (equally spaced in $\log T$) are chosen, and squared-spectra $\langle \varepsilon\, \varepsilon^* \rangle, \langle \iota\, \iota^* \rangle$ as well as cross-spectra $\langle \varepsilon\, \iota^* \rangle, \langle \iota\, \varepsilon^* \rangle$ are estimated for the 15 periods using a robust approach (ε^* is the complex conjugate of ε, and $\langle \ldots \rangle$ denotes summation over all L realizations). We then derived an estimate of $Q_1(\omega) = \varepsilon(\omega)/\iota(\omega)$ from $\sqrt{\frac{\langle \iota\, \varepsilon^* \rangle \langle \iota\, \iota^* \rangle}{\langle \varepsilon\, \varepsilon^* \rangle \langle \varepsilon\, \iota^* \rangle}}$ (this is an unbiased estimate if there is equal relative noise in q_1^0 and g_1^0), as well as squared coherency $\mathrm{coh}^2 = \frac{\langle \iota\, \varepsilon^* \rangle \langle \varepsilon\, \iota^* \rangle}{\langle \varepsilon\, \varepsilon^* \rangle \langle \iota\, \iota^* \rangle}$; cf. OLSEN (1999*b*) for details. Finally, Q_1 is transformed to the C-response by means of

$$C = \frac{a}{2} \frac{1 - 2Q_1}{1 + Q_1}. \tag{3}$$

Squared coherency coh^2, shown in the upper part of Figure 3, is above 0.9 for periods between 2 days and 1 month, but decreases especially at shorter periods. The minimum value near $T = 1$ day is probably due to internal contributions which are not purely induced. Since the satellites revisit a specific region after one day, this may indicate insufficient removal of Earth-fixed internal (e.g., core and crustal) contributions.

The lower part of the figure compares the C-response found in this study with previous estimates based on data from Magsat and from European observatories, respectively. In addition to covering a larger range of periods, the estimated errors are smaller for the results of this study, compared to Magsat. However, the uncertainty of the satellite response still exceeds that of the observatory response, due to the much shorter time series of the satellite data.

5 Conclusion

We presented a method that allows to combine different data sets (satellite and/or ground-based) for modelling the time-space structure of large-scale magnetospheric sources. The method has been applied to observatory data alone, and to the combined magnetic data from three satellites. In future we plan to use the method for the joint analysis of ground-based and satellite data. This requires considering (and perhaps co-estimating) observatory biases (a practice introduced by Langel et al. (1982) for combining observatory and satellite data in main field modeling) and will probably improve the estimation of electromagnetic transfer functions.

References

Holme R, Olsen N, Rother M & Luehr H (2002) CO2: A CHAMP magnetic field model, in *Proceedings of the First CHAMP Science Meeting*

Langel RA & Estes RH (1985) The near-Earth magnetic field at 1980 determined from MAGSAT data, *J Geophys Res*, **90**, 2495–2509.

Langel RA, Estes RH & Mead GD (1982) Some new methods in geomagnetic field modelling applied to the 1960-1980 epoch, *J Geomagn Geoelectr*, **34**, 327–349.

Olsen N (1999*a*) Induction Studies With Satellite Data, *Surveys in Geophysics*, **20**, 309–340.

Olsen N (1999*b*) Long-period (30 days – 1 year) electromagnetic sounding and the electrical conductivity of the lower mantle beneath Europe, *Geophys J Int*, **138**, 179–187.

Olsen N (2002) A Model of the Geomagnetic Field and its Secular Variation for Epoch 2000 Estimated From Ørsted Data, *Geophys J Int*, **149**, (2), 454–462.

Unraveling the Magnetic Mystery of the Earth's Lithosphere: The Background and the Role of the CHAMP Mission

Dhananjay Ravat[1] and Michael Purucker[2]

[1]Dept. of Geology MS 4324, Southern Illinois University at Carbondale, Carbondale, IL 62901-4324, USA.
[2]Raytheon STX/Geodynamics Branch, Code 921, Goddard Space Flight Center, Greenbelt, MD 20771, USA.

Summary. The utility of high-resolution aeromagnetic surveys in geologic mapping is well known. As the wavelength of anomalies increases, geologic interpretation becomes less direct and the magnetic response of geology more complex. The long wavelength magnetic anomalies observed at the altitudes of near-Earth satellites represent the collective effect of many contiguous regional-scale geologic units. The variations also reflect the effect of the Curie isotherm that governs the thickness of the magnetic crust, and therefore may be used to constrain the models of regional variations of heat flow. This information in turn can be useful in deriving lithospheric strength variations and also constraining models of global seismological parameters. The challenge of the CHAMP magnetic experiment is to unravel this complex information beyond the level of Magsat and extract from it new knowledge regarding the Earth's lithosphere that could be used by geologists and geophysicists. There are a number of fronts along which improvements appear possible in the anomaly field and their interpretation from the overall mission characteristics of the CHAMP satellite. They are: improvements due to data redundancy and attitude determinations, better external field models, increased anomaly resolution during late-phase lower altitude part of the mission, new direct interpretation techniques and, taking advantage of gravity field from the CHAMP mission, improved interpretations resulting from correlations with gravity anomaly data through Poisson's relation. Magnetic anomaly data provide a unique perspective on geologic processes as evident from the well known example of magnetic stripes associated with the seafloor spreading—a process that could not be inferred readily through other geophysical quantities. Similarly, information on bulk geologic variations of size greater than a few hundred km is difficult to obtain through routine geologic techniques. This is the scale at which satellite magnetic and gravity data from CHAMP can find its niche in the lithospheric interpretation.

Key words: magnetic anomalies, gravity anomalies, Proterozoic geology, CHAMP, magnetic interpretation.

1 Introduction

Just a few years ago, Bob Langel and Bill Hinze commented in their treatise on the Earth's magnetic field measured by satellites that we have reached a plateau in

the progress of the new subdiscipline of satellite geomagnetism. The plateau was largely reached due to lack of new data since Magsat (c. 1980) (Langel and Hinze 1998). New data were needed to define the secular changes in the geomagnetic field, but also to improve the quality and the wavelength bandwidth of the presumed "static" lithospheric magnetic anomalies. The situation has changed dramatically in the last few years due to significant coordination and commitment of geomagnetists worldwide, leading to three high quality magnetic field satellites currently in orbit: Ørsted, SAC-C, and CHAMP. Of these, data from the CHAMP satellite, especially in its anticipated late low-altitude phase (< 300 km), will be most beneficial for improving the resolution of the lithospheric anomaly field because the field is attenuated as an inverse distance function and the minimum meaningful wavelength obtained from these data is closely associated with the observation altitude (e.g., see an example of increased amplitudes and improved resolution of anomaly data at 250 km altitude over the U.S. in Ravat and Purucker 1999). Although the orbits of Ørsted and SAC-C are higher than ideal for the analyses of lithospheric fields, their contribution toward deriving reliable lithospheric field signal from the CHAMP data cannot be overstated. Through synergistic utilization of data from these satellites, models of the external fields and the main field of the Earth of unprecedented precision may be derived—ultimately reducing the "noise" envelope of the CHAMP lithospheric magnetic anomaly data.
The role of CHAMP in shaping the bright future of the lithospheric magnetic field studies is clear: we anticipate many improvements in the determination of lithospheric magnetic field and its interpretation. Expectations may be summarized as improvements due to data redundancy and attitude determinations (for vector data), better external field models, increased anomaly resolution during the late-phase lower altitude part of the mission, development of new direct interpretation techniques and, taking advantage of the gravity field from the CHAMP mission, improved interpretations resulting from correlations with gravity anomaly data through the application of Poisson's relation. It is not our intention to develop fully all these ideas here, but to give only a flavour how these data could be utilised by geomagnetists, geophysicists, and also geologists in different parts of the world to their advantage.

2 The Background

For satellite altitude magnetic data to contribute substantively to our understanding of geology, they must provide information that other geological and geophysical data sets cannot. Or, at least do so more readily or inexpensively than other techniques. For many years, this very ability—the utility of the anomaly data—had been questioned because the data did not always agree with people's conception of geology and geophysics of a particular region. In terms of wide applicability of concepts in interpretation, on the continents, Michael Mayhew has demonstrated the significance of the inverse correlation of the anomalies or the derived equivalent source susceptibilities with the regions of high heat flow (e.g., Mayhew 1982). With Peter Wasilewski, he has also made important contributions in our understanding of magnetic petrology of the lower continental crust and shown anomaly/susceptibility correlations with the thickness of the crust (Mayhew et al. 1991; Wasilewski and Mayhew 1992). But generally satellite anomaly interpretations have been tentative and always seeking confirmation in other data sets—perhaps because there are indeed a large number of conditions that control the magnetisation of the Earth's lithosphere. However, as we reason below, the idea of finding confirmation with near-surface geologic knowledge developed from areally small geologic features itself is flawed. We will show in this paper that the apparent discrepancy in some regions of the satellite altitude magnetic anomalies and the near-surface geological and geophysical knowledge may be actually pointing to new, previously overlooked (or not understood) geologic information that is readily apparent in the satellite magnetic anomaly data.

Where on the Earth should we look for these new geologic insights? In order to improve our chances of finding useful interpretations, we must focus on geologic sources that primarily contribute in the anomaly wavelength band of 400-500 km to about 3000 km (the wavelengths of present satellite altitude anomaly data). The advantage here is that these are not the wavelengths normally retained accurately by compilations of marine magnetic and aeromagnetic surveys. And, with the exceptions of global-scale earthquake seismology, satellite gravity, and topography data, this is certainly not the scale of geologic mapping or the collection of many other geologically-oriented geophysical data sets. The utility of satellite altitude anomalies can therefore be demonstrated better for geological problems in this bandwidth.[1]

Another common difficulty with the satellite magnetic anomalies is that they represent effects from a multitude of geologic sources. Indeed, the geologic interpretation of these anomalies is less direct than what one is used to with high-resolution aeromagnetic surveys (e.g., Ravat 2000; Reeves 2001). But what is not appreciated is that only rarely do areally small geological magnetic effects coalesce to form large satellite altitude magnetic anomalies. One can demonstrate this easily by continuing upward to the altitude of satellite magnetic observations(about 400 km) the aeromagnetic anomalies of wavelengths lower than 500 km.[2] The upward continued anomalies from such a procedure result in amplitudes smaller than the noise envelope of present satellite altitude anomaly maps (i.e., < 1 nT) (see Ravat et al. 2002). Such examples demonstrate the futility of the direct comparison of small-scale near-surface geologic sources. Yet, upon the

---

[1] The concept of satellite data contributing to geologic knowledge in the wavelength band of 400 to 3000 km appears obvious now, but previously this was not a very well-understood concept because there is a large amount of literature on the interpretation of these anomalies through correlation of areally small near-surface geology.

[2] This cutoff includes the contribution of all *regionally small* geological sources in the Earth's lithosphere above the Curie isotherm—from igneous plutons in the upper crust to most ancient rifts, including the large Midcontinent rift in the U.S.A.

suggestion "take a look at what the satellite magnetic field suggests", the most common reaction of geoscientists is to seek confirmation in what they already know—that is, in the near-surface geologic and geophysical variations. Educating geoscientists is necessary in the proper use of these data to learn more about the subsurface.

There were also other issues that inhibited the due recognition of the utility of the satellite-derived anomalies for geologic interpretation, and the most prominent among them was noise—both observational and due to external fields. This situation has changed substantially with the verification of these anomalies independently from two recent high quality satellites: Ørsted (Purucker et al. 2002) and CHAMP (Maus et al. 2002).

In the remainder of the paper, we show the untapped potential of the magnetic anomaly data in understanding the regional-scale geologic makeup of the continents using an example of Middle Proterozoic provinces in the U.S. In retrospect, this was the most important case which undermined the utility of satellite altitude magnetic anomalies for geologic mapping in the minds of prominent geoscientists in the U.S., and therefore it is fitting to illustrate through this very example how the tide may be turning. Finally, we discuss the advantages of combined gravity and magnetic anomaly analysis and interpretation. But first, we discuss which one of the available POGO and Magsat satellite altitude magnetic anomaly maps likely preserve the most complete spectrum of the long-wavelength magnetic anomalies and, at the same time, is compatible with the regional-scale aeromagnetic anomalies. This aspect is important because certain direct interpretation techniques require high precision uninterrupted wavenumbers in an anomaly under investigation.

3 Completeness of the Long-wavelength Anomaly Spectrum and its Significance

Just as the reliability of the derived anomalies is important, so is its completeness. Having an incomplete or partial spectrum is like seeing only part of the picture and that can only add to the difficulties associated with the theoretical ambiguity in the interpretation of potential-field data. In the last few years, we have developed methods to address testing the issues of reliability and completeness using joint inversion of aeromagnetic and satellite altitude data (Ravat et al. 2002). Here, we show which one of the commonly available satellite altitude anomaly tations is likely to be most reliable using comparisons with Canadian long aeromagnetic profiles collected from 1969 to 1976, and what is the cause of the problems in the other representations compared. With the approach outlined in Ravat et al. (2002), we compared four satellite magnetic anomaly data sets over Canada: POGO anomalies of Langel (1990), and Magsat anomaly representations of Cain et al. (1990), Ravat et al. (1995), and Alsdorf et al. (1998). We have not had the opportunity to compare with this technique the satellite anomaly models from any of the recent comprehensive models from the Sabaka et al. (2000) approach or the CHAMP anomaly model (Maus et al. 2002).

Fig. 1.1 shows coherence (the degree of association between two data sets in a band of wavenumbers about a central wavenumber) between the Canadian aeromagnetic data and the above satellite anomaly maps using identical multi-taper parameters. It is clear from the comparison that the model of Cain et al. (1990, here with spherical harmonics up to degree 49) compares well with the aeromagnetic data above the wavelength of about 650 km with coherence of > 0.85. The other three satellite data sets compare poorly below the wavelength of 1000 km. The main reason for the poor performance of the data sets of Ravat et al. (1995) and Alsdorf et al. (1998) is that they have applied notch-filtering at multiple junctures and have removed all information from wavenumbers that did not correlate between subsets of their data (e.g., dawn and dusk or different altitude bins). The principal focus of the Ravat et al. (1995) study was to derive the most robust anomaly features and alleviate doubts regarding the reliability of the anomalies. Consequently, the study relied heavily on filtering the data when subsets did not highly correlate. These missing wavenumbers in both the above studies have been poorly reconstructed in a later gridding step, either the equivalent-source altitude normalisation or the collocation. Very likely, patchwork derivation of POGO anomalies and their reconstruction by the equivalent-source method, in addition to a higher original altitude of these data, is the cause of the poor performance of the POGO model. Since the Arkani-Hamed et al. (1994) model (not compared here) is constructed by combining above maps by Langel (1990) and Ravat et al. (1995),

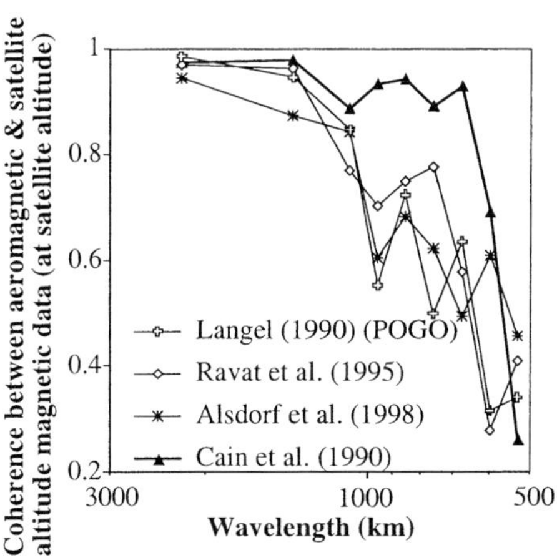

Fig. 1.1 Coherence in the range of wavenumbers from 2500 km to 500 km of the Canadian high-altitude aeromagnetic data and four different satellite altitude data sets computed at the satellite altitude. Based on studies of Ravat et al. (2002) and Wang et al. (2001), the Cain et al. (1990) anomaly models to spherical harmonic degree 60 retain all wavelengths down to about 425 km and also agree with long-profile aeromagnetic data.

it must also perform badly below 1000 km and should not be used to rigorously compare satellite magnetic anomalies below the wavelength of 1000 km.

We have made similar comparisons over the U.S. with Cain et al.'s (1990) Magsat anomaly model extending to spherical harmonic degree 60 and found that high coherence between that model and long aeromagnetic profile Project Magnet data over the U.S. extends to wavelengths of about 425 km (Wang et al. 2001). Clearly, if the Canadian and the U.S. long aeromagnetic profile data are correct, and our method of comparison considered different sources of errors in both aeromagnetic and satellite altitude data sets, the present yardstick for the *global* satellite magnetic anomaly data should be the degree 60 model of Cain et al. (1990), and not any other model. It would be interesting and valuable to compare the anomalies over North America from the recent versions of the comprehensive model and also CHAMP.

4 Delineation of a Mid-Proterozoic Continental Building Block in the Midcontinent of the U.S.

The igneous crust of North America has been analysed geochemically and geochronologically for a long time using surface and borehole samples. Through this work, a concept has emerged of different age provinces that have accreted to or collided with the earlier assemblies of North America (Hoffman 1989). In the U.S. particularly, the oldest province is the Archean Wyoming craton to which crusts formed elsewhere have accreted (Fig. 1.2a). Many of the rocks found in the reachable parts of these crusts have geochemical signatures of primary mantle melts and also reworking of their earlier crusts, and they have intruded the upper crust in widespread igneous events that gave rise to the Eastern Granite-Rhyolite province (1,470±30 Ma) and the Southern Granite-Rhyolite province (1,370±30 Ma) (EGR and SGR, respectively, in Fig. 1.2a). In the early 1980s, the problem for the interpretation of the satellite magnetic anomalies was that the boundaries of the magnetisation contrasts derived from the anomalies (one example of which is shown in Fig. 1.2b) and the boundaries of the Granite-Rhyolite provinces did not coincide—and for that matter, the southeastern boundary of these magnetisation highs did not coincide with anything known at that time.

In the last decade, however, new insights developed from the analyses of Nd isotopic data have identified crustal formation ages. With dense data in the regions of Precambrian basement outcrops and sparse data elsewhere, Van Schmus et al. (1996) have been able to distinguish regions of crustal formation of age older and younger than 1.55 Ga across a line known as the T_{DM} line (the dashed lines in Fig. 1.2). The boundary marked by the T_{DM} line is likely a reactivating fault because one can observe alignment of topographic features in Michigan, USA, and alignment of parts of Wabash river in Indiana, USA, in commonly available high-resolution shaded-relief topographic maps. There is a strong correlation between the T_{DM} line and the gradient of magnetisation contrast indicated by Magsat in this region. In fact, one could continue the interpretation of the T_{DM} line further to the southwest very easily on the basis of Magsat magnetisation model. The western

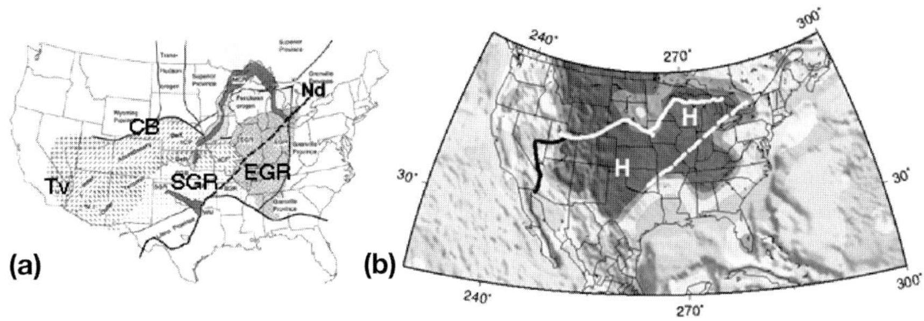

Fig.1.2 (a) Middle Proterozoic (1.9-1.6 Ga) crustal provinces in the U.S. are bounded on the north and west by Mojave Province (Tv), Cheyenne Belt (CB), and their northwestern limit continues up to the Killarney region (west of the letters Nd) of Ontario (Canada). SGR and EGR are Southern Granite-Rhyolite and Eastern Granite-Rhyolite provinces. Dashed line (Nd) represents inferred southeastern limit as defined by Nd isotopic data which bounds the Middle Proterozoic crustal provinces on its southeastern margin and divides SGR and EGR provinces (from Van Schmus et al. 1996). (b) Magnetic contrasts over the U.S. (see Ravat and Purucker (1999) for a color version of the global map). Hs mark the relative highs in the depicted susceptibility variation. Heavy lines show the Middle Proterozoic provinces in the U.S. Tv from part "a" is excluded from the boundary because presently this region is governed by the high heat flow related to the Basin and Range province which has decreased the magnetic thickness of the crust and hence it does not appear as a magnetic high.

and northern boundaries of this high magnetisation zone are also marked in part by Nd isotopic crustal formation age boundaries and, where the provinces have not been substantially affected at a later stage (e.g., high heat flow in the Rio Grande rift regions in New Mexico, USA, or the Basin and Range province in Nevada, USA), these isotopically determined boundaries also correspond well with the magnetisation boundaries. The magnetization of the remaining parts of the U.S. has been discussed in Langel and Hinze (1998) and will not be repeated here. The most important point about this example is not the new interpretation of the original crusts in the U.S., but it is about utilisation of these two pieces of information (isotopic model crustal formation ages and satellite altitude magnetic contrasts) for deciphering inexpensively the nature and evolution of crustal assemblies in large parts of the world. A refined picture from the CHAMP magnetic anomaly data from its late-phase lower altitude mission could substantially improve the present day resolution of this fundamental knowledge of the outer layers of the Earth.

5 Combined Gravity and Magnetic Anomaly Analyses

Perhaps the most important aspect of CHAMP for the interpretation the Earth's crust is the availability of both gravity and magnetic anomalies with the same near-Earth orbital characteristics. For certain geologic sources, the additional information will reduce (not eliminate) the non-uniqueness of potential-field interpretation. Since only a few studies have used extensively the correspondence (or lack thereof) between satellite-derived global gravity and magnetic fields to interpret regional-scale geologic features (e.g., von Frese et al. 1982, 2002), this section is somewhat speculative. We feel that because of the long-wavelength nature of the data, correlations may be used for the interpretation of a geologic province, and not identification of the causative source as is possible sometimes with the combined analysis of near-surface gravity and magnetic anomaly data. What kind of correlations (or lack thereof) may we expect? Perhaps it may be instructive to look at some scenarios. A regional-scale deep sedimentary basin under prevailing isostatic equilibrium may have no significant long-wavelength gravity anomaly (low densities of the sedimentary rocks are compensated by a mantle upwarp), but because the sedimentary rocks and the upwarped mantle are both non-magnetic (Wasilewski and Mayhew 1992), there will be a magnetisation low over such a basin. On the other hand, if such a basin were compensated by high density igneous rocks within the crust (e.g., assuming that the basin was initially formed by a large meteoritic impact and significant amount of mafic magmatism ensued due to decompression melting), the magnetisation and gravity lows of the sedimentary basin might be substantially offset by the high magnetic property and density of the intrusives. In another example, high mantle heat flow of active continental rifts would reduce magnetisable thickness of the crust and, if the lithosphere is thinned, there will likely be a long-wavelength gravity low. In a yet another example, when lower crustal high density–high seismic velocity mafic layers have magnetisation contrasts, but little density contrast, as in the example of 1.6-1.9 Ga provinces in the central U.S. discussed in the previous section, once again there will be non-correlated anomalies.

Finding geologic information that can provide new insights means we must first figure out what information is in agreement with the existing geophysical data sets. In the recent years, a number of other geological and geophysical global databases have become available (e.g., surface geology, sedimentary thickness, heat flow, seismic velocities and thicknesses in the crust and upper mantle). Interpretation of satellite potential-field data can significantly benefit from these constraints. The approach of starting out with *a priori* global crust and upper mantle models to assess what is in agreement has been applied separately with both global magnetic and gravity data sets (Purucker et al. 1998; Kaban et al. 1999).

Moreover, taking advantage of the above global gravity and magnetic models, one can explore the commonality of gravity and magnetic sources and presence of remanence in the magnetic sources using Poisson's relation and computing psuedo-gravity anomalies (see, e.g., Blakely 1995) from the derived magnetisation models. Comparison of observed gravity anomalies and psuedo-gravity anomalies will allow us to better isolate the lithospheric gravity anomaly field from deeper sources of gravity anomalies. Secondly, the non-correspondence of the isolated "crustal" gravity anomaly and the psuedo-gravity field might indicate the presence of remanently magnetised sources. Thus, there are many unexplored avenues in the area of joint interpretation of gravity and magnetic anomalies and these will significantly advance our knowledge of large-scale geologic variations on the Earth.

6 Conclusion

The most important contribution of CHAMP to geological knowledge will be recognising regional-scale sources of density and magnetic contrasts (with anomaly wavelengths > 400 km) that are difficult to investigate inexpensively through routine geologic, geochemical, and geophysical mapping either because of their scale, difficult terrain, or the great depth of the sources. The example of the utility of satellite magnetic data in delineating Middle Proterozoic provinces in the U.S., identified independently by Nd isotope model ages, demonstrates the untapped potential of these data in demarcating boundaries of such regional-scale deep crustal geologic units worldwide. Another advantage of these data is that they will contribute toward a more reliable fuller spectrum of anomalies that can be helpful in investigating deep crustal and upper mantle sources of geodynamical interest. Finally, it is hoped that these data will spur the development of new techniques to answer old questions, raise new important questions related to geodynamics, and contribute to our knowledge regarding the changing Earth.

Acknowledgements

We thank Prof. Colin Reeves for the review of the manuscript. DR also thanks Prof. Hermann Luehr for his invitation for a talk at the 1st CHAMP meeting. NASA provided the funding for this research.

References

Alsdorf D, Taylor P, von Frese RRB, Langel R, Frawley J (1998) Arctic and Asia Lithospheric Satellite Magnetic Anomalies. Phys Earth Planet Inter 108: 81-99

Arkani-Hamed J, Langel RA, Purucker M (1994) Scalar magnetic anomaly maps of Earth derived from POGO and Magsat data. J Geophys Res 99: 24075-24090

Blakely RJ (1995) Potential theory in gravity and magnetic applications. Cambridge University Press, Cambridge

Cain JC, Holter W, Sandee D (1990) Numerical Experiments in Geomagnetic Modeling. J Geomag Geoelectr 42: 973-987

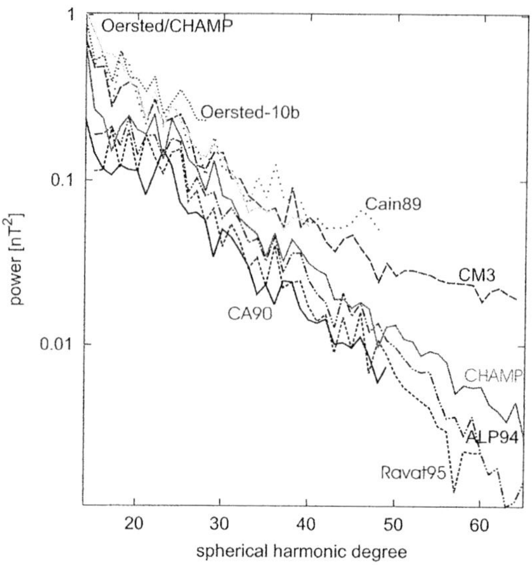

Fig. 1. Power Spectra of the total intensity anomaly (Arkani-Hamed et al., 1994) at 450 km altitude of several Magsat and POGO, Ørsted and CHAMP lithospheric field models

2 Class I: Models based on along-track filtered data

Cohen and Achache (1990) derive the potential of the lithospheric field using Magsat data up to spherical harmonic degree and order 14-50 as shown in Fig. 2. Instead of the index K_p, a 3 hour global index A_m, based on 24 observatories is used. To obtain the residual field, a main field model is subtracted from the measured field and then each component is averaged over 1-s interval to reduce high-frequency noise and the effect of external magnetic disturbances. A mean equatorial anomaly (MEA) is computed by averaging all corrected profiles plotted at constant dip latitude to correct for equatorial electrojects(EEJs). North-south filtering is done on the data set. After the least squares inversion of the spherical harmonic coefficients of the potential, the resulting discrepancies between dawn and dusk averages are less than 1 nT.

Arkani-Hamed et al (1994) present a scalar magnetic anomaly map(ALP94) from combined data sets of the POGO and Magsat satellites. Magsat scalar anomaly map used here was derived by Ravat et al. (1995) using Magsat dawn and dusk and POGO scalar anomaly map at an altitude of 400 km using POGO data. Two anomaly maps are derived based on two selection criteria: stringent and less-string ent. The stringent criteria seek to isolate anomaly signal of lithospheric origin by eliminating signals of remaining unmodeled main field, its secular variation and of external origin (ring currents), but at the expense of suppressing some lithospheric anomaly signal. The map shown in Fig. 3 is represented by spherical harmonic of degree 15-60. The map based on less-stringent criteria seeks to encompass maximum signals of lithospheric origin. The map is not shown here. Accuracy for the maps is estimated by the authors to be ± 2 nT.

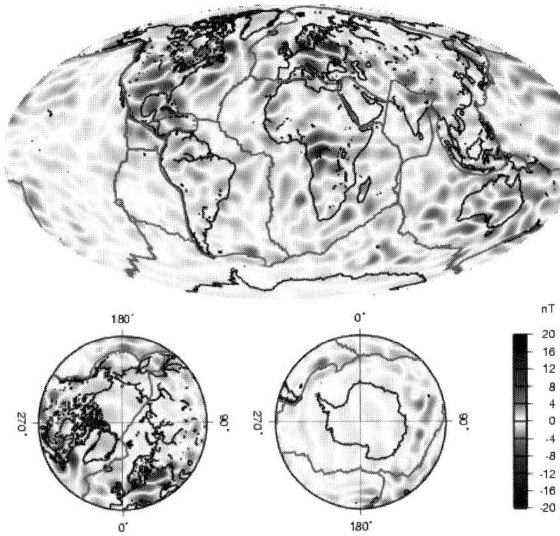

Fig. 2. Global scalar anomaly map prepared using Magsat vector data (Cohen and Achache, 1990)

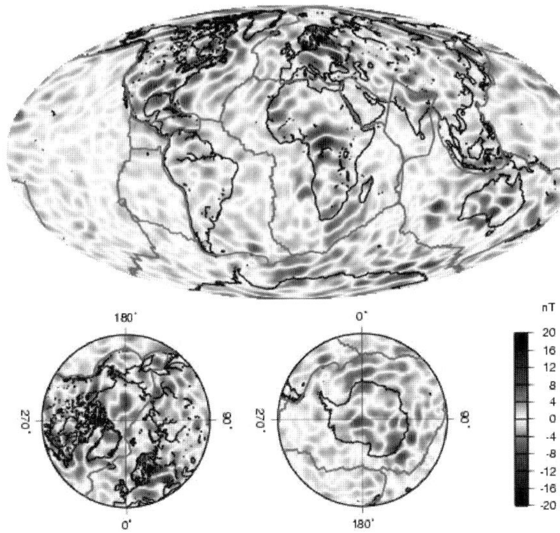

Fig. 3. Global scalar anomaly map prepared using Magsat dusk and dawn data and POGO satellite data (Arkani-Hamed et al, 1994)

Ravat et al (1995) Scalar anomaly map shown in Fig. 4 is derived using scalar Magsat dawn and dusk data for spherical harmonic degrees 15-65. The data selection for magnetic quiet days uses index K_p less than 2^+ for equatorial and mid-latitudes and AE index less than 50 nT for polar latitudes while the variance for the residual (ΔB) is kept at less than 80 $(nT)^2$. High pass Kaiser filter with a cut-off wavelength of 4000 km is applied, to remove the magnetic fields due to ionospheric current systems. To remove the inconsistencies between the near-by passes and between dawn and dusk data, pass-by-pass correlation followed by crossover analysis is carried out. Equivalent source dipole representations are used to generate the scalar anomaly map. Covariant spherical harmonic analysis is used to isolate the 'common features' of dawn and dusk data.

Maus et al (2002) The CHAMP satellite data were processed in 120^0 track segments at the mid-equatorial latitudes and 100^0 track segments at the poles. Mid-latitude data is selected between 22:00 – 6:00 LT. A planetary activity index K_p less than 2^+ is used for the selection of magnetically quiet days. After subtracting a recent Ørsted main and external field model, the remaining unmodeled large-scale external contribution is removed by fitting a homogeneous field and subtracting it on a track by track basis. In order to preserve N/S trending features, the data are not filtered further along-track. At the poles, a subset of quiet data with minimum root mean square (rms) is chosen for the analysis. Spherical harmonic coefficients 15-65 of the magnetic potential are estimated by a least squares minimum norm procedure. The anomaly map is shown in Fig. 5.

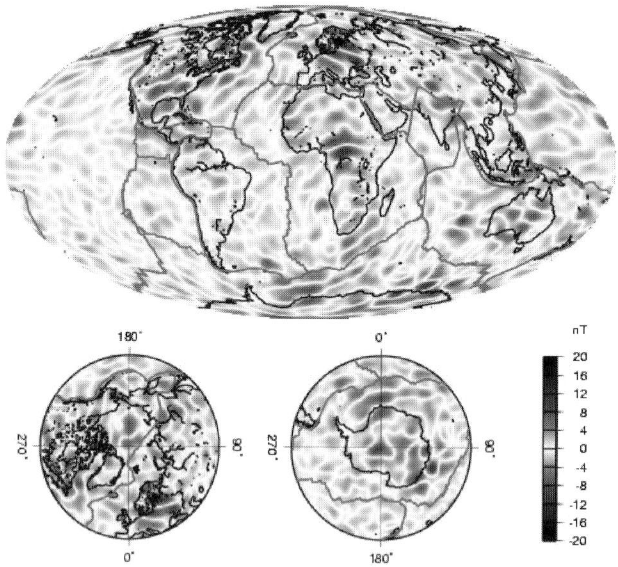

Fig. 4. Global scalar anomaly map prepared using Magsat scalar data (Ravat et al, 1995)

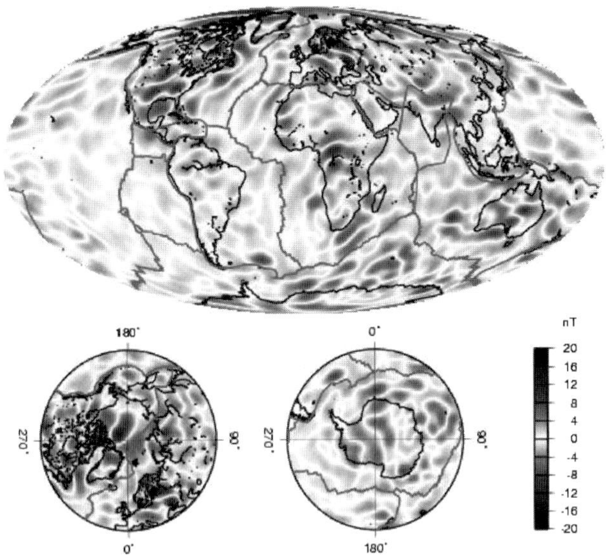

Fig. 5. Global scalar anomaly map prepared using CHAMP scalar data (Maus et al, 2002)

3 Class II: No filtering

The following set of scalar anomaly maps applies no filtering techniques to arrive at their final models.

Cain et al (1989a) Scalar magnetic anomaly map shown in Fig. 6 is derived for spherical harmonic degrees 14-63. A selection of observed Magsat vector data is taken and averaged over data blocks of 3^0 x 3^0 in size. Data are selected for intervals with K_p less than 2^+. After D_{st} correction, residuals are corrected for ionospheric currents. The spherical harmonic coefficients are derived up to n=63, however the spectrum computed by Cain et al (1989b) is claimed to be correct up to n=59, beyond which it is said to be plagued with noise. The average altitude of the anomaly map is 450 km., however no altitude correction to the data has been performed and this can be a source of error.

CM3 Model (2000) A new model as shown in Fig. 7, called CM3 (Compre- hensive Model: Phase 3) has been derived using the Magsat and Pogo satellite data and incorporating observatory hourly and annual means data. The spherical harmonic expansion is done to degree/order 14-65 in order to account for fields from the Earth's lithosphere at satellite altitude. Data selection procedure considers the index K_p less than 1^- and the Magsat dawn and dusk data are selected with $|D_{st}|$ less than 20 nT. Special quasi-dipole (QD) conforming harmonic functions have been used, including terms accounting for seasonal variation and variation with solar activity. Least squares inversion is performed to arrive at the comprehensive model. The fit to the data is much better than in its predecessor models.

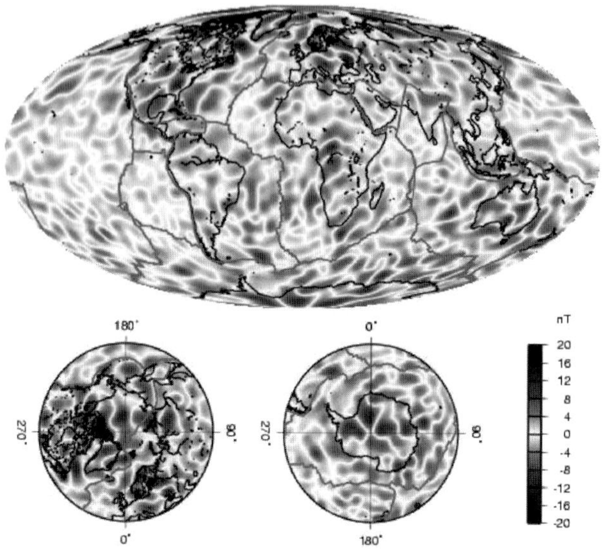

Fig. 6. Global scalar anomaly map prepared using Magsat vector data (Cain et al, 1989a)

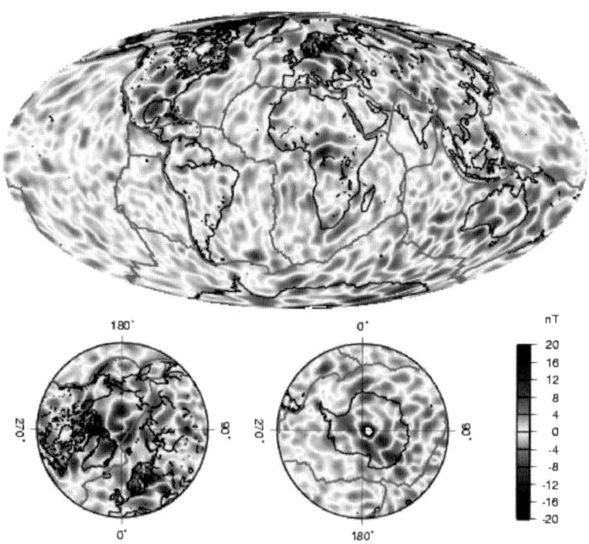

Fig. 7. Global scalar anomaly map prepared using Magsat vector and Pogo scalar data (Sabaka et al, 2000).

Olsen (2001) A spherical harmonic model, Ørsted_01c_o2, of the lithospheric field up to degree 41 and of the secular variation up to degree 13 is derived using Ørsted satellite data as shown in Fig. 8. The quiet data selection is based on index K_p equal to 1^+. At the polar caps data recorded with IMF (Inter-planetary magnetic fields) field $|B_y|$ greater than 3 nT is rejected. A modified D_{st} index derived from data of ground-based observations was used. Magnetospheric contributions are modeled up to degree/order 2 with zonal terms varying with annual and semi-annual periodicity. The model is estimated using Iteratively Re-weighted Least Squares with Huber weights to account for the non-Gaussian data error distribution. RMS misfit to the data is 3 nT. The model predicts scalar observations from CHAMP satellite very well with an rms misfit of 3.5nT at non-polar latitudes and about 6 nT at polar latitudes.

Over the coming years, a continuous stream of high quality magnetic data from several near Earth satellites will allow increasingly accurate simultaneous modeling of the main, lithospheric and ionospheric fields. These modeling results will help us remove the effects of the external sources and the main field contribution to the internal field to produce a more accurate global lithospheric field model. The new field models will thus provide a new basis for the study of crustal composition, dynamics and heat flow.

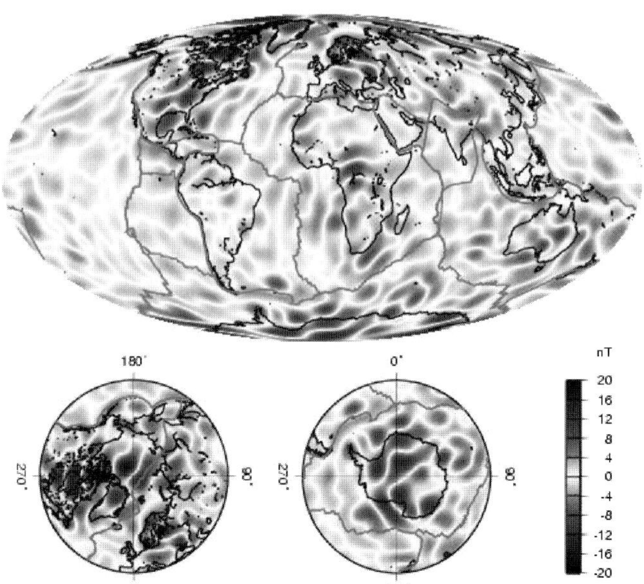

Fig. 8. Global scalar anomaly map prepared using CHAMP and Ørsted vector data (Olsen, 2001).

References

Arkani-Hamed J, Langel RA, Purucker ME (1994) Magnetic anomaly maps of Earth derived from POGO and Magsat data. J Geophys Res 99: 24075-24090.

Cain JC, Schmitz DR, Muth L (1984) Small-scale features in the earth's magnetic field observed by Magsat. J Geophys Res 89: 1070-1076.

Cain JC, Wang Z, Kluth C, Schmitz DR (1989a) Derivation of a geomagnetic model to n = 63. Geophys J 97: 431-441.

Cain JC, Wang Z, Schmitz DR, Meyer J (1989b) The geomagnetic spectrum for 1980 and core-crustal separation. Geophys J 97: 443-447.

Cohen Y, and Achache J (1990) New global vector anomaly maps derived from Magsat data. J Geophys Res 95: 10783-10800.

Langel RA (1990) Global magnetic anomaly maps derived from POGO spacecraft data. Phys. Earth Planet. Inter. 62: 208-230.

Maus S, Rother M, Holme R, Luhr H, Olsen N, Haak V (2002) First CHAMP satellite magnetic data resolve uncertainty about strength of the lithospheric magnetic field. Geophys Res Letts (in press)

Olsen N (2001) A model of the Geomagnetic Main Field and its Secular Variation for Epoch 2000 Estimated from Ørsted Data. Geophys J Int(in press)

Ravat D, Langel RA, Purucker M, Arkani-Hamed J, Alsdorf DE (1995) Global vector and scalar Magsat magnetic anomaly maps. J Geophys Res 100: 20111-20136.

Sabaka TJ, Olsen N, Langel RA (2000) A comprehensive model of the near-Earth Magnetic Field: Phase 3. NASA/TM-2000-209894, National Aeronautics and Space Administration, Goddard Space Flight Center, Greenbelt, Maryland.

Schmitz DR, Meyer J, Cain JC (1989) Modelling the Earth's geomagnetic field to high degree and order. Geophys J R Astr Soc 97: 421-430.

Mapping the Lithospheric Magnetic Field from CHAMP Scalar and Vector Magnetic Data

Stefan Maus, Kumar Hemant, Martin Rother and Hermann Lühr

GFZ, Div. 2, Telegrafenberg, 14473 Potsdam, *smaus@gfz-potsdam.de*

Summary. Eighteen months into the CHAMP mission there is now a seizable data set of scalar and vector magnetic measurements. With this data the lithospheric magnetic field can be mapped far more accurately than in previous missions.

Key words: geomagnetic field, magnetic anomalies, field models

1 Data Selection

We estimate two separate field models from the CHAMP scalar magnetic data of 9 Aug. 2000 to 30 Sep. 2001, and the vector data of 1 Jun. 2001 to 21-Dec. 2001. From all of the polar latitude data we select a 5% subset of tracks with particularly low residuals. For the lower latitudes, night time data are selected from 22:00 to 6:00 local time (LT), excluding the period of 19:00 to 22:00 LT in order to avoid the sporadic disturbances caused by F region currents [4]. Figure 1 shows the behaviour of mean scalar residuals with the K_p index, which is a three-hour-range index derived from 13 magnetic observatories quantifiyng the effect of solar particle influx on the magnetic field. We use only data at magnetically quiet times, quantified by $K_p \leq 2$.

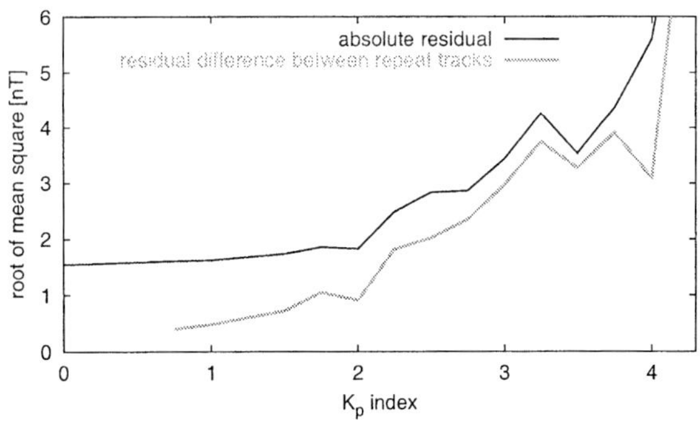

Fig. 1. Along track residual rms for night time low-latitude scalar data against the K_p index. We use only data with $K_p \leq 2$. The residual difference between repeat tracks can be extrapolated to around 0.1 nT for $K_p = 0$, indicating the extraordinary precision of the measurements. In contrast, the corresponding curves for the vector components (not shown here) tend to 1 nT for $K_p = 0$, due to star camera noise.

2 Correction for magnetospheric fields

After the standard main and external field corrections the magnetospheric signal remaining in the data is still of the order of the strength of the lithospheric field (Figure 2a). This is a key issue in lithospheric field mapping. The significant disagreement between earlier lithospheric field models is partly a consequence of using data which still contain significant external field signal. We remove this contribution by fitting and subtracting track by track an external field with the corresponding induced internal fields coupled by Langel's empirical factor of 0.27 to the external fields. This is done to degree-1 for the scalar and to degree-2 for the vector data. Subsequently, the corrected signal is clearly dominated by the lithospheric field (Figure 2b).

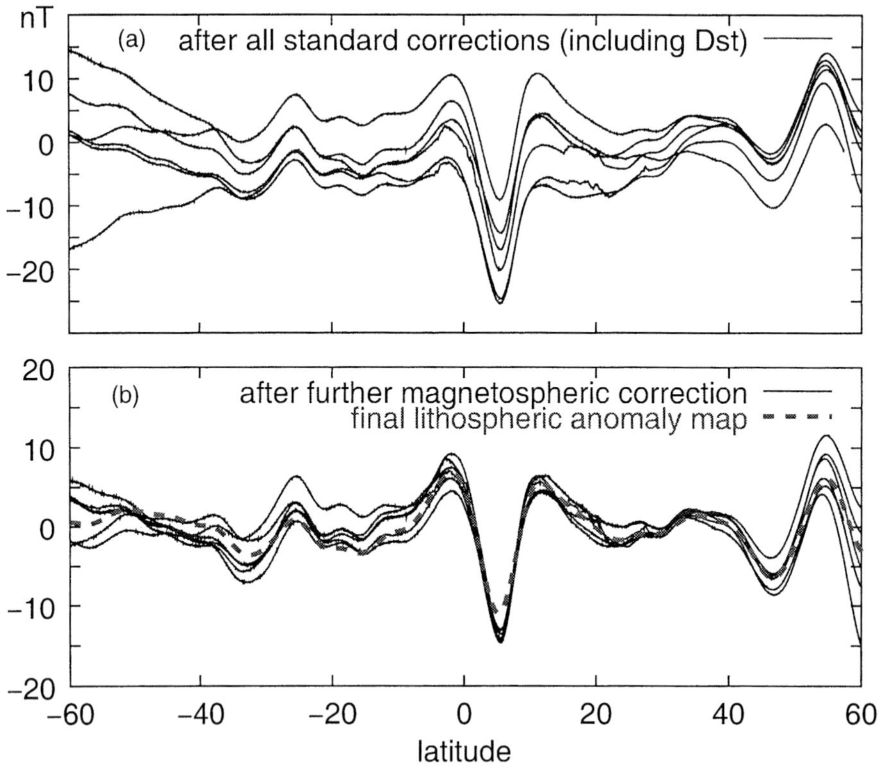

Fig. 2. Night time sample tracks at 20.0°-20.7° longitude. After all standard corrections the remaining residuals (a) are still dominated by external fields. In order to retain as much along-track signal as possible, we only subtract a best fitting homogeneous field from each track (b). Note the signature of the well known Bangui anomaly at 5°N. Underestimation of this peak by our final model (dashed) is due to truncation at spherical harmonic degree 65.

3 Discarding noisy tracks

After the previous corrections, some noisy tracks remain. They can be identified and eliminated by plotting the along-track residual rms against the longitude at the equator crossing (Figure 3).

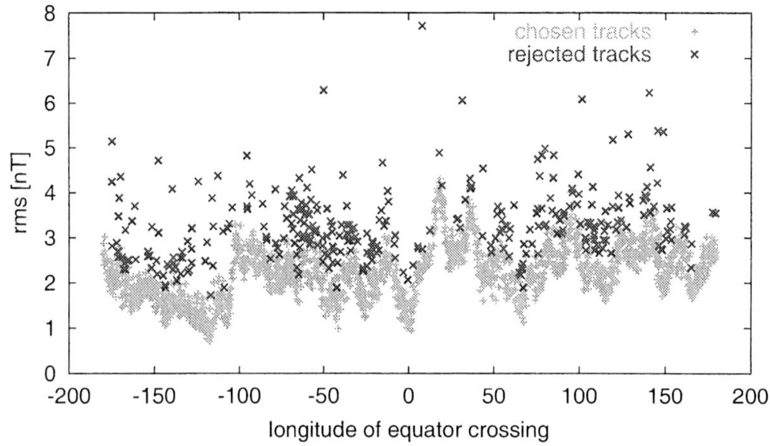

Fig. 3. Noisy tracks are identified and removed using a 4° sliding window in longitudes

4 Field model estimation

A field model is conveniently represented by the coefficients of the spherical harmonic expansion of the scalar magnetic potential. These coefficients can be estimated by a least squares inversion. We estimate separate models from the scalar and the vector data in order to get an idea of the uncertainties involved. Later, a joint model will be generated. In the case of scalar-only data, a minimum norm solution is sought, due to the ambiguity caused by the Backus effect.

5 Comparison with other field models

Apart from visual comparisons, which are given in a companion paper in this volume, there are statistical tools for gaining insight into the differences between models. The first is the spherical harmonic power spectrum (Figure 4). The strong differences in power between field models reflect the ambiguity between magnetospheric and ionospheric contamination on the one side and along-track trending lithospheric anomalies on the other. The models with higher power rely on carefully selected but

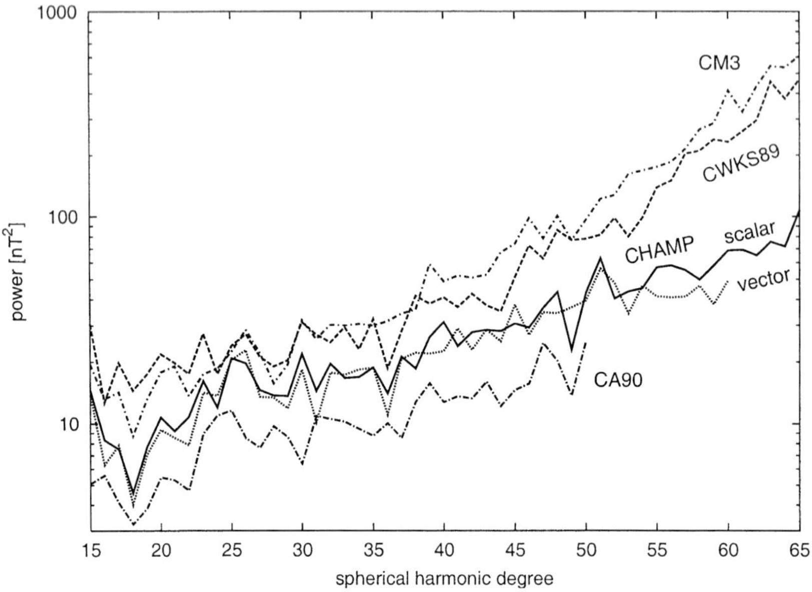

Fig. 4. Magnetic power spectra of CA90 [3], CWKS89 [2], CM3 [7], and our CHAMP scalar and vector field models. The main reason for the upward slope in the power spectra is the peculiar definition of the Mauersberger/Lowes spectrum, which slopes upward for a spatially uncorrelated magnetic field [6]

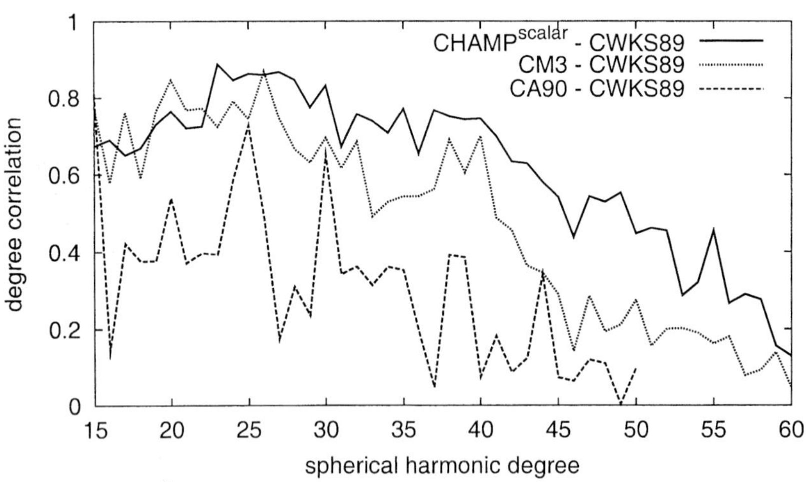

Fig. 5. Correlations of model CWKS89 with models CM3, CA90 and our scalar field model.

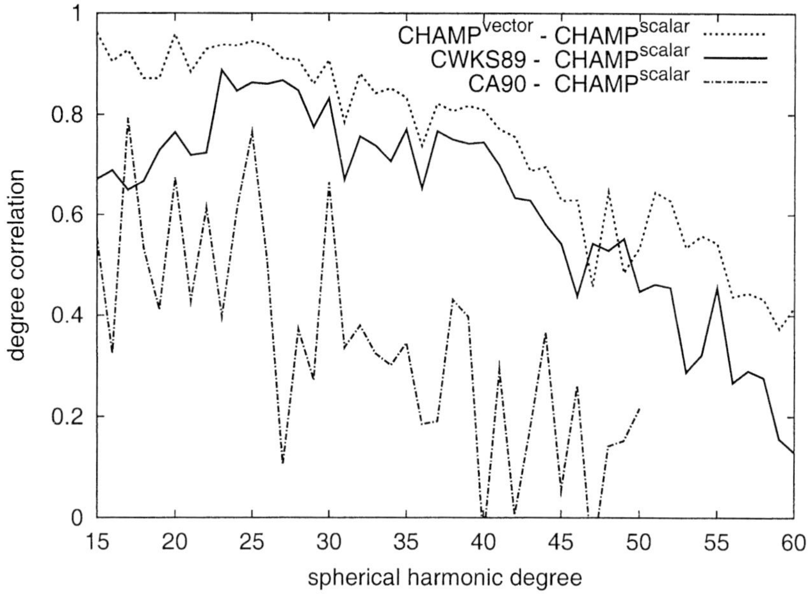

Fig. 6. Correlations of our scalar field model with CWKS89, CA90 and our vector field model.

unfiltered data, while the lower power models rely on various techniques of track-by-track external field removal. The true strength of the lithospheric field must lie somewhere in between.

Further insight is provided by the average correlation between the coefficients of two models, separately for each degree [1]. Two sets of degree correlation curves are shown in Figures 5 and 6. As may have been expected from the power spectra, the correlation between different models is rather poor. Interestingly, our model from scalar-only CHAMP data has the highest correlation with CWKS, despite the uncertainty known as Backus' effect (Figure 5). The higher correlation among the CHAMP scalar and vector models (Figure 6) is partly a consequence of using a similar processing scheme for both kinds of data. But it also shows that significant gains in accuracy are possible with the new CHAMP data.

6 A map of B_z at 450 km altitude

Since the scalar-only model has already been published elsewhere [5], we show our preliminary vector-only model in Figure 7.

Acknowledgement. This study was carried out under the grant of the German Federal Ministry of Education and Research (BMBF) No. 03F0333a.

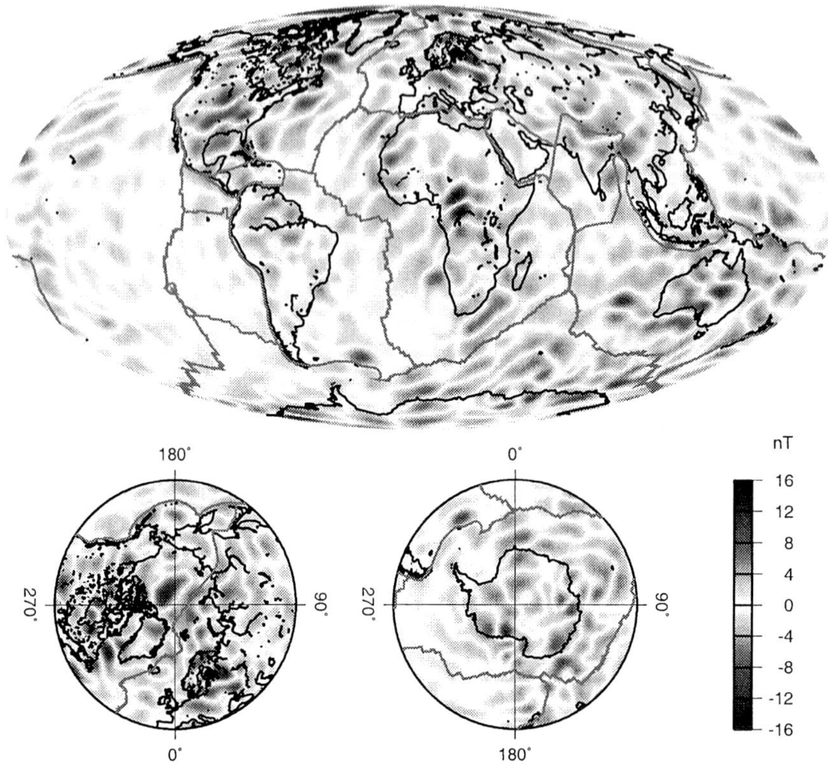

Fig. 7. Downward (B_z) component at 450 km altitude of our preliminary lithospheric field model derived from vector data only.

References

1. Arkani-Hamed J, Langel RA, and Purucker M (1994) Scalar magnetic anomaly maps of Earth derived from POGO and Magsat data, *J Geophys Res*, *99*, 24,075–24,090.
2. Cain JC, Wang Z, Kluth C and Schmitz DR (1989) Derivation of a geomagnetic model to n = 63, *Geophys J Int*, *97*, 431–441
3. Cohen Y, and Achache J (1990) New global vector magnetic anomaly maps derived from Magsat data, *J Geophys Res*, *95*, 10,783–10,800.
4. Lühr H, Maus S, Rother M and Cooke D (2002) First in situ observation of night time F region currents with the champ satellite, *Geoph Res Lett, in print*.
5. Maus S, Rother M, Holme R, Lühr H, Olsen N, and Haak V (2002) First scalar magnetic anomaly map from CHAMP satellite data indicates weak lithospheric field, *Geoph Res Lett, in print*.
6. Maus S (2001)*New statistical methods in gravity and magnetics*, Habilitation thesis, University of Braunschweig, http://www.gwdg.de/~smaus/habil.pdf.
7. Sabaka TJ, Olsen N, and Langel RA (2000) A comprehensive model of the near-Earth magnetic field: phase 3, *Technical Report, NASA/TM-2000-209894*.

Improving the Definition of Cratonic Boundaries Utilizing the Lithospheric Magnetic Field derived from CHAMP Observations

Michael E. Purucker[1] and Nils Olsen[2]

[1]Raytheon ITSS at Geodynamics Branch, NASA/GSFC, Greenbelt, MD 20771 USA
[2]Danish Planetary Institute, Juliane Maries Vej 30, DK-2100, Copenhagen, Denmark

Summary. We confirm the utility of a global non-satellite-based magnetic model of induced magnetizations in the terrestrial lithosphere by comparison in the North American region. We discuss the extension of the model to other regions, pointing out where inadequacies exist in our knowledge. Finally, we suggest some possible improvements that would make the model less subjective and more automatic.

Key words: Geomagnetism, lithosphere

1 Introduction

The launch of CHAMP in July of 2000 represented the first return of a magnetic field satellite to sub-500 km altitudes since Magsat flew in 1979 and 1980. The low altitude of the satellite makes it ideal for studies of the lithospheric magnetic field, and we expect that models will improve dramatically over the course of the mission.

2 The Global Magnetization Model

The global magnetization model that we utilize is identical with the one in Purucker et al. (2002). To review, the model contains both an induced and remanent magnetization component. The induced component is based on an estimate of the volume of magnetic crust from the 3SMAC (3-dimensional Seismological Model A Priori Constrained, Nataf and Ricard, 1996) and a sediment thickness model of Laske (1997). Magnetic susceptibilities of 0.035 SI are assigned to continental crust and 0.04 SI to oceanic crust. Figure 1 (top left) shows the scalar or total magnetic field (the magnetic field in the direction of the Degree 1-13 field) associated with the induced model. The remanent component of the model is based on the oceanic lithospheric model of Dyment and Arkani-Hamed (1998), calibrated in the South Atlantic (Purucker and Dyment, 2000). Figure 1 (top right) shows an example of the scalar magnetic field associated with the remanent model. The sum of the scalar field from the induced and remanent models is shown in Figure 1 (center left) in a region centered over North America. Because the field from the core obscures the lithospheric magnetic field at low degrees, we discard degrees between 1 and 14, as well as degrees unresolved by the satellite observations (Figure 1, center right).

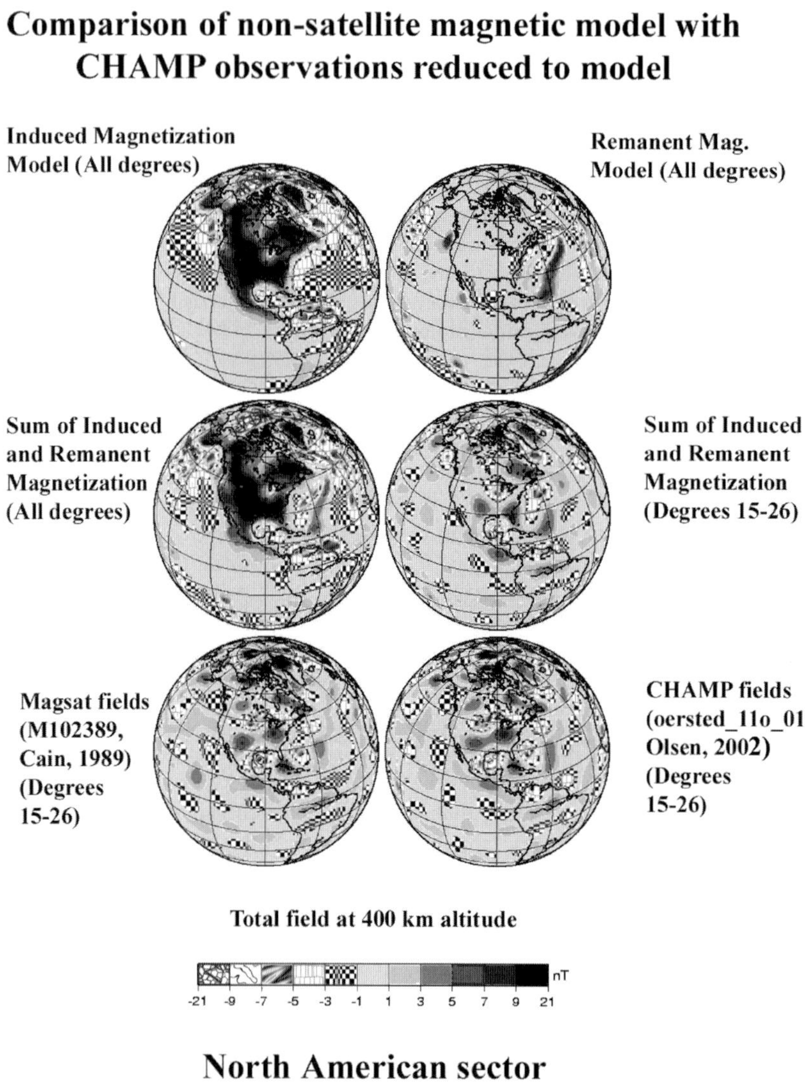

Fig. 1. Maps of the scalar or total magnetic field (the magnetic field in the direction of the inducing field) shown at an altitude of 400 km using an orthographic projection centered at 90 degrees West, 30 degrees North

3 New satellite magnetic field observations

The new CHAMP observations have been reduced to a spherical harmonic model (Figure 1, bottom right) utilizing techniques described in Olsen (2002). Geomagnetically quiet, night-time vector and scalar observations from CHAMP and Ørsted were utilized in making the model (named Oersted (11o(01)). Data selection criteria are described in Olsen et al (2000). For comparison, a map of the magnetic field from Magsat (Years : 1979-1980) derived by Cain et al (1990) is shown. As was the case for the Ørsted observations (Purucker et al., 2002), the CHAMP observations reduced to model show an almost one-to-one correspondence to the model's features. The model is closest to the observations over North America and the North Atlantic Ocean.

4 Extension to other cratonic regions

South America and adjacent South Atlantic Ocean

The pattern of magnetic anomalies in the South American model (Figure 2, top row) is created by a pair of cratons centered in Venezuela and Brazil, separated by an Amzonian region of younger, thinner crust. The weakness of the total field anomalies, compared with North America, is a result of 1) the factor of two difference in the inducing field between the two regions, and 2) smaller cratonic blocks. The weaker features in southern South America, where thick cratonic blocks are absent, are not as well fit by the model. Sea-floor spreading anomalies in the South Atlantic Ocean, previously recognized by Purucker and Dyment (2000), are again recognized in the CHAMP observations reduced to model.

Southern Africa

The pattern of magnetic anomalies in Africa (Figure 2, second from top row), in contrast to the situation in the adjacent oceans, shows a poor match to the model. The observations show substantial anomalies while the predicted field is very weak. This may be do the presence of a substantial remanence component (Whaler and Langel, 1996), which our global model does not account for on the continents. It is also probably due to inadequacies in the 3SMAC model of crustal thickness, which do not correspond well to our knowledge of cratonic regions in Africa (Kevin Burke, personal communication). For example, the 3SMAC model has included no information on the ancient cratonic terranes of southern Africa and only limited information on cratonic regions in central Africa.

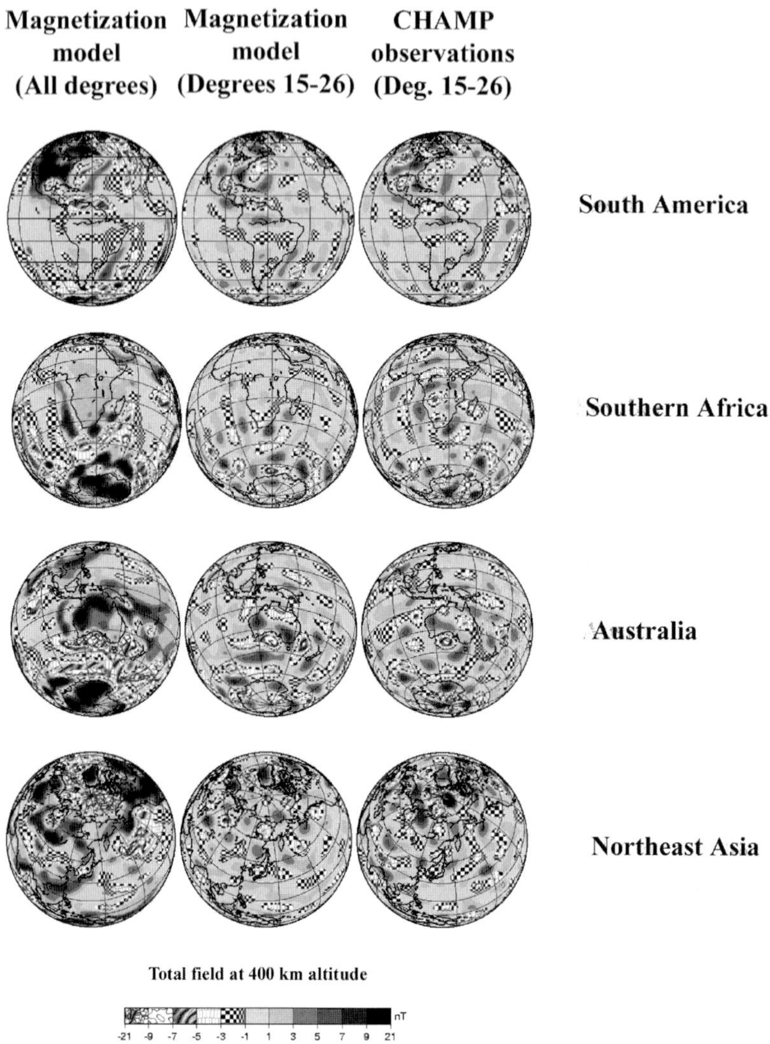

Fig. 2. Maps of the total magnetic field predicted from the model (left two columns) with the CHAMP observations reduced to model (right column) for four cratonic regions. The maps are shown at 400 km using orthographic projections

Australia

The pattern of magnetic anomalies in Australia (Figure 2, third from top row) represents a good match to the model, probably because of the maturity of our geological knowledge there and the incorporation of that information in the 3SMAC model. The comparison in the oceanic realm is not as satisfactory, particularly south and west of Australia, where our knowledge of the pattern of seafloor spreading is incomplete.

Northeast Asia

Neither the continental nor oceanic regions adjacent to northeast Asia (Figure 2, bottom row) shows a good agreement with the model. In part this may be due to anomaly control by other features, such as metamorphic grade (Sobolev, 1978 ; Rajaram and Purucker, in preparation), and the presence of remanence. Although some first-order oceanic features, like the western edge of the Cretaceous Quiet Zone near 30 degrees North, 180 East, correspond well, many others do not.

5 Discussion

In order to better fit the satellite observations to the global non-satellite magnetic model, we have previously utilized an iterative inverse approach, combined with hand-editing of the model in areas where the wavelength components of the observation and model do not match. A more automatic and objective approach would be to utilize the rubber-sheeting approach common in photogrammetry (Clarke, 1990), using geometric transformations to match 'control' points. This would begin with the assignment of individual features in the model to the satellite observations, and the subsequent generation of a surface that maps one into the other. This surface could then be propagated to the underlying model and the new model could be tested against the observations. Because of the non-linear nature of the parameterization, several iterations would probably be required. The presence of two layers (induced and remanent magnetization) in the model will complicate the application of this method, and it may only be possible to apply the method where one or the other layer is clearly dominant, such as in the interior of continents or oceans.

References

Cain JC, Holter B, Santee D (1990) Numerical experiments in geomagnetic modeling. J Geomag Geoelectr 42. 973-987.

Clarke KC (1990) Analytical and Computer Cartography, Prentice-Hall, Englewood Cliffs, New Jersey

Dyment J, Arkani-Hamed J (1998) Contribution of lithospheric remanent magnetization to satellite magnetic anomalies over the world's oceans. J Geophys Res 103. 15423-15441.

Laske G, Masters G (1997) A global digital map of sediment thickness (abstract). EOS Trans. AGU. 78, Fall Meeting Suppl, F483.

The relationship between the magnetic field due to the crust measured at satellite altitude and crustal magnetisation is expressed by the following integral (Parker et al., 1987):

$$\Delta B^{(\eta)}(\mathbf{r}_j) = -\hat{\mathbf{I}}_j^{(\eta)}.\nabla_{\mathbf{r}_j} \int_V \mathbf{H}(\mathbf{r}_j, \mathbf{s}).\mathbf{M}(\mathbf{s})dV, \tag{1}$$

where $\Delta B^{(\eta)}(\mathbf{r}_j)$ is a satellite anomaly datum at position $\mathbf{r}_j$, $^{(\eta)}$ denotes either the X-, Y- or Z- component, $\mathbf{H}(\mathbf{r}_j, \mathbf{s}) = \frac{\mu_0}{4\pi}\nabla_s \frac{1}{|\mathbf{r}_j - \mathbf{s}|}$, and $\mathbf{s}$ is position within the volume V of the magnetized lithosphere. $\hat{\mathbf{I}}_j^{(\eta)}$ is a unit vector in the direction of the appropriate orthogonal anomaly field component, and the subscript on the del operator indicates which set of variables differentiation is with respect to.

This relationship can be linearised by noting that the crustal field is of small amplitude compared to that originating in the core, giving

$$\Delta B = \frac{1}{B_c} \sum_{\eta} \Delta B^{(\eta)} B_c^{(\eta)}, \tag{2}$$

where the subscript c denotes the main field, and summation is over the three orthogonal components. ΔB is the intensity of the scalar field, and B_c that of the core field. By inserting (1) into (2) for each $\Delta B^{(\eta)}$, we can form an integral relationship between our scalar anomaly data, and the quantity of interest, the magnetisation of the crust, which I write symbolically as

$$\Delta B(\mathbf{r}_j) = \int_V \mathbf{G}(\mathbf{r}_j, \mathbf{s})\mathbf{M}(\mathbf{s})dV,$$

where $\mathbf{G}$ is the known data kernel. We invert these data using minimum norm modelling, following the work of Parker et al. (1987), Jackson (1990) and Whaler and Langel (1996), to find the best fitting magnetisation minimising $\int_V \mathbf{M}^2 dV$. The solution is expressed as

$$\mathbf{M}(\mathbf{r}) = \sum_{j=1}^{N} \alpha_j \mathbf{G}(\mathbf{r}, \mathbf{r}_j), \tag{3}$$

where there are N data. $\boldsymbol{\alpha}$ is the solution of

$$(\Gamma + \lambda \mathbf{I})\boldsymbol{\alpha} = \Delta \mathbf{B}, \tag{4}$$

where $\Delta\mathbf{B}$ is the data vector (length N), Γ is the $N \times N$ Gram matrix whose elements are given by

$$\Gamma_{ij} = \int_V \mathbf{G}(\mathbf{r}_i, \mathbf{s})\mathbf{G}(\mathbf{r}_j, \mathbf{s})dV, \tag{5}$$

and λ is a damping parameter controlling the relative importance of fit to the data and smoothness of the model.

The inversion makes no assumption as to magnetisation direction. Most previous methods assume magnetisation is in the direction of the main field. The extent to which this is a good approximation depends on the relative importance of induced and remanent magnetisation, on the amount of subsequent movement and rotation of crustal rocks after their remanent magnetisation has been frozen in, and on the extent to which the main field has changed since this happened. It is an inappropriate assumption for large areas of the oceanic basins, where magnetisation is dominated by magnetic stripes. However, the severe non-uniqueness in deducing magnetisation from external measurements of the magnetic field, as exemplified by Runcorn (1975), means that interpretation of all models must be undertaken with caution.

For datasets of typically tens of thousands of points, direct implementation of this method is not practical. Whaler and Langel (1996) followed the depleted basis approach of Parker and Shure (1983), in which the solution is expressed as a linear combination of data kernels at only a subset of data points. This reduces the numerical problem from solving an $N \times N$ linear system to solving one of dimension the number of depleted basis points. Whaler and Langel (1996) chose to concentrate basis points over a continental-sized region. However, the positions of the depleted basis points are subjective, and involve a loss of resolution. To avoid this problem, and enable global solutions to be produced, I have used the method of Purucker and Whaler (1999). Only the largest (in magnitude) elements of the $N \times N$ Gram matrix given by (5) are treated as non-zero; the sparse linear system given by (4) can then be solved using the iterative conjugate gradient algorithm. This is possible because satellite data are predominantly sensitive to the crustal magnetisation in a small diameter disk centred on the point directly below the satellite. The result of this is that the Gram matrix is numerically banded, with large elements concentrated on and near the leading diagonal. A similar approach was adopted by Purucker et al. (1996) to calculate global equivalent dipole solutions from satellite magnetic data.

Langel and Whaler (1996), using a depleted basis to model scalar anomaly data, were able to express the solution as a linear combination of any set of convenient basis functions (O'Brien and Parker, 1994). For mathematical simplicity, they chose to expand the solution in terms of the data kernels for vertical component data. Here, I must use the appropriate basis functions for scalar data, expressions for which, in terms of the basis functions for vector data, are given by Langel and Whaler (1996).

2 Data and numerical implementation

I used a $1^\circ \times 1^\circ$ grid calculated from Maus et al.'s (2002) spherical harmonic expansion of the total intensity anomaly between degrees 14 and 65 at a mean altitude of 438km., shown elsewhere in this volume. This gave 64619 data (excluding the points at the poles); they were assumed to have equal uncertainty, and thus the data were not normalised to unit variance prior to inversion. The core field model used was that of Olsen (2002).

I employed Jacobi preconditioning to improve convergence of the conjugate gradient algorithm, whereby each Gram matrix element Γ_{ij} was normalised by $\sqrt{(\Gamma_{ii}\Gamma_{jj})}$, such that all diagonal elements are unity. The data are also appropriately normalised, and the preconditioning factor is removed from the solution vector after its calculation.

The majority of the computational effort is expended on calculating the Gram matrix. Fortunately, this part of the calculation parallelises well: for a given datum point $\mathbf{r}_i$, the program then calculates the Gram matrix element Γ_{ij} for every datum point $\mathbf{r}_j$, and this can be distributed across a number of processors. The inefficiency of recalculating elements whose values could be determined from the symmetry properties of the Gram matrix is far outweighed by the advantage of parallelisation. Once a row of the Gram matrix has been calculated, the program returns to serial mode, and those elements above the specified threshold are retained in compressed row storage form (see Press et al., 1992, for details). The conjugate gradient algorithm has also been parallelised, although this is not a particularly time-consuming part of the calculation. The code was parallelised using OpenMP. On a Sunfire 6800 system, efficiencies of up to 0.9 were achieved, though speed-up drops off significantly beyond about 8 processors.

3 Results

Figure 1 shows preliminary results, presented as plots of the component of magnetisation parallel to the main field, perpendicular to the main field in the meridian plane, and perpendicular to the meridian plane, at the Earth's surface. The picture at depth is almost identical, with a slow decrease in amplitude (as found by Whaler and Langel, 1996). This is imposed by the form of the data kernels; satellite data are unable to resolve depth variations in such a thin layer from heights of several hundred kilometres. The thickness of the magnetised layer was set at 40km; previous studies show that altering it changes the amplitude of the magnetisation such that the vertically-integrated magnetisation is the same. The model shown in figure 1 was obtained by retaining 19.3M, or 0.46%, of the Gram matrix elements. Typically, the conjugate gradient algorithm converged within 100 iterations (it is guaranteed to converge within N). I have investigated the effect of changing the damping parameter, λ, in (4). The primary effect is to alter the amplitude of the deduced magnetisation; the pattern is robust. For the solution shown in figure 1, the root-mean-square (rms) magnetisation is 2.8 Am^{-1}, and the rms misfit 1.7 nT. The results compare favourably with those obtained over Africa and the continental USA by Whaler and Langel (1996), and the global models of Purucker and Whaler (1999). Note that the amplitudes of the components of magnetisation parallel to the main field and perpendicular to the

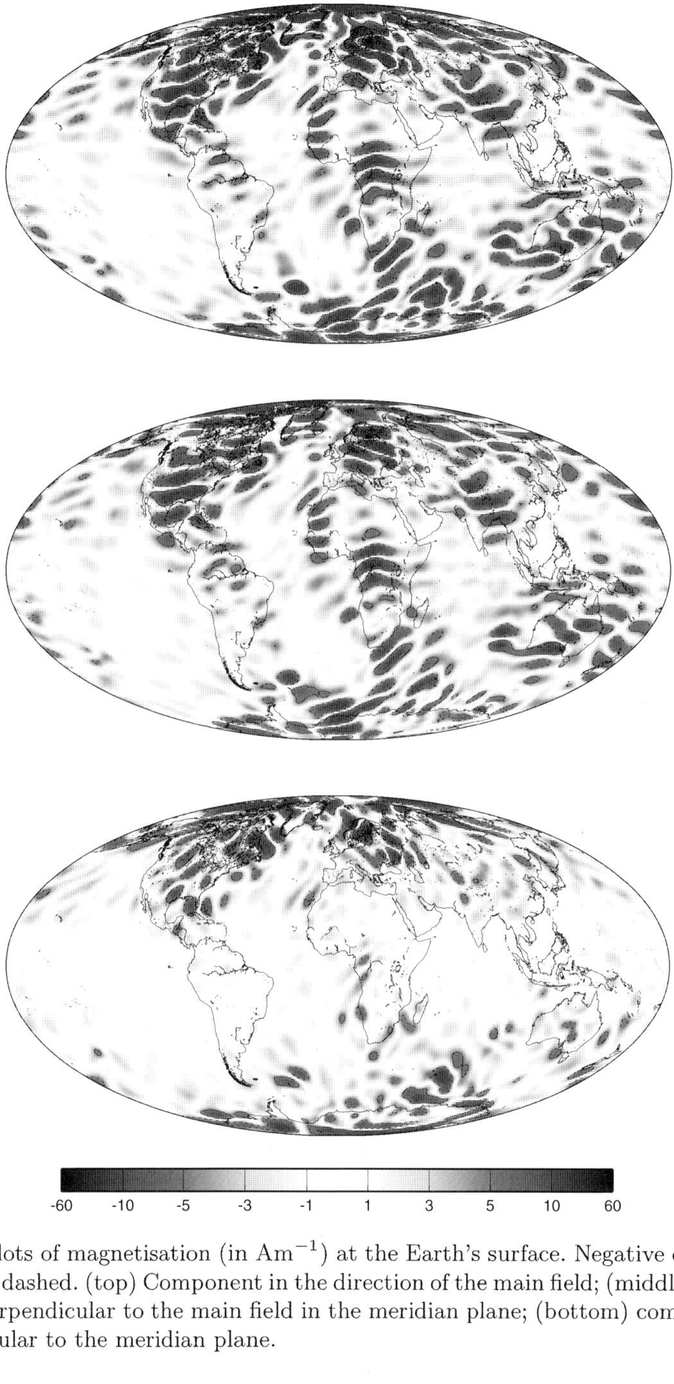

Fig. 1. Plots of magnetisation (in Am^{-1}) at the Earth's surface. Negative contour values are dashed. (top) Component in the direction of the main field; (middle) component perpendicular to the main field in the meridian plane; (bottom) component perpendicular to the meridian plane.

main field in the meridian plane are roughly equal, so that magnetisation for this model is not predominantly induced. The observation by Whaler and Langel (1996) that the component of magnetisation perpendicular to the meridian plane is small over Africa is true over much of the globe, including over oceanic regions.

Figure 1 shows high-amplitude anomalies near the poles, particularly prominently for the component of magnetisation perpendicular to the meridian plane that is otherwise small, for which there is no evidence in the data. Their amplitude varies little with damping parameter, so they are less noticeable for lightly damped models whose mean amplitude is higher. Similar high amplitude 'features' appear near the poles in magnetisation maps for Mars deduced from vector Mars Global Surveyor data inverted using a similar scheme. Thus, they are unlikely to be the result of linearisation to treat scalar data, and I think the most likely cause is retaining insufficient elements of the Gram matrix. Future work will test this. Other planned developments include the use of binned CHAMP scalar anomaly data, rather than data synthesised from Maus et al.'s (2002) spherical harmonic model, recently provided by S Maus (personal communication). The binning process gives a standard deviation on each scalar anomaly datum that can be used to weight it during inversion, and a mean altitude for each bin.

4 Acknowledgements

The application of the conjugate gradient algorithm has been developed in collaboration with Michael Purucker of NASA's Goddard Space Flight Center, and the parallelisation in conjunction with the University of Edinburgh Parallel Computing Centre, with contributions from Magnus Hagdorn, Adam Carter and Josephine Beech-Brandt. Stefan Maus provided the grid of scalar anomaly values. Nils Olsen provided the coefficients of his Ørsted 10B model in advance of publication.

References

Jackson A (1990) Accounting for crustal magnetization in models of the core magnetic field. Geophys J Int 103:657-673

Langel RA, Whaler KA (1996) Maps of the magnetic anomaly field at Earth's surface from scalar satellite data. Geophys Res Lett 23:41-44

Maus S, Rother M, Holme R, Lühr H, Olsen N, Haak V (2002) First scalar magnetic anomaly map from CHAMP satellite data indicates weak lithospheric field. Geophys Res Lett, accepted for publication.

O'Brien MS, Parker RL (1994) Regularized geomagnetic field modelling using monopoles. Geophys J Int 118:566-578

Olsen N (2002) A model of the geomagnetic field and its secular variation for epoch 2000 using Ørsted data. Geophys J Int 149:454-462

Parker RL, Shure L (1982) Efficient modeling of the Earth's magnetic field with harmonic splines. Geophys Res Lett 9:812-815

Parker RL, Shure L, Hildebrand JA (1987) The application of inverse theory to seamount magnetism. Rev Geophys 25:17-40

Press WH et al (1992) Numerical Recipes. Cambridge University Press, Cambridge, England

Purucker ME, Sabaka TJ, Langel RA (1996) Conjugate gradient analysis: A new tool for studying satellite magnetic data sets. Geophys Res Lett 23:507-510

Purucker M, Whaler K (1999) Global models of lithospheric magnetization from satellite magnetic anomaly data sets. EOS Trans AGU 80:F281

Runcorn SK (1975) On the interpretation of lunar magnetism. Phys Earth planet Inter 10:327-335

Whaler KA, Langel RA (1996) Minimal crustal magnetizations from satellite data. Phys Earth planet Inter 98:303-319

Multiscale Downward Continuation of CHAMP FGM-Data for Crustal Field Modelling

Thorsten Maier, Carsten Mayer

Geomathematics Group, University of Kaiserslautern, PO-Box 3049, 67653 Kaiserslautern

Summary. Crustal field determination and downward continuation from CHAMP-FGM-data will be presented within a multiscale framework using scalar as well as vectorial radial basis functions. The basic idea is to formulate the problem in terms of integral equations relating the radial or tangential projections of the geomagnetic field at satellite height with the magnetic field at the Earth's surface. In both, the radial as well as the tangential case, crustal field downward continuation turns out to be an exponentially ill-posed problem. As an appropriate solution method multiscale regularization in terms of scalar and vectorial wavelets is presented based on the knowledge of the singular systems corresponding to the aforementioned integral operators. Finally, first wavelet results of crustal field contributions at Earth level, calculated from CHAMP-FGM-data, will be presented.

Key words: multiscale regularization, wavelets, crustal field, downward continuation

1 Introduction

Lithospheric magnetisation gives rise to a geomagnetic field contribution that can be mapped by appropriately low-flying satellites. The German geoscientific satellite CHAMP is a recent low-earth-orbiter designed, among other tasks, for such lithospheric studies. CHAMP provides us with highly accurate scalar (Overhauser magnetometer) as well as vectorial (Fluxgate magnetometer, FGM) data on an almost circular orbit with almost global coverage. While the scalar data are commonly used to derive magnetic anomaly maps at satellite height, the vectorial data are usually used to derive magnetic anomaly maps at Earth's level, a process commonly referred to as downward continuation.

We present a multiscale technique for the downward continuation of vectorial data. To do so, the connection of the geomagnetic field at Earth's level as well as at satellite height is described using integral equations with kernel functions providing us with explicitly known corresponding singular systems. In fact, this has been done before (e.g. [1]). However, the authors did not take advantage of the explicit knowledge of the singular systems but numerically performed a truncated singular value decomposition. From a mathematical point of view downward continuation is an exponentially ill-posed problem and the numerical solution of the respective integral equations requires sophisticated means of regularization. We use the known singular systems to construct regularization wavelets in order to solve the integral equations within a scalar and vectorial multiscale framework, i.e. in a locally adaptive manner. This, however, enables us to make use of the localization properties of the wavelets and, therefore, gives us the possibility to perform local computations with spatially varying resolution. In more detail, our approach can retain finer details where the data distribution warrants it, without introducing artefacts in areas of poorer data coverage. Consequently, the method can be applied to data sets with heterogeneous distribution as well as to comparatively evenly distributed data. This, however, is not the subject of this article (cf. e.g. [3] and [2] for more details). Additionally, the multiscale framework enables us to use fast tree-algorithms for the efficient computations of the occuring convolutions. While in [2] a general approach to spaceborne geodata can be found, further reading and a complete treatment of the present subject will be available in the upcoming [5]. Details on the implementation and first numerical results can be found in [8].

2 Mathematical Formulation

For the sake of brevity we will assume the spherical approximation to be valid, i.e. we suppose the Earth's surface Ω_{ρ_1} as well as the satellite's orbit Ω_{ρ_2} to be spheres of fixed radii $\rho_1, \rho_2 \in \mathbb{R}^+$, $\rho_1 < \rho_2$, centered around the origin. We furthermore suppose the pre-Maxwell equations to be valid, i.e.

$$\nabla \cdot b = 0,$$
$$\nabla \wedge b = \mu_0 j,$$

where b denotes the magnetic induction, j is the electric current density and μ_0 is the vacuum permeability. If the data used for modeling are suitably chosen and carefully preprocessed (see e.g. [6] and Section 3) we might also neglect the electric currents between the Earth's surface and the satellite's orbital sphere, i.e. $j \approx 0$ in the spherical shell $\Omega_{(\rho_1,\rho_2)}$. This assumption, however, results in the existence of a scalar potential U with $\Delta U = 0$, such that $b = -\nabla U$ in $\Omega_{(\rho_1,\rho_2)}$.

In what follows, V_{ρ_1} will denote a (scalar or vectorial) function on the Earth's surface Ω_{ρ_1}, representing those quantities which we want to calculate from the given satellite data. For instance, V_{ρ_1} may be chosen to be the product of the geomagnetic vertical component (i.e. Z component) by the radius, it may be the geomagnetic potential U or $\nabla^* U$ (∇^* denotes the surface gradient), i.e. a linear combination of the horizontal geomagnetic (i.e.

North (X), East (Y)) components. In the following, the function G_{ρ_2} will denote a (given) function on the satellite's orbit Ω_{ρ_2} and will represent, as far as this article is concerned, the product of the geomagnetic vertical component by the radius. It should be remarked that we can also formulate the problem for given horizontal components but we omit this approach for the sake of brevity. Finally, we suppose V_{ρ_1} and G_{ρ_2} to be elements of separable Hilbert spaces $\mathcal{H}(\Omega_{\rho_1})$ and $\mathcal{K}(\Omega_{\rho_2})$, respectively, of square-integrable functions (scalar or vectorial, see Theorem 1). A connection between V_{ρ_1} and G_{ρ_2} can be established via an integral equation:

Theorem 1. *Let $V_{\rho_1} \in \mathcal{H}(\Omega_{\rho_1})$ and $G_{\rho_2} \in \mathcal{K}(\Omega_{\rho_2})$ be given as specified above. Then the integral operator*

$$\Lambda : \mathcal{H}(\Omega_{\rho_1}) \to \mathcal{K}(\Omega_{\rho_2})$$

connecting V_{ρ_1} and G_{ρ_2} via

$$(\Lambda V_{\rho_1})(x) = G_{\rho_2}(x) = \int_{\Omega_{\rho_1}} V_{\rho_1}(y) K(y,x) \; d\omega_{\rho_1}(y), \quad x \in \Omega_{\rho_2} \tag{1}$$

is characterized by the kernel function

$$K(\cdot,\cdot) = \sum_{n=0}^{\infty} \sum_{k=1}^{2n+1} \sigma_n H_{n,k}(\cdot) K_{n,k}(\cdot).$$

$\{H_{n,k}\}$ and $\{K_{n,k}\}$, $n \in \mathbb{N}_0$, $k = 1, \ldots, 2n+1$, are complete orthonormal systems in $\mathcal{H}(\Omega_{\rho_1})$ and $\mathcal{K}(\Omega_{\rho_2})$, respectively. $\{\sigma_n\}$, $n \in \mathbb{N}_0$, are the singular values of Λ. The system

$$\{\sigma_n, H_{n,k}, K_{n,k}\}$$

is called the singular system of Λ. The resulting Hilbert spaces and singular systems are given as follows:

V_{ρ_1}	$\mathcal{H}(\Omega_{\rho_1})$	$H_{n,k}$	$\mathcal{K}(\Omega_{\rho_2})$	$K_{n,k}$	σ_n
Z *on* Ω_{ρ_1}	$\mathcal{L}^2(\Omega_{\rho_1})$	$Y^{\rho_1}_{n,k}$	$\mathcal{L}^2(\Omega_{\rho_2})$	$Y^{\rho_2}_{n,k}$	$\left(\frac{\rho_1}{\rho_2}\right)^n$
$\nabla^* U$ *on* Ω_{ρ_1}	$l^2_{(2)}(\Omega_{\rho_2})$	$y^{(2),\rho_2}_{n,k}$	$\mathcal{L}^2(\Omega_{\rho_1})$	$Y^{\rho_1}_{n,k}$	$\left(\frac{\rho_1}{\rho_2}\right)^n \sqrt{\frac{n+1}{n}}$
U *on* Ω_{ρ_1}	$\mathcal{L}^2(\Omega_{\rho_1})$	$Y^{\rho_1}_{n,k}$	$\mathcal{L}^2(\Omega_{\rho_2})$	$Y^{\rho_2}_{n,k}$	$\left(\frac{\rho_1}{\rho_2}\right)^n \frac{n+1}{\rho_2}$

A proof of the previous theorem can be found in [5], similar results are presented in [4]. For the sake of completeness it should be mentioned, that $\left\{Y^{\rho_1}_{n,k}\right\}$ and $\left\{Y^{\rho_2}_{n,k}\right\}$ are scalar spherical harmonics normalized on Ω_{ρ_1} and Ω_{ρ_2}, respectively. The system $\left\{y^{(2),\rho_2}_{n,k}\right\}$ is a system of vector spherical harmonics given by $y^{(2),\rho_2}_{n,k} = (n(n+1))^{-1/2} \nabla^* Y^{\rho_2}_{n,k}$, $n \in \mathbb{N}$, $k = 1, \ldots 2n+1$. The space $l^2_{(2)}(\Omega_{\rho_2})$ is the completion of the span of the system $\left\{y_{n,k}^{(2),\rho_2}\right\}$ with respect to the l^2-norm. For more details about vector spherical harmonics and the corresponding Fourier theory see [3].

In our setting, the process of downward continuation or, more explicitly, the calculation of the geomagnetic field vector (or potential) at Earth's level from the vertical component at satellite height, is equivalent to solving equation (1) for V_{ρ_1}. The solution of (1) can, formally, be found using the Moore-Penrose inverse

$$\Lambda^{-1}G_{\rho_2} = \sum_{n=0}^{\infty}\sum_{k=1}^{2n+1} (\sigma_n)^{-1} G_{n,k} H_{n,k}, \tag{2}$$

with $G_{n,k} = (G_{\rho_2}, K_{n,k})_{\mathcal{K}(\Omega_{\rho_2})}$, where $(\cdot,\cdot)_{\mathcal{K}(\Omega_{\rho_2})}$ denotes the inner product in $\mathcal{K}$. For arbitrary functions $G_{\rho_2} \in \mathcal{K}(\Omega_{\rho_2})$ (note that, for example, the product $Z\rho_2 = G_{\rho_2}$ obtained from satellite measurements might be affected by observational errors) the right hand side of (2) is not necessarily convergent. Since we know from Theorem 1 that the values of σ_n^{-1} diverge exponentially with n, the problem is exponentially ill-posed. To force convergence of (2) we have to replace the series by a filtered version. We will do this by means of regularization wavelets (note that we just present concrete examples and omit the general definitions and results), i.e.:

Theorem 2. *Let G_{ρ_2} be defined as above. Let the decomposition and reconstruction regularization wavelets at scale j, i.e. Ψ_j and $\tilde{\Psi}_j$, be defined by*

$$\Psi_j(\cdot,\cdot) = \sum_{n=0}^{\infty}\sum_{k=1}^{2n+1} \Psi_j^{\wedge}(n) H_{n,k}(\cdot) K_{n,k}(\cdot),$$
$$\tilde{\Psi}_j(\cdot,\cdot) = \sum_{n=0}^{\infty}\sum_{k=1}^{2n+1} \Psi_j^{\wedge}(n) H_{n,k}(\cdot) H_{n,k}(\cdot).$$

For a bandlimited variant let the corresponding symbols be given by (truncated singular value decomposition filter):

$$(\Psi_j)^{\wedge}(n) = \begin{cases} 0 & \text{for } n = 0, \ldots 2^j - 1 \\ \sigma_n^{-1/2} & \text{for } n = 2^j, \ldots, 2^{j+1} - 1. \\ 0 & \text{for } n > 2^{j+1} - 1 \end{cases}$$

For a non-bandlimited variant let the symbols be given by (Tikhonov filter):

$$(\Psi_j)^{\wedge}(n) = \left(\frac{\sigma_n}{\sigma_n^2 + \gamma_{j+1}^2} - \frac{\sigma_n}{\sigma_n^2 + \gamma_j^2} \right)^{1/2},$$

with $\lim_{j\to\infty}\gamma_j = 0$. Then, for the J-level approximations

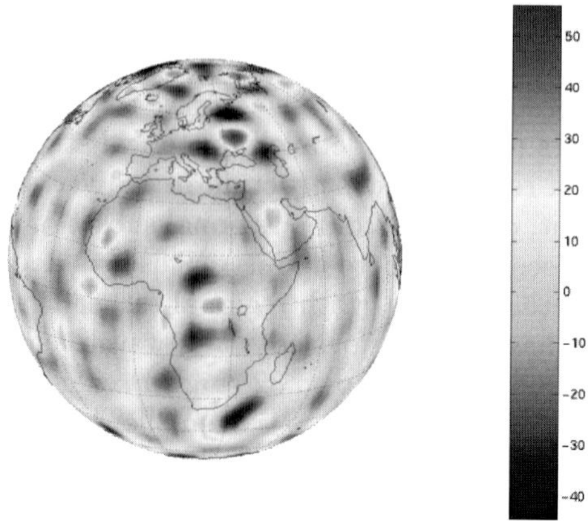

Fig. 2. Z component of crustal field at mean Earth radius [nT]. Calculated from a synthetic FGM data set

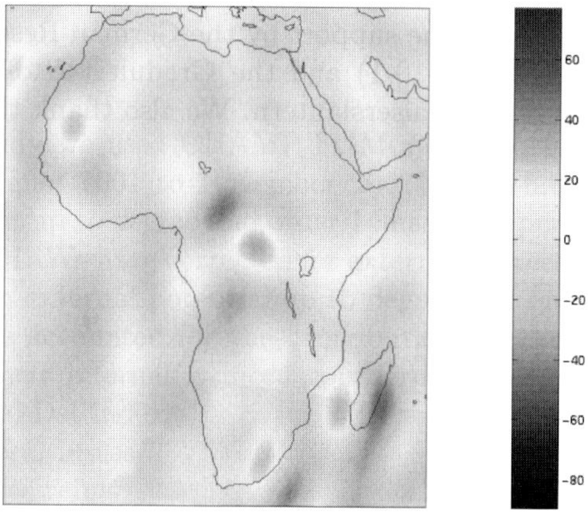

Fig. 3. Regional Z component of crustal field at mean Earth radius [nT]. Calculated from CHAMP FGM-data

References

1. Achache J, Abtout A, Le Mouel J-L (1987) The Downward Continuation of MAGSAT Crustal Anomaly Field over Southeast Asia. J Geophys Res, 92, pp 11.584-11.596
2. Freeden W (1999) Multiscale Modelling of Spaceborne Geodata. BG Teubner, Stuttgart, Leipzig
3. Freeden W, Gervens T, Schreiner M (1998) Constructive Approximation on the Sphere (With Applications to Geomathematics). Oxford Sciences Publication. Clarendon Press

4. Freeden W, Glockner O, Thalhammer M (1999) Multiscale Gravitational Field Recovery from GPS-Satellite-to-Satellite Tracking. Studia Geoph Geod, 43, pp 229-264
5. Maier T (2002) Multiscale Geomagnetic Field Modelling from Satellite Data: Theoretical Aspects and Numerical Applications. Doctoral Thesis, Geomathematics Group, University of Kaiserslautern (in preparation)
6. Maus S, Rother M, Holme R, Lühr H, Olsen N, Haak V (2002) First CHAMP Satellite Magnetic Data Resolve Uncertainty About Strength of the Lithospheric Magnetic Field. Geophys Res Lett (submitted)
7. Olsen N (2002) A Model of the Geomagnetic Field and its Secular Variation for Epoch 2000 Estimated From Ørsted Data. Geophys J Int (in press)
8. Zimmermann S (2001) Multiskalenanalyse des Krustenfeldes aus Satellitendaten. Diploma Thesis, Geomathematics Group and Computational Material Science Group, University of Kaiserslautern

CHAMP Enhances Utility of Satellite Magnetic Observations to Augment Near-Surface Magnetic Survey Coverage

Hyung Rae Kim[1], Ralph R.B. von Frese[1], Patrick T.Taylor[2], and Jeong Woo Kim[3]

[1]Dept. of Geological Sciences, The Ohio State University, Columbus, OH 43210 USA,
[2]Geodynamics Branch, NASA/GSFC, Greenbelt, MD 20771 USA,
[3]Dept. of Earth Sciences, Sejong University, Seoul 143-747, ROK

Summary. Regional to continental scale magnetic anomaly maps are becoming increasingly available from airborne, shipborne, and terrestrial surveys. Satellite data can be used to help fill the coverage gaps in regional compilations of these near-surface surveys. For the near-surface Antarctic magnetic anomaly map being produced by the Antarctic Digital Magnetic Anomaly Project (ADMAP), we show that near-surface magnetic anomaly estimation is greatly enhanced by the joint inversion of the near-surface data with the CHAMP observations relative to the data from the Magsat and Ørsted missions that respectively suffer from lower measurement accuracy and higher orbital altitudes. The analysis also suggests that considerable new insights on the magnetic properties of the lithosphere may be revealed by a further order-of-magnitude improvement in the accuracy of the magnetometer measurements at minimum orbital altitude.

Key words: Magnetic anomaly continuation, joint inversion, Antarctica, ADMAP.

1 Introduction

We investigate the utility of satellite magnetometer observations to augment coverage gaps in near-surface anomaly compilations of terrestrial, airborne and shipborne magnetic survey data. The nature of the problem is shown by the near-surface magnetic anomalies compiled by the Antarctic Digital Magnetic Anomaly Project (ADMAP) in Figure 1 [3]. Here regional coverage gaps are evident that presently are only constrained by satellite magnetic observations.

In the next section we develop an approach for using the satellite magnetic data to fill in coverage gaps in the near-surface data based on the joint inversion of the two data sets. Specifically, by the use of simulations we study the relative utilities of the Magsat, Ørsted, and CHAMP observations for this application. Our analysis finds that the CHAMP data will be particularly suitable because the measurement accuracy is an order-of-magnitude greater than Magsat's and the orbital altitudes are much lower than Ørsted's.

2 Simulations

For our analysis, we considered the near-surface magnetic anomalies of the Wed-

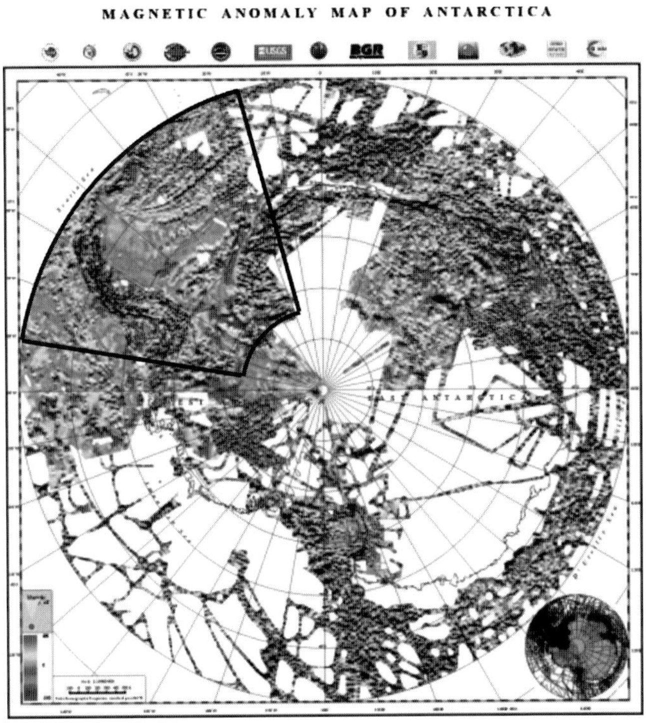

Fig. 1. Near-surface total field magnetic anomalies (nT) of the Antarctic with the Weddell Sea study area highlighted. Lower right insert shows the survey line coverage

dell Sea sector in Figure 1. These anomalies were low-pass filtered for 400 km wavelengths and larger that are likely to be detected at satellite altitudes of 350 km and higher. The low-pass filtered anomalies were then related by least squares matrix inversion to the volume magnetic susceptibilities of a series of crustal blocks based on the Antarctic crustal thickness model from [11] using spherical coordinate Gauss-Legendre quadrature integration [9]. Figure 2.A gives the inversion predictions that match the input low-pass filtered magnetic anomalies with negligible error.

With this inversion model we now consider the problem of estimating near-surface anomaly values within the coverage hole or gap outlined by the white border in Figure 2. For example, using a conventional interpolation based on minimum curvature [1; 8], we obtain the results shown in Figure 2.B. Quantitative measures of the fit of the interpolated values to the 'true' values within the gap from the crustal magnetic model include the root-mean-squared (RMS) difference and correlation coefficient (CC) that are listed in Table 1.

For the 400-km altitude Magsat simulation in Figure 2.C, the inversion model was evaluated to nearest 3 nT to simulate the magnetometer measurement errors [5]. The Magsat anomalies were then combined with the near-surface anomalies

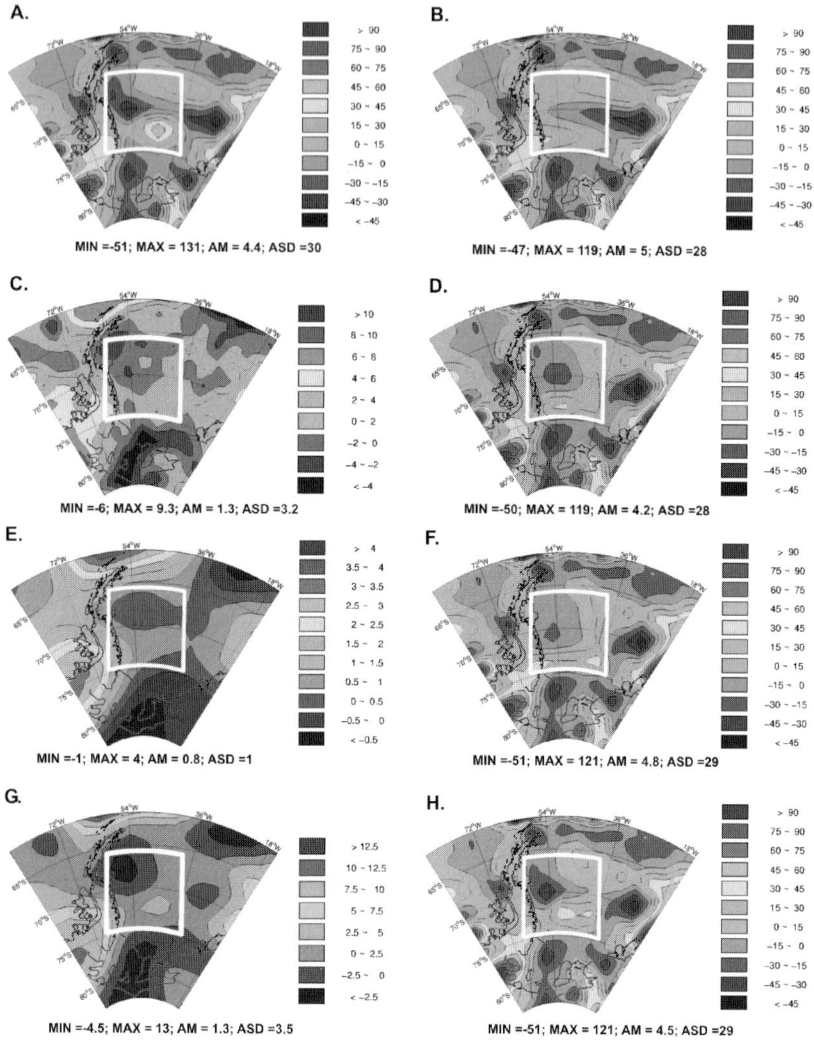

Fig. 2. Map A shows the 'true' near-surface total field anomaly values (nT) at 2 km above sea level from our crustal magnetic model. Map B gives the near-surface anomalies where the estimates within the regional coverage gap marked by the white border were obtained by minimum curvature from the surrounding near-surface anomalies. Simulated Magsat (map C at 400 km), Ørsted (map E at 700 km), and CHAMP (map G at 350 km) satellite magnetic data from our crustal magnetic model were used to estimate the near-surface anomalies within the gap in maps D, F, and H, respectively. Listed attributes in nT for each map include the minimum (MIN) and maximum (MAX) amplitudes, the amplitude mean (AM), and amplitude standard deviation (ASD).

outside the gap by joint inversion to obtain the gap predictions shown in Figure 2.D with the RMS difference and correlation listed in Table 1.

Figures	2.B	2.D	2.F	2.H
RMS	108.7	98.5	74.5	32.1
CC	0.34	0.51	0.81	0.93

Table 1. Performance statistics for using miminum curvature (2.B) and Magsat (2.D), Ørsted (2.F), and CHAMP (2.H) magnetic anomalies to fill a gap in aeromagnetic anomaly coverage. The prediction statistics include the root-mean-square (RMS) difference in nT and the correlation coefficient (CC).

The joint inversion, formulated after [7], is based on the matrix system $\mathbf{AX} = \mathbf{B}$. Here, the design matrix $\mathbf{A} = [a1\ a2]^T$ involves coefficients a1 and a2 related to the geometric properties between the crustal source blocks and the near-surface and satellite altitude magnetic observations, respectively. The observation vector $\mathbf{B} = [b1\ b2]^T$ includes the near-surface and satellite anomalies b1 and b2, respectively. The weighted mean-square-error solution vector $\mathbf{X}$ of magnetic susceptibilities for the crustal source blocks is given by $\mathbf{X} = [\mathbf{A}^T\mathbf{PA} + (EV)\mathbf{I}]^{-1}\ [\mathbf{A}^T\mathbf{PB}]$, where $\mathbf{I}$ is the identity matrix, $\mathbf{P}$ is the weight matrix, and the scalar EV is error variance to enhance the stability of the inversion [10]. For this study, the working precision of our algorithms was sufficiently great that no weights were required and we set $\mathbf{P} = \mathbf{I}$. However, trade-off diagrams are necessary to establish optimal values of EV for a solution $\mathbf{X}$ that will yield effective anomaly predictions in the coverage gap.

Figure 3 illustrates the EV trade-off diagram for the simulated Magsat anomaly predictions obtained in Figure 2.D. Here, the correlation coefficients and RMS differences from the predictions of the solutions for various values of EV are compared. These results are plotted for the near-surface anomalies outside and within the gap by solid and dashed curves, respectively. In a real application, we will only be able to estimate the trade-off diagram for the near-surface anomalies outside the gap, but these results pretty much mirror the performance of the solution within the gap as suggested in Figure 3. Specifically, the predictions in Figure 2.D with minimal RMS difference and maximum CC relative to the near-surface anomalies were obtained using $EV = 10^{6}$ as indicated by the trade-off diagram in Figure 3. Comparable trade-off diagrams were also observed for predicting near-surface anomaly values in the gap from simulated Ørsted and CHAMP observations.

Figure 2.E gives the simulated 700-km altitude Ørsted anomalies from the original inversion model evaluated to the nearest 0.3 nT to simulate the measurement errors [7]. The simulated satellite anomalies were then combined with the near-surface anomalies outside the gap by EV-optimized joint inversion to obtain the gap predictions shown in Figure 2.F with the RMS difference and correlation listed in Table 1.

Ørsted's magnetometers are also being carried by CHAMP, but at significantly lower altitudes. Figure 2.G shows the simulated 350-km CHAMP anomalies evaluated to the nearest 0.3 nT from the original inversion model. Combining these CHAMP anomalies with the near-surface anomalies outside the gap by EV-

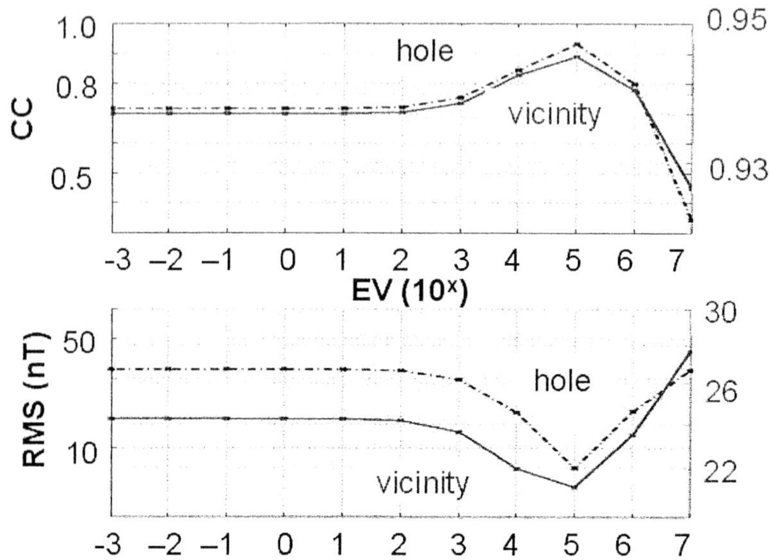

Fig. 3. Trade-off diagrams for obtaining an "optimal" value of error variance (EV) in terms of the CC and RMS difference of the predictions within the hole (dashed line refers to left vertical axis) and the surrounding vicinity (solid line refers to right vertical axis).

optimized joint inversion yields the gap predictions shown in Figure 2.F with the RMS difference and correlation listed in Table 1.

3 Conclusions

The results in Table 1 clearly favor the use of CHAMP observations for supplementing near-surface magnetic surveys with regional coverage gaps such as encountered by the ADMAP compilation in Figure 1. The enhanced correlation of gap predictions suggested in Table 1 for the use of CHAMP data can be broadly interpreted in terms of improvements in the signal-to-noise ratio (e.g., [2; 4]) from $(N/S) \approx \sqrt{[|CC|^{-1} - 1]}$. Accordingly, gap predictions using CHAMP data can have noise levels reduced by 72% relative to Magsat-based predictions simply by virtue of the order-of-magnitude increase in measurement accuracy reflected by the CHAMP data. Indeed, our analysis suggests that increasing the measurement accuracy in the magnetic observations a further order-of-magnitude (i.e., 0.03 nT) could reduce noise levels by nearly 99% relative to the Magsat predictions.

Of course these results are limited in practice by the errors in reducing magnetic observations for their lithospheric components. These reduction errors can be especially severe in the polar regions where strong and highly dynamic external magnetic fields operate. However, improving measurement accuracy can greatly facilitate the reduction of magnetic observations for non-lithospheric effects because of the enhanced correlation that can result between near-surface and satellite magnetic anomalies of the lithosphere.

Acknowledgement

Elements of this research were supported by NASA grant NAG5-7645.

References

1. Briggs IC (1974) Machine contouring using minimum curvature. Geophysics, **39**, 39-48.
2. Foster MR, Guinzy NJ (1967) The coefficient of coherence: Its estimation and use in geophysical prospecting. Geophysics, **32**, 602-616.
3. Golynsky AV, Chiappini M, Damaske D, Finn C, Ishihara T, Morris P, Nogi Y and von Frese RRB (2001) ADMAP – A digital magnetic anomaly map of the Antarctic. EOS (AM. Geophys. Union Trans.) **82**, S132.
4. Kim J-H (1995) Improved Recovery of Gravity Anomalies from Dense Altimeter Data. Ph.D. Dissertation, The Ohio State University, Columbus.
5. Langel RA, Hinze WJ (1998) The Magnetic Field of the Earth's Lithosphere: The Satellite Perspective, Cambridge Univ., USA.
6. Neubert T, Mandea M, Hulot G, von Frese RRF, Primdahl F, Jørgensen JL, Friis-Christensen E, Stauning P, Olsen N, Risbo T, (2001) Ørsted satellite captures high-precision geomagnetic field data. EOS (Am Geophys Union Trans), **82**, 81, 86-88.
7. Ravat DN, Purucker M, (1999) The future of satellite magnetic anomaly studies is bright!. The Leading Edge, **18**, 326-329.
8. Smith WHF, Wessel P, (1990) Gridding with continuous curvature splines in tension. Geophysics, **55**, 293-305.
9. von Frese RRB, Hinze WJ, Braile LW, Luca AJ, (1981) Spherical earth gravity and magnetic anomaly modeling by Gauss-Legendre quadrature integration. J Geophys, **49**, 234-242.
10. von Frese RRB, Ravat DN, Hinze WJ, McGue CA, (1988) Improved inversion of geopotential field anomalies. Geophysics, **53**, 375-385.
11. von Frese RRB, Tan L, Kim JW, Bentley CR, (1999) Antarctic crustal modeling from the spectral correlation of free-air gravity anomalies with the terrain. J Geophys Res, **104**, 25275-25296.

Comparing Magsat, Ørsted and CHAMP crustal Magnetic Anomaly Data over the Kursk Magnetic Anomaly , Russia

Patrick T. Taylor[1], James J. Frawley[2], Hyung Rae Kim[3], Ralph von Frese[3], and Jeong Woo Kim[4]

[1]Geodynamics Branch, NASA/GSFC, Greenbelt, MD 20771 USA
[2]Herring Bay Geophysics, Tracy's Landing, MD 20779 USA
[3]Dept. of Geological Science, The Ohio State University, Columbus, OH 43210 USA
[4]Dept. of Earth Sciences, Sejong University, Seoul 143-747, RoK

Summary. Prior to the launch of Magsat, it was not obvious that a satellite altitude magnetometer mission would record crustal magnetic fields, despite the POGO and COSMOS-49 results. Subsequent studies, however, soon revealed that Magsat was able to map crustal magnetic anomalies. One of these investigations was centered on the Kursk Magnetic Anomaly (KMA) of Russia. Not only did this earlier Magsat study of the KMA record its magnetic signature at satellite altitude but also it was revealed that the anomaly displayed a significant component of remanent magnetization. With the subsequent launch of the higher altitude Ørsted satellite a comparison between the KMA and these newer magnetic anomaly data was made. It was determined that Ørsted could indeed record the crustal magnetic field and with a greater signal/noise ratio than Magsat since at a higher altitude the spacecraft was further from the ionosphere. It is not surprising, therefore, that after the successful launch of CHAMP we continued our initial study of the crustal anomaly field over the KMA with these newer magnetic observations. The KMA appears to be well represented in these CHAMP data as well.

Key words: Comparing satellite magnetic anomalies, Kursk magnetic anomaly, remanent magnetization.

1 Introduction

Recording magnetic fields at satellite altitude produced by variations of magnetization of the crust of the Earth has only been accomplished within the last thirty-eight years with the launch of Cosmos 49. Since these signals are far weaker than either the Earth's main magnetic dipole field or the omnipresent time-varying external fields their representation was not universally acknowledged and their presence has to be continually verified. In order to establish that these short, time-invariant magnetic signatures represented crustal units and therefore are useful in regional geologic and tectonic studies it was necessary to make satellite magnetic anomaly maps over well known magnetic features or to confirm a correspondence between the satellite measurements and ground-truth. For the case of Magsat this was done when a satellite altitude (350 km) map was made over the well-known

Kursk Magnetic Anomaly region (40^0N-65^0N and 15^0E-45^0E) of Russia (Taylor and Frawley 1987). When the more recent Ørsted satellite was launched in February 1999 (Neubert et al. 2001) its higher altitude (850 by 613 km) than Magsat lead some to believe that crustal magnetic anomalies would not be resolved; however, another map over the KMA using these data (Taylor et al. 2000) revealed a distinct and similar crustal magnetic signal. Therefore with the launch of CHAMP in July 2001 we used the initial data from this mission to make yet another crustal anomaly map of the KMA. In this report we will discuss how these different satellite magnetic maps were made, compare them and discuss the geologic significance of our studies.

2 Data Analysis

Magsat had a sun-synchronous orbit with an inclination of 97^0 and therefore the dawn (descending) and dusk (ascending) orbits crossed with a relatively high angle. We made use of this property to construct our initial magnetic anomaly map since the static magnetic field, at the point where these orbits intersected at common altitude should have the same value. We removed a low-order polynomial from each orbit to produce a minimum difference (in a least-square sense) at these orbital intersection positions (Taylor and Frawley 1987). This method is also used to reduce data from other geopotential field missions with crossing orbits. In the case of our KMA study we restricted the orbits used to vary in altitude between 340 to 360 kilometers and produced a field at 350 km (Fig. 1). There is a 27 nT

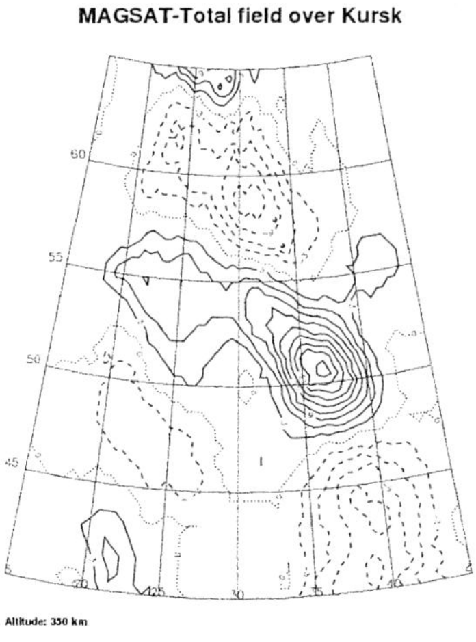

Fig. 1. Total field (scalar) magnetic anomaly map of the region of Eastern Europe centered on the Kursk Magnetic Anomaly computed at 350 km altitude (Taylor and Frawley 1987). Contour interval is 3 nT, zero contour is dotted and labeled, solid contours represent positive values and dashed negative. Owing to the nature of our mapping technique the zero-level is arbitrary.

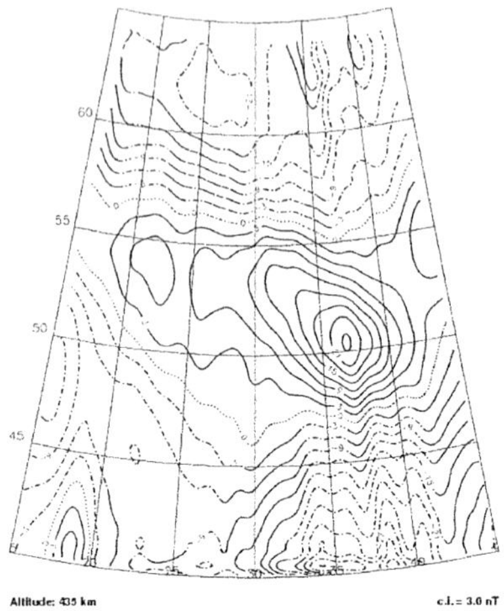

Fig. 4. Total field (scalar) magnetic anomaly map of the same region as Fig. 1 computed at 410 km altitude data from the CHAMP mission magnetic data. Contour interval is 3 nT with zero contour dotted and positive and negative values given by solid and dashed contours respectively.

/akima interpolation algorithm. The resulting grid was then low-pass filtered with a cutoff wavelength of 100-250 km. Mean altitude of orbits in the survey area was 410 km (Fig. 4).

The resultant field compares favorably with that computed from both MAGSAT and Ørsted (see Figs. 1, 2 and 4). Differences between these anomaly fields are due to the greater number of orbit profiles and different ranges of altitude and orbital eccentricities. To demonstrate the correspondence between the Magsat and CHAMP maps we plotted one on top of the other (Fig. 5). Note that these anomaly contours over the KMA are very similar, having a generally northwest-southeast trend across this region and mirroring the iron-ore formations (Krutikhouskaya et al. 1964). There is a general agreement with the anomalies over the other significant magnetic features of these two maps.

3 Interpretation

The Kursk Magnetic Anomaly (KMA) of Russia (51^0 N, 37^0 E) has long been recognized as one of the largest magnetic anomalies on Earth. It is associated with the massive quartz iron-ore formations; however, model studies have revealed that the relationship between the two is complex since this magnetic anomaly cannot be produced by the thickness of iron formation observed at the surface. There are large anomalies in all three-satellite maps over the center of the KMA (approxi-

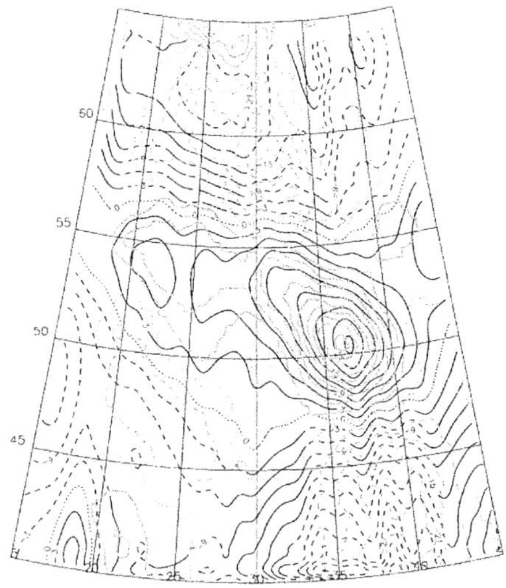

Fig. 5. Anomaly map of Figs. 1 and 4 plotted together. Magsat contours are in a gray shade and the CHAMP contours are bolder.

mately 50^0 N, 35^0 E); see and compare these data with the recently compiled magnetic anomaly map of Europe (Wonik et al. 2002). Correlating the reduced-to-the-pole Magsat data with crustal thickness data derived from deep seismic refraction studies (Belayevsky et al. 1968) revealed that the greater than 50 km thick crust in this region is coincident with the Magsat anomaly high (Taylor and Frawley 1987).

Magsat was the first of the near-Earth magnetic satellite missions to record vector field components. However, due to an abrupt transition between the star camera attitude determinations and the satellite attitude motion model there were frequent offsets in these measurements. In the selection of orbits over the region of the KMA our data were uncharacteristically free of this particular type of noise and therefore we were able to extract enough component data to map these vectors (Taylor and Frawley 1987). From these component data we derived a directional total magnetization vector for the KMA of east-northeast (~ 66^0) compared with the main magnetic field declination of some 5^0 east. Krutikhouskaya et al. (1964) reported measurements on the rock magnetic properties, including direction of magnetization, from thousands of samples from the region of the KMA. These rock samples were not subjected to magnetic cleaning processes and were therefore more indicative of the values reflected in the magnetic anomaly field data.
Despite the scatter in these data they are in general agreement with a vector of magnetization significantly east of the main field declination. This ground truth result encourages us to compute magnetic component maps using CHAMP data

and compute directions of magnetization. These directions of magnetization results will also indicate the degree of remanent magnetization contributing to the magnetic anomaly since significant variations from an induced anomaly would indicate the importance of remanence.

4 Conclusions

Both the currently operating Ørsted and CHAMP geopotential satellites as well as the earlier Magsat are able to record magnetic anomalies from sources within the Earth's crust and therefore can be used for geologic and tectonic studies. From the magnetic vector components being recorded by the Ørsted and CHAMP mission we should be able to compute a directions of magnetization for the larger anomalies and from this to evaluate the role of remanent magnetization in the formation of these anomalies, this procedure could tell us the nature of the dominant crustal magnetic minerals.

References

Alsdorf DE, von Frese RRB, Arkani-Hamed J, and Noltimier H (1994) Separation of lithospheric, external, and core components of the polar geomagnetic field at satellite altitude. Journal of Geophysical Research 99: 4655-4667

Beliayevsky NA, Borisov AA, Volvovsky IS and Schukin YuK (1968) Transcontinental crustal sections of the USSR. Canadian Journal of Earth Science 5: 1067-1078

Kim JW, Kim JH, von Frese RRB, Roman DR and Jezek KC (1998) Spectral attention of track-line noise. Geophysical Research Letters 25: 187-190

Krutikhouskaya ZA, Zavoyskyi VN, Podolyanko CM and Ya Savenko B (1964) Remanent rock magnetization of iron formations from the big Krivoy Rog and KMA [In Russian]. Ukrainian Institute of Geophysics, Science House, Kiev

Neubert T *et al* (2001) Ørsted Satellite Captures High-Precision Geomagnetic Field Data. *Eos*, Transactions, American Geophysical Union 82: 81 and 87-88

Olsen N *et a l*(2000) Ørsted Initial Field Model. Geophysical Research Letters 27: 3607-3610

Taylor PT and Frawley JJ (1987) Magsat anomaly data over the Kursk region, USSR. Physics of the Earth and Planetary Interiors 45: 5-15

Taylor PT, von Frese RRB, Kim HR (2000) Ørsted and Magsat: A comparison over the Kursk magnetic anomaly. In: Neubert T, Ultré-Guérard (eds) Proceedings, 3rd International Science Team Meeting, Grasse, pp 45-48

Wonik T, Trippler K, Geipel H, Greinwald S, Pashkevitch I (2001) Magnetic anomaly map for Northern, Western, and Eastern Europe. Terra Nova 13: 203-213

CHAMP, Ørsted, and Magsat Magnetic Anomalies of the Antarctic Lithosphere

Ralph R.B. von Frese[1], Hyung Rae Kim[1], Patrick T.Taylor[2], and Jeong Woo Kim[3]

[1]Dept. of Geological Sciences, The Ohio State University, Columbus, OH 43210 USA,
[2]Geodynamics Branch, NASA/GSFC, Greenbelt, MD 20771 USA,
[3]Dept. of Earth Sciences, Sejong University, Seoul 143-747, ROK

Summary. We processed CHAMP, Ørsted and Magsat mission data for the Antarctic using advanced spectral correlation theory to separate spatially and temporally static components from the dynamic ones. Ignoring measurement noise and processing errors, the dynamic components include strong external field effects, whereas the static components reflect lithospheric and core field effects. Consistency is apparent in the correlation of all the anomaly fields at degree and order 13 and higher. However, the longer wavelength components are corrupted by errors in the core field reductions that do not account for the crustal thickness magnetic effects. To assess these magnetic effects, we developed an Antarctic crustal thickness model from free-air and computed terrain gravity effects. The pseudo magnetic effects evaluated from the crustal model indicate quite strong crustal anomaly contributions for degrees lower than 13 in the satellite magnetometer observations. The Antarctic gravity field from which the crustal thickness effects were inferred is limited by the general lack of terrestrial gravity observations. However, CHAMP is providing significant new insight on the magnetic and terrestrial gravity fields for advancing our understanding of the Antarctic lithosphere and its tectonic evolution

Key words: Lithospheric magnetic anomalies, Antarctica, CHAMP, Ørsted, Magsat

1 Degree 13 and higher Lithospheric Magnetic Anomalies

Figure 1 compares the total magnetic field anomalies of degree 13 and higher that we obtained from the Magsat [1], Ørsted [5], and CHAMP magnetometer observations for the Antarctic south of 55°S. The anomaly features between the maps are fairly well correlated and thus support the veracity of the lithospheric anomalies first mapped by the six-month Magsat mission during austral summer and fall when the strongly corrupting effects of the auroral external fields were maximum.

These lithospheric anomaly estimates were obtained using advanced spectral correlation theory [10] to separate the spatially and temporally static components, which include the lithospheric effects, from the dynamic non-lithospheric components. We divided the orbital observations into subsets of ascending and descending orbit passes over the Antarctic. Common lithospheric anomaly components will be registered on neighboring passes separated by distances that are small compared to their altitude, while data components that are not correlated between these passes will reflect non-lithospheric effects [1].

The correlative and non-correlative data components can be extracted using the correlation spectrum given by $CC(k) = \cos(\Delta\theta_k)$, where $CC(k)$ is the correlation coefficient between the k-th wavenumber components of the two signals and $\Delta\theta_k$ is the phase difference in these components. Here the fast Fourier transform can be used with absolutely no loss of generality because the correlation spectrum depends only on the phase differences between co-registered and orthogonally gridded data.

Figure 2 gives a spectral correlation filtering application for two Magsat ascending (i.e., dusk) orbits separated by an average distance of about 7 km at roughly 330 km altitude across the Arctic from northeastern Greenland to southwestern Finland. Figure 2.A shows the raw data after removal of a core field model using the adjustment procedures of [1]. The relatively low correlation coefficient (= 0.479) between the data passes reflects none-correlative features from auroral external field effects, measurement errors, and other non-lithospheric sources.

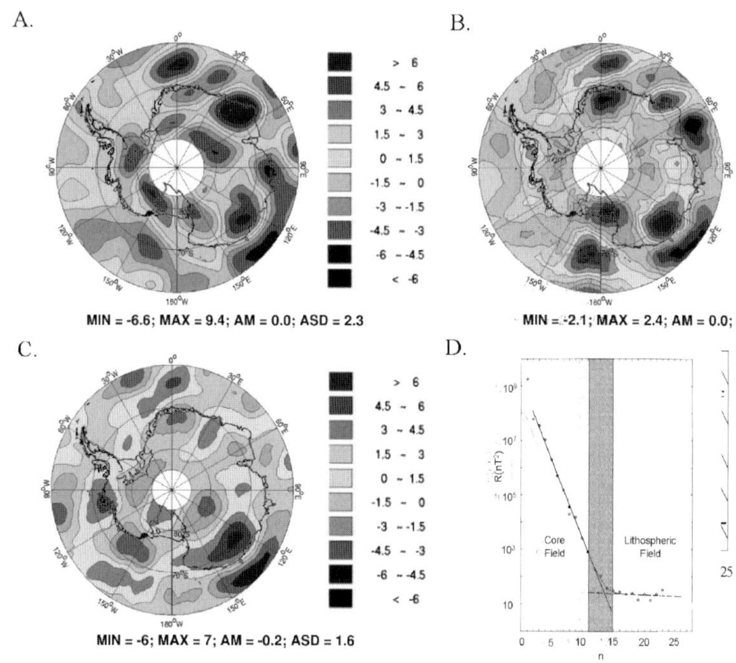

Fig. 1. Degree 13 and higher magnetic anomalies from A) Magsat at 430 km, B) Ørsted at 700 km, and C) CHAMP at 450 km. Attributes listed for each map in nT include the maximum (MAX) and minimum (MIN) amplitudes, the amplitude mean (AM), and amplitude standard deviation (ASD). D) The generalized geomagnetic field spectrum [7] of power, R, versus spherical harmonic degree, n, includes the shaded transition zone where slopes merge and are difficult to differentiate.

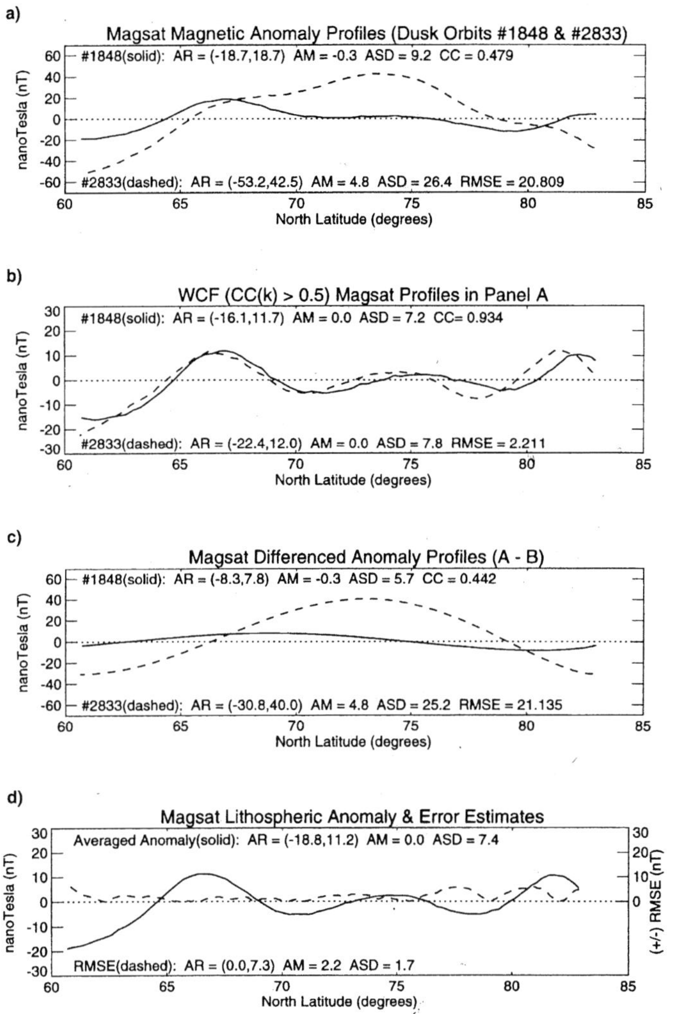

Fig. 2. Wavenumber correlation filtering (WCF) of two nearest-neighbor orbits from the Magsat mission. Additional data attributes listed include the amplitude range (AR) of (MIN, MAX)-values, the correlation coefficient (CC), and the root-mean-square error (RMSE).

Using the correlation spectrum for the signals in 2.A, we inversely transformed all wavenumber components with $CC(k) \geq 0.5$ for the estimates in Figure 2.B of their lithospheric components. The enhanced correlation (= 0.934) obtained by these results can be broadly interpreted for improvements in the signal-to-noise ratio (e.g., [2; 4]) from $(N/S) \approx \sqrt{[|CC|^{-1} - 1]}$. Accordingly, relative to the raw residuals in Figure 2.A, the correlation filtered signals reflect roughly a 75% reduction in the non-lithospheric noise effects.

Figure 2.C shows the components of the raw residuals that were rejected by the correlation filters. Although partly correlated, they reveal little apparent sensitivity for the regional magnetic effects of the crustal geology (e.g., ocean-continent margins, ocean ridges, basins, etc.). They appear to be more readily ascribable to non-lithospheric effects (e.g., external field effects, core field reduction errors, etc.) and hence were discarded. Accordingly, Figure 2.D gives the least squares estimate of lithospheric anomalies from the point-by-point averages of the coherent signals in Figure 2.B. Here, we also give the differences between the coherent signals as root-mean-square-errors (RMSE) to constrain interpretations of the lithospheric anomaly estimates.

From the correlation filtered ascending and descending passes, we constructed maps at common altitude and spherical coordinates by least squares collocation [3]. These co-registered anomaly maps were correlation filtered for their common components to further reduce non-lithospheric noise. For the Magsat anomaly map in Figure 1.A, the correlation filtered ascending and descending anomalies were essentially averaged and smoothed by low-pass filtering to remove inappropriate higher frequency components [1]. For the Ørsted and CHAMP anomalies in Figures 1.B and 1.C, respectively, the correlation filtered ascending and descending anomaly maps were combined by spectrum reconstruction [6] to minimize track-line or corrugation noise due to along-track processing effects.
Although greatly reduced in non-lithospheric noise effects, Figures 1.A, 1.B, and 1.C may be substantially incomplete estimates of the magnetic anomalies of the Antarctic lithosphere at satellite altitudes. The geomagnetic field spectrum generalized in Figure 1.C from [7] shows a transition zone roughly between degrees 11 and 15 where the slopes for the core field and regional lithospheric effects are difficult to differentiate. Hence, regional lithospheric effects may corrupt the core field estimates at degree 13 and lower so that the core field reduction may also remove significant lithospheric signal [8; 11]. However, the Antarctic gravity field can help to recover these regional magnetic anomalies of the lithosphere.

2 Degree 13 and lower Lithospheric Magnetic Anomalies

Figure 3 illustrates our approach to separating degree 13 and lower core field and lithospheric anomaly components. In Figure 3.A, for example, we obtained the Magsat anomalies relative to the core field estimates through degree 11 that we infer from the geomagnetic field spectrum of [7] to be largely uncontaminated by lithospheric effects. Subtracting the degree 13 and higher anomaly field of Figure 1.A from these results leaves residual degree 11 through 13 anomalies dominated by core field and regional lithospheric components that include the magnetic effects of crustal thickness variations.

To estimate the crustal thickness effects in the residual degree 11-13 anomalies,

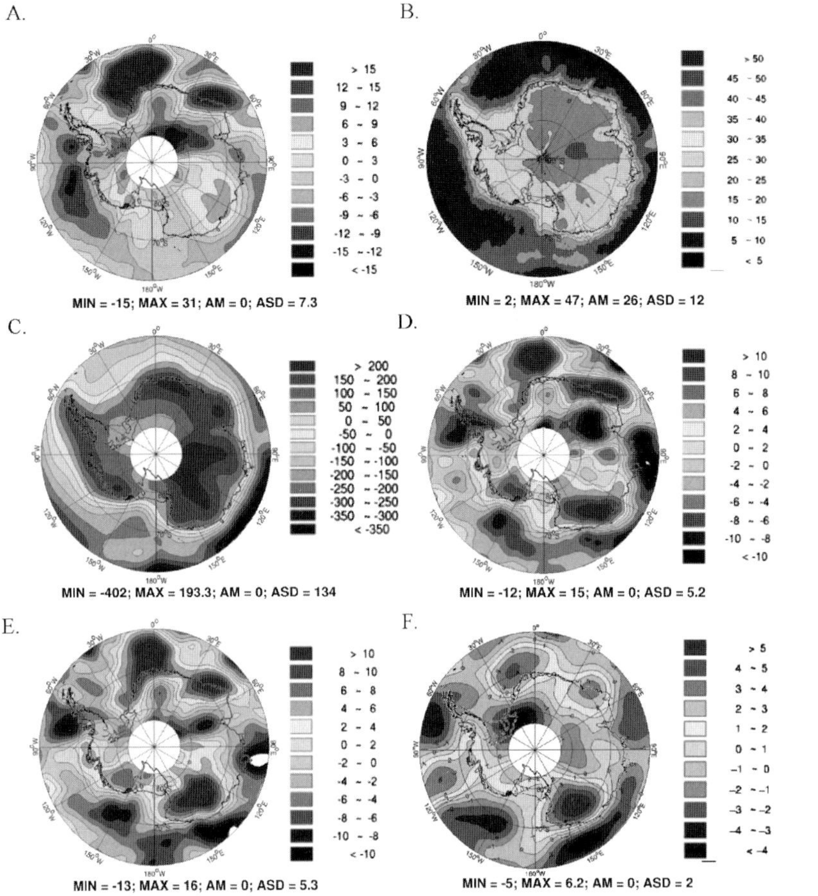

Fig. 3. To estimate the Magsat comprehensive lithospheric magnetic anomalies (nT) at 430 km in map E that include the degree 11-to-13 lithospheric components (nT) in map D, we consider the Magsat anomalies (nT) at 430 km relative to the degree 11 core field in map A, the Antarctic crustal thickness variations (km) in map B, and their pseudo magnetic total field effects at 430 km in map C. A similar analysis gives the Ørsted comprehensive lithospheric magnetic anomalies at 700 km in map F.

we used the crustal thickness model in Figure 3.B from [12]. This model was obtained from a spectral correlation analysis at satellite altitude of EGM96 free-air anomaly estimates and the gravity effects of the rock, water, and ice components of the terrain. The pseudo magnetic total field anomalies of this model obtained via Poisson's relation for correlative magnetic and gravity potentials (e.g., [9]) are shown in Figure 3.C. Spectral correlation analysis of these pseudo magnetic effects and the residual degree 11-13 anomalies reveals the phase coherent residual anomalies in Figure 3.D that we interpret for degree 13 and lower Magsat magnetic anomalies of the Antarctic lithosphere. These results suggest that considerable lithospheric signal may be missing in satellite anomaly maps where the data have been reduced for core field estimates through degree 13. Our estimates may be limited, however, because the Antarctic gravity predictions from EGM96 are poorly constrained due to the paucity of terrestrial observations [13]. The gravity observations from CHAMP will provide the highest resolution gravity field mapped to date for the Antarctic, and hence important new insight for magnetic and geologic studies of the south polar lithosphere. Combining Figures 1.A and 3.D yields the comprehensive lithospheric anomaly map for the Magsat data given in Figure 3.E. A similar analysis for the Ørsted data gives the comprehensive lithospheric anomaly field in Figure 3.F. We are currently integrating these results with the improved magnetic and gravity observations from CHAMP for further insights on the Antarctic lithosphere.

Acknowledgement. Elements of this research were supported by NASA grant NAG5-7645.

References

1. Alsdorf DE, von Frese RRB, Arkani-Hamed J, Noltimier HC, (1994) Separation of lithospheric, external, and core components of the south polar geomagnetic field at satellite altitudes. J Geophys Res, **99**, 4655-4667.
2. Foster MR, Guinzy NJ (1967) The coefficient of coherence: Its estimation and use in geophysical prospecting. Geophysics, **32**, 602-616.
3. Goyal HK, von Frese RRB, Hinze WJ, (1990) Statistical prediction of satellite magnetic anomalies. Geophys J Int, **102**, 101-111.
4. Kim J-H (1995) Improved Recovery of Gravity Anomalies from Dense Altimeter Data. PhD Dissertation, The Ohio State University, Columbus.
5. Kim HR, von Frese RRB, Kim JW, Taylor PT, Purucker ME, Neubert T (2002) Ørsted verifies regional magnetic anomalies of the Antarctic lithosphere. Geophys Res Lett (**in-press**).
6. Kim JW, Kim J-H, von Frese RRB, Roman DR and Jezek KC, (1998) Spectral attenuation of track-line noise. Geophys Res Lett, **25**, 187-190.
7. Langel RA, Estes RH (1982) A geomagnetic field spectrum. Geophys Res Lett, **9**, 250-253.
8. Meyer J; Hufer JH, Siebert M, Hahn A (1985) On the identification of Magsat anomaly charts as crustal part of internal field. J Geophys Res, **90**, 2537-2542.
9. von Frese RRB, Hinze WJ, Braile LW (1982) Regional North American gravity and magnetic anomaly correlations. Geophys JR Astr Soc, **69**, 745-761.
10. von Frese RRB, Jones MB, Kim JW, Kim H-J (1997) Analysis of anomaly correlations. Geophysics, **62**, 342-351.
11. von Frese RRB, Kim HR, Tan L, Kim JW, Taylor PT, Purucker ME, Alsdorf DE, Raymond CA (1999) Satellite magnetic anomalies of the Antractic crust. Annali di Geofisica, **42**, 309-326.
12. von Frese RRB, Tan L, Kim JW, Bentley CR (1999) Antarctic crustal modeling from the spectral correlation of free-air gravity anomalies with the terrain. J Geophys Res, **104**, 25275-25296.
13. von Frese RRB, Roman DR, Kim J-H, Kim JW, Anderson AJ (1999) Satellite mapping of the Antarctic gravity field. Annali di Geofisica, **42**, 293-307.

Separation of External Magnetic Signal for Induction Studies

Monika Korte, Stephen C. Constable and Catherine G. Constable

Institute of Geophysics and Planetary Physics, Scripps Institution of Oceanography, University of California at San Diego, 9500 Gilman Drive, La Jolla, CA 92093-0225. *mkorte@igpp.ucsd.edu*

Summary. Magnetic satellites offer highly improved capabilities for studies of Earth's mantle electrical conductivity, but the necessary separation of the external forcing signal from all other field contributions requires good models of the main, crustal, and ionospheric fields. We use the updated comprehensive field model [5] to remove the unwanted field contributions and fit the remaining magnetospheric ring current signal with a simple axial dipolar structure. These preparatory steps for an induction study show that while the comprehensive model is generally well suited for the separation, there are still limitations in both the model and the CHAMP data.

Key words: geomagnetic induction, reference fields

1 Introduction

Earth's mantle electrical conductivity has traditionally been studied by using geomagnetic observatory records. Magnetic satellites like CHAMP offer a vastly superior spatial coverage of the globe. Like observatories, however, satellites record the sum of all geomagnetic field contributions, while the desired signals for conductivity studies are only variations external to Earth and their induced fields. We plan to use the external forcing signal asssociated with the recovery phase of magnetic storms in particular, with the ultimate goal of developing a 3D model of global mantle conductivity distribution. Here we present the first step, an attempt to separate the desired signal from CHAMP data.

2 Using the Comprehensive Field Model to Extract the Dst Signal

A model of the different contributions to the geomagnetic field is necessary to extract a specific signal like the inducing and induced fields from all the other contributions that have to be regarded as noise, in this case the core field and crustal and ionospheric fields. *Constable and Constable* [2] successfully used *Sabaka et al.'s* [4], [5] comprehensive field model (CMP3) for that task when

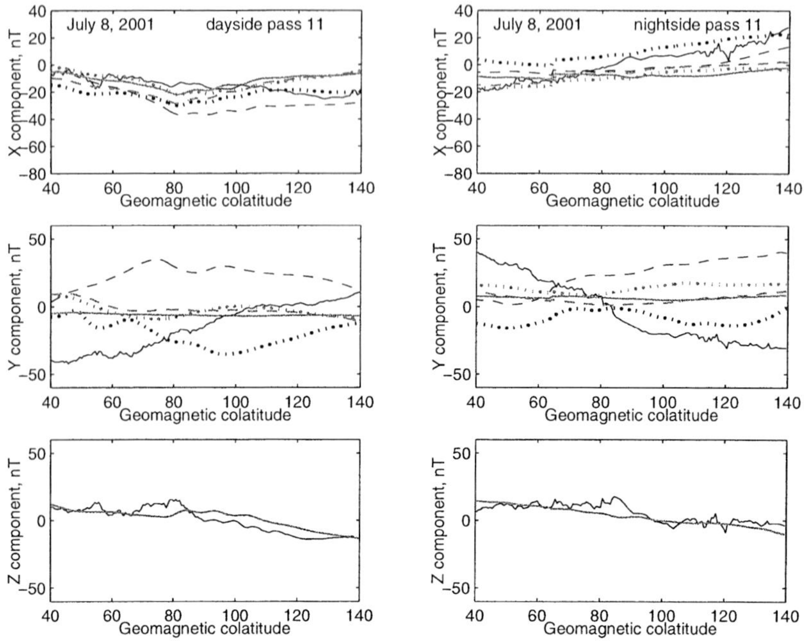

Fig. 1. External and induced geomagnetic field and model predictions. Solid black line: CHAMP data, CM3d core and crustal field predictions subtracted. Other: CM3d predictions of all field contributions external to Earth and induced, with the different possibilities of coupling current predictions, dawn (black dotted), dusk (black dashed), 10 am (gray dotted), 4 pm (gray dashed) and coupling currents turned off (gray solid line).

analyzing Magsat data. The original CMP3 only covered a timespan up to 1985, but Sabaka et al. have updated their model (*Sabaka, pers. comm.*) with Oersted data so that we were able to use that version (CM3d) with CHAMP data of summer 2001. First, we determined the model predictions for the core and crustal fields (up to spherical harmonic degree 65) and subtracted those from the data. Shape and order of magnitude of the remaining residuals suggest that the CM3d predictions successfully served to extract the external and induced field parts (Fig. 1). We are, however, not interested in all external variations, but only in the ring current or Dst signal. To verify whether CM3d meets our demands of extracting that, we compared our residual data to the external and induced model predictions, that include primary and induced ionospheric fields, primary and induced magnetospheric fields and fields from ionospheric coupling currents. The latter are slightly problematic: due to the limitation of the Magsat data to local dawn and dusk orbits, the original CMP3 could predict the coupling currents only for those two local times. CM3d involves a first attempt to include Oersted dayside data in the model, but so far only can predict the coupling currents for 4 local times, which are dawn, dusk, 10 am and 4 pm. Figure 1 compares data and model predictions. For the data we chose a CHAMP orbit that falls approximately into local dawn and dusk (6 pm and 6 am) and we have plotted the model predictions using the coupling current prediction for every one of the four possible local times. Keep in mind that all other field contributions are modeled continuously in time by CM3d. Generally, the predicted effects of coupling currents are small for 10 am and 4 pm and they have no effect on the Z component. The strongest differences are seen between the dawn and dusk predictions of the Y component. In Fig. 1 we would expect the black dashed model prediction to fit best for dusk, the black dotted one for dawn. Obviously, this is not always the case, as this example shows. While the respective model prediction might fit the data slightly better than the other ones for the dawn pass Y component, it fits worse than most others in the dusk pass Y component. We conclude that the prediction of coupling currents by CM3d is not suitable for our data. We turned off the option to predict coupling currents (solid gray line in Fig. 1), in order not to generate artificial variations in the data by subtracting inadequate predictions. The general fits of the remaining external and induced predictions of CM3d, however, are satisfying. Note in particular the remarkably good fit for the Z component. The small step in the model dawn X component predictions at about 65° geomagnetic colatitude in Fig. 1 is due to the use of the non-interpolated hourly Dst index in the model evaluation. For a smooth model we would have to smoothly interpolate that index in time. For our study, however, we will not use that part of the model anymore as we are trying to extract just that ring current signal. To do so, we now subtract the CM3d predictions for all field contributions except for the coupling currents and the Dst contribution. The latter is achieved by setting the Dst-index to zero when evaluating the model. Figures 2 and 3 show examples of the remaining Dst residual signal. The

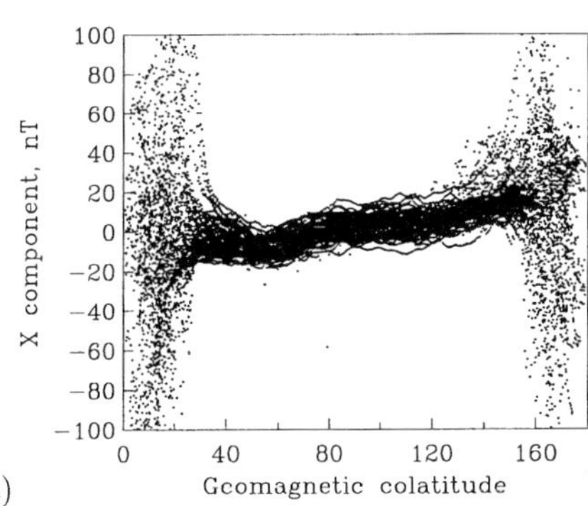

a)

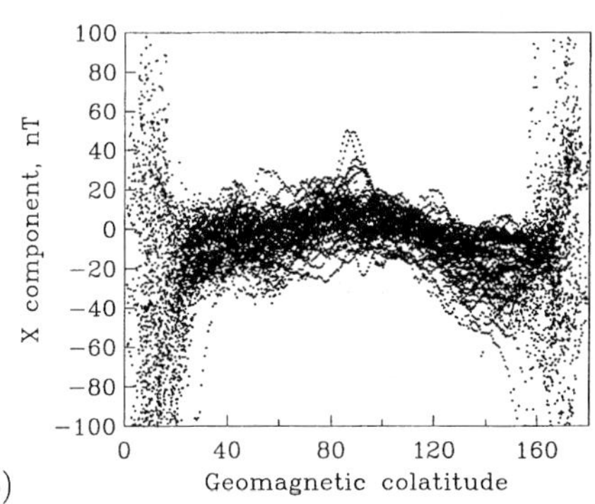

b)

Fig. 2. Dst residuals, ring current signal extracted from CHAMP data, for nightside passes (a) and dayside passes (b) of the 5 magnetically most quiet days of August 2001, north component.

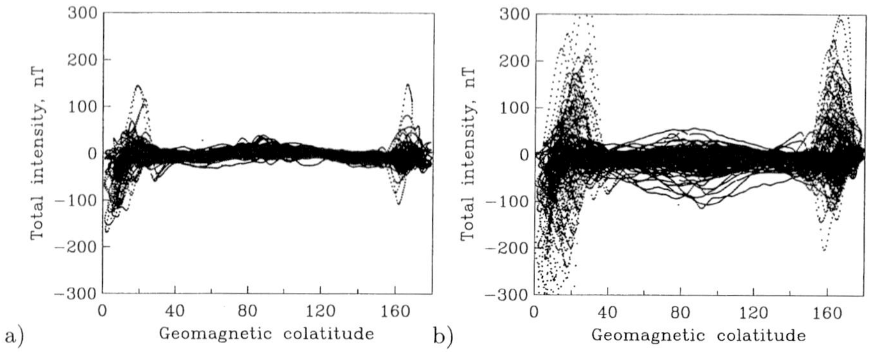

Fig. 3. Dst residuals for the 5 magnetically most quiet (a) and most disturbed (b) days of August 2001, total intensity

figures display all CHAMP passes of the five magnetically most quiet or most disturbed days of August 2001. Figure 2 shows that the dayside Dst-residuals are much more scattered than the nightside data. There are two possible reasons for that: First, although the difference is visible in the total intensity, it is stronger in the vector components. This suggests that orientation errors are larger during daytimes, compatible with *Holme et al.* [3]. Second, the model predictions of the variations to be removed might be less complete for daytimes, when more local time dependent effects occur. However, Fig. 3 shows that we have extracted the Dst signal to a good approximation: Except for high latitudes there is not much signal in the data of the quiet days, whereas for the disturbed days a clear bow-shaped signal as expected from a ring current source is visible. At high latitudes, within about 40° of the poles, the Dst-residuals are dominated by fields associated with field-aligned currents, which could not be eliminated by the CM3d model predictions. To avoid those, we use observations from the colatitude range between 40 and 140° only.

3 Fitting Dst

As the next step on our way to an induction study we need to characterize the inducing (external to Earth) and the induced (internal) parts of the Dst signal from our Dst-residual data. *Banks* [1] showed that variations in the Earth's magnetic field at periods shorter than one year are dominated by a simple P_1^0 spherical harmonic geometry. We assume this to be a valid approximation in particular for the ring current in mid-latitude regions and write the magnetic field **B** as the negative gradient of a scalar potential ψ_1^0,

$$\mathbf{B}(r,\theta,\phi) = -\nabla\psi_1^0(r,\theta,\phi) \tag{1}$$

in geomagnetic coordinates r, θ, ϕ. This potential can be written as a spherical harmonics expansion of only the associated Legendre polynomial of degree 1 and order 0, P_1^0:

$$\psi_1^0(r,\theta) = a\left\{i_0^1(t)\left(\frac{a}{r}\right)^2 + e_1^0(t)\left(\frac{r}{a}\right)\right\} P_1^0(cos\theta) \tag{2}$$

where a is the mean Earth radius and i_1^0, e_1^0 are the Schmidt quasi-normalized spherical harmonic coefficients for the internal (induced) and external (inducing) ring current signal. The magnetic components then are given by

$$B_r = \left[-e_1^0 + 2i_1^0\left(\frac{a}{r}\right)^3\right] cos(\theta) \tag{3}$$

$$B_\theta = \left[e_1^0 + i_1^0\left(\frac{a}{r}\right)^3\right] sin(\theta) \tag{4}$$

$$B_\phi = 0. \tag{5}$$

We determine the coefficients by a least-squares fit to our Dst-residuals and compare this fit to the ring current predictions of CM3d, which utilizes the global Dst-index. Figure 4 gives some examples for magnetically more quiet (a) and more disturbed (b) conditions. The shapes of our P_1^0-fit and the Dst-index fit are similar, but often the amplitudes of our fit agree better with the data, both for quiet and disturbed conditions, as can be seen in the bottom panels of Fig. 4 where we have subtracted both fits respectively from the data. In particular the fit for B_θ mostly is worse for the CM3d prediction. We conclude that the satellite estimates more reliably quantify the magnitude of ring current excitations than the Dst-index obtained from ground based observations. Compared to the Magsat results of *Constable and Constable* [2], our fits for CHAMP data generally are slightly worse. We might not have eliminated fields of sources other than the ring current well enough. It seems reasonable that CMP3 can predict Magsat data better than CM3d can predict CHAMP data, as CMP3 was obtained using Magsat data whereas CM3d is extrapolated to the observation times of our data. Moreover, the CHAMP data might not yet be definitive, as we expect revisions and improvements as the mission continues.

4 Conclusion

We have tried to extract the ring current or Dst signal from the CHAMP recordings of all geomagnetic field contributions. We tested the updated comprehensive model CM3d for that task and came to the conclusion that this model is well-suited to remove the core and crustal fields. For the external and induced fields it also does a reasonably good job, except for the prediction of field-aligned coupling currents, which at the moment only can be predicted

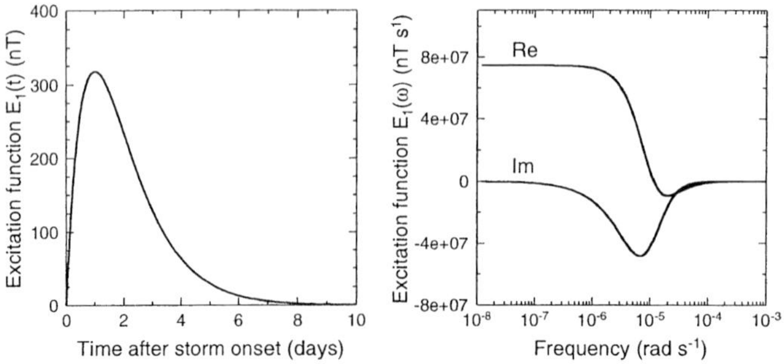

Fig. 1. Transient ring-current excitation function $E_1(t)$ with ring-current decay constant $\tau = 1$ day and scale factor $A = 0.01$ nT s^{-1} (*left*) and its Fourier image $E_1(\omega)$ (*right*). These functions are used to simulate a geomagnetic storm in the time and frequency domain, respectively.

with a $P_{10}(\cos\vartheta)$ spatial structure, where ϑ is colatitude in spherical coordinates. Consequently, all spherical harmonics of the external scalar magnetic potential are equal to zero except the first-degree coefficient $E_1(t)$. The ring current has a finite build-up time. After the onset at time $t = 0$ it is assumed to evolve according to

$$E_1(t) = A\, t\, e^{-t/\tau}, \tag{1}$$

where A is a scale factor and τ governs the decay of the ring current. Values of τ typical for a moderate to large geomagnetic storm range from 1 day to 1 week [5]. We will consider $A = 0.01$ nT s^{-1} and $\tau = 1$ day in all numerical examples. The Fourier image of $E_1(t)$ has an analytical form [1]:

$$E_1(\omega) = \frac{A}{2\pi} \frac{\alpha^2 - \omega^2 - 2i\alpha\omega}{(\alpha^2 + \omega^2)^2}, \tag{2}$$

where ω is the angular frequency and $\alpha = 1/\tau$. The external excitation function (1) and its Fourier image (2) are shown in Fig. 1.

3 Transient responses

In this paper, we study satellite-recorded induction signals due to ring-current excitation in three conductive spheres: a) a homogeneous sphere with radius $a = 6371$ km and electrical conductivity $\sigma = 0.1$ S/m, b) a two-layer sphere with the inner-sphere radius $b = 3500$ km and conductivity $\sigma_2 = 10$ S/m and the outer-sphere radius $a = 6371$ km and conductivity $\sigma_1 = 0.1$ S/m, and c) an axisymmetrically nested-sphere configuration with the same geometrical and physical parameters as for the two-layer sphere, but the inner sphere translated along the z-axis northwards by an offset of $d = 2700$ km. The last configuration is perhaps the simplest 2-D spherical conductor for which a semi-analytical solution to the ring-current excitation is available in the frequency domain [2]. A related investigation in the

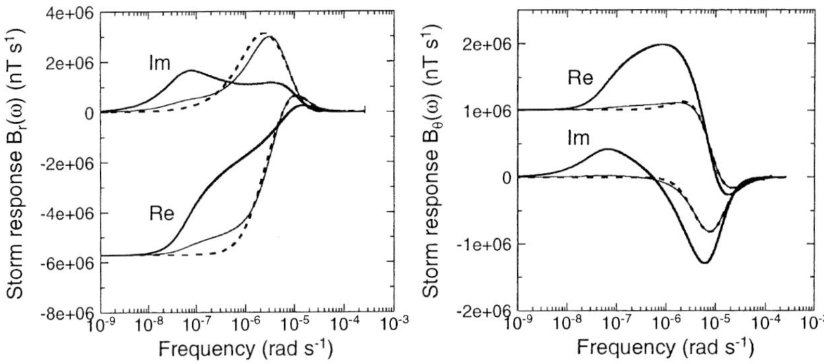

Fig. 2. The frequency magnetic responses $B_r(\omega)$ and $B_\vartheta(\omega)$ of a homogeneous conductive sphere (*dashed lines*), the two-layer sphere (*thin solid lines*) and the axisymmetrically nested-sphere model (*thick solid lines*) to ring-current excitation. See the text for the description of sphere configurations.

time domain has been carried out by [4] using the time-domain, spectral-finite element approach

Figure 2 shows the real and imaginary parts of the radial $B_r(\omega)$ and polar $B_\vartheta(\omega)$ magnetic storm responses at the outer-sphere surface $r = a$ and the colatitude $\vartheta = 10°$ as a function of angular frequency ω for the homogeneous sphere (dashed lines), the two-layer sphere (thin solid lines) and the axisymmetrically nested-sphere configuration (thick solid lines). The figure allows us to make an immediate interpretation in terms of the basic physics of ring-current induction in a conductive sphere. For instance, the low frequency responses of all conductive spheres approach the primary field, namely, $B_r(\omega) = -A\cos\vartheta/2\pi\alpha^2$ and $B_\vartheta(\omega) = A\sin\vartheta/2\pi\alpha^2$ for $\omega \to 0$. This is expected since a sufficiently low frequency excitation cannot detect the presence of a conductive sphere of finite conductivity and radius. On the other hand, at sufficiently high frequencies, the responses of the two-layer sphere and axisymmetrically nested sphere approach the response of the homogeneous sphere since the induced eddy currents are confined to a thin layer near the surface of the conductive sphere as if the sphere were perfectly conductive, and cannot detect the inner sphere.

Figure 2 also shows that the magnetic response of the two-layer sphere is similar to that of the homogeneous sphere with the conductivity of the outer sphere. In this case the frequency with the largest response corresponds to that which makes the skin depth roughly equal to the radius of the inner sphere. Hence, the response of the two-layer sphere is nearly unaffected by the presence of the inner sphere. We can also see that the magnetic response due to the presence of the axisymmetrically nested sphere is more pronounced in the polar component than in the radial component and is larger at lower frequencies than that for the homogeneous sphere.

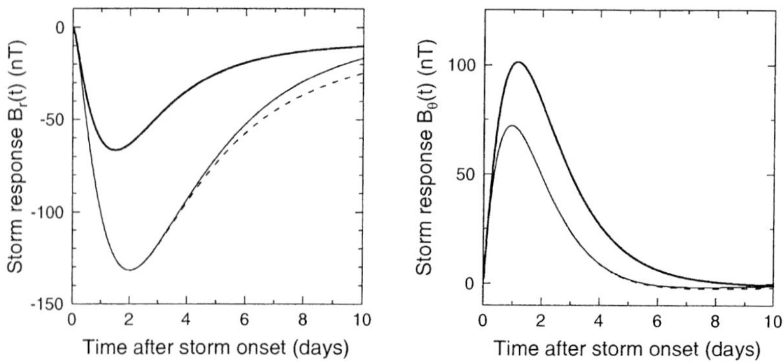

Fig. 3. The transient magnetic responses $B_r(t)$ and $B_\vartheta(t)$ of a homogeneous conductive sphere (*dashed lines*), the two-layer sphere (*thin solid lines*) and the axisymmetrically nested-sphere model (*thick solid lines*) to ring-current excitation.

The anomalous response (the difference between the magnetic responses of the axisymmetrically nested spheres and the two-layer sphere) decreases with increasing colatitude (not shown here), which is expected since the inner sphere is displaced to northern latitudes. In the southern hemisphere, the anomalous response is insignificant.

The magnetic response in the frequency domain is transformed to the time domain by the inverse Fourier transform. Since the solutions in the time domain must be real and causal functions, the inverse Fourier transform can be computed by the cosine transform. The time-domain responses were checked using the more general time-domain, 2-D spectral-finite element approach [4] in order to validate the numerical procedure. We obtained excellent agreement between the responses computed using the inverse Fourier transform and the time-domain, spectral-finite element approach. For more details, see [4].

The transient responses of the homogeneous sphere, the two-layer sphere and the axisymmetrically nested spheres at the outer-sphere surface $r = a$ and colatitude $\vartheta = 10°$ are shown in Fig. 3 for the ring-current excitation described by eq. (1). The comparison of Figs. 1 and 3 shows that the induced eddy currents persists for longer times after the storm onset than does the ring current. Figure 3 also shows that the effect of the inner sphere in the two-layer sphere model on both the radial, $B_r(t)$, and the polar, $B_\vartheta(t)$, responses in the two-layer sphere is small because of the reason discussed above. On the contrary, the effect of the inner sphere in the axisymmetrically nested-sphere model is dominant, particularly at later times. The strong late-time signature of the inner sphere suggests that the extraction of information about mantle electrical heterogeneity from satellite-recorded induction signals may be possible. However, the interpretation of satellite-recorded induction signals is, in principle, ambiguous.

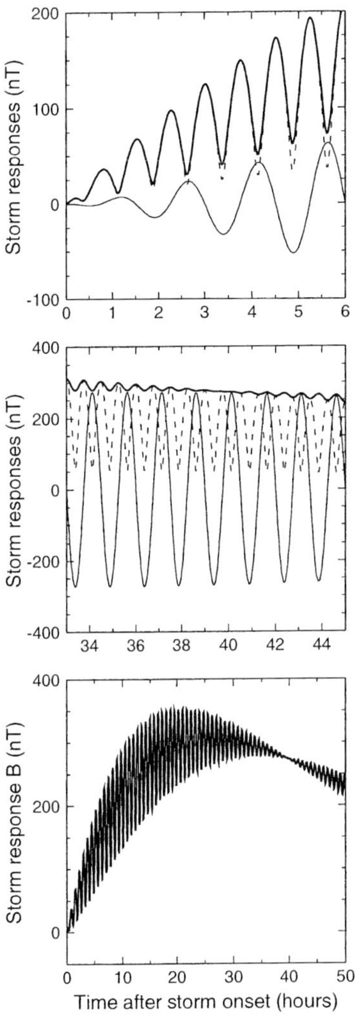

Fig. 4. The radial, $B_r(t)$, (*thin solid lines*), polar, $B_\vartheta(t)$, (*dashed lines*) and scalar intensity, $B(t)$, (*thick solid lines*) spatiotemporal induction signals measured by a satellite orbiting a homogeneous conductive sphere with the conductivity $\sigma = 0.01$ S/m along the circular trajectory with the orbital declination $\Omega = 10°$ and the orbital period $T = 90$ min.

4 Spatiotemporal responses

The 90-minute period of near-Earth satellites [6] is much shorter than the 1-day to 1-week periods characteristic for a magnetic storm. Thus, the satellite will orbit many times around the Earth during the course of a single storm. As the satellite orbits, its onboard magnetometer samples the magnetic signals from the induced eddy currents. At the same time, the ring-current intensity evolves, increases during the main phase of the storm and decays during the recovery phase. Quantitative interpretations of the satellite induction signal involve deconvolving the spatial and temporal variations recorded by a satellite magnetometer. This requires the development of spatiotemporal geomagnetic induction forward modelling. Here we propose a modelling procedure based on highly simplified orbital dynamics, ring-current excitation, and 1-D or 2-D electrical conductivity models.

The satellite trajectory is assumed to be circular. The time-varying spatial position of a satellite in Cartesian coordinates is described by

$$\begin{aligned} x(t) &= R\sin(2\pi t/T)\sin\Omega, \\ y(t) &= -R\cos(2\pi t/T), \\ z(t) &= R\sin(2\pi t/T)\cos\Omega, \end{aligned} \tag{3}$$

where R is the orbital radius, T is the orbital period, and Ω is the orbital declination measured from the z-axis. The satellite is assumed to be at the ascending node at time $t = 0$, which corresponds to the onset of the magnetic storm. Note that the ascending node is the point on the equator that is occupied by the satellite when it passes northward.

Besides the radial, $B_r(t)$, and polar, $B_\vartheta(t)$, response components, we simulate a scalar response $B(t)$ by

$$B(t) = \sqrt{B_r^2(t) + B_\vartheta^2(t)}, \tag{4}$$

which would be measured by a scalar magnetometer on a satellite during

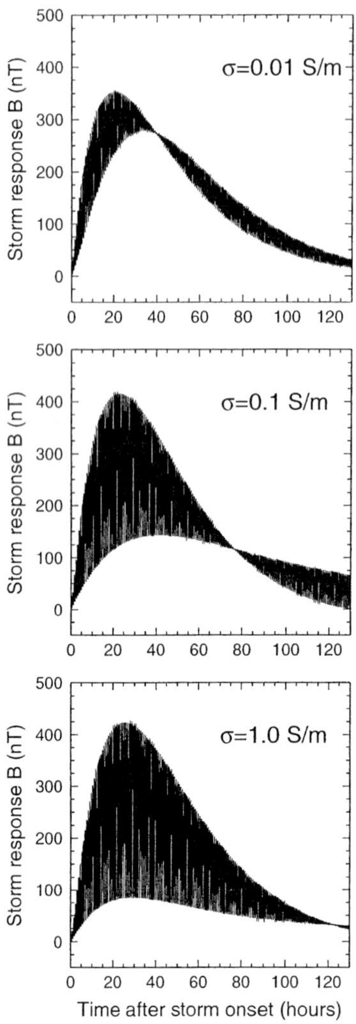

Fig. 5. The effect of electrical conductivity of a homogeneous sphere on the scalar intensity $B(t)$ of satellite spatiotemporal induction signal.

a ring-current excitation. The responses $B_r(t)$, $B_\vartheta(t)$ and $B(t)$ are shown in Fig. 4 for the orbital period $T = 90$ min and the declination $\Omega = 10°$ (a trajectory which is close to the polar orbit). The top panel in Fig. 4 shows that the polar response $B_\vartheta(t)$ undergoes two complete oscillations in amplitude during each complete 90-minute orbit, while the radial response $B_r(t)$ experiences only one. Moreover, the radial response changes the sign, while the polar component stays positive. This behaviour is a consequence of the spherical geometry of the problem. The radial component $B_r(t)$ is negative at the north pole and positive at the south pole, while the polar component $B_\vartheta(t)$ is positive at both poles. This explains the oscillatory character of the spatiotemporal response shown in Fig. 4. The oscillations are damped with a time scale that is controlled by τ, the decay constant of the ring-current excitation.

The top panel in Fig. 4 also shows that, at earlier times, the amplitude of the polar response $B_\vartheta(t)$ is larger than the amplitude of the radial response $B_r(t)$. However, at later times, the amplitude of the radial response is larger than that of the polar response (see the middle panel in Fig. 4). There is an intermediate time for which the scalar response $B(t)$ does not oscillate. In Fig. 4, this time roughly corresponds to $t = 39.5$ hours and is manifested by the "node" of the scalar response. This node can be clearly seen at the bottom panel in Fig. 4.

The appearance of the node in the spatiotemporal scalar response is caused both by the spherical geometry of the global induction problem and by the fact that the radial component of the induced magnetic field grows and decays over a time scale different from that of the polar component. Figure 5 shows the spatiotemporal scalar response $B(t)$ for a homogeneous sphere with conductivities $\sigma = 0.01$, 0.1 and 1.0 S/m, respectively. We can see that for a more conductive sphere, the occurrence of the node is shifted to later times. Thus, the time of occurrence of the node is a good indicator of the conductivity of the sphere. A node also appears in

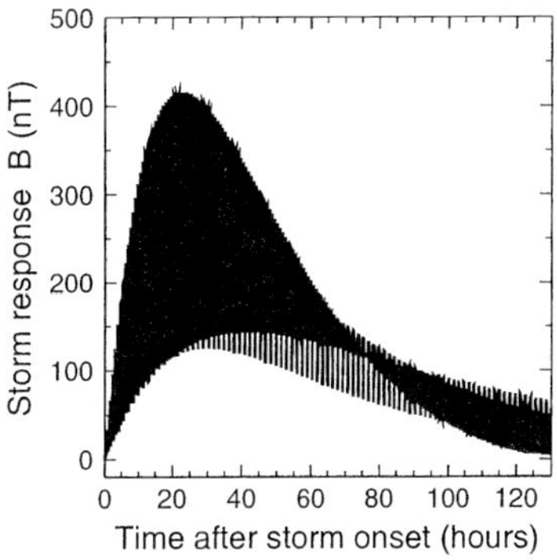

Fig. 6. The scalar intensity $B(t)$ of satellite spatiotemporal induction signal for the axisymmetrically nested-sphere model.

the spatiotemporal response $B(t)$ of the axisymmetrically nested-sphere model. In Fig. 6, this time roughly corresponds to $t = 75$ hours. The time of occurrence of the node is proportional to the electrical conductivity of the sphere. This provides a useful tool for interpreting transient satellite-recorded induction data. However, its detection in actual satellite data remains to be proved.

5 Conclusion

Calculations presented in this paper show the transient electromagnetic responses of homogeneous and heterogeneous conductive spheres to ring-current excitation. The computed response curves reveal that the information about the spheres' electrical conductivity is contained in the intermediate-time and late-time magnetic responses. The magnetic response evolves such that the polar response is largest at early times, while at later times the radial response becomes larger. At the cross-over time when the radial and polar components are of similar strength, the scalar magnetic intensity becomes independent of colatitude. This is manifested by the occurrence of a "node" in the intensity of the spatiotemporal response. The occurrence time of the node is governed by the geometry and electrical conductivity of the sphere configuration. Its appearance provides useful insight into the transient geomagnetic induction process, but its detection in actual satellite-recorded induction data remains to be shown.

Acknowledgements. I would like to thank H. McCreadie and D. Wolf whose reviews of an earlier version of the manuscript are very much appreciated.

References

1. Ben-Menahem A, Singh SJ (1981) Seismic Waves and Sources. Springer-Verlag, New York.
2. Everett ME, Schultz A (1995) Geomagnetic induction in eccentrically nested spheres. Phys Earth Planet. Inter, **92**, 189–198
3. Hultqvist B (1973) Perturbations of the geomagnetic field. In: Egeland A, Holter O, Omholt A (eds) Cosmical Geophysics. Universitetsforlaget, Oslo
4. Martinec Z, Everett ME, Velímský J (2002) Time-domain, spectral-finite element approach to transient two-dimensional geomagnetic induction in a spherical heterogeneous earth. Submitted to Geophys J Int
5. McPherron RL (1995) Magnetospheric dynamics. In: Kivelson MG, Russell CT (eds) Introduction to Space Physics. Cambridge Univ Press, 400–458
6. Neubert T, Mandea M, Hulot G, von Frese R, Primdahl F, Jorgensen JL, Friis-Christensen E, Stauning P, Olsen N, Risbo T (2001) Oersted satellite captures high-precision geomagnetic field data. Eos, Trans AGU, (82), 81, 87–88

Night-time Ionospheric Currents

Hermann Lühr

GeoForschungsZentrum Potsdam (GFZ), Division 2, Solid Earth Physics and Disaster Research (*hluehr@gfz-potsdam.de*),

Summary. During night-time the electron density in the ionosphere is significantly reduced. As a consequence of this, it is expected that most of the currents in the ionosphere should die out. The validity of this assumption has been checked by scanning the literature for reports on current observations on the night side. The search was focused on events during magnetically quiet times as defined by the K_p index. Nighttime current observations have been published from high, middle and low latitudes. These findings demand a reconsideration of the term 'magnetically quiet condition' for field modeling purposes.

Key words: ionosphere, conductivity, ionospheric currents, electron density

1 Introduction

It is good practice in main and lithospheric field modeling to limit the data selection to the nightside orbital arc on magnetically quiet days. Ionospheric currents flowing predominantly below the satellite orbit generate magnetic fields which can hardly be separated from the internal field sources. During the dark hours the conductivity in the ionospheric E region – the layer where most of the horizontal currents are flowing – is reduced by about two orders of magnitudes. There has, however, never been a comprehensive study on the occurrence of currents on the nightside and their implication on field modeling. Besides these more practical aspects there is also a considerable plasma physical interest in the currents at night-time. Which processes drive these currents? Are there plasma instabilities involved in supporting these currents?

In the literature ionospheric currents on the nightside have been reported from various latitudes, even on magnetically quiet days. Detrick and Lanzerotti [1] present large-amplitude Sq variations in the southern polar cap during summer solstice. The systematic appearance of field-aligned currents (FAC) at mid-latitudes during post-midnight hours is reported by Saito et al. [7]. Rather recently Lühr et al. [5] have found evidence for F region currents in the equatorial region during the post-sunset hours.

In the sections to follow we briefly introduce the prime ionization processes then present current observations from the various latitudes and finally discuss their specific features and their relevance for field modeling purposes.

2 Ionization Processes

Electric currents require two ingredients to flow, conductivity and a force providing a differential motion of the oppositely charged particles. Here we will not address the question of driving forces, but concentrate on the ionization of the E region.

Dominant processes are the photo ionization and the collision-induced ionization. Most effective for the ionization is the extreme ultra violet (EUV) part of the solar spectrum The E region can to good approximation be described as a Chapman layer. From which we deduce a height-dependent ionization rate

$$q(h) = q_m \exp\left(1 - y - \frac{1}{\cos\chi} e^{-y}\right) \tag{1}$$

with y = $(h-h_m)/H$, where h_m is the height of the peak ionization rate, q_m and H is the atmospheric scale height. χ denotes the zenith distance of the sun, ranging from 0° to 90°. More important than the ionization rate is the electron density, n_e, which is depending on production and loss processes. At the altitude of the E region the electron/ion recombination is the governing loss process. For slowly varying production rates the electron density will adjust at an equilibrium

$$n_e = \sqrt{\frac{q}{\alpha}} \tag{2}$$

where α is the recombination coefficient. When considering day-time height-averaged E region electron density, which is the controlling parameter for the conductivity, a simple relation reveals, suitable for practical purposes

$$n_e = a\sqrt{\mathrm{I}\cos\chi} + b \tag{3}$$

where I is the EUV intensity which can be approximated by the solar flux index F10.7, a and b are the coefficients of the linear regression.

At higher altitudes, e.g. in the F region, electron attachment is the main loss process. Further more, charge transport becomes a prominent factor here. Thus no simple relations can be given for the electron density at this level.

Ionization by collision is important at auroral latitudes. The efficiency varies strongly with altitude and depends on the energy and the pitch angle of the precipitating particles. A maximum yield is reached at the altitude h_m

$$q_e(h_m) \propto \frac{F_e W_e}{H} \tag{4}$$

where F_e is the flux and W_e is the energy of the precipitating particles.

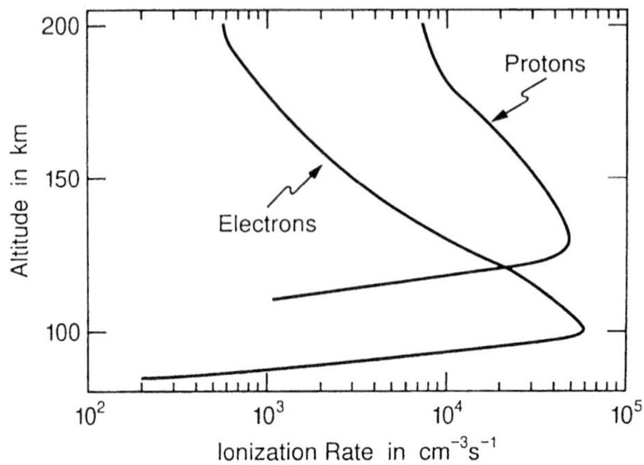

Fig. 1. Ionization due to precipitating of 10 keV particles

In a realistic atmosphere the peak ionization rate is reached at about 130 km and 80 km for electrons with energies of 1 keV and 300 keV, respectively. Figure 1 shows the height distribution of the ionization rate for precipitating particles with an energy of 10 keV. Obviously, electrons penetrate deeper into the atmosphere than ions. For an enhancement in the E region conductivity precipitating electrons in the energy range 1 to 10 keV are most effective. Such electron populations are also responsible for the visible auroras. Figure 2 shows the average location of the auroral oval. Along this ribbon a channel of enhanced conductivity and thus sizable currents, e.g. the polar electrojets (PEJ), are experienced during the dark hours.

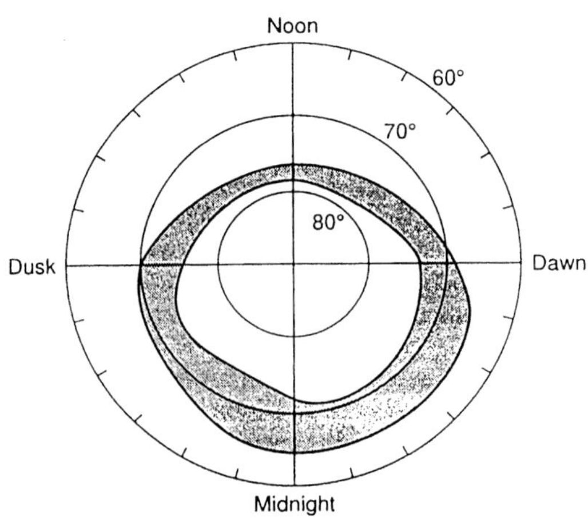

Fig. 2. Average location of the auroral oval for an activity level of K_p=2$^+$

3 Quiet-time high-latitude magnetic signature

The polar regions are known to be magnetically very active. It is thus particularly challenging to select an unperturbed data set from these regions for magnetic field modeling. The definition of night-time is not straight forward here, since there are seasons with continuous sun shine or darkness. Particle precipitation can modify the conditions at any time.

For this study we will make use of ground-based observations in the polar cap. In a recent paper Detrick and Lanzerotti [1] presented data from the Automatic Geophysical Observatory (AGO/P6) obtained during the solar minimum year 1997. With an invariant latitude of 85° AGO/P6 is located well within the polar cap. Figure 3 shows recordings of the three magnetic elements for the five quietest days of each month. There is a strong dependence of the observed field deflections on the season. This is particularly clear in the Z component. During summer solstice (Dec.) the amplitudes sometimes exceed 150 nT even on days with $K_p = 0$. During the dark months the deflections are smaller by almost one magnitude.

The authors have derived the persistent diurnal variation for each month. This can be used to construct an overhead current system. In Figure 4 a possible current configuration is sketched which could explain the quiet day magnetic deflections during summer solstice. Across the polar cap we find currents flowing from the evening side towards the 08 MLT sector.

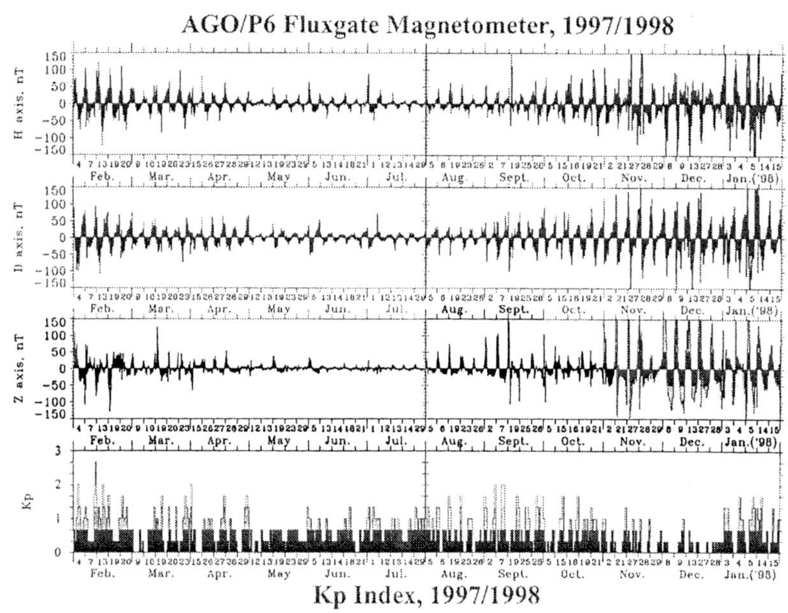

Fig. 3. Magnetic deflections recorded at the Antarctic station AGO/P6 for the five quietest days of the months (from [1])

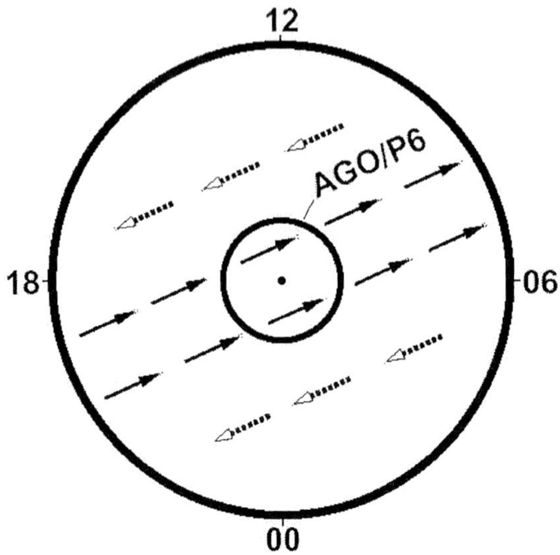

Fig. 4. Sketch of average quiet-day polar current system

Equatorward of the station some return currents are required to account for the ratio between horizontal to vertical deflections. Such a configuration is reminiscent of the well-known two-cell pattern, but the symmetry axis is rotated towards morning. In the paper by Friis-Christensen [3] such a tendency can be found for IMF $B_z > 0$. Further studies should be devoted to the properties of high-latitude current configurations during quiet times to confirm our suggestions.

4 Mid-latitude field-aligned current pattern

As reported by Saito et al. [7], the DE-2 satellite observed electric field fluctuations at the topside of the night-time mid-latitude ionosphere. Such fluctuations were often detected at magnetically conjugate points in the northern and southern hemisphere. Figure 5 shows the geographical distribution of its footprints in the E region. The two bands appear symmetrically on both sides of the dip equator and the occurrence rate peaks at 30° invariant latitude. Mid-latitude electric field fluctuations (MEF) occur preferably past midnight as can be seen in Figure 6. The earlier, less pronounced peak at 20 MLT could be caused by a different process e.g. the F region dynamo as described by Rishbeth [6].

The authors have tried to estimate the field-aligned current (FAC) density which is associated with the MEFs. Due to the coarse resolution of the vector magnetometer on DE-2 (3 nT digitization step size) the results are somewhat questionable. They report a FAC density of 0.2 to 0.3 $\mu A/m^2$. The source region of the FACs is sometimes found in the northern and sometimes in the southern hemisphere. The conjugate ionosphere is then acting as a load in the circuit.

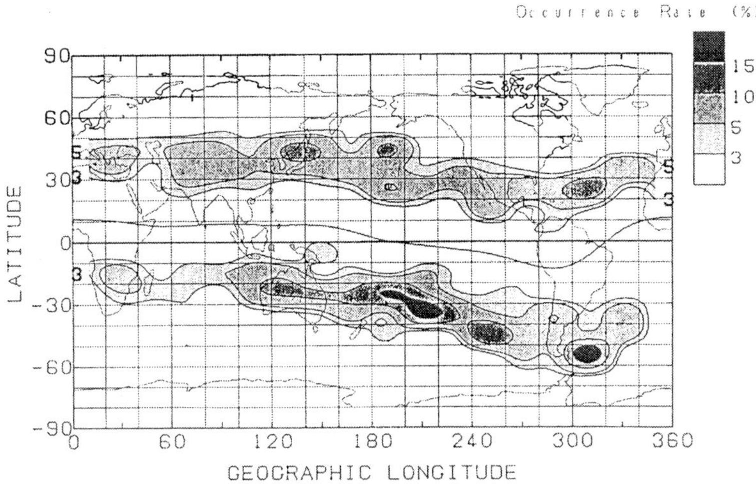

Fig. 5. Geographic distribution of the occurrence of MEFs (from [7])

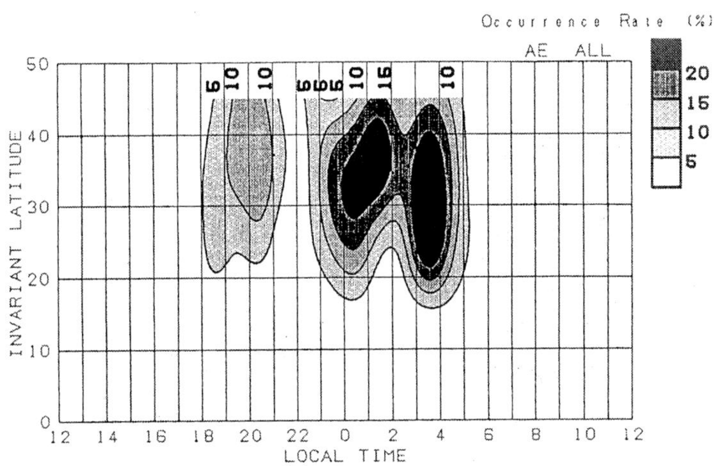

Fig. 6. Occurrence of MEFs with respect to local time and latitude (from [7])

The appearance of MEFs shows no clear dependence on the level of magnetic activity. They are assumed to be stimulated by medium-scale traveling ionospheric disturbances [8]. During the night hours the rate of decrease in electron density is higher at lower altitudes due to the faster recombination rate. The resulting vertical gradient in electron density lowers the threshold for launching plasma instabilities which in return stimulate the MEFs.

5 F region currents at equatorial latitudes

Currents in the F region can become important at night-time when the E region conductivity is small. At other times these currents will be by-passed through the lower E layer. It was now possible for the first time to perform in-situ observations of such F region currents with the CHAMP satellite [5].

On its polar, low-Earth orbit (<450 km) CHAMP is a well suited platform to study F region phenomena. The high precision scalar and vector magnetic field instruments allow to detect even faint current densities. A technique which has been applied rather successfully in many respects, is to compare the magnetic field readings from tracks where CHAMP has sampled the same area on ground. When looking at the differences between repeat tracks only the time varying parts will stand out. With this method ionospheric current features can be detected very reliably.

The upper frame of Figure 7 shows scalar magnetic field residuals of four repeat tracks from the longitude strip 321.5°±0.5° East. The main field was removed from the total field readings by subtracting a recent model. Furthermore, the ring current effect was eliminated by fitting a suitable function on an orbit-by-orbit approach. For a better readability the curves in Fig. 7 have been separated by 2nT. The two lower curves are very similar in shape. The persistent variations can be attributed to magnetic signatures of the crust. The two upper curves exhibit clear deviations from the common undulations in the equatorial region. Prime differences between the two pairs of tracks is the local time at which they have been taken.

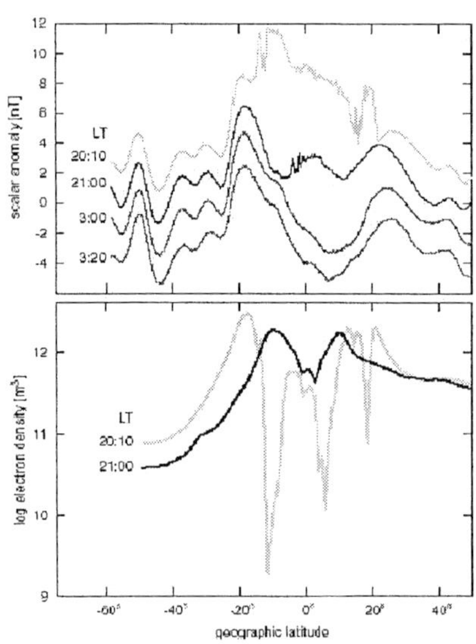

Fig. 7. CHAMP repeat tracks at 321.5° ± 0.5° longitude, (top) deflections of the magnetic field magnitude (bottom) in-situ electron density measurements (from [5])

In the lower frame of Figure 7 the electron density is shown, measured on board CHAMP by means of a Langmuir probe. The two curves were taken simultaneously with the two upper magnetic field traces. A typical feature are the deep density depletions observed during the 20:10 LT event. The depletions are accompanied by deflections in the magnetic field. During the 21:00 LT event there is only a small depletion close to the equator. Accordingly the magnetic perturbation is limited to a smaller amplitude and a more confined area.

After having seen the effect in a single longitude band, it was of interest to examine the global distribution of this phenomenon. Figure 8 shows the geographic distribution of the occurrence of scalar field disturbances. In the upper map, containing observations from before midnight, we find two bands of disturbances centered on the dip equator. The distribution with longitude exhibits high occurrence rates in the Atlantic sector, fewer over the Pacific and almost a void in the Indian sector. As can be seen in the lower map, no magnetic disturbance events were detected during the post-midnight hours.

We interpret these magnetic field fluctuations as indications for F region currents. Firstly, the magnetic deflections are accompanied by short-period fluctuations which represent scale sizes of about 100 km. As a rule of thumb, the distance to the source is comparable to the smallest scale size observed. With CHAMP at an altitude of about 450 km this brings the currents layer into the evening time F2

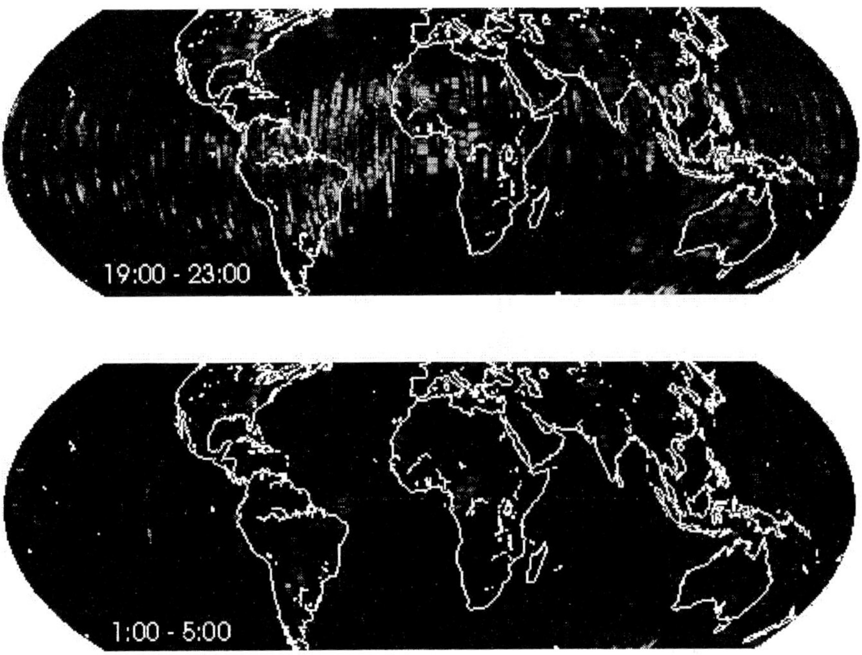

Fig. 8. Global distribution of F region current features (top) before midnight (bottom) after midnight (from [5])

2. Fejer BG (1997) The electrodynamics of the low-latitude ionosphere: Recent results and future challenges. J Atmos Sol Terr Phys 59: 1465-1482
3. Friis-Christensen E (1984) Polar cap current systems. In: *Magnetospheric Currents* (ed T A Potemra): 86-95, AGU Geophys Monogr 28, Washington D C
4. Huang CY, Burke WJ, Machuzak JS, Gentile LC and Sultan PJ (2001) DMSP observations of equatorial plasma bubbles in the topside ionosphere near solar maximum. J Geophys Res 106: 8131-8142
5. Lühr H, Maus S and Rother M (2002) First in-situ observation of night-time F region currents with the CHAMP satellite. Geophys Res Lett 29: (in press)
6. Rishbeth H (1971) Polarisation fields produced by winds in the equatorial F region. Planet. Space Sci 19:357-369
7. Saito A, Iyemori T, Sigiura M, Maynard NC, Aggson TL, Brace LH, Takeda M and Yamamoto M (1995) Conjugate occurence of the electric field fluctuations in the nighttime mid-latitude ionosphere. J Geophys Res 100: 21439-21451
8. Saito A, Iyemori T, Blomberg LG, Yamamoto M and Takeda M (1998) Conjugate observations of mid-latitude electric field fluctuations with the MU radar and the Freja satellite. J Atmos Sol Terr Phys 60:129-140
9. Whalen JA (2000) An equatorial bubble: Its evolution observed in relation to bottom side spread F and the Appleton anomaly. J Geophys Res 105: 5303-5315

Multiscale Determination of Radial Current Distribution from CHAMP FGM-Data

Carsten Mayer, Thorsten Maier

Geomathematics Group, University of Kaiserslautern, PO-Box 3049, 67653 Kaiserslautern

Summary. A multiscale method using spherical vectorial wavelets for the computation of radial current distributions from CHAMP-FGM data is presented. The considerations are essentially based on the Mie representation of solenoidal vector fields in terms of toroidal and poloidal parts as well as the Helmholtz decomposition of spherical vector fields. It will be shown how a vectorial wavelet representation of the geomagnetic field at the satellite's orbit can be used to determine a multiscale approximation of the radial current distribution at that height by considering the connection of the toroidal magnetic field scalar with the radial part of the current distribution via the Beltrami differential equation. The applicability and efficiency of the multiresolution technique will be illustrated and first results obtained from CHAMP-FGM data will be presented.

Key words: Mie representation, Helmholtz decomposition, Vector wavelets, Radial current distributions

1 Introduction

We start our considerations with a short note on notation: The standard notation in this paper is such that capital letters F, G etc. denote scalar functions. Vector fields are represented by lower-case letters f, g etc. . A related notation is used for function spaces, i.e. $\mathcal{C}^{(i)}(\Omega)$ and $c^{(i)}(\Omega)$ are the spaces of $i-$times continuously differentiable scalar respectively vector functions on the unit sphere Ω.

The theoretical basis for dealing with macroscopic electrodynamics are Maxwell's equations for polarized media. Neglecting the influence of Maxwell's displacement current density (cf. [2]) results in the decoupling of the equations for the magnetic case and it suffices to look at the static equations (i.e. the pre-Maxwell equations)

$$\nabla \cdot b = 0, \tag{1}$$

$$\nabla \wedge b = \mu_0 j, \tag{2}$$

where b is the magnetic induction, i.e. the magnetic field, and μ_0 is the vacuum permeability, $\mu_0 = 4\pi \cdot 10^{-7} VsA^{-1}m^{-1}$. Applying the divergence operator to (2) we get

$$\nabla \cdot j = 0 \ . \tag{3}$$

Theorem 3. *Let Ω_R be the (spherical) satellite's orbit. Then the current system on Ω_R is given by*

$$\begin{aligned} j(R\xi) = & \left[\frac{1}{\mu_0}\left(\xi \cdot \frac{\partial}{\partial r} \wedge L^*_\xi\right) F^r_3(\xi)\right]_{r=R} \\ & + \frac{1}{\mu_0 R}\left(\xi \cdot \Delta^*_\xi F^R_3(\xi) - \nabla^*_\xi F^R_3(\xi) - L^*_\xi F^R_1(\xi)\right) \\ & - \frac{1}{\mu_0 R^2} L^*_\xi \left[\frac{\partial}{\partial r} r^2 \frac{\partial}{\partial r} P_b(r\xi)\right]_{r=R}, \quad \xi \in \Omega \end{aligned}$$

with

$$\nabla^*_\xi F^R_2(\xi) = \left[\left(\xi \cdot \frac{\partial}{\partial r} \wedge L^*_\xi\right) P_b(r\xi)\right]_{r=R} - \frac{1}{R}\nabla^*_\xi P_b(R\xi).$$

Hence, it can easily be seen that

$$\xi \cdot j(R\xi) = \frac{1}{\mu_0 R}\, \Delta^*_\xi F^R_3(\xi), \quad \xi \in \Omega. \tag{6}$$

This theorem and equation (6) tell us that the radial part (radial projection) of the current system at the satellite's orbit Ω_R is totally determined by the Beltrami operator applied to the toroidal scalar of the magnetic field. Or vice versa radial currents crossing a sphere induce toroidal magnetic fields on that sphere. This result has already been determined in [9], where the author uses a spherical harmonic expansion to model the toroidal scalar of the magnetic field and gets the radial current distribution by simply applying the Beltrami operator to this series expansion which is a straightforward calculation.
We choose a different approach for modelling the toroidal Helmholtz scalar in terms of scaling functions and wavelets. Details of this techniques are presented in [5] and [7]. We will just recapitulate them now in a short manner. Let $l^2(\Omega)$ and $\mathcal{L}^2(\Omega)$ be the space of square-integrable vector respectively scalar fields on the unit sphere Ω. Let $\{Y_{n,l}\}, n \in \mathbb{N}_0, l = 1, \ldots, 2n+1$, be a system of $\mathcal{L}^2(\Omega)$–orthonormal (scalar) spherical harmonics.

Definition 1. The operators $o^{(i)} : C^{(1)}(\Omega) \to c^{(0)}(\Omega), i = 1, 2, 3$ are defined by

$$\begin{aligned} o^{(1)}_\xi F(\xi) &= \xi F(\xi), \\ o^{(2)}_\xi F(\xi) &= \nabla^*_\xi F(\xi), \\ o^{(3)}_\xi F(\xi) &= L^*_\xi F(\xi) = \xi \wedge \nabla^*_\xi F(\xi). \end{aligned}$$

For $i = 1, 2, 3$ we define a system of *vector spherical harmonics of type i* by

$$y^{(i)}_{n,l}(\xi) = \frac{1}{\sqrt{\mu^{(i)}_n}}\, o^{(i)}_\xi Y_{n,l}(\xi), \quad \xi \in \Omega$$

for $n = 0_i, 0_i + 1, \ldots l = 1, \ldots, 2n + 1$, where $\mu_n^{(1)} = 1$, $\mu_n^{(i)} = n(n+1)$ for $i = 2, 3$, $0_i = 0$ for $i = 1$ and $0_i = 1$ for $i = 2, 3$.
The use of the symbols 0_i is necessary, since there exist no constant tangential spherical vector fields but constant radial spherical vector fields.

The system of vector spherical harmonics is a complete system in $l^2(\Omega)$, i.e. every function $f \in l^2(\Omega)$ can be represented by its Fourier series in terms of vector spherical harmonics

$$f = \sum_{i=1}^{3} \sum_{n=0_i}^{\infty} \sum_{l=1}^{2n+1} f_{n,l}^{(i)} y_{n,l}^{(i)}$$

in the sense of the $\|\cdot\|_{l^2(\Omega)}$ −norm, where $f_{n,l}^{(i)} = (f, y_{n,l}^{(i)})_{l^2(\Omega)}$.
Next we want to define $l^2(\Omega)$-kernel functions, which we use to recover a function according to the Helmholtz theorem.

Definition 2. Let $k_J^\wedge(n), n \in \mathbb{N}_0$ be a sequence dependend on a parameter $J \in \mathbb{N}$ satisfying

$$\begin{array}{lll} (i) \sum_{n=0}^{\infty} ((2n+1)/4\pi) \left|k_J^\wedge(n)\right|^2 < \infty, & & \text{for all } J \in \mathbb{N}, \\ (ii) & k_{J'}^\wedge(n) \le k_J^\wedge(n) & \text{for } J' \le J \text{ and all } n \in \mathbb{N}, \\ (iii) & \lim_{J\to\infty} k_J^\wedge(n) = 1, & \text{for all } n = 0_1, 0_i + 1, \ldots . \end{array}$$

Then we call the function

$$k_J^{(i)}(\xi, \eta) = \sum_{n=0_i}^{\infty} \sum_{l=1}^{2n+1} k_J^\wedge(n) Y_{n,l}(\eta) y_{n,l}^{(i)}(\xi), \quad \xi, \eta \in \Omega$$

an $l^2(\Omega)$ *radial basis function of type* i.

These kernel functions can simply be understood as an approximation of the vectorial delta distribution. Relation (iii) of Definition 2 guaranties the convergence of the radial basis functions to the delta distribution. This fact also shows that the radial basis functions defined above are space localizing kernel functions. For increasing scale J the kernel functions localize more and more. Examples for such types of kernels can be found in [4] or in [5]. The easiest example is the Shannon scaling function given by $k_J^\wedge(n) = 1$ for $n = 0, \ldots, 2^J$. This kernel function has strong oscillations in the space domain which can be reduced by smoothing the edge in the frequency domain resulting in other scaling functions. An example of a radial basis function of type 3 on the sphere is shown in Figure 1 for two different scales. The property of space localization can clearly be seen.
The kernel functions defined above fit perfectly to the Helmholtz theorem, i.e. we are able to approximate each part of the Helmholtz components of a given vector field $f \in l^2(\Omega)$ separately, since

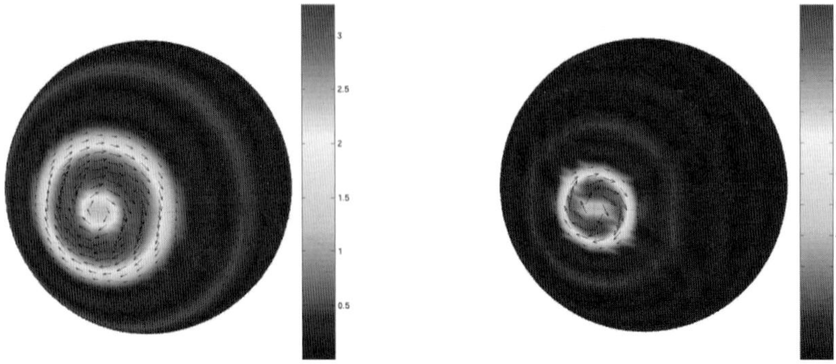

(a) scale $J = 3$ (b) scale $J = 4$

Fig. 1. CUP radial basis functions of type 3 (for the definition of CUP vectorial scaling functions and wavelets see [4]). Radial basis functions of type 3 are surface divergence free. They are desinged to approximate the toroidal part of a spherical vector field and thus can be used to compute the radial current distribution on a sphere. The space localizing property of this type of radial basis functions is clearly visible.

$$\lim_{J\to\infty} \int_\Omega \left(\int_\Omega f(\zeta) \cdot k_J^{(i)}(\zeta,\eta) d\omega(\zeta) \right) k_j^{(i)}(\,.\,,\eta) d\omega(\eta) = f^{(i)} \tag{7}$$

holds in the sense of the $\|\cdot\|_{l^2(\Omega)}$ −norm. So let $k_J^{(3)}$ be a radial basis function of type 3 and let $J \in \mathbb{N}$ be large enough, then we get from (5) and (7)

$$\begin{aligned} b^{(3)}(R\xi) = L_\xi^* F_3^R(\xi) &\simeq \int_\Omega \left(\int_\Omega b(R\xi) \cdot k_J^{(3)}(\zeta,\eta) d\omega(\zeta) \right) k_J^{(3)}(\xi,\eta) d\omega(\eta) \\ &= L_\xi^* \underbrace{\int_\Omega \left(\int_\Omega b(R\xi) \cdot k_J^{(3)}(\zeta,\eta) d\omega(\zeta) \right) K_J(\xi,\eta) d\omega(\eta)}_{F_3^R(\xi)}, \end{aligned}$$

where the scalar kernel K_J is just the former vector radial basis function with the vector spherical harmonic of type 3, $y_{n,l}^{(3)}$, replaced by $(\mu_n^{(3)})^{-1/2} Y_{n,l}$. The symbol $\simeq$ means that the error is negligible since the scale $J \in \mathbb{N}$ is chosen large enough. By applying the Beltrami operator to F_3^R we get

$$\Delta_\xi^* F_3^R(\xi) \simeq \int_\Omega \left(\int_\Omega k_J^{(3)}(\zeta,\eta) \cdot b(R\xi) d\omega(\zeta) \right) C_J(\xi,\eta) d\omega(\eta), \quad \xi \in \Omega \tag{8}$$

with the *radial current approximating kernel* defined by

$$C_J(\xi,\eta) = \sum_{n=1}^{\infty} \sum_{k=1}^{2n+1} -\sqrt{n(n+1)}\, k_J^\wedge(n)\, Y_{nk}(\eta)\, Y_{nk}(\xi), \quad \xi,\eta \in \Omega.$$

Thus by comparison of (6) and (8) we easily get

$$\xi \cdot j(R\xi) \simeq \frac{1}{\mu_0 R} \int_\Omega \left(\int_\Omega b(R\zeta) \cdot k_J^{(3)}(\zeta, \eta) \, d\omega(\zeta) \right) C_J(\xi, \eta) \, d\omega(\eta) \ . \tag{9}$$

Using appropriate integration rules on the sphere (see e.g. [5]), the convolutions in (9) can be discretized and be computed. Fast methods of computing the kernel functions and the convolutions in (9) can be achieved by using the addition theorem for scalar and vector spherical harmonics. Details about the numerical implementation can be found in [6].

3 Examples of Radial Current Distributions

As an example of our computations we show in Figure 2 the reconstructions of radial current distributions computed from CHAMP-FGM data. For the present study we took data of quiet magnetic days ($K_p \leq 2$) from August 2001 to September 2001 from which we subtracted the main field model Ørsted-10b-01 up to degree 13 (for details about the model see [10]). For applying our methods we averaged main-field residuals to a regular grid using an inverse-distance method (Related techniques have already been used by [8] for crustal field determination and [9] for the determination of radial current distributions). It should be pointed out that this is not the optimal method of averaging disturbed measurements to a regular grid. In [1] many algorithms are presented and the conclusion is that there is no best one. The method we use is very sensitive to data points lying close to the grid point at which the average is being calculated, but is not very robust for outliers. It should be combined with a method which takes the distribution of the data into consideration. This will be done in the future.
Since the radial basis functions we use are space localizing, we are also able to determine the current systems in a local framework, i.e by using only local data. This method is not presented here but will be applied in future work. It gives us the possibility to cater to the data situation. If on the one hand the coverage of "good" data is locally very dense we can determine also high frequent perturbations in the current systems. If on the other hand the data coverage is more or less scattered we can only reconstruct on a low scale level. This fact is a crucial difference to a reconstruction of the toroidal part of the magnetic field with vector spherical harmincs as applied in [9]. Spherical harmonic reconstruction needs a globally dense, uniformly distributed coverage of data.

Acknowledgements: We gratefully acknowledge the Graduiertenkolleg "Mathematik und Praxis", University of Kaiserslautern, and the German Research Foundation (DFG contract No. FR 761-10-1) for financial support. Furthermore we thank the GFZ Potsdam for having made CHAMP magnetic-data available.

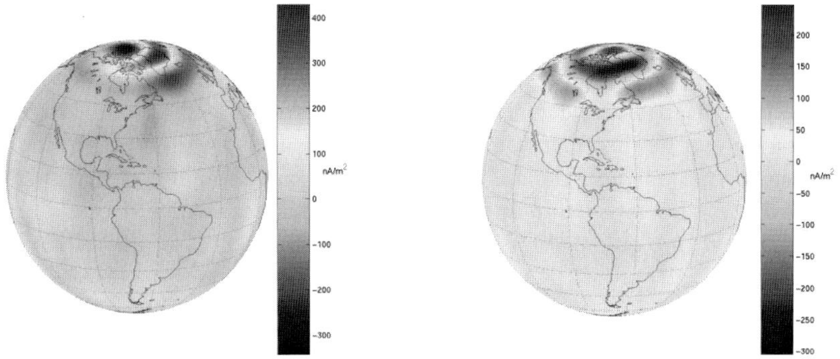

(a) day data (10:00-16:00 local time)

(b) night data (22:00-6:00 local time)

Fig. 2. Radial current distribution at the satellite's orbit computed with CHAMP main field residuals

References

1. Amm O (1998) Direkte Bestimmung der flächenhaften Verteilung ionosphärischer elektrodynamischer Parameter aus Bodenmessungen: Theorie und Anwendung in sphärischen Koordinaten. Doctoral Thesis, TU Braunschweig
2. Backus G (1986) Poloidal and Toroidal fields in geomagnetic field modelling. Reviews in Geophysics 24: 75-109
3. Backus G, Parker R, Constable C (1996) Foundations of Geomagnetism. Cambridge University Press, Cambridge
4. Bayer M, Freeden W, Maier T (2001) A vector wavelet approach to iono- and magnetospheric geomagnetic satellite data. J Atmosph and Solar-Terrestrial Physics 63: 581-597
5. Freeden W, Gervens T, Schreiner M (1998) Constructive Approximation on the Sphere (With Application to Geomathematics). Oxford Science Publications, Clarendon Press, Oxford
6. Maier T (1999) Multiscale analysis of the geomagnetic field. Diploma Thesis, Department of Physics, University of Kaiserslautern
7. Maier T, Bayer M (1998) Multiscale Analysis of the Geomagnetic Field. In: Freeden W (Ed.), Progress in Geodetic Science at GW 98. Shaker-Verlag, Aachen, pp 274-283
8. Maus S, Rother M, Holme R, Lühr H, Olsen N, Haak V (2001) First CHAMP satellite magnetic data resolve uncertainty about strength of the litospheric magnetic field, Geophys Res Lett submitted
9. Olsen N (1997) Ionospheric F region currents at middle and low latitudes estimated from Magsat data. J Geophys Res 102: 4563-4576
10. Olsen N (2002) A Model of the Geomagnetic Field and its Secular Variation for Epoch 2000. Geophys J Int 149: 454-462

Ionospheric Currents from CHAMP Magnetic Field Data - Comparison with Ground Based Measurements

Patricia Ritter[1], Ari Viljanen[2], Hermann Lühr[1], Olaf Amm[2], and Nils Olsen[3]

[1] GeoForschungsZentrum Potsdam (GFZ), Division 2, Solid Earth Physics and Disaster Research *pritter@gfz-potsdam.de*
[2] Finnish Meteorological Institute Geophys. Res. Div., Helsinki, Finnland,
[3] Danish Space Research Institute, Copenhagen, Denmark

Summary. One important contribution to the magnetic field measured at satellite height is due to the external current systems. Above the polar regions the scalar data sampled by the Overhauser Magnetometer on CHAMP are used to study the horizontal Ionospheric Hall Current system. A series of line currents, placed at high latitudes at a height of 110 km is fitted to the spatial variations of the residual magnetic field across these regions. The observed current distribution depends on the local time of the satellite track. At dawn/dusk orbits the Polar ElectroJets (PEJs) can be detected clearly in the northern and southern auroral regions. Its variation with time is evidently correlated with the AE activity index. For a period of six months the results obtained from CHAMP measurements are compared to the currents determined from ground based observations of the Scandinavian IMAGE magnetometer network. During periods of increased activity the two current estimates are in good agreement. At quiet times uncorrected lithospheric magnetic fields cause spurious current features in the satellite results.

Key words: Magnetic field, Polar ElectroJet, satellite, ground measurements

1 Introduction

The most intense current system in the ionosphere is that of the polar electrojets (PEJ) in the auroral oval. The strength and latitudinal position of these current flows depend on the solar wind activity and related magnetospheric processes. The primary PEJs are directed from the day to the night side, i.e. eastwards at dusk and westwards at dawn in both hemispheres. The ionospheric currents at high latitudes are determined from the scalar magnetic field data of the Overhauser magnetometer on CHAMP. Above the polar regions the horizontal currents are approximated by a series of infinite line currents [3]. A model is fitted to the data using Biot Savart's law.

For the inversion the currents are positioned at a height of 110 km above the surface of the Earth. They are separated horizontally by 1° of latitude (also approx. 110 km). For the inversion algorithm an orbit segment of ±60°

centered at the closest approach to the geographic pole is used. The scalar readings are available at a rate of 1 Hz, which is equivalent to 16 measurements per degree of latitude. The data processing includes subtraction of the main field (OIFM, [4]) and elimination of the ring current effect (D_{ST} correction), trend correction over the entire orbit segment and parabolic tapering of the currents at the edges of the interval. In section 2 we present the temporal and geographical dependency of the currents. The comparison with ground based measurements over a period of half a year is given in section 3.

2 Currents at northern and southern polar regions

In October 2000 CHAMP was in a dawn/dusk orbit, thus crossing the PEJs perpendicularly. Fig.1 shows the currents determined in the polar regions of the northern and southern hemisphere. The time period includes all 31 days of the month. The satellite track can be imagined as crossing the contour plots from top to bottom for each orbit. The PEJs are prominent along two ribbons of enhanced current density at geomagnetic colatitudes of about ±20°. The positive amplitudes mark anti-sunward directed currents. The currents in the polar cap region are weaker and directed oppositely as expected. Strength and latitudinal extension of the current flow are correlated clearly with the amplitudes of the AE index.

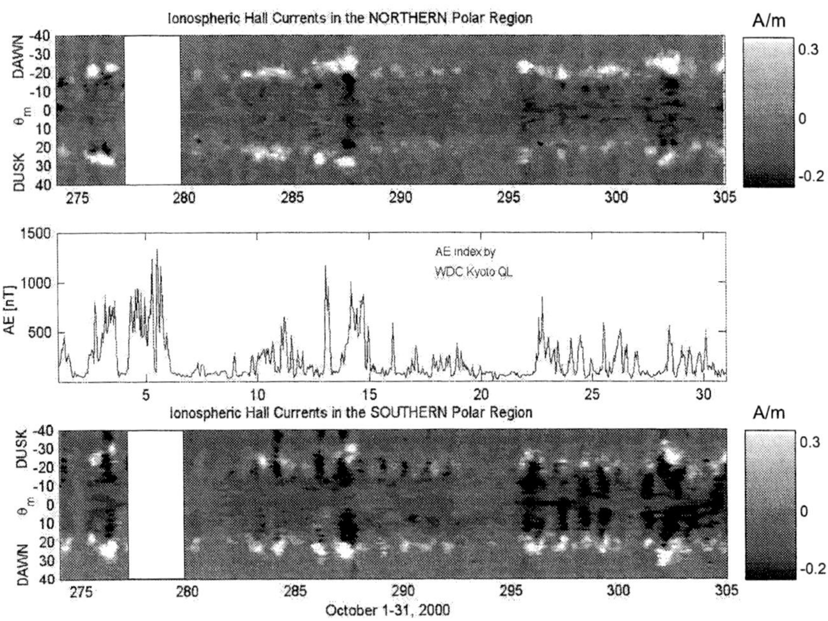

Fig. 1. Ionospheric currents in the polar regions of the northern and southern hemispheres (geomagetic colatitudes). For comparison the AE index is plotted in the middle graph. Horizontal scale one month, given as day numbers in MJD2000.

In the subauroral regions (equatorwards of the PEJs) the amplitude levels of the currents show a systematic difference for dawn and dusk. This is no edge effect of the model, as the considered orbit arcs are considerably longer than shown in the contour plots, and tapering is applied in the omitted parts. The different amplitudes might rather hint to a non-sufficient D_{ST} correction due to the asymmetric ring current.

The orbital plane of CHAMP is changing only slowly with local time and also the PEJ is organised primarily in a geomagnetic local time frame. Scanning the PEJ over a certain period reveals its average properties in a given local time sector. Fig.2 shows the current density distribution separately for dawn and dusk and for both hemispheres for a period of three days on Oct. 10-12, 2000. The AE index during those days was moderate, hence average conditions can be assumed.

As the Earth is rotating below the orbit during the considered 45 overflights, the mean location of the PEJ for a given time emerges. The rings of enhanced current density at an angular distance of approx. 20° from the corrected geomagnetic poles mark the auroral ovals. Differences in longitude could either be due to the asymmetry of the geomagnetic field, differences in

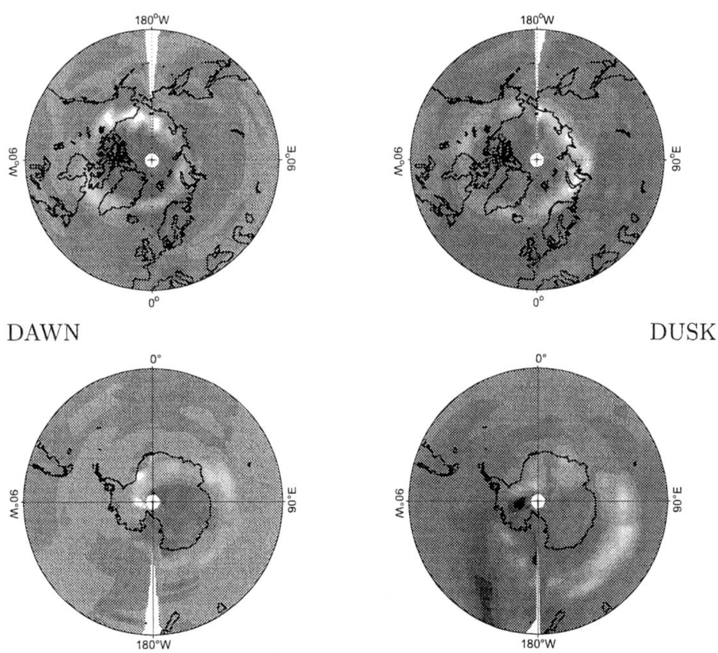

Fig. 2. Scan plots of the ionospheric currents over three days in October 2000 (dawn/dusk orbit) in the polar regions of the northern and southern hemispheres. Gray scale see Fig.1; the positive currents are directed westward on the dawn side and eastward on the dusk side.

the solar aspect angle of the relevant parts of the ionosphere, difference in the magnetic activity, or different contributions from the induction effects. Much larger studies are needed for deriving statistically significant results in order to distinguish between all these effects.

3 Comparison with ground based measurements

During a period of 6 months we identified 126 satellite overflights across the region of the IMAGE magnetometer array [1]. The currents determined from CHAMP data were compared to currents computed from the ground based measurements. The site locations are shown in Fig.3.

For the IMAGE data, the eastward current densities (y-direction) were determined at ionospheric height (100km) by 1D upward continuation of the north (x-) component of the magnetic field [2]. Fig.4 shows the current density curves of 16 overflights as examples.

The agreement of the curves is good in most examples. The CHAMP curves (solid lines) show a persistent minimum at approx. 25° co-latitude, which is not confirmed by the ground data (dashed lines) - a clear hint that the CHAMP data used for the inversion procedure are influenced by another source, probably crustal field anomalies. The minimum is evident in all curves. It is also visible in the contour plots of Fig.5 which include all overflight

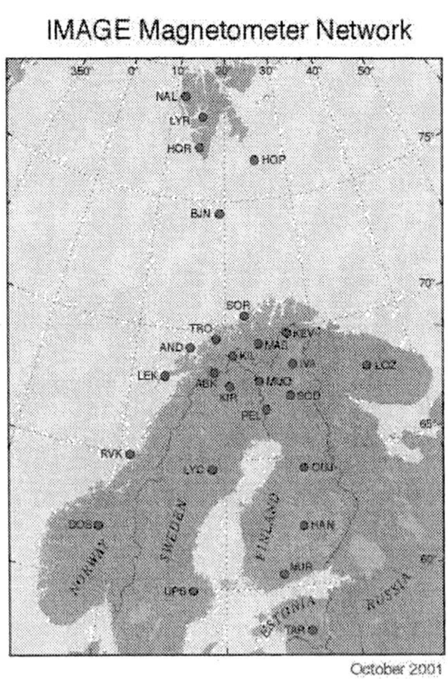

Fig. 3. The IMAGE magnetometer array (see http://sumppu.fmi.fi/image).

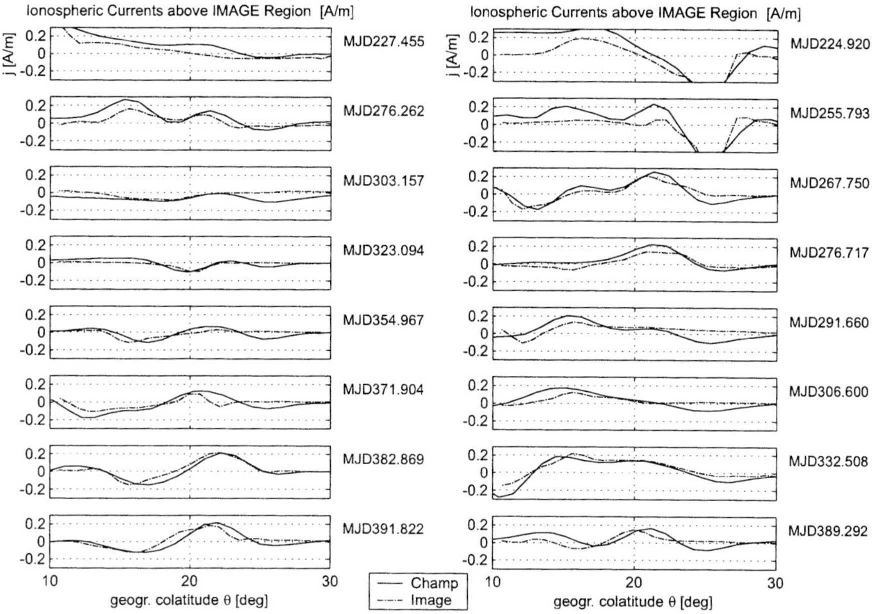

Fig. 4. Comparison of the current densities determined by CHAMP (solid lines) and IMAGE ground measurements (dashed lines) between Aug 2000 (MJD 227) - Feb 2001 (MJD 390) on a N/S profile across the IMAGE region ($10° - 30°$ geographic co-latitudes). Left side: the satellite crossed the region on ascending tracks; right side: only descending tracks.

events sorted by ascending and descending tracks on the left and right side, respectively.

Currents determined from CHAMP magnetic field data are displayed at the top, those from the IMAGE stations are presented in the two bottom frames. Times when data were corrupted or not available are blanked out. During periods of increased activity both current estimates are in good agreement. At quiet times uncorrected lithospheric magnetic fields cause spurious current features in the satellite results.

4 Conclusions and outlook

We used CHAMP magnetic field data to estimate ionospheric currents in the polar regions. For the validation of the results estimates from ground based observations were employed. Generally a good agreement was found. There are, however, systematic discrepancies apparent, which need further consideration. Among them are a better correction of the ring current effect and a removal of the crustal magnetic signatures. At the ground-based end

2 Data Processing

We use Ampère-Maxwell's law, $\mu_o \boldsymbol{I}=\nabla\times\boldsymbol{B}$, to calculate FAC intensities. A single satellite can measure the magnetic field variations in one dimension only; hence additional information is needed. Field-aligned current (FAC) structures in the auroral regions are often sheet-like. The FAC may thus be determined with a fair degree of accuracy from the gradient of the magnetic perturbations associated with the currents as observed by a satellite crossing through the structure. However, in the central polar cap the currents structures are not expected to be sheet-like, hence other algoritms for the data analysis are needed.

The basic assumption used here, like, e.g., in *Weimer* [2001], is that the FAC distribution is deterministic. Hence repeated scanning by differently oriented orbits during the same specific conditions will in due time provide a complete mapping of the related 2-D spatial distribution of magnetic perturbations over the polar regions. Here we have avoided a global functional representation of the magnetic perturbations or the FAC currents. Instead we use bi-variate interpolation [*Akima*, 1978] to derive for a fine grid over the central polar regions numerical values for the magnetic perturbations associated with the selected conditions.

The available data have been sorted on the basis of strict criteria to ensure steady interplanetary magnetic field (IMF) and solar wind (SW) plasma conditions. For the present *NBZ* study it is, furthermore, required that B_z is the dominant IMF component. These criteria are specified in Table 1.

The magnetic perturbations are found by subtracting model field values from the measured data. For the Magsat measurement the GSFC(12/83) geomagnetic reference field model for epoch 1980 [*Langel*, 1987] was used. In addition to the internal field contributions, the contributions from the magnetospheric ring current have been included through the *Dst*-dependent terms given by *Langel* [1987]. For the Ørsted magnetic vector data the Oersted (10c/99) field model for epoch 2000 [*Olsen et al.*, 2000] has been used to calculate the model field. After subtraction of the *Dst*-corrected model field from Magsat (0.5s or 5s samples) and Ørsted (1s samples) all perturbation data are converted to 5 sec averages. This temporal resolution corresponds to a spatial resolution of around 35 km along the orbit. The positions of the Magsat and Ørsted satellites are defined in standard altitude adjusted corrected geomagnetic (aacgm) coordinates and magnetic local time.

Case	IMF Bx	IMF By	IMF Bz	ΔB[1)]	Brms[2)]	Unit
Z1	$-2<Bx<2$	$-1<By<1$	$0<Bz<2$	2	1	nT
Z2	$-5<Bx<5$	$-2<By<2$	$2<Bz<5$	3	2	nT
Z3	$-10<Bx<10$	$-3<By<3$	$5<Bz<10$	5	3	nT
Z4	$-20<Bx<20$	$-5<By<5$	$10<Bz<20$	10	5	nT

[1)] ΔB is max extension of range for previous set of IMF parameters

[2)] Brms is max RMS value of IMF parameter during total time range

Table 1. Selection criteria for IMF parameters for specific *NBZ* cases.

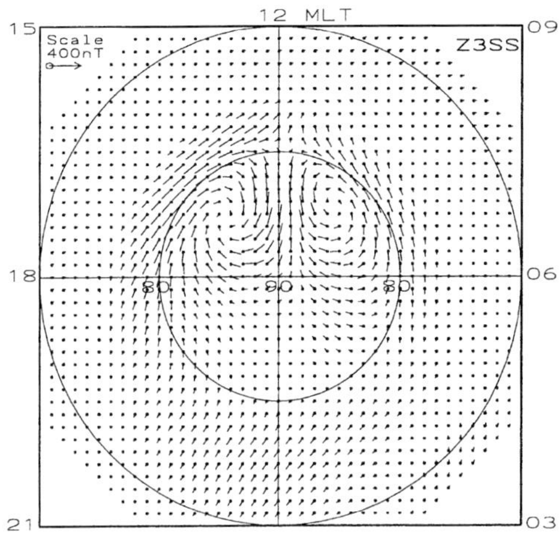

Fig. 1. Magnetic perturbation vector field for the southern polar cap referred to 120 km altitude in aacgm magnetic latitude-magnetic local time coordinates. Magsat and Ørsted data have been combined for summer passes occurring during cases of +5 < IMF B_z

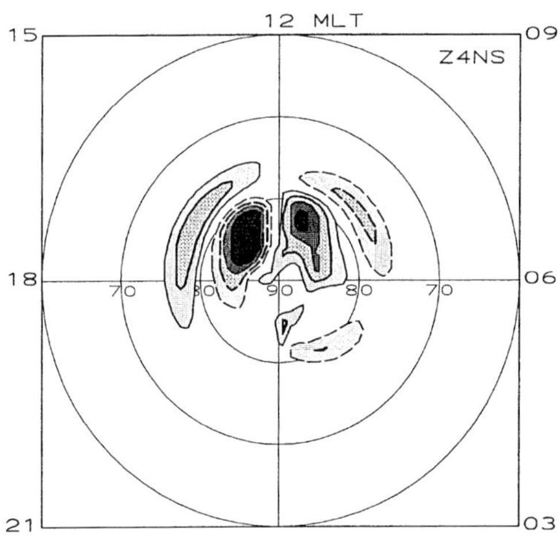

Fig. 2. Derived FAC distribution for summer conditions, for the northern polar cap and for the strongest available case, +10< IMF B_z<+20 nT (*Z4NS*). The contour lines drawn with solid line indicate upward FAC while broken lines indicate downward FAC. The step between contours is 0.25 μA. The outer FAC structures are *R1* currents, the inner are *NBZ* currents.

The resulting distribution of magnetic perturbations over the polar cap for specific conditions defined by *Z3* conditions (c.f. Table 1) for Southern Summer (i.e. *Z3SS* conditions) are shown in Fig. 1.

From such data the resulting distribution of upward and downward FAC is easily derived using Ampère-Maxwell's law. We show in Fig. 2 the FAC distribution derived from the strongest available FAC case, +10< IMF B_z<+20 nT (*Z4NS*).

4 Seasonal Variations

Based on the available data we have examined, among other, the seasonal distribution of FAC. In this study we have handled the two polar caps separately. The results are displayed in Fig. 3 for the *Z3* case, that is, for +5 < IMF B_z < +10 nT. The figure displays the marked seasonal variation. For both hemispheres the *NBZ* FAC currents are sizeable during local summer conditions only.

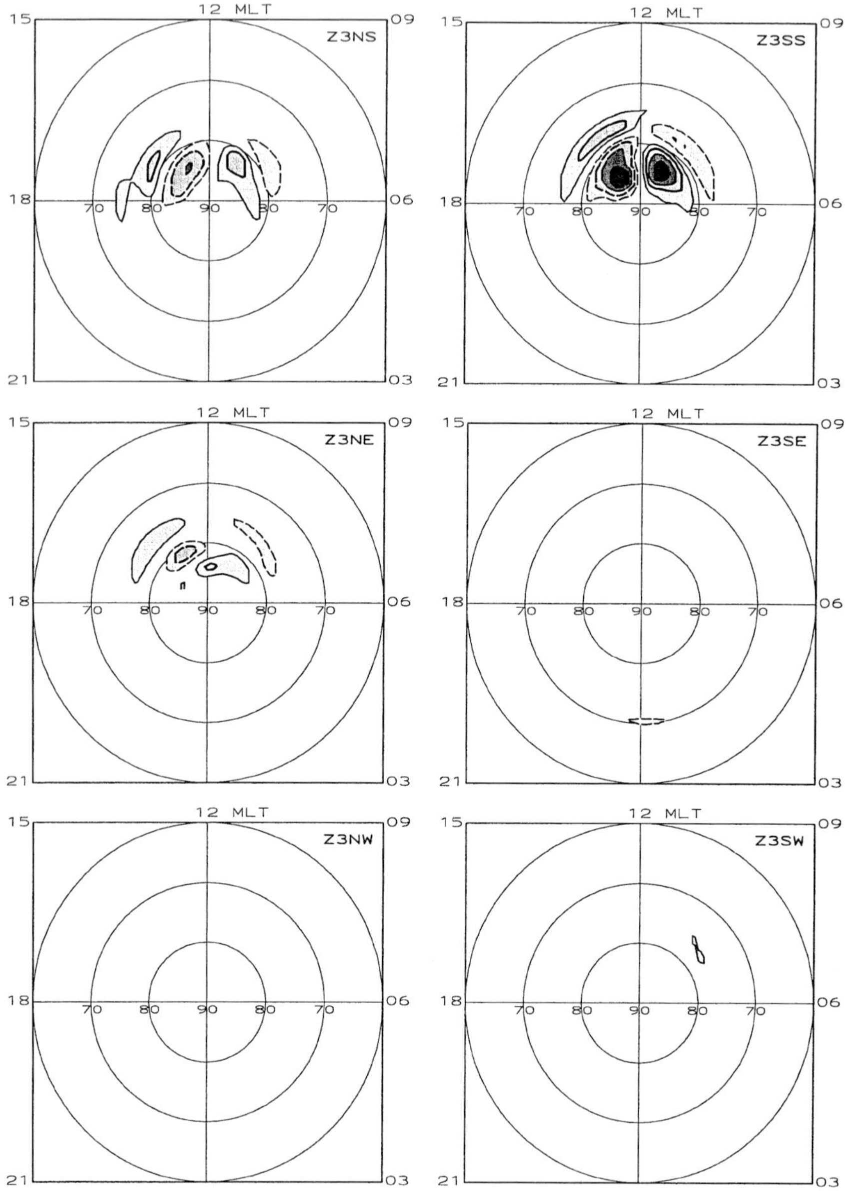

Fig. 3. Seasonal variations in NBZ field-aligned currents. Solid line contours depict up· ward FAC, broken line contours indicate downward FAC. Contour interval 0.25 $\mu A/m^2$ Zero contours are not shown. Top fields display Arctic (left) and Antarctic (right) summe conditions. Middle fields display equinox, and bottom field winter conditions.

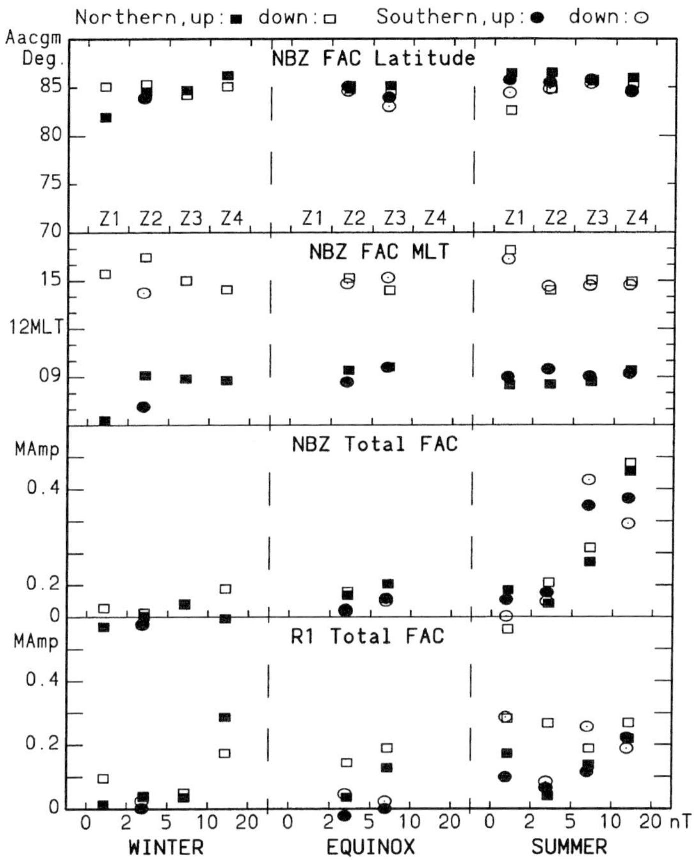

Fig. 4. Synthesis of variations in NBZ FAC currents. The figure is divided in three vertical panels for Winter, Equinox, and Summer conditions, respectively. Within each panel there is a scale marking IMF B_z values from 0 to +20 nT. The points in the top field represent the latitude of the center of NBZ currents which seems to be very steadily located at 85° (aacgm) latitude. The next field presents the MLT time for the center of FAC distribution. The two centers are positioned close to 0900 and 1500 LMT, respectively, through all conditions. The third field displays total NBZ FAC upward or downward current for the daytime polar cap region. The most remarkable feature is the shifted rise during summer conditions. The NBZ FAC currents are low up to a threshold of around +5 nT for IMF B_z and then rise strongly. The Region-1 (R1) total FAC currents (bottom) are much stronger during summer than during winter. They are much stronger for IMF B_z <0 (not shown).

5 Delayed effects from IMF Bz changes

The various effects from changes in the interplanetary magnetic field are sometimes delayed by substantial amounts. An example is shown in Figure 5. Here we show the convection response to a step-like change in IMF B_z from -10 nT to +10 nT. Note the gradual change in convection clock angle seen in bottom field.

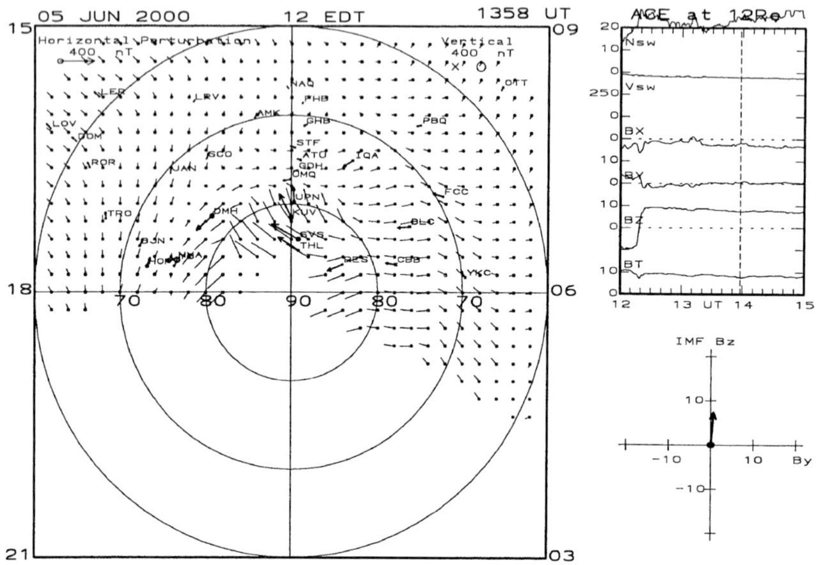

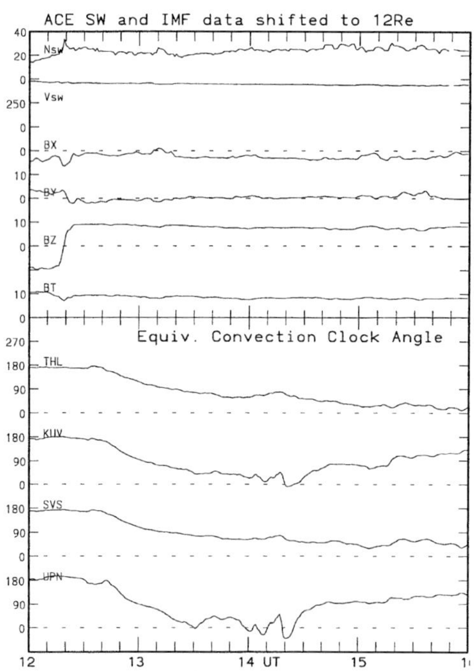

Fig. 5. (a) Top field displays equivalent convection at 1358 UT on 5 June 2000 derived from polar magnetometer recordings. Top right field displays ACE SW and IMF data shifted to 12 Re. The IMF is also shown in the dial plot. Note that the field is almost straight northward and has a strength close to 10 nT.

(b) The upper part of left field displays (again) the SW and IMF data from 1200 to 1600 UT. The lower part displays the clock angle for the equivalent convection at the polarmost stations from Thule (top) to Upernavik (bottom). Note the sudden step in IMF B_z from negative (-10 nT) to positive level (+10 nT) at 1220 UT. The convection clock angles start to change 15 min later. However, the change is very gradual. It takes around 1 hour to complete the shift from antisunward (180°) to sunward (0°) flow.

6 Discussions

The mapping of *NBZ* currents obtained in this study confirms the results of *Weimer* [2001] and some of the earlier works like, for instance, *Araki* et al. [1984]. The extensive data base now available adds to define further characteristic features of the *NBZ* FAC system. The seasonal behavior with vanishing *NBZ* FAC currents during winter conditions, no doubt, relate to the extremely low conductivity present in the central polar cap ionosphere during this season. At the high latitudes in question, solar UV ionization processes are weak during local winter.

Furthermore, as the combined result of the high latitudes in question and the infrequent occurrence of substorms during *NBZ* conditions the energetic particle fluxes are very weak; hence the precipitation-induced ionization is low. The resulting ionospheric conductivity in the central polar caps could well be more than an order of magnitude lower during winter than during summer conditions. This feature impedes the closure of FAC in the winter ionosphere.

Another specific observation noted above is the steady locations of the centers of the *NBZ* FAC system at 85° aacgm latitude and 0900 and 1500 MLT, respectively. The *NBZ* FAC regions extend poleward of the cusp/cleft latitudes defined, for instance, by *Newell et al.* [1989]. Hence the *NBZ* FAC regions must be characterized as the ionospheric projections of mantle/lobe regions. Their constant positions agree with the observation that the cusp/cleft location changes little with varying IMF B_z when $B_z>0$ [*Newell et al.*, 1989].

For the possible source mechanism for the *NBZ* FAC currents the complementary characteristics of *R1* and *NBZ* currents with respect to location, direction and intensity variation with IMF B_z indicate that their generation mechanisms are similar while their geometries are complementary. The dayside *R1* currents are presumably generated in the low latitude boundary layer (LLBL) just inside the magnetopause at the front and along the flanks of the magnetosphere [*e.g., Potemra*, 1996]. In the low latitude *R1* source regions the magnetospheric field direction is northward. The *R1* currents, are enhanced during increasingly negative IMF B_z since this orientation favors merging processes.

The equivalent FAC generation mechanism in the high-latitude boundary layers (HLBL) have reversed geometry since the magnetospheric field in the possible source region (the frontal mantle and lobe regions) has a southward component. The *NBZ* FAC, accordingly, should flow opposite of *R1*, that is, up at the morning side and down at the evening side, they should be located poleward of the *R1* FAC, and they should increase with increasing positive IMF B_z. These anticipated features agree well with observations.

A particularly interesting feature is the temporal development of the polar convection patterns following a sudden IMF B_z shift to a positive level. The slow and gradual change could be related to a reconfiguration of the magnetosphere during *NBZ* conditions to give a more inflated shape at the upper and lower mantle which, in turn, would further enhance the *NBZ* FAC generation processes.

7 Conclusions

Using the combined data base from the Magsat and Ørsted missions with strict criteria to select cases of steady NBZ conditions we have noted these points.

- The *NBZ* FAC system comprises a pair of upward and downward currents flowing along field lines extending from rounded or bean-shaped regions of the polar ionosphere. The regions are centered at around 85° aacgm magnetic latitude and at 0900 MLT for the upward and at 1500 MLT for the downward currents. The location of the *NBZ* FAC currents are invariant to season and to IMF B_z level.
- The *NBZ* FAC currents are of substantial magnitude at the dayside polar cap and during the summer season only. They vanish at the night side and in winter.
- The total downward and upward *NBZ* FAC currents are about equal and amounts typically to $0.4{\cdot}10^6$ A at IMF B_z=+10 nT and in the summer season.
- The total *NBZ* currents are strongly non-linear functions of the IMF B_z level. They are vanishing small for IMF B_z<+5 nT, rise strongly for +5<B_z<+10 nT, and level out for +10 nT<IMF B_z.
- The characteristics found for the *NBZ* FAC system suggest that these currents constitute a competing high-latitude boundary layer counterpart to the low-latitude boundary layer *R1* FAC system. A positive IMF B_z level is required to reconfigure the magnetosphere thus imposing an IMF B_z threshold on forming *NBZ* FAC. Following sudden IMF changes the reconfiguration processes may last 30-60 min to let the coupled magnetosphere-ionosphere system reach steady conditions.

Acknowledgements. We gratefully acknowledge the use of ACE data supplied from the ACE Science Center (ACS), IMP8 and Magsat satellite data made available by the National Space Science Data Center (NSSDC), and Ørsted satellite data provided by the Oersted Science Data Center (OSDC) at DMI.

References

Akima H (1978) A method for bivariate interpolation and smooth surface fitting for irregularly distributed data points, *Assoc Comput Trans Math Software*, *4* (2), 148.

Araki T, Kamei T, and Iyemori T (1984) Polar cap vertical currents associated with northward interplanetary magnetic field, *Geophys Res Lett*, *11*, 23.

Langel RJ (1987) The Main Field, in: Jacobs JA (ed), *Geomagnetism*, Vol 1, Ch4, Academic Press.

Neubert T et al (2001), Ørsted satellite captures high-precision geomagnetic field data, *EOS*, *82* (7), 81.

Newell PT, Meng C-I, Sibeck DG, and Lepping R (1989) Some low-altitude cusp dependencies on the interplanetary magnetic field, *J Geophys Res*, *94*, 8921.

Olsen N et al (2000) Oersted Initial Field Model, *Geophys Res Lett*, *27*, 3607.

Potemra T (1994) Sources of large-scale Birkeland currents, in: Holtet JA and Egeland A (eds), *Physical Signatures of Magnetospheric Boundary Layer Processes*, p3, Kluwer Academic Publishers.

Weimer DR (2001) Maps of ionospheric field-aligned currents as a function of interplanetary magnetic field derived from Dynamics Explorer 2 data, *J Geophys Res*, *106*, 12,889.

Field-aligned Currents Inferred from Low-Altitude Earth-Orbiting Satellites and Ionospheric Currents Inferred from Ground-Based Magnetometers — Do They Render Consistent Results?

Jurgen Watermann[1], Freddy Christiansen[1], Vladimir Popov[2], Peter Stauning[1], Ole Rasmussen[1]

[1] Danish Meteorological Institute, Lyngbyvej 100, Copenhagen, Denmark
[2] IZMIRAN, Troitsk, Moscow Region, Russia
jfw@dmi.dk, fch@dmi.dk, vpopov@izmiran.rssi.ru, pst@dmi.dk, or@dmi.dk

Summary. Field-aligned currents (FACs) and ionospheric Hall currents were inferred from spatially and temporally coincident vector magnetometer measurements made along the Greenland west coast onboard the Ørsted satellite and on the ground, respectively. We selected satellite passes which were closely aligned with the ground-based magnetometer chain, and investigate whether the trajectory-integrated FAC (TIFAC) is consistent with the ionospheric Hall current across the line of magnetometers. Our analysis builds on the assumption that FACs are closed in the ionosphere via Pedersen currents driven by an electric field which also drives ionospheric Hall currents. Solar radiation conductance models suggest that the Hall/Pedersen conductance ratio, and consequently the Hall/Pedersen current ratio, and eventually the Hall/TIFAC ratio, should be slightly greater than 1. We examined 236 cases covering the midnight–morning–noon sector and found at times small and at other times large deviations from this ratio. We suggest that the latter cases represent conditions where the infinite current sheet assumption is violated or the ionosphere exhibits strong conductivity gradients or conductances are largely caused by charged particle precipitation. Deviations from a ratio of 1 appear to occur predominantly equatorward of $\approx 74°$ corrected geomagnetic latitude (CGML) in the midnight sector and poleward of $\approx 72°$ in the noon sector.

Key words: high-latitude ionosphere, ionospheric electric currents

1 Introduction

The existence of field-aligned currents (FACs) has been suggested by Birkeland as early as 1908, but direct observation of the magnetic field associated with FACs has only been possible since earth-orbiting spacecraft were equipped with suitable magnetometers. The first rather comprehensive high-latitude FAC distributions were derived from TRIAD observations [17, 1]. A considerable amount of high-precision magnetic field data from geospace has subsequently been acquired, not least during the MAGSAT mission [16], and more recently, with Ørsted [4] and CHAMP [this volume]. Field-aligned currents are now a well-established feature of magnetosphere–ionosphere coupling, and FAC dependence on the solar wind and interplanetary magnetic field has been mapped and modeled in great detail [5, 6, 8, 15].

Concurrently with satellite observations, a large amount of ground-based magnetometer data has been acquired at all latitudes particularly in the northern hemisphere. Ground-based magnetometers are sensitive to the divergence-free part of the horizontal current system which is primarily a Hall current flowing horizontally in the ionosphere. They do not normally detect the magnetic field of the curl-free part, i.e., the Pedersen current, and of the FAC [2]. However, induced telluric currents can contribute substantially to the observed magnetic field [9].

The region poleward of some 55°–60° geomagnetic latitude (from where field lines typically map to the ring current zone or to magnetospheric boundary layers) is of specific interest because this is the region where the ionosphere provides for the low-altitude closure of magnetospheric FACs *via* Pedersen currents. FAC and ionospheric Pedersen current are therefore closely related. And since the same electric field that drives the Pedersen current also drives a Hall current we must expect that FAC and Hall current are also closely related. This has indeed been confirmed in specific cases through simultaneous measurements in space and on the ground [11, 13].

It is the purpose of this paper to expand on these studies by carrying out a survey which involves all available magnetic field observations made simultaneously along the Greenland west coast and onboard the Ørsted satellite when it passed over western Greenland. Observations from Ørsted are particularly suited to study this topic because its polebound trajectory is under optimal conditions almost perfectly aligned with the best-fit great circle through the western Greenland magnetometer chain.

2 Method

2.1 Data Selection

The Ørsted orbit inclination of 96.5° is almost identical to the best-fit great circle through the Greenland west coast magnetometer stations (96.4° inclination). Mapping the satellite trajectory along the geomagnetic field down to the ionospheric electrojet layer has little influence since the satellite orbits at only 650–865 km altitude. Nevertheless, we applied accurate field line tracing based on the IGRF2000 model [3]. For the rest of this paper, all references to Ørsted parameters, including position, trajectory, trace, FAC density, integrated FAC, etc., are those resulting from mapping down to 110 km altitude.

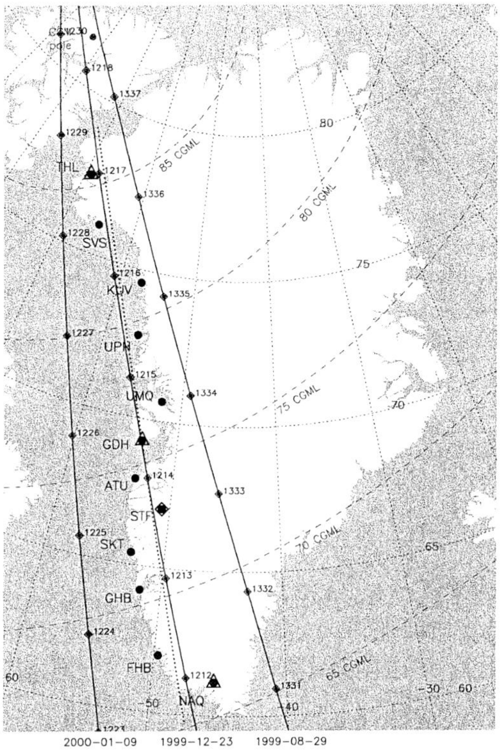

Fig. 1. Ørsted sample passes (solid lines) over west Greenland with UT marks along the traces; full circles – magnetometer sites; dotted straight line – best-fit great circle; dotted grid – geogr. longitude and latitude; dashed lines – CGML contours

Fig. 1 shows a map of Greenland with the ground magnetometer stations (full circles with 3-letter codes), the best-fit great circle (dotted line) and the traces of three polebound Ørsted sample passes (solid lines). The eastern and western traces border the strip within which the satellite trajectories must fall in order to be considered for inclusion in our data set. The full width of the strip is ≈700 km at the southern tip of Greenland. We inspected all orbits from the years 1999–2001 which fell within that strip, and retained those which had sufficiently continuous vector magnetometer coverage. Passes where gaps of more than five consecutive samples (≈42 km) occurred over Greenland were discarded. A total of 236 passes fulfilled the criteria and were thus used in our analysis.

2.2 Single Pass FAC Density Inference

We assume that the FACs form a sequence of parallel current sheets consisting of two components, oriented parallel and perpendicular to the satellite trajectory, respectively. In this paper we consider only the latter. The magnetic field gradient along the projection of the satellite trajectory on a plane perpendicular to the main geomagnetic field is measured as the satellite travels along. The gradient perpendicular to the trajectory in the same projection plane vanishes because it is aligned with the uniform current sheets under consideration. The field-aligned gradient vanishes per definition as we consider only currents which are field-aligned. Thus we have only one single component of the gradient vector which is different from zero. The FAC density (for sheets perpendicular to the satellite trajectory) is then derived from $\nabla \times \mathbf{B}$ and mapped down to 110 km altitude.

For each satellite pass we integrated the FAC density along the trajectory, starting at 55° CGML, in order to obtain a trajectory-integrated perpendicular FAC (TIFAC) for every pass at every point along the trajectory. Inspection of Ørsted magnetic field data revealed that a substantial magnetic field perturbation was practically never observed equatorward of 55° CGML.

2.3 Ionospheric Hall Current Inference

The equivalent ionospheric current, assumed to form a thin horizontal sheet 110 km above the ground, was determined by applying an inversion scheme ([7] with minor modifications) after the quiet-time magnetic field had been subtracted from the measurements. The algorithm requires two vector components of the magnetic variations along the Greenland west coast chain, oriented horizontally along the best-fit great circle and vertically down, respectively. Spatial Hilbert transforms of the two components render, under certain conditions, a magnetic field variation from which the internal (induced) contribution is removed [14]. The remaining field (solely from external sources) is then used to determine the equivalent ionospheric current perpendicular to the magnetometer chain. We assume this current to be a Hall current driven by an electric field oriented parallel to the magnetometer chain.

3 Comparison Between TIFAC and Hall Current

Uniform FAC sheets which extend perpendicularly to the magnetometer chain must be closed in the ionosphere *via* Pedersen currents flowing parallel to the chain because the current system in its entity is divergence-free. Under the assumption that FACs are negligible below 55° CGML the local Pedersen current must match the local TIFAC everywhere along the trajectory poleward of 55° CGML. The electric field that drives the Pedersen current along the chain also drives the ionospheric Hall current across the chain. The Hall/Pedersen current ratio, and consequently the Hall/TIFAC ratio, need to be identical to the Hall/Pedersen conductance ratio (Σ_H/Σ_P ratio) if the assumption of infinitely extended uniform current sheets and smoothly varying conductances were fulfilled.

Fig. 2 demonstrates this for a sample Ørsted pass on April 14, 2000. Both panels show the Hall current distribution along the magnetometer chain, between about 66° and 85.5° CGML, as it develops over time (gray-coded). The trace of the polebound satellite, between 1035 and 1040 UT, is tagged with the local FAC density (upper panel) and the TIFAC (lower panel), respectively. The local FAC alternates on short spatial scale between up (white) and down (black) which makes comparison with the ionospheric current difficult. The TIFAC intensity, however, shows a rather steady rise to an upward maximum and then changes to a downward maximum. The flip occurs near 77.5° CGML where the highest local FAC density is observed. The upward TIFAC all the way up to the flip must be fed by a Pedersen coming from the poleward side (according to our assumption no currents flow equatorward of 55°). It has thus to be associated with a westward Hall current. The argument holds *vice versa* for the downward TIFAC poleward of the flip which must be associated with an eastward Hall current. Our view is confirmed (within ≈0.5°) by the observed Hall current distribution which maximizes in westward direction near 77.5° and, after a shear at ≈78°, maximizes in eastward direction near 79°. The FAC maxima of ±0.12 A/m should be compared to the Hall

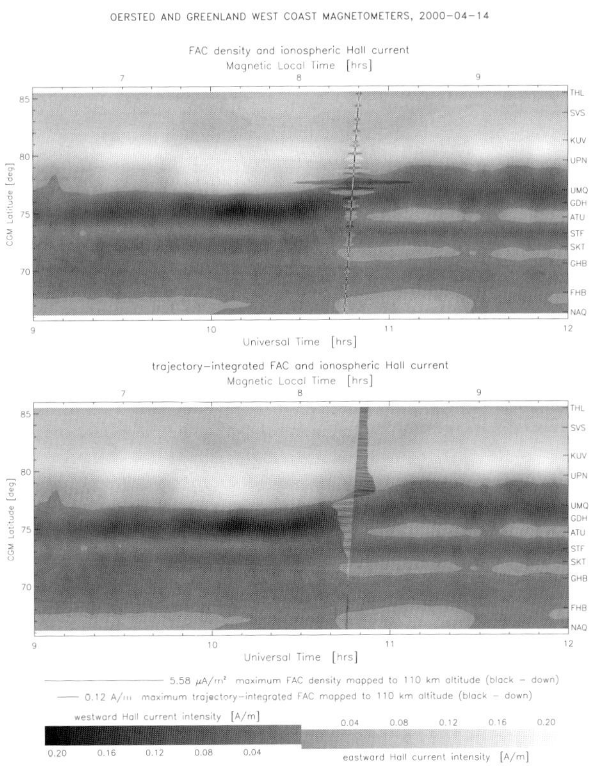

Fig. 2. Distribution of ionospheric equivalent current across the magnetometer chain *vs* CGML and time (gray-coded); Ørsted trajectory with FAC density (top panel) and TIFAC (bottom panel) overlaid (white – upward, black – downward FAC)

current maxima encountered during the satellite pass, 0.11 A/m westward and 0.10 A/m eastward, respectively. The TIFAC/Hall current ratio and, following our arguments from above, the Σ_H/Σ_P ratio, is thus ≈0.9.

Ultraviolet sunlight and energetic particle precipitation constitute the primary sources for ionization of the upper atmosphere. During this event the ionosphere above western Greenland was barely sunlit at solar zenith angles slightly below 90°. The Σ_H/Σ_P ratio resulting from solar radiation is about 1.4 at solar zenith angles between 60° and 90° according to statistical studies [12]. However, large variations in the Σ_H/Σ_P ratio may occurr if the conductivity is enhanced by particle precipitation, because the Σ_H/Σ_P ratio depends critically on the energy spectrum of the auroral electron flux [10].

In order to compare TIFAC and Hall current quantitatively we determined for every data point on each satellite pass a quasi-normalized difference. It is essentially the difference between local TIFAC and local Hall current divided by the arithmetic mean of the largest difference between upward and downward TIFAC and eastward and westward Hall current, respectively, on that specific satellite pass. We then binned the quasi-normalized differences into four MLT intervals, midnight (22–02 MLT), pre-dawn (02–06 MLT), post-dawn (06–10 MLT), and noon (10–14 MLT). The results are shown in Fig. 3 where positive values indicate a TIFAC larger than the Hall current and negative values the opposite. If our arguments from above were correct this would mean that positive values are, in principle, equivalent to $\Sigma_H/\Sigma_P < 1$ and negative values to $\Sigma_H/\Sigma_P > 1$.

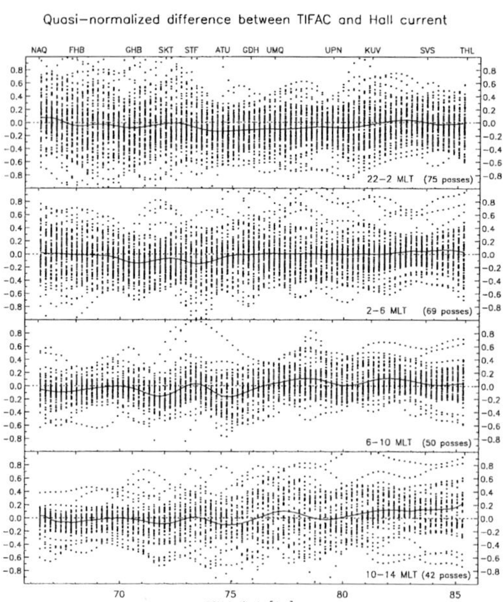

Fig. 3. Quasi-normalized difference between TIFAC and Hall current from all passes, binned according to different MLT intervals; the solid lines follow the mean values

It appears that the quasi-normalized difference is most of the time within the ± 0.2 range. It exhibits a dependence on magnetic local time and geomagnetic latitude. The distributions tend toward slightly negative values at lower latitudes (auroral zone) and are close to zero in the midnight and pre-dawn sector at high latitudes (polar cap). They become positive in the high-latitude sections of the post-dawn and noon sectors. The largest spreads are observed in the midnight sector below 74° CGML and in the noon sector above 72°, and the smallest spread is observed in the entire post-dawn sector and in the noon sector below 72°. The large spread in the midnight auroral zone likely reflects the fact that here conductances are practically solely determined by particle precipitation and their ratio thus varies considerably, and that we find here the Harang discontinuity, the majority of auroral break-ups, substorm surges etc. The smaller spread in the post-dawn and lower latitude noon sectors reflect the fact that conductances are here dominated by solar radiation and thus relatively stable. The large spread and predominantly positive values in the high-latitude noon sector may be due to the localized cusp current system which differs much from our model of infinitely extended current sheets. It may also be influenced by the annual variation of the dipole axis which moves the high-latitude noon sector ionosphere into the dark at winter time and into the sunlight at summer time.

4 Conclusion

We examined a number of cases when coincident Ørsted and ground-based magnetometer measurements over western Greenland were available. In some cases the patterns of TIFAC and Hall currents match well assuming nearly equal Pedersen and Hall conductances, in other cases they do not. We suggest the latter cases to represent situations in which our assumptions were at least partially violated. Such situations may include

- Hall and/or Pedersen conductances exhibit substantial spatial gradients
- The assumption of infinite uniform current sheets is invalid
- The upper atmosphere is ionized by charged particle precipitation

We suggest that a comparison between ground-based and satellite vector magnetometer measurements provides for a method to test whether the FAC densities inferred from single pass satellite observations are consistent with the assumption of sheet currents.

So far we have only covered the midnight–morning–noon sector and did not separate our data according to season. However, our sample size will grow and permit a more detailed comparison as more observations from the Ørsted, CHAMP and SAC–C satellites will become available. We further intend to examine Sondrestrom incoherent scatter radar data in order to obtain Σ_H and Σ_P values above Sondrestrom and thus have an independent measurement for reference and comparison.

GPS Radio Occultation with CHAMP

Jens Wickert, Georg Beyerle, Torsten Schmidt, Christian Marquardt, Rolf König, Ludwig Grunwaldt, and Christoph Reigber

GeoForschungsZentrum Potsdam (GFZ), Division 1, Kinematics & Dynamics of the Earth, Potsdam, Germany, *wickert@gfz-potsdam.de*

Summary. The GPS (Global Positioning System) radio occultation experiment onboard the German CHAMP (CHAllenging Minisatellite Payload) satellite was activated in February 2001. By the end of 2001 about 36,500 occultations were recorded. We review the occultation data processing at the GFZ and discuss selected first results. 1) CHAMP allows for atmospheric sounding with high accuracy despite of the Anti-Spoofing (A/S) mode of the GPS. 2) There are advantageous consequences for the GPS data processing due to the termination of the Selective Availability (SA) mode of the GPS. It is possible to reduce the GPS ground station acquisition rate for double difference processing. The application of space-based single differencing technique for precise occultation data processing became feasible. 3) The state-of-the-art GPS flight receiver onboard CHAMP combined with high-gain occultation antenna allows for atmospheric sounding deep into the lower troposphere. Possible applications and improvements of occultation data analysis in the lower troposphere are discussed.

Key words: CHAMP, GPS, Radio occultation, remote sensing, Selective Availability, Space-based single differencing, GFZ Potsdam

1 Introduction

The CHAMP GPS radio occultation experiment is an important step towards establishing an innovative remote sensing method for sounding of the Earth's atmosphere and ionosphere.

The GPS radio occultation technique is based on precise dual-frequency phase measurements (L-band) of a GPS receiver in a Low-Earth-Orbit tracking a setting or rising GPS satellite. Combining these measurements with the satellites' position and velocity information the small phase path increase due to the atmosphere during the occultation event can be derived. It is converted to atmospheric bending angles. Assuming spherical symmetric atmosphere vertical profiles of refractive index can be determined and converted to atmospheric parameters as pressure, temperature and, using independent knowledge of temperature, also to water vapor within the lower troposphere.

The main advantages of the calibration-free GPS occultation technique are global coverage, high vertical resolution and all-weather-capability combined with high accuracy. These properties allow various applications in atmospheric/ionospheric research, weather forecast and climate change detec-

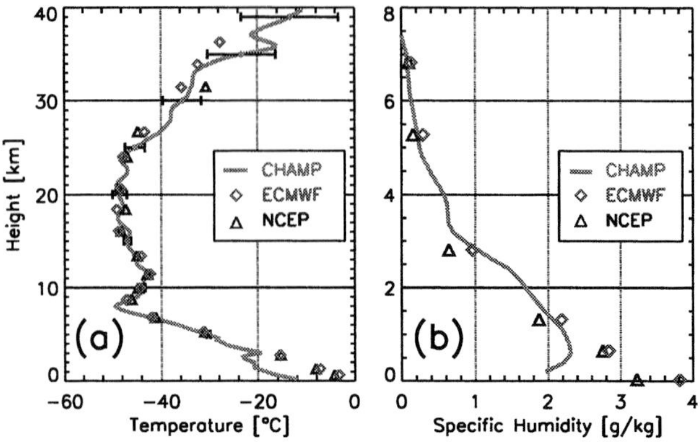

Fig. 1. Vertical profiles of (a) dry temperature and (b) specific humidity, derived from one of CHAMP's first occultation measurements (No. 5; 0.5°W, 53.2°S; 19:43 UTC) on February 11, 2001, compared to analyses data from ECMWF (European Centre for Medium-Range Weather Forecasts and NCEP (National Centers for Environmental Prediction).(from [9])

tion [1, 2]. For the first time the promising potential of the GPS radio occultation technique was demonstrated by the GPS/MET-Experiment [3]. In relation to GPS/MET, the CHAMP satellite possesses enhanced GPS flight receiver (the state-of-the-art GPS receiver 'BlackJack' provided by JPL, Jet Propulsion Laboratory) and high-gain occultation antenna.

Currently, beside CHAMP also the Argentinean SAC-C satellite (launched on November 21, 2000) performs GPS radio occultation measurements. Furthermore, on March 17, 2002 the American-German GRACE satellites (GRAvity and Climate Experiment, twin satellite mission) were successfully launched. Both satellites possess the capability to perform GPS occultation measurements. Further missions flown with GPS occultation receivers onboard are the Danish Ørsted [4] and the South African SUNSAT [5]. The Ionosphere Occultation Experiment (IOX) onboard the PICOSAT satellite (launched on September 30, 2001) provides GPS occultation measurements of the ionosphere. IOX is part of the U.S. American Department of Defense Space Test Program executed by the Air Force [6].

We review the occultation data processing at the GFZ and present selected first results. A detailed evaluation of the quality of CHAMP atmospheric data products is performed in another paper of this issue [7].

2 First Measurements

The first 7 occultation measurements with CHAMP were performed on February 11, 2001 between 19:04 and 20:04 UTC. Vertical profiles of dry temperature and specific humidity were derived (an example is shown in Fig. 1). Already these measurements showed that, in spite of the activated anti-spoofing (A/S) mode, CHAMP allows for atmospheric sounding with high accuracy and vertical resolution [8, 9].

3 Status of the experiment

Occultation measurements were performed during 217 days in 2001 giving a total of 36,514 events. Vertical profiles of atmospheric parameters were derived for 24,658 occultations (67.5 %). For the remaining profiles (32.5 %) the atmospheric excess phase calibration failed.

One reason for that relatively large percentage is the reduced data quality due to on-board-software modifications of the GPS flight receiver for parallel performing of occultation measurements in the atmosphere and ionosphere [10]. These modifications were implemented at the end of July 2001. As a result the number of daily atmospheric occultations was significantly reduced from ~220-240 daily events temporary down to ~100 (see Fig. 2). Furthermore, the data analysis software for the automatic calibration of the atmospheric excess phase needs to be improved, to detect and correct for for all types of irregularities in the GPS occultation and ground station data. Furthermore, the influence of the ionosphere during solar maximum as an additional factor for failure has to be investigated in more detail.

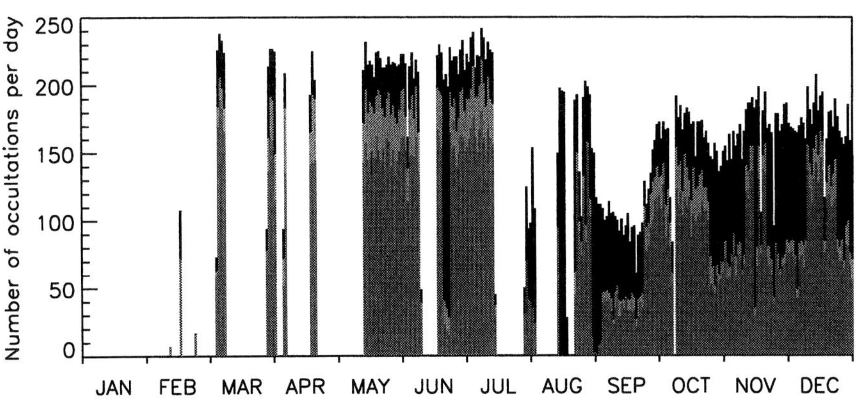

Fig. 2. Number of daily CHAMP occultations (duration > 20 s) for 2001. The total height of the columns corresponds to the number of daily measurements. The black and light grey color indicate occultations with calibration failure and insufficient data quality, respectively. The height of the dark grey columns corresponds to the number of vertical atmospheric profiles provide to the CHAMP data center at GFZ.

4 Occultation data processing and infrastructure at GFZ

The main components of the GFZ infrastructure for data generation, transfer, analysis and archiving are: the GPS receiver onboard the CHAMP satellite (provided by JPL) and the ground segment (see Fig. 3). It consists of the downlink station at Neustrelitz, Germany (53.2°N, 13.0°E) including the Raw Data Center (RDC), the fiducial GPS ground network (Fiducial Network), the Precise Orbit Determination (POD) facility, the Occultation Processing System [Atmospheric Profiling (AP)-Processor] and for archiving and distribution the CHAMP Information System and Data Center (ISDC). The downlink station and the Raw Data Center are operated by the German Aerospace Center (DLR/DFD). The GPS ground network is operated in cooperation between GFZ and JPL, the other components are maintained by GFZ.

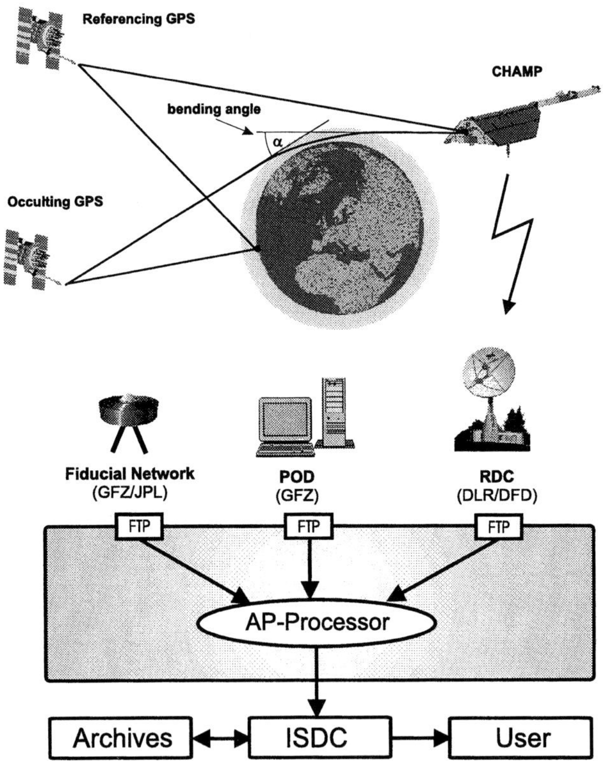

Fig. 3. Infrastructure of the CHAMP GPS radio occultation experiment (Overview).

Within the framework of the German GASP (GPS Atmosphere Sounding Project, see e.g. [11]) a pre-operational atmospheric data product delivery is aimed at. This requires a maximum time delay of not more than 3 hours between measurement and product delivery to weather services. To reach this objective, every component of the occultation infrastructure (Fig. 3) has to satisfy this 3-hour requirement. We characterize in more detail the Precise Orbit Determination, the Fiducial Network and the occultation data analysis and processing system at GFZ and focus to operational aspects.

4.1 Precise Orbit Determination

The GPS and CHAMP orbit ephemerides (Rapid Science Orbits, RSO) are provided by the orbit and gravity component within GFZ's Science Data System for CHAMP [12]. The orbits are available in the afternoon of the following day. The arc lengths of the CHAMP orbits are chosen at 14 hours. Thus, one day is covered by two arcs with 2 hours overlap.

Until November 2001, 3-D position accuracy of the RSO sized at $\sim$20 cm which was already sufficient for precise occultation processing [13]. Improved POD modelling increased then the accuracy further to $\sim$10 cm [14].

A current project is the demonstration of near-real-time orbit delivery with 3 hour rhythm (Ultra rapid Science Orbits, USO).

4.2 Fiducial Ground Network

To compensate for CHAMP and GPS clock errors, the precise phase measurements of the onboard GPS receiver are differenced with those from a fiducial ground network receiver [9, 15]. To reach global coverage of occultation measurements a global ground network is necessary (High Rate Low Latency GPS Ground Tracking network [16]). This network consists of 38 globally distributed stations (as of January 2002) and is operated jointly by JPL and GFZ with cooperating partners.

Redundancy for occultation processing Simulation studies in preparation of the CHAMP occultation experiment using a planned configuration of 28 stations have shown that an average redundancy of $\sim$4 stations for occultation processing is reached [15, 16].

To confirm this with real data we have processed $\sim$5,500 CHAMP occultations during day of year 135-160, 2001 (May 15 - June 9) applying the double difference technique using data of 24 ground stations. The average data availability of these ground stations was 91.2 %. The main reasons for loss of data are hardware problems at the ground station sites or data transmission failures from the ground stations to the Level-1 processing center at GFZ.

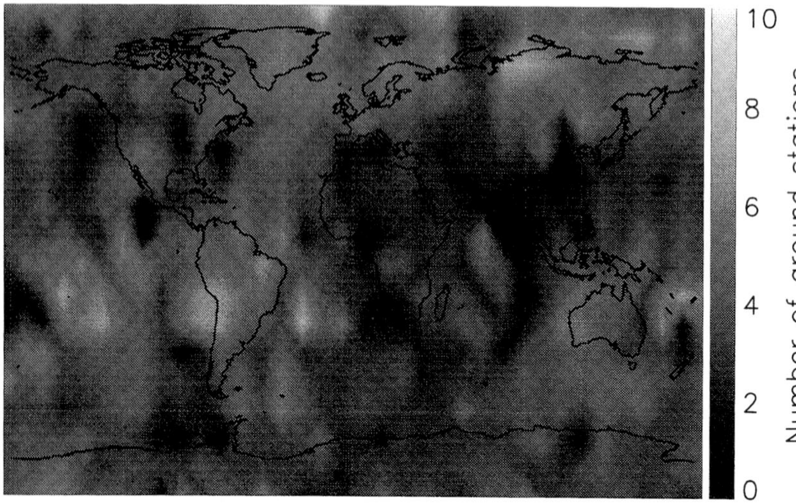

Fig. 4. Global distribution of ground station redundancy for CHAMP occultation processing during May 15 - June 9, 2001 (from [17]).

Fig. 4 shows the resulting global distribution of ground station redundancy for occultation processing. On average ~4 ground stations were available for processing, e.g. over North Atlantic or Europe. Areas with lower redundancy were found as well, e.g. India or South Africa with ~2 ground stations per occultation. Areas with higher-than-average redundancy with up to 7 ground stations per occultation exist within the Equatorial Pacific region.

We conclude that the established network configuration allows for processing of globally distributed occultation events with average redundancy of ~4 ground stations per event.

Data flow The GPS ground data are recorded with an acquisition rate of 1 Hz (Pseudoranges and phases on both GPS frequencies). The data (15 min binary files, ~540 MByte daily) are transmitted via FTP (File Transfer Protocol) to the Level-1 processing center at GFZ. Here, 1 hourly RINEX (Receiver INdependent EXchange format) files are generated and transferred to the CHAMP Information System and Data Center. The time delay between measurement and data provision is currently between ~30 min and ~70 min. Due to the use of 15 min intervals for the data transmission from the ground stations to the Level-1 processing center, these time delays can be further minimized down to about 15-30 min.

4.3 Occultation data analysis and processing system

The basics of the standard analysis of CHAMP occultation data are described in [9], a more detailed description can be found in [13]. The standard data analysis is based on the double difference method for deriving the atmospheric excess phase, the geometric optics approximation and the assumption of spherical symmetry for the derivation of the atmospheric refractivity, respectively. For details on the GPS radio occultation method in general see e.g. [2].

The analysis software is embedded in a dynamically configurable and extendable system for operational data processing and product generation [18], which also provides an interface to GFZ's data archive (ISDC). The analysis and processing software is implemented in FORTRAN, C++ and Perl programming languages.

The main processing computer is a SUNFIRE 280R server, 2 UltraSparc III processors (750 MHz), 4 GB RAM.

5 Consequences of the termination of Selective Availability

Selective Availability was part of the intentional degradation of the user's navigation solution. SA implied that GPS transmitter clocks varied by $\sim 10^{-7}$ s (tens of meters in units of length) over timescales of 100 s, with rates on the order of 1 m/s. To correct for these clock rates, which are unacceptable for precise atmospheric profiling [15], the double difference technique is used to process GPS occultation data [9, 19].

Due to the termination of SA on May 2, 2000 the amplitude of the described errors in the transmitter clocks was reduced by orders of magnitude [20]. This made the clocks more predictable and has advantageous consequences for occultation data processing, which are briefly described here.

5.1 Acquisition rate of the ground station data

To correct for the clock variations (induced by SA) a ground station acquisition rate of 1 Hz was required. This rate is currently used for standard occultation processing (double differencing) at GFZ.

However, already during the preparation of the CHAMP occultation experiment it was shown, that after the termination of SA a reduction of the ground station acquisition rate to 1/10 Hz would not influence the accuracy of the calibration of the atmospheric excess phase for occultation measurements [15].

Here we investigate the influence of reduced (in relation to the standard 1 Hz rate) ground station acquisition rates (1/5, 1/10 and 1/30 Hz) on the quality of vertical dry temperature profiles derived from CHAMP measurements. A set of 436 occultations during April 19-21, 2001 was processed with average redundancy of ~ 3.2 ground stations per event using acquisition rates of 1, 1/5, 1/10 and 1/30 Hz, respectively. The resulting 1,400 dry temperature profiles were compared with corresponding ECMWF profiles.

The 1/5 Hz profiles exhibit identical deviations (mean and standard) from ECMWF as the reference profiles, which were processed using 1 Hz. Even a reduction of the acquisition rate down to 1/30 Hz results in statistically identical profiles compared to ECMWF as the reference set [20].

Our study shows that a reduction of the ground station acquisition rate down to 1/30 Hz will be possible. However, further investigation of larger data sets is required to confirm this result.

5.2 Space-based single differencing

As an alternative to double difference processing the application of less-differencing techniques became feasible with termination of SA. First results of GPS occultation data processing utilizing a single difference technique, i.e., clock correction using 5 min GPS clock solutions without directly involving GPS ground station data, were presented in [20]. Fig. 5 shows one of the results, the statistical comparison of two sets of 436 vertical profiles of refractivity and dry temperature, generated by the double and single difference method, respectively. The mean deviation in refractivity is <0.1 % up to 30 km. The standard deviation is <0.2 % up to heights of 20 km and reaches 0.4% at 30 km. A mean deviation of temperature is not discernible over the entire height interval. The standard deviation is <0.6 K at heights up to 20 km and ~1.2 K at 30 km.

This suggests that after termination of SA the application of the single difference technique provides nearly identical refractivity and temperature profiles compared to profiles processed using double differencing. For the first time it was demonstrated, that future GPS clock correction may have no

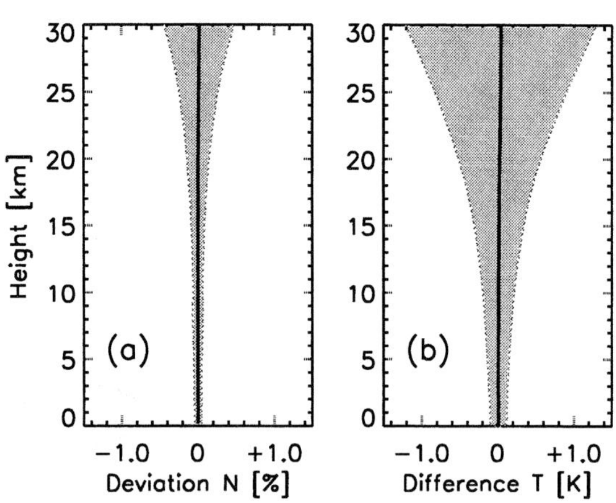

Fig. 5. Statistical comparison between 436 CHAMP refractivity (a) and dry temperature profiles (b), derived using single and double difference technique during April 19–21, 2001.(from [20])

need for direct use of auxiliary ground station data. But, further examination is needed to find the reasons for the statistical deviations between single and double difference profiles [20]. These are not insignificant for GPS radio occultation measurements with sub-kelvin accuracy potential below 40 km and 0.2 K at the tropopause [2].

The application of single difference methods would not only significantly simplify processing systems, but also entirely avoid measurement error contributions from ground-to-satellite links (e.g. phase path noise, uncorrected ionosphere/troposphere or multipath at the ground station sites). We note however, that the GPS ground station data (1/30 Hz) also are used for the Precise Orbit Determination of the GPS and CHAMP satellites, including the derivation of the precise 5 min GPS clock solutions.

6 Lower troposphere and surface reflections

The analysis of the first occultation measurements indicated that the amount of profiles reaching the Earth's surface will be significantly larger compared with the GPS/MET-Experiment [9]. Therefore the CHAMP data set could be valuable especially for improving the retrieval techniques in the lower troposphere, a task of current occultation research (e.g. [21]) We give two examples, where advanced retrieval techniques were applied to CHAMP radio occultation data in the lower troposphere.

6.1 Surface reflections

Using the radioholographic technique (e.g. [22]) ~3,800 occultation events observed between May 14 and June 10, 2001 were analysed. About 1,200 (~30 %) of these analysed occultations exhibit components due to GPS signal reflection on Earth's surface. Characteristic frequency shifts of direct and reflected GPS signals have been determined in observations and radio propagation simulations. It was shown that these frequency shifts can be exploited to obtain information on ground elevation at the reflection point and ground-level refractivity [23].

Fig. 6 shows one result of the radioholographic analysis of ~ 10,000 occultations recorded between August 26, 2001 and February 1, 2002. Confirming the result of [23] ~30 % of the analysed occultations exhibit reflected components.

We note that CHAMP has the capability to receive left-hand circular polarized GPS signals, reflected from Earth's surface by a nadir viewing GPS antenna. In the future a feasibility study for GPS based altimetry/scatterometry (see e.g. [24]) is planned.

the improvement of numerical weather prediction by the provision of accurate observations of the Earth's atmosphere in a global scale with low latency. This will be realized by routine operated science data reception with a downlink station located at Ny Alesund (Spitsbergen, 78.9°N, 11.9°E), which was successfully tested already during end of 2001, and provision of ultra rapid science GPS and CHAMP orbits.

Furthermore, the successful launch of the American-German twin-satellite GRACE mission with occultation capability for both satellites on March 17, 2002 can provide up to 500 for- and aft-looking occultations per day in the near future.

Acknowledgement. We are grateful to U. Foelsche and K. Hocke for critical and constructive review of the paper. We thank our colleagues at GFZ and JPL: T. Meehan for GPS receiver operation and control, R. Galas, Ch. Förste, Wolfgang Köhler and Bernd Ritschel for occultation and ground station data handling, preprocessing and archiving. The German Weather Service (DWD) provided ECMWF analysis via the Stratospheric Research Group, Berlin. We thank K. Labitzke and K. Schulz-Schöllhammer for the ECMWF data preparation. NCEP data were obtained through NCEP's Climate Prediction Center (CPC). This study was carried out in the GASP (GPS Atmosphere Sounding Project) strategy fonds program under the grant of the German Federal Ministry of Education and Research (BMBF) no. 01SF9922/2.

References

1. Hajj GA, Lee LC, Pi X, Romans LJ , Schreiner WS, Straus PR, and Wang Ch (2000) COSMIC GPS Ionospheric Sensing and Space Weather, Terrestrial, Atmospheric and Oceanic Science, **11**, 235-272
2. Kursinski ER, Hajj GA, Hardy KR, Schofield JT, and Linfield R, (1997) Observing Earth's atmosphere with radio occultation measurements using the Global Positioning System, J Geophys Res, **102**, 23429-23465
3. Ware R et al (1996) GPS Sounding of the atmosphere from low Earth orbit: Preliminary results, Bull Am Meteorol Soc, **77**, 19-40
4. Escudero A, Schlesier AC, Rius A, Flores A, Rubek F, Larsen GB, Syndergaard S, and Høeg P (2001) Ionospheric tomography using Ørsted GPS measurements - Preliminary results, Phys Chem Earth (A), **26**, 173-176
5. Mostert S, and Koekemoer JA (1997) The science and engineering payloads and experiments on SUNSAT, Acta Astronautica, **41**, 401-411
6. PICOSAT web pages (2002): http://www.losangeles.af.mil/SMC/PA/Fact_Sheets/PICOSat.htm
7. Marquardt C, Schöllhammer K, Beyerle G, Schmidt T, Wickert J, and Reigber Ch (2002) Validation and Data Quality of CHAMP Radio Occultation Data, this issue
8. Wickert J, Schmidt T, Marquardt C, Reigber Ch, Neumayer K-H, Beyerle G, Galas R, and Grunwaldt L (2002) GPS radio occultation with CHAMP: First results and status of the experiment, Proc of IAG Scientific Assembly 2001,

Vistas for Geodesy in the New Millenium, September 2-7, Budapest, Hungary, Springer Series, **125**, 273-278

9. Wickert J, Reigber Ch, Beyerle G, König R, Marquardt C, Schmidt T, Grunwaldt L, Galas R, Meehan TK, Melbourne WG, and Hocke K (2001) Atmosphere sounding by GPS radio occultation: First results from CHAMP, Geophys Res Lett, **28**, 3263-3266
10. Jakowski N, Wehrenpfennig A, Heise S, Reigber Ch, Lühr H, Grunwaldt L, and Meehan T (2002) GPS Radio Occultation Measurements of the Ionosphere from CHAMP: Early Results, Geophys Res Lett, in press
11. Reigber Ch (1998) GPS Atmosphere Sounding An innovative approach for the recovery of atmospheric parameters, HGF Strategy fund proposal, Potsdam, Germany
12. Reigber Ch (2002) CHAMP home page, Internet: http://www.gfz-potsdam.de/champ
13. Wickert J (2002): The CHAMP radio occultation experiment: Algorithms, Processing system, and First results, PhD Thesis (in German), University of Graz, Scientific Technical Report 02/07 GFZ Potsdam, Germany
14. Michalak G, Baustert G, König R, and Reigber Ch (2002) CHAMP rapid science orbit determination - Status and future prospects, this issue
15. Wickert J, Galas R, Beyerle G, R. König, and Ch. Reigber (2001) GPS ground station data for CHAMP radio occultation measurements, Phys Chem Earth (A), **26**, 503-511
16. Galas R, Wickert J, and Burghardt W (2001), High Rate low latency GPS ground tracking network for CHAMP, Phys Chem Earth (A), **26**, 649-652
17. Wickert J, Galas R, Schmidt T, Beyerle G, Reigber Ch, Förste C, and Ramatschi M (2002) Atmospheric sounding with CHAMP: GPS ground station data for occultation processing, subm to Phys Chem Earth (A)
18. Wehrenpfennig A, Jakowski N, and Wickert J (2001) A Dynamically Configurable System for Operational Processing of Space Weather Data, Phys Chem Earth (C), **26**, 601-604
19. Rocken C, Kuo Y-H, Schreiner WS, Hunt D, Sokolovskiy S, and McCormick Ch (2000) COSMIC System description, Terrestrial, Atmospheric and Oceanic Science, **11**, 21-52
20. Wickert J, Beyerle G, Hajj GA, Schwieger V, and Reigber Ch (2002) GPS radio occultation with CHAMP: Atmospheric profiling utilizing the space-based single difference technique, Geophys Res Lett, **29**, 8, 10.1029/2001GL013982
21. Gorbunov M (2001) Radioholographic methods for processing radio occultation data in multipath regions, DMI Report 01-02, DMI Copenhagen, Denmark
22. Pavelyev AG, Volkov AV, Zakharov AI, Krutikh SA, and Kucherjavenkov AI (1996) Bistatic radar as a tool for Earth investigation using small satellites, Acta Astronautica, **39**, (9-12), 721-730
23. Beyerle G, Hocke K, Wickert J, Schmidt T, Marquardt C, and Reigber Ch (2002) GPS radio occultations with CHAMP: A radio holographic analysis of GPS signal propagation in the troposphere and surface reflections, J Geophys Res, in press
24. Garrison JL, and Katzberg SJ (2000) The application of reflected GPS signals to ocean remote sensing, Remote Sensing and Environment, **73**, 175-187
25. Vorob'ev VV, and Krasil'nikova TG (1994) Estimation of the accuracy of the atmospheric refractive index recovery from Doppler shift measurements at frequencies used in the NAVSTAR system, Phys Atmos Ocean, **29**, 602-609

Validation and Data Quality of CHAMP Radio Occultation Data

Christian Marquardt[1,2], Kathrin Schöllhammer[3], Georg Beyerle[1], Torsten Schmidt[1], Jens Wickert[1], and Christoph Reigber[1]

[1] GFZ Potsdam, Division 1, Kinematics and Dynamics of the Earth
[2] Current address: Met Office, Bracknell, UK
[3] Freie Universität Berlin, Institut für Meteorologie, Berlin, Germany

Summary. Since the first successful radio occultation measurement taken by the GPS receiver onboard CHAMP, more than 40,000 vertical soundings of stratospheric temperature and tropospheric humidity have been processed at GFZ Potsdam. During the first year of the radio occultation experiment, several multi–week batches of quasi–continuous measurements have been collected, allowing for detailed validation against other meteorological measurements. This paper presents the first validation results and discusses some conclusions regarding the data quality of CHAMP radio occultation data drawn from these.

Key words: CHAMP, GPS radio occultation, validation, canonical transform

1 Introduction

In February 2001, the first radio occultation measurements were taken by the GPS receiver onboard the German satellite CHAMP (Wickert et al., 2001). Since then, the automated processing system at GFZ Potsdam has successfully processed more than 40,000 occultations (Wickert et al., 2002). Beginning in September 2001, the measurements have been quasi–continuous; before September, two four–week periods during May/June and June/July 2001 are available. In this paper, we present first results of validation studies for CHAMP radio occultation data.

Section 2 briefly reviews the data processing strategy implemented at GFZ. In section 3, we show statistical comparisons of CHAMP temperature profiles in the lower and mid stratosphere against meteorological analyses provided by ECMWF and nearby radiosonde observations. The quality of tropospheric refractivity measurements is discussed in section 4, where we show that refractivity exhibits a negative bias against meteorological analysis data. This translates into a large dry bias in humidity. We present evidence that the Canonical Transform (CT) method (Gorbunov, 2001, 2002) does not provide a significant improvement of the retrievals when applied to CHAMP data. We further speculate that the internal functioning and tracking algorithms implemented in the GPS receiver are an important factor for the data quality of radio occultation data in the mid and lower troposphere which has not yet been addressed properly. Conclusions in section 5 close the paper.

2 Data Processing

The operational data processing for CHAMP radio occultation data follows the scheme known from the GPS/MET data (see, e.g., Vorob'ev and Krasil'nikova, 1994; Kursinski et al., 1997; Hocke, 1997; Rocken et al., 1997; Hajj et al., 2002). Using high precision orbits of CHAMP and the occulting GPS satellite, bending angle vs. impact parameter profiles are derived from measured excess path delays. Clock errors are removed by means of a double differencing technique; this involves the use of additional data obtained from a referencing GPS satellite and an appropriate GPS ground station. Ionospheric effects are removed by a linear combination of the bending angles calculated from the two GPS frequencies (Vorob'ev and Krasil'nikova, 1994). An Abel transform (Fjeldbo et al., 1971) allows for the calculation of neutral atmospheric refractivity N. This, in turn, is related to atmospheric pressure p, temperature T and partial water vapor pressure e via

$$N = \kappa_1 \frac{p}{T} + \kappa_2 \frac{e}{T^2} \tag{1}$$

with empirically known coefficients κ_1 and κ_2 (Smith and Weintraub, 1953).

Note that the derived bending angles exhibit a large amount of noise at altitudes of 50 km and above, representing residual ionospheric errors. Thus, some smoothing has to be applied to the raw bending angle data prior to the Abel transform. At GFZ, a statistically optimal interpolation (Sokolovskiy and Hunt, 1996; Healy, 2001b) is applied to the raw bending angles, using a climatological bending angle profile as *a priori*. Error estimates for measured and *a priori* data are calculated as in Sokolovskiy and Hunt, neglecting vertical error correlations. The role of the *a priori* data in the upper level treatment of raw bending angles will be discussed in section 3.

If water vapor can be neglected (as in the upper troposphere and the stratosphere), the hydrostatic equation is used in order to calculate pressure and temperature by downward integrating the refractivity profile. At GFZ, the downward integration is initialized using climatological data at a high altitude (at the top of the measurements, or at most at 90 km altitude).

An example of a "dry temperature" retrieval is shown in Fig. 1, along with nearby radiosonde and ECMWF data. The radiosonde was launched from the German Meteorological Observatory Lindenberg 90 minutes after the occultation took place; the distance between the location of the occultation and the radiosonde launch site is only 27 km. Note the good agreement of temperatures between all three profiles in the altitude range from 400 to 30 hPa (5 to 25 km). The high vertical resolution of the CHAMP data in the vicinity of the tropopause – CHAMP agreeing better with the radiosonde than the analysis does – is also noteworthy.

If humidity does provide a significant contribution to refractivity, additional external data is required to seperate the dry and moist contribution. This is apparent in the temperature profile in Fig. 1, where dry temperature

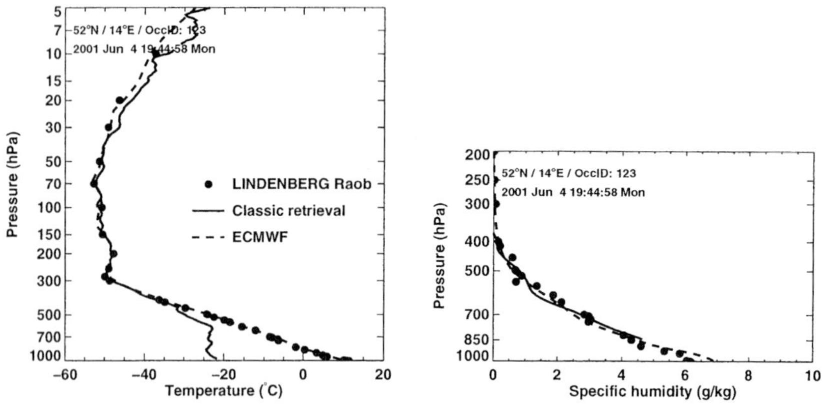

Fig. 1. Comparison of a CHAMP profile taken on June 4, 2001, located close to the German Meteorological Observatory Lindenberg. Shown are dry a temperature (left) and specific humidity (right) profiles from CHAMP, radiosonde data (bullets), and ECMWF analyses (dashed).

starts to be significantly colder than analysis or radiosonde temperatures at 400 hPa and below. We follow Gorbunov and Sokolovskiy (1993) and use a temperature profile obtained from ECMWF in order to solve for the partial water pressure in equation (1). At GFZ, the retrieval of water vapor begins 1.5 km below the lapse rate tropopause, as determined from the dry temperature profile. Fig. 1 (right) shows the specific humidity profile obtained for the occultation near Lindenberg. Again, the CHAMP profile agrees well with both analysis and radiosonde.

As an alternative to the traditional retrieval of bending angles, a Canonical Transform (CT) method (Gorbunov, 2001, 2002) has also been implemented. The profile shown in Fig. 1 has been calculated with the traditional approach; experiences with CT retrievals are discussed in section 4.

After the processing, all profiles undergo an automated quality control procedure which filters out physically implausible temperature and refractivity profiles. Specifically, profiles exhibiting stratospheric temperature deviations above 25 K and relative tropospheric refractivity deviations above 10 % against ECMWF are discarded. Typically, this affects 15–20% of the profiles.

3 Stratospheric Temperatures

Figure 2 show zonal mean (bias) and root mean square (rms) differences between CHAMP dry temperatures and ECMWF in the upper troposphere and stratosphere between May 14 and June 10, 2001. Analysis data was obtained by interpolation from 6 hourly data sets, and the statistics have

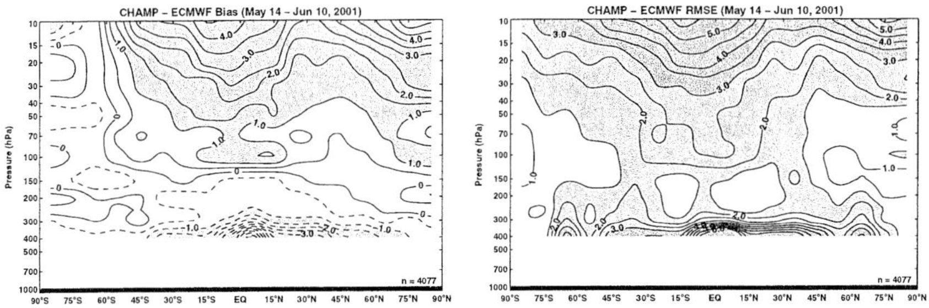

Fig. 2. Mean (left) and root mean square (right) deviations between CHAMP dry temperatures and ECMWF. Light shading denotes regions with a bias (rmse) below 1 (1.5) K, dark shading regions with a rmse above 3 K.

been calculated seperately for 10° wide latitude bins. Note that data below 400 hPa is not shown, as humidity is no longer negligible. The apparent cold bias of CHAMP temperatures around the lower boundary of the plotted data is also related to neglecting humidity in the retrieval of dry temperature.

The mean deviation in the upper troposphere and lower stratosphere (200 to 30 hPa) is less than 1.5 K everywhere outside the tropics, with rms deviations below 1.5 - 2 K in the same regions. The increase of error variance in the tropical stratosphere is related to convectively generated gravity waves. These are resolved by CHAMP, but not by the analyses. For a discussion of gravity wave characteristics obtained from radio occultation data, see, e.g., Tsuda et al. (2000).

Above 30 hPa (25 km), however, CHAMP retrievals show a considerable bias of up to 4 K against ECMWF temperatures, accompanied by an increase in rms deviations. This bias is related to details of the upper level data handling, of which the raw bending angle smoothing is part of. To illustrate this, Fig. 3 shows the same statistics as before, but calculated from data obtained from a slightly different retrieval setup. In this case, *a priori* bending angles used for optimal interpolation were calculated from exponentially extrapolated ECMWF refractivity at stratopause levels (around 50 km altitude) rather than from a climatology; all other processing options remained unchanged. Apparently, a "bias free" retrieval against ECMWF has been obtained up to 10 hPa everywhere except in high latitudes of the Southern Hemisphere (where a negative bias has been introduced). Rms deviations have also been reduced in the Northern Hemisphere. It is interesting to note that the difference in the temperature retrievals can be tracked down to the tropical tropopause region, where the "zero bias" retrieval is warmer by 0.5 K than the original one. In high latitudes of the Northern Hemisphere the modified retrieval is by up to 1 K colder than the original one at 100 hPa. This illustrates that *a priori* information used in the upper level treatment of radio occultation data does have a discernible influence on stratospheric tempera-

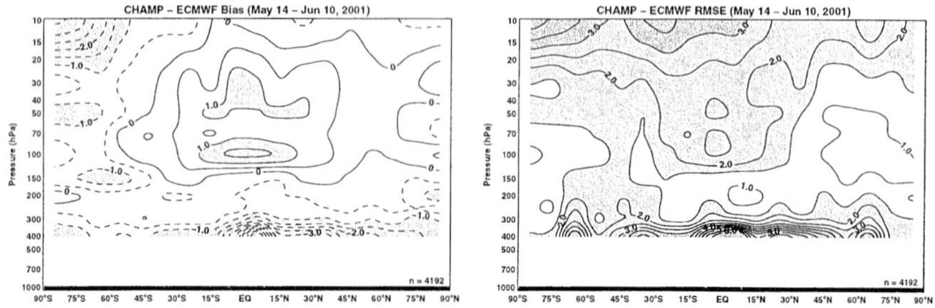

Fig. 3. As Fig. 2, but for a different implementation of the upper level smoothing of raw bending angle data (see text for details).

ture retrievals even in the lower stratosphere. Also note that radio occultation measurements as those provided by CHAMP do not carry enough information at altitudes above 30 hPa to allow for the highly accurate retrieval of temperatures without the proper use of *a priori* information. This is consistent with findings on the information content of radio occultation data in a 1DVar context (e.g., Healy, 2001b).

Regardless of the upper level bending angle smoothing, *a priori* data also enters the retrieval in the initialisation of the downward integration of the hydrostatic equation. Thus, one way of "tuning" temperature retrievals towards a given analysis is to to use analysis data to initialise the vertical integration at, e.g., at 10 hPa (30 km). This will, of course, lead to "unbiased" temperature retrievals when compared to the same analysis. The retrieved temperature at (and below) the chosen level, however, is determined by the *a priori*, and does not constitute an independent measurement.

Statistical comparisons of standard CHAMP dry temperatures with nearby radiosonde ascents for the periods May/June, June/July and November 2001 are shown in Fig. 4. Global statistics were calculated on standard pressure levels. Profiles were taken into account if the distance between the occultation and the radiosonde launch site was less then 300 km and 3 hours.

Again, there is an excellent agreement with biases < 1 K and rms deviations < 2–3 K between 150 and 30 hPa. At 10 hPa, a warm bias exists against radiosondes and reaches 1.5–3 K, depending on season. Note that the bias is smaller than against ECMWF analyses. This is consistent with the fact that ECMWF suffers from a cold bias the mid stratosphere, which is most likely related to the use of ATOVS radiances without applying a bias correction to upper stratospheric and mesospheric channels (T. McNally, pers. comm.). This, as the existence of large scale systematic biases between state–of–the–art analyses of the stratosphere as found by Schöllhammer et al. (2002), underlines that the tuning of parameters and/or smoothing methods in order to obtain "bias free" retrievals against a single specific analysis (like the one shown in Fig. 3) should be avoided.

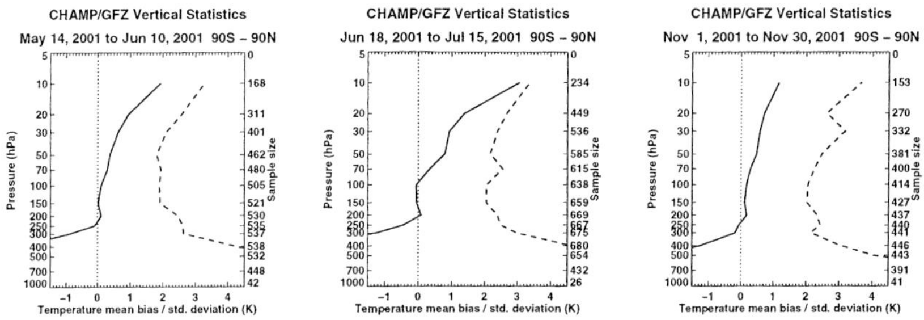

Fig. 4. Bias and RMSE (dashed) between standard CHAMP dry temperatures and nearby radiosondes for May/June (left), June/July (middle) and November 2001 (right). Note that data below 150 hPa exhibits a cold bias, as humidity is neglected in the dry temperature retrieval. Figures on the right axis denote the number of radiosonde matches for each pressure level.

4 Refractivity and Humidity in the Troposphere

Figure 5 (left) shows the zonal mean relative deviation of CHAMP refractivity with respect to refractivity calculated from ECMWF for the May/June 2001 period. The traditional retrieval based on geometrical optics has been used. At all latitudes and altitudes below 300 hPa, a negative bias in refractivity is apparent, though most pronounced in the lower troposphere and the tropics and subtropics. The bias is below 1 % in the free troposphere above 500 hPa. It reaches 5% near the ground in mid latitudes of both hemispheres, and exceeds 10% in the tropical lower troposphere[1]. A negative bias has also been reported on for GPS/MET data (Rocken et al., 1997), although no information on its meridional distribution is available.

It has been pointed out in the literature that the refractivity bias found in GPS/MET data is (indirectly; see below) related to atmospheric multipath effects (e.g., Rocken et al., 1997; Gorbunov and Gurvich, 1998; Gorbunov, 2001, 2002; Sokolovskiy, 2001a,b). Multipath is expected to be of major importance in the tropical troposphere where large variations in the water vapor field occur. The meridional distribution of the CHAMP refractivity bias is, by and large, consistent with this interpretation, suggesting that multipath effects also effect the CHAMP measurements.

The traditional retrieval method used for the derivation of refractivity shown in Fig. 5 is not able to deal with data affected by multipath. Thus, several algorithms have been developed in the past aiming at solving the multipath problem (e.g., Gorbunov et al., 1996; Mortensen and Hoeg, 1998; Hocke et al., 1999; Sokolovskiy, 2001b). The most promising method currently

[1] Note that for the purpose of the discussion of refractivity biases in the troposphere, profiles exhibiting more than 10% relative deviation against ECMWF (which are usually removed during the automated quality control) have been kept.

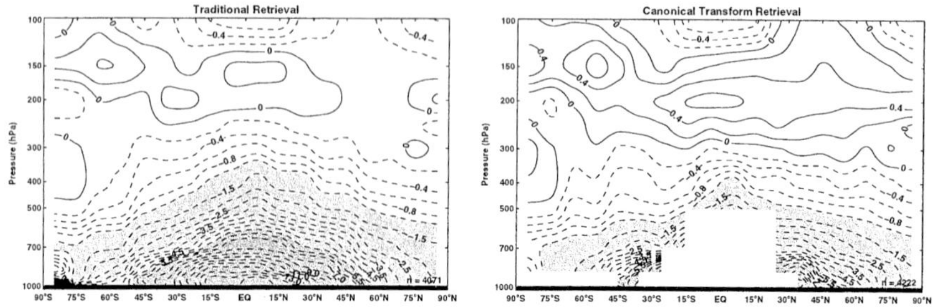

Fig. 5. Relative difference between CHAMP and ECMWF refractivity with respect to ECMWF, for the traditional (left) and an experimental CT retrieval (right). Light (dark) shading denotes areas where differences are larger than ± 1 (5)% (see text for details).

seems to be the CT method (Gorbunov, 2001, 2002) which has been shown to have a remarkable capability to reconstruct bending angle profiles from simulated GPS phase and amplitude data based on tropical radisonde ascents.

While performing well in simulations, the application of CT (and some of the other retrieval methods cited above) to real data obtained from GPS/MET measurements has only shown significant improvements for few selected profiles (Gorbunov and Gurvich, 1998; Gorbunov, 2002). The authors claim that "corrupted" data caused by problems in the receivers signal tracking under low signal–to–noise (SNR) conditions are responsible for the lack of improvement in the majority of cases. However, no immediate relationship between occurances of tracking failures of the receiver and problems in data quality (e.g., as judged by properties of the "CT amplitude"; see below) have been presented. Sokolovskiy (2001b) points out that changes in the GPS receiver tracking firmware onboard GPS/MET significantly influenced both the penetration depth of occultations and the magnitude of the refractivity bias. This supports the view that details in the receiver's tracking algorithms significantly affect the data quality in the troposphere.

The state–of–the–art BlackJack receiver onboard CHAMP is known to have improved tracking capabilities compared to GPS/MET; radio occultation profiles obtained from this instrument apparently reach further down into the lower troposphere (Wickert et al., 2001). Thus, it has been hoped that CHAMP data will provide a better suited data set for the application of improved retrieval methods than GPS/MET. Applying the CT rather than the traditional retrieval to CHAMP data should give a noticeable improvement in the quality of bending angles and refractivities, and especially a significant reduction of the refractivity bias.

Unfortunately, this is not the case. Fig. 5 (right) shows the zonal mean deviation between the experimental CT retrieval and ECMWF. There is a slight reduction of the refractivity bias at altitudes above 500 hPa, suggesting that atmospheric multipath does play a non-negligible role even in the upper troposphere. Below 500 hPa, however, biases are only slightly reduced and still prevail in the tropics and subtropics.

A possible atmosphere related explanation for the refractivity bias remaining in the CT retrievals – which would also apply to GPS/MET data – is that horizontal gradients (see, e.g., Healy, 2001a) might be a more dominating factor than previously anticipated. The CT may fail significantly if large horizontal gradients are present (see, e.g., Fig. 15 of Gorbunov, 2001). Gorbunov (2002), however, argues that cases with sufficiently strong gradients occur only very rarely. Thus, they should not dominate the statistics.

Alternatively, the CHAMP receiver in its current configuration may suffer from similer problems as the GPS/MET receiver. Under low SNR conditions, the CHAMP receiver switches into a "flywheeling" (F/W) mode [*Meehan*, pers. comm.], similar to the one experimented with during the GPS/MET mission. If the SNR of the received GPS signal drops below a certain threshold value, the receiver, instead of declaring to have lost the signal, propagates the phase information onwards in time using an extrapolation based on the last few available measurements. The receiver may operate in F/W mode for several seconds (corresponding to several km height difference in the atmosphere), before it either stops tracking, or continues its standard phased locked tracking mode provided SNR has regained sufficiently large values again. As low SNR occurs during the strong amplitude oscillations caused by atmospheric multipath, F/W events can be viewed as a further indicator for the occurance of multipath. However, they also indicate that the receiver's standard PLL has failed; the receiver has to bridge periods of data outage. Note that loosing cycles during F/W events probably remains unnoticed, as the receiver still delivers apparently smooth and continuous phase "measurements".

As mentioned above, the CT provides a diagnostic quantity to judge on the quality of the measured amplitude and phase data. The "CT amplitude" represents a hypothetical amplitude of the GPS signal after the removal of multipath effects by the CT. Gorbunov (2002) interprets the occurrance of large, chaotic fluctuations in transformed amplitude as indicator for "corrupted" data, speculating they might be linked to phase lock loop failures of the GPS/MET receiver.

Fig. 6 shows the CT amplitude for three different CHAMP profiles. For the one located over Antarctica, the CT amplitude remains fairly constant over most of the altitude range shown. It rapidly drops to zero when the occulting GPS satellite sets behind the Earth's horizon. This is consistent with the expectation that multipath effects do not occur in the dry polar regions. Ideally, measurements undergoing multipath conditions should exhibit qualitatively similar CT amplitude profiles, provided the CT was successful in removing the multipath effects. However, in contrast to the Antarctic profile, the subtropical and tropical occulations exhibit strong and rapid oscillations

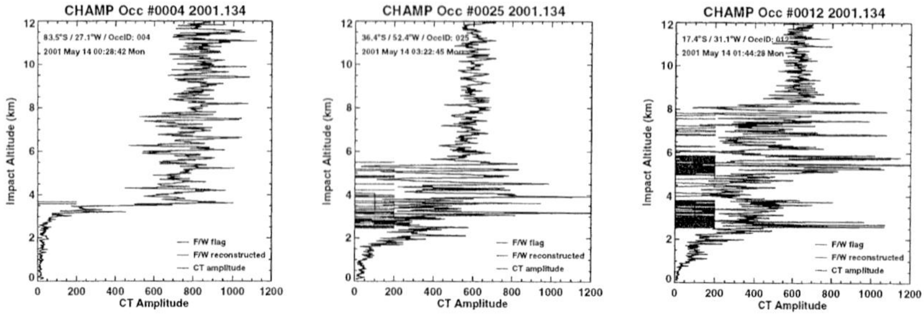

Fig. 6. CT amplitude of three arbitrarily selected CHAMP profiles from May, 14, 2001 over Antarctica (left), the Southern Hemisphere subtropics (middle) and tropics (right). Shown are the transformed amplitude (line) and reconstructed F/W flag (spikes on the left vertical axis; see text for details).

below 6 and 8 km impact altitude, respectively, indicating "corrupted" data according to Gorbunov.

In order to check for a possible relation to the receiver's tracking mode states, we have reconstructed likely F/W events for the profiles shown. The reconstruction is based on L1 SNR using a set of parameters provided by T.K. Meehan (JPL). Periods with active flywheeling are indicated on the left vertical axis of the CT amplitude plots; height referencing was provided by means of the traditional phase–only retrieval. While the receiver apparently did not activate F/W during the antarctic occultation (apart from the last moment of the occultation where the line of sight touches the Earth's surface), it did so several times during the other occultations. In the case of the tropical occultation, F/W was actually active during most of the tropospheric measurement. We note that the regions of poor data quality (according to CT) are closely linked to periods when the receiver is in F/W mode. Manual inspection of several hundred CHAMP profiles indicates that the close relation between active F/W and large fluctuations in the transformed amplitude is a generic feature of the vast majority of CHAMP profiles.

This suggests that poor data quality (from point of view of the CT) in the troposphere is linked to the CHAMP receiver's internal functioning under low SNR conditions. Note that we do not claim that F/W or any other property of the receiver indeed causes any degradation of the signal's quality; the complex signal structure is, in first place, related to the complex humidity distribution in the troposphere. We also do not claim that the CT fails in resolving multipath problems; it has shown its ability in numerous simulations. We do believe, however, that the role of the receiver and its internal working for the receiver's output signal structure are currently not sufficiently understood in order to be used appropriately in present day retrieval algorithms. As with GPS/MET, the data which is available from the CHAMP receiver in its current tracking configuration is not meeting the CT's implicit idealised assumptions about GPS measurements.

The simplest way to avoid problems with flywheeled data is to exclude them from the retrievals. The data stream output by the receiver offers a F/W flag, though only sampled at full seconds (thus, not all F/W data will be caught). In the experimental CT retrieval shown in Fig. 5, data has been discarded after the receiver has signalled the activation of F/W for the first time. As a result, a considerable part of the tropospheric data has been removed from the CHAMP data set. Thus, the implementation of F/W or any other open loop tracking capability within the receiver is clearly required if radio occultation measurements in the tropical and lower troposphere are intended. Questions that need to be addressed, however, are to what degree such data can be used in algorithms like the CT, and how large the uncertainties of the derived parameters like bending angles are.

We already noted that systematic errors in refractivity in mid latitude lower tropospheric refractivity are in the order of 5%. This seems to be an acceptable error, especially as humidity measurements with traditional instruments like radiosondes are often quoted to have random errors of 20–30% (Elliott and Gaffen, 1991) or more. It should be kept in mind, though, that refractivity depends only weakly on humidity. Rearranging equation (1) and taking the first variation with respect to N gives

$$\delta e = \frac{T^2}{\kappa_2}\,\delta N \; .$$

A 5% error in refractivity at a 15°C, 1000 hPa and a relative humidity of 50% (corresponding to a partial water vapor pressure of 8.4 hPa) results in a partial water vapor pressure error of 3.4 hPa, i.e. a systematic error of 40%. This is consistent with comparisons between retrieved water vapor profiles and nearby radiosondes, shown in Fig. 7: mean relative deviations of specific humidity against radiosondes are between 30 and 60%, with rms differences between 60 and 100%. Also note that the correlation between CHAMP and radiosonde humidities is below 0.6 at all altitudes, indicating that the current humidity retrievals cover less than 40% of the humidity variability present in the radiosonde data set.

We finally note that the traditional retrieval suffers from a high sensitivity to errors in the *a priori* temperature used for the derivation of humidity from refractivity. Marquardt et al. (2001) demonstrated that the use of different temperature analyses can cause local differences in specific humidity of up to 50 to 70% at the 500 hPa level. Thus, even if the refractivity bias issue can be solved, the traditional water vapor retrieval is at most suitable for a qualitative demonstration of the potential usefulness of radio occultations in NWP and climate research. Early attempts to use variational retrieval schemes like those proposed by Healy and Eyre (2000) and Palmer et al. (2000) have been very promising, and are likely to become the standard way of water vapor retrieval at GFZ. For variational retrievals, however, as for variational data assimilation, a full error characterisation of the measurements including the receiver's role is required, but currently missing.

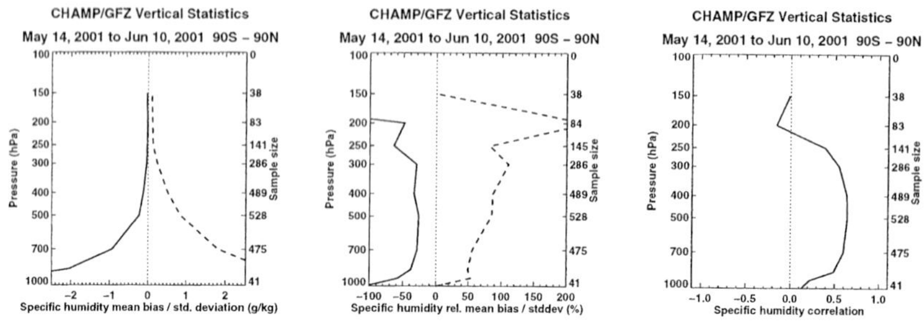

Fig. 7. Absolute (left) and relative (middle) bias and rms (dashed) differences as well as correlations (right) between radiosonde and CHAMP specific humidities during May/June 2001. Figures on the right axis denote the number of radiosonde matches for each pressure level.

5 Conclusions

We have shown that CHAMP radio occultation retrievals of stratospheric temperatures provided by GFZ agree well with both ECMWF analyses and radiosondes. Between the upper troposphere and 30 hPa (about 25 km), mean deviations are below 1 K against radiosondes, with rms deviation below 2–3 K. A warm bias is apparent above 30 hPa, which is more pronounced against ECMWF (4 K) than against radiosondes (2 K). The bias in the mid stratosphere can be significantly influenced by changing implementation details of the upper level bending angle smoothing in the retrieval; the information content of radio occultation soundings as provided by CHAMP is not sufficient to provide independent temperature retrievals without the use of *a priori* data or assumptions.

Tropospheric refractivity obtained from CHAMP soundings suffers from a systematic bias of up to 10% in the tropical lower trosposphere; in mid latitudes, biases in the order of 5% are found near the ground. These biases translate into mid latitude dry biases in the order of 40% in partial water vapor pressure. Similar large biases (and rms deviations in the order of 60–100%) have been found in specific humidity comparisons with nearby radiosonde data. Systematic biases of this magnitude question the usefulness of lower tropospheric radio occultation data for NWP and climate research applications. Thus, understanding the causes of and finding workarounds for these biases are crucial in the future.

The application of more elaborate retrieval algorithms like the Canonical Transform method has not improved the CHAMP retrievals significantly; large biases in the lower troposphere still exist. This is surprising, as various simulations indicate that the CT has a remarkable ability to reconstruct bending angles even under strong multipath conditions. By using the internal diagnostics of the CT, we have provided evidence that degradations (from point of view of the CT) in the quality of amplitude and phase measurements are linked to details of the tracking algorithms of the GPS receiver under low SNR conditions. Therefore, CHAMP, with its current signal aquisition and tracking configuration,does not provide an enhanced quality when compared to GPS/MET. We suggest that the role of the GPS receiver for the GPS signal's structure requires more attention than previously anticipated. Error estimates taking these issues into account are required for the use of radio occultation data in variational retrieval and data assimilation systems, but not yet available.

Acknowledgement. The GFZ part of this study was carried out under the grant of the German Federal Ministry of Education and Research (BMBF) No. 01SF9922/2.

References

Elliott W, Gaffen D (1991) On the utility of radiosonde humidity archives for climate studies. Bull Amer Meteor Soc 72:1507–1520

Fjeldbo G, Kliore G, Eshleman V (1971) The neutral atmosphere of Venus as studied with the Mariner V radio occultation experiments. Astron J 76:123–140

Gorbunov M (2001) Radioholographic methods for processing radio occultation data in multipath regions. Scientific Report 01-02, Danish Meteorological Institude, Copenhagen

Gorbunov M (2002) Radio-holographic analysis and validation of Microlab-1 radio occultation data in the lower troposphere. J Geophys Res 107:10.1029/2001JD000,889

Gorbunov M, Gurvich A (1998) Microlab–1 experiment: Multipath effects in the lower troposphere. J Geophys Res 103:13.819–13.826

Gorbunov M, Sokolovskiy S (1993) Remote sensing of refractivity from space for global observations of atmospheric parameters. Report 119, Max–Planck–Institut für Meteorologie, Hamburg

Gorbunov M, Sokolovskiy S, Bengtsson L (1996) Space refractive tomography of the atmosphere: Modelling of direct and inverse problems. Report 210, Max–Planck–Institut für Meteorologie, Hamburg

Hajj G, Kursinski E, Romans L, Bertiger W, Leroy S (2002) A technical description of atmospheric sounding by GPS occultation. J Atmos Sol-Terr Phys 64:451–469

Healy S (2001a) Radio occultation bending angle and impact parameter errors caused by horizontal refractive index gradients in the troposphere: A simulation study. J Geophys Res 106:11,875–11,889

Healy S (2001b) Smoothing radio occultation bending angles above 40 km. Ann Geophys 19:459–468

Healy S, Eyre J (2000) Retrieving temperature, water vapor and surface pressure information from refractive–index profiles derived by radio occultation: A simulation study. Quart J Roy Meteorol Soc 126:1661–1683

Hocke K (1997) Inversion of GPS meteorology data. Ann Geophys 15:443 – 450

Hocke K, Pavelyev A, Yakovlev O, Barthes L, Jakowski N (1999) Radio occultation data analysis by the radioholographic method. J Atmos Sol-Terr Phys 61:1169–1177

Kursinski E, Hajj G, Schofield J, Linfield R, Hardy K (1997) Observing earth's atmosphere with radio accultation measurements using the Global Positioning System. J Geophys Res 102:23.429–23.465

Marquardt C, Labitzke K, Reigber C, Schmidt T, Wickert J (2001) An assessment of the quality of GPS/MET radio limb soundings during February 1997. Phys Chem Earth 26:125–130

Mortensen M, Hoeg P (1998) Inversion of GPS occultation measurements using Fresnel diffraction theory. Geophys Res Lett 25:2441–2444

Palmer P, Barnett J, Eyre J, Healy S (2000) A nonlinear optimal estimation inverse method for radio occultation measurements of temperature, humidity, and surface pressure. J Geophys Res 105:17.513–17.526

Rocken C, Anthes R, Exner M, Hunt D, Sokolovsky S, Ware R, Gorbunov M, Schreiner W, Feng D, Herman B, Kuo YH, Zou X (1997) Analysis and validation of GPS/MET data in the neutral atmosphere. J Geophys Res 102:29.849–29.866

Schöllhammer K, Marquardt C, Labitzke K (2002) Comparison of three different meteorological datasets (ECMWF, Met Office and NCEP). This issue

Smith E, Weintraub S (1953) The constants in the equation for atmospheric refractivity index at radio frequencies. In: Proc. IRE, volume 41, 1035–1037

Sokolovskiy S (2001a) Modeling and inverting radio occultation signals in the moist troposphere. Radio Sci 36:441–458

Sokolovskiy S (2001b) Tracking tropospheric radio occultation signals from low Earth orbit. Radio Sci 36:483–498

Sokolovskiy S, Hunt D (1996) Statistical optimization approach for GPS/MET data inversions. URSI GPS/MET Workshop, Tucson, Arizona

Tsuda T, Nishida M, Rocken C, Ware R (2000) A global morphology of gravity wave activity in the stratosphere revealed by the GPS occultation data (GPS/MET). J Geophys Res 105:7257–7273

Vorob'ev V, Krasil'nikova T (1994) Estimation of the accuracy of the atmospheric refractive index recovery from doppler shift measurements at frequencies used in the NAVSTAR system. USSR Phys Atmos Ocean, Engl Transl 29:602–609

Wickert J, Beyerle G, Schmidt T, Marquardt C, König R, Grunwaldt L, Reigber C (2002) GPS radio occultation with CHAMP. This issue

Wickert J, Reigber C, Beyerle G, König R, Marquardt C, Schmidt T, Grunwaldt L, Galas R, Meehan T, Melbourne W, Hocke K (2001) Atmosphere sounding by GPS radio occultation: First results from CHAMP. Geophys Res Lett 28:3263–3266

Global Climate Monitoring based on CHAMP/GPS Radio Occultation Data

Ulrich Foelsche, Gottfried Kirchengast, and Andrea K. Steiner

Institute for Geophysics, Astrophysics, and Meteorology (IGAM), University of Graz, Graz, Austria.

Summary. The global coverage, all-weather capability, high accuracy, and self-calibrated nature of radio occultation (RO) data suggests them as a near-ideal resource to build global climatologies of fundamental atmospheric variables such as temperature and humidity. Such climatologies are not available yet, but some heritage exists from an on-going project on a climate observing system simulation experiment using realistically simulated RO data (GNSS-CLIMATCH, a joint project of IGAM/Univ. of Graz and MPI for Meteorology, Hamburg). The CHAMP (Challenging Minisatellite Payload) RO data will provide the very first opportunity to create real RO based climatologies. We aim at using the complete CHAMP RO data flow for month-to-month, season-to-season, and year-to-year global climate monitoring via temperature, geopotential height, and humidity fields.

The RO based climatologies will be built in two modes: 1) fully independent based on statistical binning and averaging techniques applied to the dataset of original profiles, 2) weakly background-dependent but higher resolved based on optimal fusion (3DVAR assimilation) of RO data into averaged ECMWF atmospheric analysis background fields. Here we describe the set-up for the first approach and show preliminary results based on realistically simulated data. These suggest that reliable season-to-season temperature climatologies resolving scales > 1000 km can be obtained even with RO data received from a single satellite.

Key words: radio occultation technique, CHAMP satellite mission, climate monitoring, forward modeling and inversion, atmospheric remote sensing, anthropogenic climate change

1 Introduction

Over the coming decades climate change monitoring will be of high relevance since there exists global concern (and increasing evidence) that the Earth's climate is significantly influenced by human activities (e.g., IPCC 2001). High quality observations of the atmosphere are particularly required (but not sufficiently available) in this context. The Global Navigation Satellite System (GNSS) radio occultation (RO) technique has the potential to substantially contribute to this scientific challenge.

The GNSS currently consists of the U.S. GPS and the Russian GLONASS with nominal constellations of 24 satellites each. A European system with some thirty additional satellites (GALILEO) is under development and scheduled to be operational by 2008.

Radio signals from these GNSS satellites, received onboard satellites in Low Earth Orbit (LEO), can be used to probe the Earth's atmosphere in a limb sounding mode, since the propagation of the signals is influenced by the atmospheric (and ionospheric) refractivity field. The principle observable is the excess phase path, which can be measured with millimetric accuracy. Such RO phase data are the basis for the calculation of key climate variables such as temperature.

The RO technique (for a detailed description see, e.g., the reviews of Kursinski et al. 1997 and Steiner et al. 2001) provides an encouraging combination of:

- Global coverage,
- All weather capability,
- High accuracy and vertical resolution, and
- Long-term stability.

In contrast to many other conventional (non-satellite) data sources (like radiosondes) observation density over oceans is as high as over land. The use of L-band signals with wavelengths of 19.0 and 24.4 cm, respectively, makes the method almost insensitive to clouds and aerosols (a major advantage given the fact that about half of the Earth is on average covered by clouds). Temperature profiles, for example, can be retrieved with an accuracy of < 1 K in the upper troposphere and lower stratosphere with a height resolution of 1-2 km (Kursinski et al. 1997). The long-term stability of the data is of particular importance for climate studies, it is achieved since precise clocks, driven by very stable oscillators, are the basis for accurate phase measurements during each single occultation event, independent of whether two events are separated by an hour or by decades.

Sensing of the Earth's atmosphere with the RO method was first successfully demonstrated with the GPS Meteorology (GPS/MET) experiment performing measurement campaigns from April 1995 to March 1997. Analysis and validation of GPS/MET data sets confirmed most of the expected strengths of the technique listed above (Kursinski et al. 1997; Rocken et al. 1997; Steiner et al. 1999).

The climate monitoring utility of a GNSS occultation observing system has not yet been tested due to the lack of long-term measurements for such a study. The CHAMP (Challenging Minisatellite Payload) RO data will provide the very first opportunity to create real RO based climatologies. We will use the complete CHAMP RO data flow for global climate monitoring via temperature, geopotential height, and humidity.

In addition, we undertake a large-scale climate observing system simulation study, which aims at testing the climate change detection capability of GNSS occultation sensors and helps setting up the climate monitoring system for the CHAMP RO data. An overview of the main results relevant in this context is given in the following sections 2 and 3.

2 The GNSS Climate Monitoring Study

A detailed description of rationale and design of this study has been given by Steiner et al. (2001). Here we focus on new results, which are relevant for this work. The study involves five main parts of work as follows.

1. Realistic modeling of the neutral atmosphere and the ionosphere over the time period 2001 to 2025.
2. Realistic simulations of occultation observables for a small GNSS receiver constellation (6 satellites) for the same time period.
3. State-of-the-art data processing for temperature profile retrieval in the troposphere and stratosphere to establish a sufficient database.
4. Statistical analysis of temperature trends in the "measured" climatology (database of simulated measurements) and the "true" climatology (model atmosphere).
5. Assessment of how well a GNSS occultation observing system is able to detect climatic trends in the atmosphere.

2.1 Atmosphere and Ionosphere Modeling

In order to obtain reliably simulated occultation data it is of particular importance to use realistic models of the neutral atmosphere and the ionosphere. For the former we employ the ECHAM5 General Circulation Model (the successor of the ECHAM4 model; Roeckner et al. 1999) with a resolution of T42L39, the highest of the 39 model levels being at ~0.1 hPa or ~80 km. Above this height MSISE-90 climatology is used (Hedin, 1991). After successful termination of the control run, a model run for the time period 2001 to 2025 including transient anthropogenic forcings due to greenhouse gases, sulfate aerosol, and ozone is currently performed at the MPI for Meteorology in Hamburg, Germany. The atmospheric fields are stored every three hours in order to capture diurnal variations.

For the ionosphere we use the NeUoG model (Leitinger et al. 1996), a global empirical climatology for the 3D ionospheric electron density field. It is driven by day-to-day variations of the solar activity (including the ~11 year solar cycle), represented by the F10.7 solar flux index. F10.7 data from the solar cycles 21, 22, and 23 are used to obtain quasi-realistic solar activity variations for the time period 2001 to 2025.

2.2 Simulation of Occultation Data

We assumed a reasonable constellation of six LEO satellites equipped with GNSS receivers, comparable to currently planned occultation missions like ACE+ (Hoeg and Kirchengast 2002) and COSMIC (Lee et al. 2001). Under the assumption that occultation events with a high angle of incidence with respect to the LEO Orbit plane are more vulnerable to horizontal structure errors, we restricted to occulta-

2.4 Temperature Trend Analysis

The "measured" temperature profiles during the respective summer season in each of the 17 latitude bins are averaged and interpolated onto 34 height levels between 2 and 50 km, yielding a summer mean temperature field in form of a 17 x 34 matrix. For comparison, a "true" climatology matrix is computed on the same grid directly from the GCM temperature field.

As soon as all the 25 JJA mean temperature fields are computed, linear trends (and their standard deviations) in each matrix element over the 25 year period will be determined based on a multivariate weighted least-squares analysis approach. Model runs of the ECHAM4 GCM in T42L19 mode indicate that temperature trends of up to ~2 K can be expected in parts of the upper troposphere and lower stratosphere during the next 25 years with anthropogenic forcing (Roeckner et al. 1999). Given the accuracy of RO data, such trends should be clearly detectable.

3 Performance Analysis for the Test-bed Season

3.1 Observational Error

Temperature error statistics for each latitude bin have been computed as described in section 2.3. The temperature bias on the 17 x 34 grid is displayed in the left panel of Figure 3 (cf. the black lines in Figure 2). The total climatological observational error (root-mean-square error of the bias) is shown in the right panel of Figure 3; in most parts of the core region between 8 and 40 km it is smaller than 0.2 K. The largest observational errors are found in the southernmost bin ("worst case" in Figure 2).

3.2 Sampling Error

In addition to the observational error we have to consider the sampling or representativeness error. Even with perfect observations at the occultation locations the

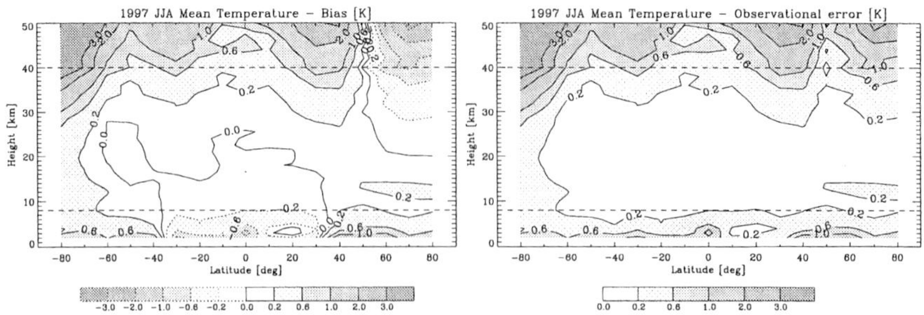

Fig. 3. Systematic error of the JJA 1997 mean temperature (left panel) and total observational error (right panel). The core region between 8 and 40 km is indicated by dashed lines.

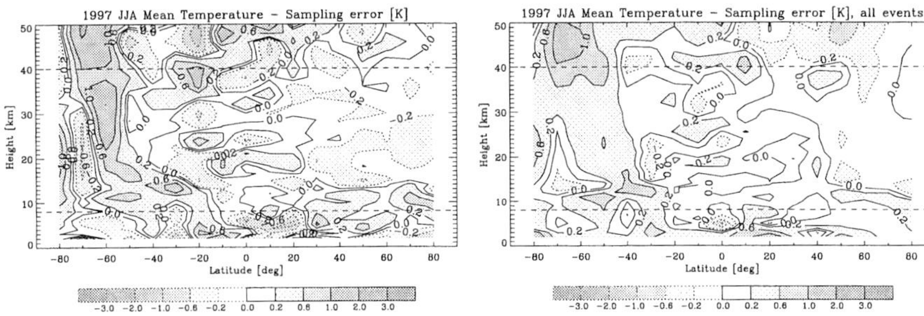

Fig. 4. Sampling error for the JJA 1997 mean temperature using the selected ~1000 events (left panel) and using all ~13,000 events (right panel).

"measured" climatology would differ from the "true" one as the sampling through occultation events is discrete and not dense enough to capture the entire spatio-temporal evolution. This error can be determined by comparison of the "true" mean temperature field and the mean field obtained from "true" profiles at the locations and dates of the occultation events. The sampling error for the 1997 summer season, using the selected ~1000 events, is shown in Figure 4 (left panel).

The largest errors of about 2 K are encountered in high latitude winter areas (around 60°S), where the latitudinal temperature gradient is particularly strong. In the core region the rms error is about 0.5 K (0.3K for the Northern Hemisphere).

For reference we computed the sampling error resulting from the use of all ~13000 events, this one is shown in the right panel of Figure 4. The error reduction is clearly visible but notedly smaller than what one might expect given an increase of occultation events by a factor of ~13. "Additional" occultation events frequently occur in close spatial and temporal vicinity to existing ones, which do not markedly improve the sampling situation.

3.3 Total Climatological Error

The total climatological error, finally, is a combination of the observational error and the sampling error; it is displayed (for JJA 1997) in Figure 5. In the core region and north of 50°S it is smaller than 0.6 K almost everywhere.

South of 50°S the coincidence of large observational and sampling errors leads to total errors of 1 to 2 K, indicating that high latitude winter areas can be a challenging region for GNSS based climatology. Above ~40 km the total error increases markedly, reaching values of 2 to 4 K.

Temperature trends over the next 25 years, however, can be expected to be larger than 1 K in important parts of the core region (Roeckner et al. 1999) and should thus have a good chance to be detected with a small constellation of GNSS receivers if the good performance during the "test-bed" season can be obtained in the "real" world over the entire 25 year period.

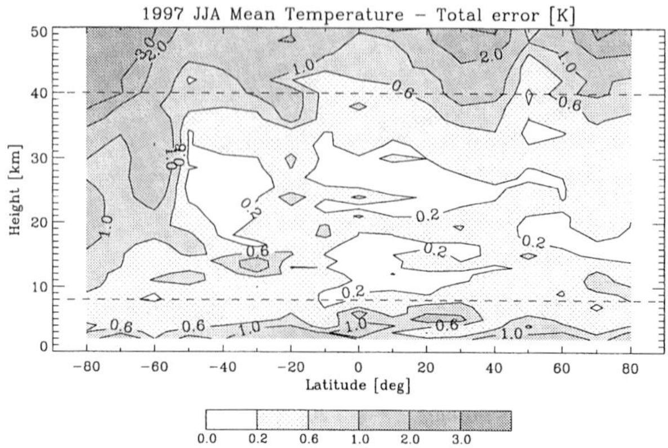

Fig. 5. Total climatological error for the JJA 1997 mean temperature.

4 Climate Monitoring with CHAMP

With the continuous stream of CHAMP RO data it will be possible to generate radio occultation based climatologies. In one approach we will use the profiles as described in section 2 and 3. While the individual profiles can be expected to be of the same quality as the ones in the simulation study (e.g., Wickert et al. 2001) the sampling error has to be considered carefully given the somewhat unfavorable single LEO satellite situation. In this context the recent results of Foelsche and Kirchengast (2002), referred to in the first paragraph of section 2.2, are quite helpful as they give us the justification to use all CHAMP occultation events within ± 45° about the LEO orbit plane if we focus on heights above ~7 km.
We computed realistic CHAMP occultation locations during a summer season (nominal GPS) and calculated the sampling error as described in section 3.2. For easing comparison we used again the JJA 1997 atmospheric fields. Even with a single satellite ~1000 (setting) occultations can be obtained during a 3 month period in the selected geographic region; the distribution is displayed in Figure 6.

Due to the high inclination of the satellite (87.3°), the event density in low latitude bins is particularly small. Comparatively small temperature variations in these bins, however, prevent the sampling error from increasing dramatically as Figure 7 shows. North of 30°N and south of 30°S the sampling error remains smaller than 0.6 K in most parts of the core region, comparable to the sampling error in the Southern Hemisphere in section 3.2.

This situation varies considerably from year to year as the two panels of Figure 7 indicate. For the results displayed in the right panel (termed "1998") we used the realistic event locations which would occur one year after JJA 1997 and computed the sampling error (based on the same atmospheric fields as used above).

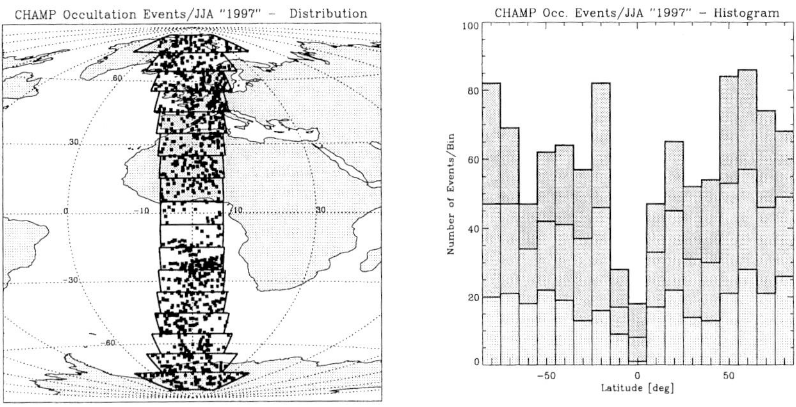

Fig. 6. Left panel: Geographic domain and locations of all ~1000 simulated CHAMP occultation events during "JJA 1997" (black squares). Right panel: Cumulative distribution of the events in the 17 equal area bins in June (light gray), June + July (plus medium gray), and June + July + August (plus dark gray).

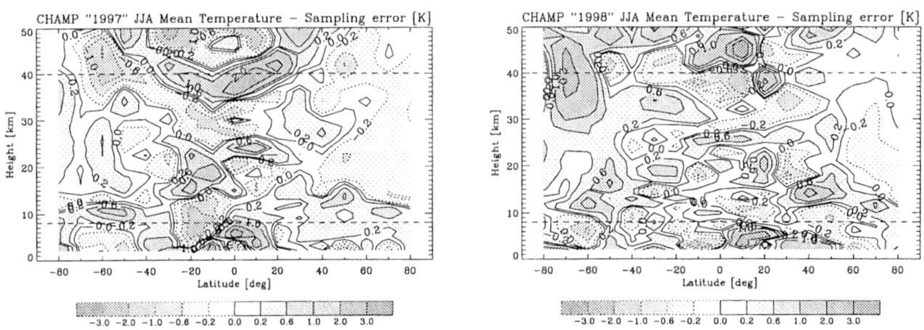

Fig. 7. Sampling error during "JJA 1997" (see text) using all ~1000 CHAMP events (left panel) and during "JJA 1998" using all available ~800 events (right panel).

Even though the total number of available events is reduced to ~800 the sampling error somewhat decreases in most parts of the core region due to the spatial and temporal sampling being better distributed during JJA 1998.

Given the encouraging performance of the GNSS receiver onboard CHAMP (Wickert et al. 2001) we are confident that reliable temperature climatologies can be achieved even with a single LEO satellite for season-to-season climatologies resolving horizontal scales > 1000 km (large-scale climatologies).

5 Summary, Conclusions, and Outlook

After a brief review of the utility of GNSS radio occultation data for climate monitoring we reported on the latest results of the GNSS-CLIMATCH study (a climate observing system simulation experiment using realistically simulated RO data) which suggest that expected temperature trends in the upper troposphere and stratosphere over the coming 25 years could be detected with a small constellation of GNSS receivers onboard LEO satellites.

Even with a single satellite like CHAMP useful temperature climatologies seem to be achievable but the sampling error, being the most important error source in this case, has to be considered carefully. As soon as a sufficient amount of CHAMP occultation data will available, we look forward to implement the described approach and to build real climatologies.

Acknowledgements. The authors thank all colleagues in the CHAMPCLIM and GNSS-CLIMATCH projects for their support. U.F. and A.K.S. were funded for this work from the START research award of G.K. financed by the Austrian Ministry for Education, Science, and Culture and managed under Program Y103-CHE of the Austrian Science Fund.

References

Foelsche U, Kirchengast G (2002), Sensitivity of atmospheric profiles retrieved from GNSS occultation data to horizontal variability in the troposphere. Tech Rep ESA/ ESTEC-2/2002, IGAM, University of Graz, Austria, 56 pp

Gobiet A, Kirchengast G (2002), Sensitivity of atmospheric profiles retrieved from GNSS occultation data to ionospheric residual and high-altitude initialization errors. Tech Rep ESA/ESTEC-1/2002, IGAM, University of Graz, Austria, 56 pp

Hedin AE (1991) Extension of the MSIS thermosphere model into the middle and lower atmosphere. J Geophys Res 96: 1159–1172

Healy SB (2001) Smoothing radio occultation bending angles above 40 km. Ann Geophys 19 (4): 459–468

Hoeg P, Kirchengast G (2002) ACE+ – Atmosphere and Climate Explorer based on GPS, GALILEO, and LEO-LEO radio occultation (ESA Earth Explorer Opportunity Mission Proposal), Wissenschaftl Ber 14, IGAM, University of Graz, Austria, 121 pp

IPCC (2001) Climate change 2001: The scientific basis. Cambridge Univ Press, 881 pp

Kursinski ER, Hajj GA, Schofield JT, Linfield RP, Hardy KR (1997) Observing Earth's atmosphere with radio occultation measurements using the Global Positioning System. J Geophys Res 102: 23429–23465

Lee L-C, Rocken C, Kursinski R (eds) (2001) COSMIC – Applications of constellation observing system for meteorology, ionosphere, and climate. Springer, Hong Kong Berlin Heidelberg

Leitinger R, Titheridge JE, Kirchengast G, Rothleitner W (1996) A "simple" global empirical model for the F layer of the ionosphere (in German; English version avail. from the authors). Kleinheubacher Ber 39: 697–704

Rocken C, Anthes R, Exner M, Hunt D, Sokolovskiy S, Ware R, Gorbunov M, Schreiner W, Feng D, Herman B, Kuo YH, Zou X (1997) Analysis and validation of GPS/MET data in the neutral atmosphere. J Geophys Res 102: 29849–29866

Roeckner E, Bengtsson L, Feichter J, Lelieveld J, Rodhe H (1999) Transient climate change simulations with a coupled atmosphere-ocean GCM including the tropospheric sulfur cycle. J Climate 12: 3004–3032

Steiner AK, Kirchengast G, Ladreiter HP (1999) Inversion, error analysis, and validation of GPS/MET occultation data. Ann Geophys 17: 122–138

Steiner AK, Kirchengast G, Foelsche U, Kornblueh L, Manzini E, Bengtsson L (2001) GNSS occultation sounding for climate monitoring. Phys Chem Earth 26: 113–124

Syndergaard S (1999) Retrieval analysis and methodologies in atmospheric limb sounding using the GNSS radio occultation technique. DMI Scient Report 99-6, Danish Meteorol Institute, Copenhagen, Denmark, 131 pp

Wickert J, Reigber C, Beyerle G, Konig R, Marquardt C, Schmidt T, Grunwaldt L, Galas R, Meehan TK, Melbourne WG, Hocke K (2001) Atmosphere sounding by GPS radio occultation: First results from CHAMP. Geophys Res Lett 28 (17): 3263–3266

Initial Results on Ionosphere/Plasmasphere Sounding based on GPS Data Obtained On Board CHAMP

Stefan Heise[1], Norbert Jakowski[1], Andreas Wehrenpfennig[1], Christoph Reigber[2], and Hermann Lühr[2]

[1] Deutsches Zentrum für Luft- und Raumfahrt (DLR), Institut für Kommunikation und Navigation, Kalkhorstweg 53, 17235 Neustrelitz, Germany
[2] GeoForschungsZentrum Potsdam (GFZ), Divisions 1 and 2.

Summary. For precise orbit determination the GPS receiver on board CHAMP tracks presently up to 8 GPS satellites. Since these navigation data are measured with a sampling rate of 0.1 Hz, the dual frequency GPS signals provide valuable information on the ionization state of the topside ionosphere and plasmasphere. Combining differential carrier and code phases, absolute values of the total electron content (TEC) along the numerous radio links are estimated after calibrating the instrumental biases of each CHAMP receiver GPS transmitter combination by applying a model assisted technique. The calibrated TEC data derived for a full CHAMP revolution are assimilated into the Parameterized Ionospheric Model (PIM) that includes also the plasmasphere. The assimilation technique is briefly described. Preliminary results that provide a 2-D electron density slice of the ionosphere/plasmasphere within the CHAMP orbit plane from the CHAMP altitude up to GPS orbit heights will be discussed. First validation checks will be presented using electron density in situ measurements by the Langmuir probe on board CHAMP and electron density profiles from digisonde and incoherent scatter radar measurements for selected coincidence cases.

Key words: Ionosphere, Plasmasphere, GPS, CHAMP

1 Introduction

Beside the GPS occultation measurements, CHAMP tracks presently up to 8 GPS satellites simultaneously for navigation, using a zenith looking antenna. The GPS measurements on board CHAMP are performed by the TRSR2 space receiver ("Black Jack") developed by the Jet Propulsion Laboratory (JPL, USA). The navigation data are measured with a sampling rate of 0.1 Hz. Due to the dispersive propagation properties of the ionosphere for radio signals, TEC can be derived from differential GPS phase measurements. Thus, the dual frequency navigation signals from CHAMP provide valuable information on the ionization state of the topside ionosphere and plasmasphere. Whereas ground based GPS measurements are well established as a tool to monitor the ionosphere [6], space borne GPS observations offer new opportunities of ionospheric sounding and monitoring on global scale. In the following we illustrate shortly the data processing at DLR IKN Neustrelitz for the reconstruction of the topside ionosphere and plasmasphere electron density distribution and present preliminary results including first validation checks.

2 Data Processing

The data processing consists of 3 main parts: the preprocessing of GPS and orbit data, the calibration of the GPS based TEC measurements and the assimilation of the TEC data into a model of the ionosphere and plasmasphere.

2.1 Preprocessing of GPS Data

Main inputs for the retrieval are GPS low rate (0.1 Hz) measurements and orbit data from CHAMP and GPS. These input data are provided by the GFZ Potsdam. During the preprocessing an algorithm as described by [1] is applied to remove outliers and to correct cycle slips. Furthermore, orbit information is allocated to every measured GPS transmitter - CHAMP receiver link. For the calculation of TEC we use differential GPS pseudoranges (P2-P1) and carrier phases (L1-L2) and work finally with pseudorange-leveled carrier phase differences. The so derived relative TEC still includes satellite and receiver differential group delay biases. Therefore, to obtain the absolute TEC information, a calibration is necessary. A detailed description on the estimation of TEC as well as GPS receiver and satellite biases is given e.g. by [7].

2.2 TEC Calibration

The GPS differential code biases are estimated in a reliable manner by different ground based GPS processing centers of the International GPS Service. A similar bias solution is regularly processed at DLR IKN Neustrelitz as well in connection with a regional European TEC monitoring. Therefore, we focus on the estimation of the unknown CHAMP receiver bias and estimate subsequently the bias for every single CHAMP-GPS combination using the known GPS biases. To estimate the CHAMP receiver bias a model assisted technique is applied. For assistance, the Parameterized Ionospheric Model (PIM) [2] is used that also includes the Gallagher model [3] of the plasmasphere. We assume that the assisting model meets the real ionospheric/plasmaspheric electron density distribution averaged over several days and estimate the CHAMP receiver bias by averaging the difference of the CHAMP TEC measurements and the corresponding modeled TEC. To keep the calibration errors small, this average is calculated from preferably small TEC values. Thus,

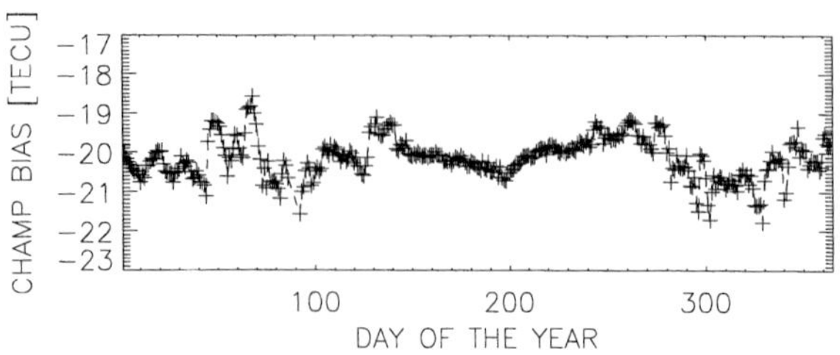

Fig. 1. Estimated CHAMP GPS receiver bias for 2001.

we consider only radio links measured under nighttime conditions in middle and high latitudes with an elevation angle of at least 50° with respect to the zenith antenna. Fig. 1 shows the estimated CHAMP bias in TEC units (TECU) for 2001, using GPS measurements from 5 consecutive days as a running mean. Because of a stable temperature on board CHAMP, which is most important for a stable receiver bias, it is possible to average the bias over several days. As Fig. 1 shows, the calculated CHAMP bias is quite stable over a considerable time period, varying around -20 TECU. It should be noted that in particular severe disturbances of the ionospheric and plasmaspheric state could violate the assumption of this calibration approach resulting in fluctuations of the estimated CHAMP bias.

2.3 Assimilation of TEC Data

After calibration, the link related TEC data derived for a full CHAMP revolution are assimilated into the PIM. For this period that takes 93 minutes, the ionospheric/plasmaspheric system is assumed to be stationary. Up to 4000 link related TEC measurements are available for a full revolution, depending on data quality and receiver settings. To give an impression of the

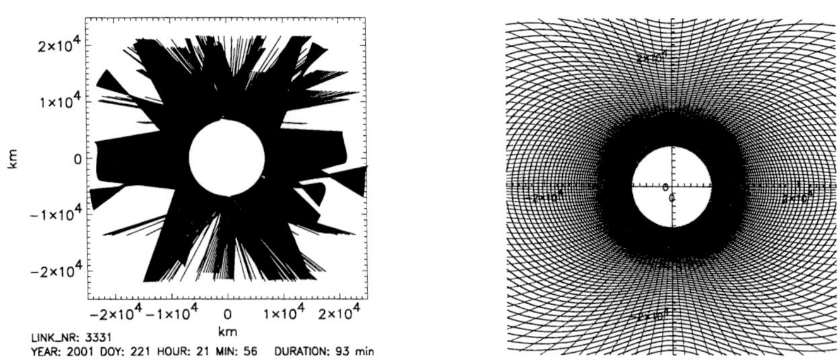

Fig. 2. Left panel: 2-D projection of a typical distribution of GPS-CHAMP radio links obtained during a full CHAMP revolution. Right panel: Voxel structure for the assimilation of link related TEC data.

typical data coverage, the left panel of Fig. 2 shows the 2-dimensional projection of the distribution of GPS-CHAMP radio links obtained during a full CHAMP revolution. To assimilate the link related TEC data into an ionospheric/plasmaspheric model (e.g. PIM) a 3-dimensional voxel structure has been defined which is shown in the right panel of Fig. 2. Because of the importance of the Earth's magnetic field for the plasmaspheric electron density distribution, this structure is handled in dipolar magnetic coordinates and its shape follows the natural (undisturbed) electron density distribution in the ionospheric/plasmaspheric system. The vertical resolution begins at 10 km near the Earth surface and grows up to 1000 km at GPS orbit height, while the horizontal resolution is 2.5 x 5° in latitude and longitude, respectively. At the beginning of the assimilation process PIM is used to initialize this voxel structure. In the following an iterative algorithm is applied that modifies the electron density of voxels crossed by the CHAMP - GPS radio links to adapt the voxel-TEC (integrated voxel electron density along a radio link) to the TEC measurements. According to the Multiplicative Algebraic Reconstruction Technique (MART, e.g. [4]), the next iteration step is reached by multiplying the crossed voxels by the ratio between the TEC measurement and the voxel-TEC. After application of this algorithm the assimilated information is still restricted to crossed voxels. Finally, to obtain a coherent 3-dimensional electron density distribution each crossed voxel influences its environment depending on distance and location. Of course, the result of this procedure is not a detailed reproduction of the real state of the ionosphere and plasmasphere. However, depending on the data coverage it is at least an improved model output. Independent on data coverage the result is always stable and physical reasonable.

3 Results

The result of the assimilation process is a 3-dimensional electron density distribution, but the data coverage and the influence of the assimilation process have their maximum inside and near the orbit plane. Other areas of the voxel structure stay completely unaffected by the assimilation process and give therefore a pure model output. Thus, we consider only a 2-dimensional slice from the assimilated voxel structure within the orbit plane. The left panel of Fig. 3 gives an impression of the reconstructed electron density distribution in the orbit plane obtained after a selected assimilation process. The right panel shows the percentage deviation of the assimilation product from the model output used for initialization. A remarkable and quite strong deviation of the assimilation output from the model input occurs in the area of the modeled plasmapause. This could be interpreted as a correction of the PIM plasmapause position by the assimilation. First validation checks were performed using in situ electron density measurements of the Planar Langmuir Probe (PLP) on board CHAMP and measurements of ionosonde stations at Athens

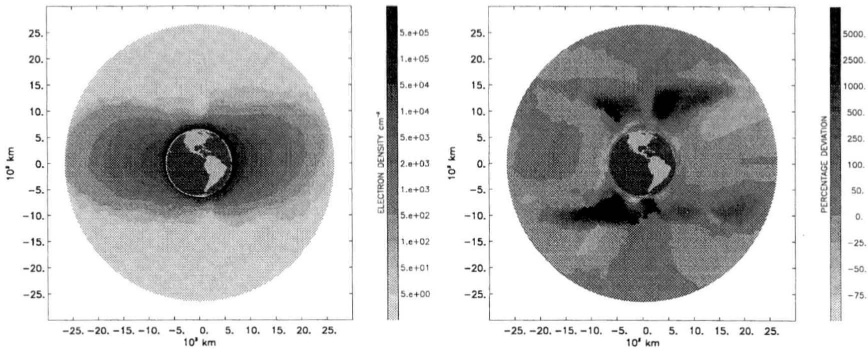

Fig. 3. Left panel: 2-D reconstruction (meridional cut) of ionospheric/plasmaspheric electron density distribution after assimilation of TEC data for a full CHAMP revolution. Right panel: Percentage deviation of the assimilation result from initial model (PIM). Assimilation for August 9, 2001, begin of assimilation: 14:08 UT, duration: 93 minutes, meridional cut (orbit plane) along 0.2° E.

and Juliusruh and the Millstone Hill incoherent scatter radar for coincidence cases with CHAMP. The in situ electron density measurements by the PLP provide a particularly valuable data source since these measurements enable the validation of the assimilation result at the location of the satellite using large data sets measured on global scale. Even if these data are not sufficient to validate the entire assimilation result, systematic errors along the satellite path would indicate deficiencies of the assimilation technique, leading to errors above the orbit height. The left panel of Fig. 4 shows a scatter plot of assimilation results versus PLP electron density measurements from 23 to 28 April 2001. The right panel shows a corresponding scatter plot of the initial

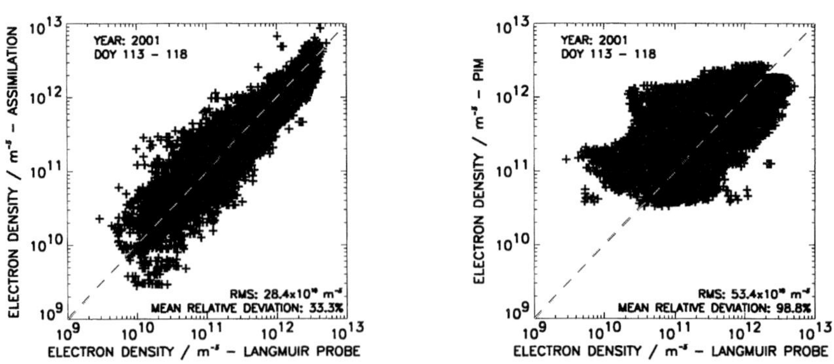

Fig. 4. Left panel: scatter plot assimilation versus corresponding Langmuir Probe measurements from 23 to 28 April 2001. Right panel: scatter plot initial model versus Langmuir Probe for the same period.

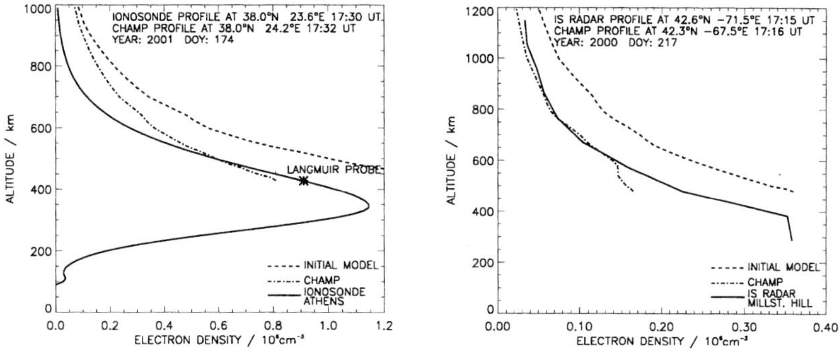

Fig. 5. Electron density profiles for selected coincidence cases between CHAMP and: Left panel: ionosonde station Athens, Jun. 23, 2001, 17:32 UT; Right panel: incoherent scatter radar Millstone Hill.

model versus PLP. It is evident that the assimilation process improves the initial model assumption significantly. The initial model coincides with PLP electron density values within an RMS spread of only 5.3 x $10^{11}m^{-3}$ (98.8 % mean relative deviation) while the assimilation result agrees within 2.8 x $10^{11}m^{-3}$ (33.3 %) for the considered period. Furthermore, the correlation between PLP and assimilation is quite consistent over a large dynamic range.

Ground based sounding of the ionosphere e.g. by incoherent scatter radar or ionosondes enables the validation of assimilation results above the orbit height if the CHAMP orbit trace comes close to the station site. The used ionosonde stations Athens and Juliusruh are equipped with the digisonde instrument that allows an estimation of the ionospheric topside profile from bottomside ionosonde measurements [5]. Fig. 5 shows electron density profiles for selected coincidence cases from digisonde station Athens (left panel) and incoherent scatter radar Millstone Hill (right panel). The initial model profile and the assimilation profile are plotted for comparison. It is clearly visible that the assimilation process improves the initial model assumption which is relatively far from reality in both cases. The corresponding PLP measurement on the left panel of Fig. 5 is in very good agreement with the digisonde and is close to the assimilation result. Probably, the upper digisonde topside profile underestimates the real electron density. In all tested coincidence cases with incoherent scatter radar and digisondes we found an essential improvement of the initial model assumption by the assimilation.

4 Conclusion

We presented preliminary results on reconstruction of ionosphere plasmasphere electron density distribution using GPS data obtained on board CHAMP. The data processing and assimilation technique to derive these

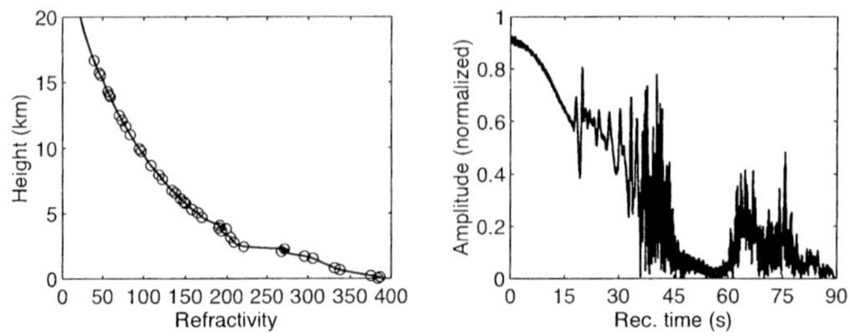

Fig. 2. Simulated occultation. *Left*: input refractivity profile shows a superrefraction layer at about 3 km in height. *Right*: received signal at the observation plane has an extended period of time where the amplitude remains small.

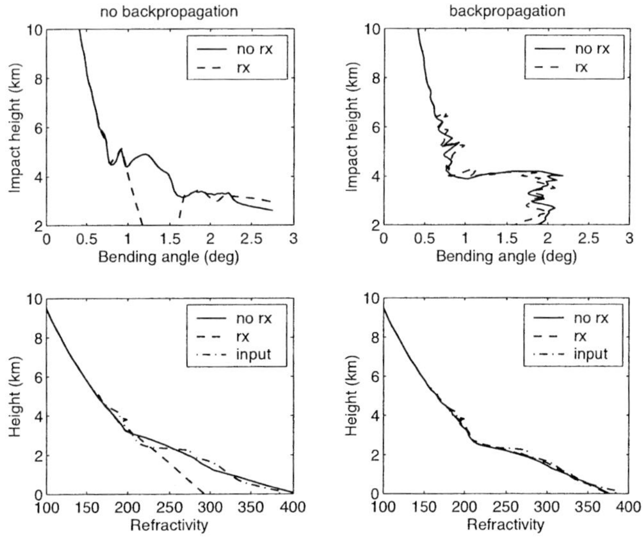

Fig. 3. Effects of receiver tracking on standard and BP retrievals as manifested in bending angle and refractivity. "no rx": MPS data with no receiver or perfect tracking; "rx": MPS data with simulated receiver tracking.

flywheeling mode. In Fig. 3, we compare the standard and BP retrievals using data with and without receiver tracking. Several observations can be made from these comparisons: (1) Without BP and with perfect tracking, multiple values of the bending are deduced, resulting in significant retrieval errors below 3 km in altitude; (2) BP reduces the multipath ambiguities but does not eliminate them; (3) Without BP, tracking errors cause a substantial negative bias in refractivity; (4) BP is *nearly unaffected* by tracking errors. The last point is at first glance rather surprising since one would expect that the back-continuation of flawed data should still result in flawed retrievals. However, it can be observed from Eq. (1) that the BP integral involves a linear weighting by the signal amplitude. Thus the regions where the *worst* error occurs would contribute the *least* to the BP field.

4 Backpropagation on CHAMP and SAC-C Data

Several complications arise when the BP algorithm is applied to real occultation data. These issues and their corresponding solutions are discussed briefly here. (1) *Transmitter motion compensation.* The diffraction integral used in the BP method assumes time-harmonic waves. The slow time dependence in the transmitting satellite position must first be removed. This is done by transforming the satellite positions such that the transmitter is fixed at all times. The phase is adjusted to account for the straight-line path differences between the original satellite positions and the transformed satellite positions. (2) *Synchronization.* Each data point is associated with a receiver time tag t, which is used in relating the occultation tangent point to its geodetic location. However, such information is lost after BP is performed since each BP point is now associated with an artificial time tag t_{bp}. To relate t_{bp} back to t, we compare the BP impact parameter $a(t_{bp})$ with the impact parameter $a(t)$ from the standard retrieval. While some errors in a are expected from the standard retrieval in multipath regions, this will not seriously affect the determination of the tangent point coordinates. (3) *BP on L1 only*. L2 measurements have lower SNR and are affected by Anti-Spoofing; moreover, its amplitude data is currently stored only once every second. Since BP depends crucially on amplitude information, we apply BP on L1 data only. Ionospheric calibration is based on the BP L1 data and the non-BP L2 data.

We present BP retrievals for CHAMP and SAC-C occultations taken on September 30, 2001. Typical examples are shown in Figs. 4 and 5. In Fig. 4, BP is able to remove all multipath ambiguities and reduces the refractivity bias with respect to the profile obtained from the National Centers for En-

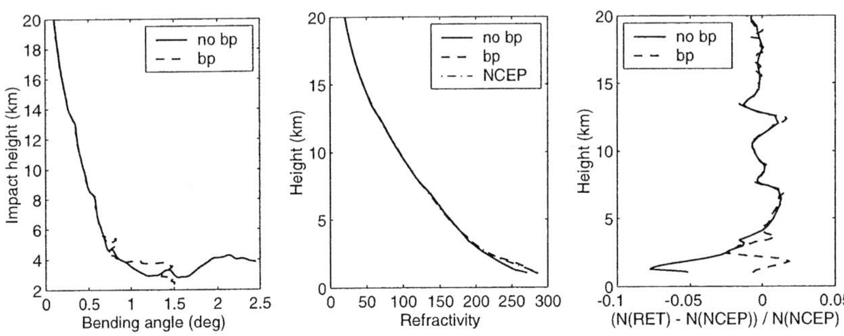

Fig. 4. An example showing improved retrieval with BP. SAC-C; occultation start time is 2001/09/30 06:14 UT.

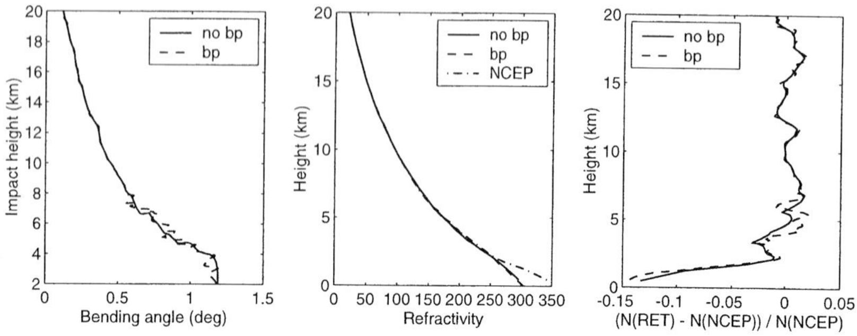

Fig. 5. An example showing no improvement in retrieval with BP. CHAMP; occultation start time is 2001/09/30 09:58 UT.

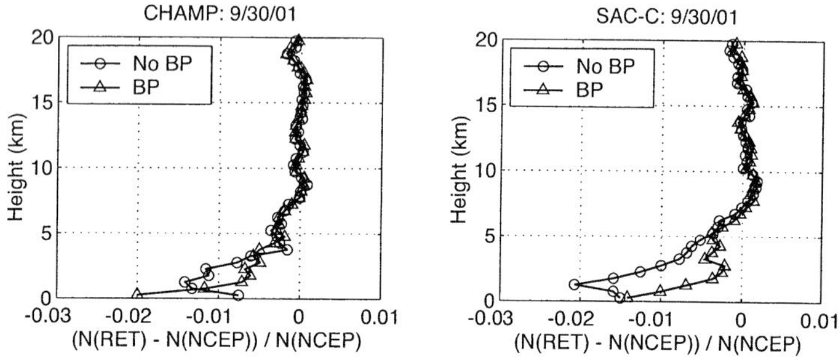

Fig. 6. Mean refractivity difference relative to NCEP based on one day of retrievals.

vironmental Prediction (NCEP). In Fig. 5, BP differs very little from the standard retrieval, both of which exhibit a negative bias with respect to the NCEP profile. Mean refractivity differences relative to NCEP are shown in Fig. 6. It can be seen that the negative refractivity bias below 5 km is generally reduced by BP, especially for the SAC-C data. The remaining bias could be due to a number of factors, including the failure of BP to remove all multipaths, limitations of Abel inversion, premature termination of occultation tracking, as well as errors in numerical weather prediction models. More work is needed to quantify these various effects.

5 Conclusions

We have studied the backpropagation (BP) method on simulated occultations which incorporated realistic receiver tracking as well as on real data from CHAMP and SAC-C. From the simulations, it was found that BP did very well (but not perfectly) in removing multipath ambiguities that often plagued lower tropospheric retrievals. Moreover, BP was found to be quite robust in the presence of receiver tracking errors. This is because the BP reconstruction is linearly weighted by the received amplitude. Regions of large tracking errors are typically associated with small SNR so that their contributions to the BP field are small. BP of CHAMP and SAC-C data also showed improved retrievals and generally reduced the negative refractivity bias with respect to numerical weather prediction models.

Acknowledgement: This research was carried out at the Jet Propulsion Laboratory, California Institute of Technology, under a contract with the National Aeronautics and Space Administration.

References

1. Fjeldbo G, Kliore AJ, and Eshleman VR (1971) The neutral atmosphere of Venus as studied with the Mariner V radio occultation experiments. *Astron J*, 76:123–140.
2. Yunck TP, Lindal GF, and Liu CH (1988) The role of GPS in precise earth observation. In *IEEE Position, Location and Navigation Symposium*, Orlando, FL, Nov 29–Dec 2.
3. Kursinski ER, Hajj GA, Schofield JT, Linfield RP, and Hardy KR (1997) Observing earth's atmosphere with radio occultation measurements using the Global Positioning System. *J Geophys Res*, 102(D19):23429–23465.
4. Hajj GA, Kursinski ER, Romans LJ, Bertiger WI, and Leroy SS (2002) A technical description of atmospheric sounding by GPS occultation. *J Atmospheric and Solar-Terrestrial Phys*, 64(4):451–469.
5. Rocken C, Anthes R, Exner M, Hunt D, Sokolovskiy S, Ware R, Gorbunov M, Schreiner W, Feng D, Herman B, Kuo YH, and Zou X (1997) Analysis and validation of GPS/MET data in the neutral atmosphere. *J Geophys Res*, 102(D25):29849–29866.
6. Sokolovskiy SV (2001) Tracking tropospheric radio occultation signals from low Earth orbit. *Radio Sci*, 36(3):483–498.
7. Gorbunov ME, Gurvich AS, and Bengtsson L (1996) Advanced algorithms of inversion of GPS/MET satellite data and their application to reconstruction of temperature and humidity. *Max-Planck Institute for Meteorology Technical Report (Hamburg)*, (211).
8. Karayel ET and Hinson DP (1997) Sub-Fresnel-scale vertical resolution in atmospheric profiles from radio occultation. *Radio Sci*, 32(2):411–423.
9. Gorbunov ME, Gurvich AS, and Kornblueh L (2000) Comparative analysis of radioholographic methods of processing radio occultation data. *Radio Sci*, 35(4):1025–1034.
10. Sokolovskiy SV (2001) Modeling and inverting radio occultation signals in the moist troposphere. *Radio Sci*, 36(3):441–458.
11. Gorbunov ME (2001) Radioholographic methods for processing radio occultation data in multipath regions. *Danish Meteorological Institute Scientific Report (Copenhagen)*, (01-02).
12. Igarashi K, Pavelyev A, Hocke K, Pavelyev D, Kucherjavenkov IA, Matyugov S, Zakharov A, and Yakovlev OI (2000) Radio holographic principle for observing natural processes in the atmosphere and retrieving meteorological parameters from radio occultation data. *Earth Planets Space*, 52:893–899.

13. Kong JA (1990) *Electromagnetic Wave Theory*. Wiley, New York, 2nd edition.
14. Levy M (2000) *Parabolic Equation Methods for Electromagnetic Wave Propagation*. The Institution of Electrical Engineers, London.

Combination of NOAA16/ATOVS Brightness Temperatures and the CHAMP Data to get Temperature and Humidity Profiles

Éva Borbás[1], Jun Li[1] and W. Paul Menzel[2]

[1]Space Science and Engineering Center, University of Wisconsin / Madison
[2]NOAA/NESDIS Office of Research and Applications, 1225 West Dayton St., Madison, WI 53706, USA

Summary. This paper examines whether the radio occultation (RO) measurements of the tropopause region are able to improve radiometric tropospheric profile retrievals, as represented by the ATOVS (Advanced TIROS Operational Vertical Sounder) temperature and humidity soundings from the NOAA16 (National Oceanic and Atmospheric Administration) satellite. First, a simulation study is presented wherein a statistical regression was used to infer temperature and humidity retrievals from ATOVS brightness temperatures and GPS (Global Positioning System) refractivity data. The ATOVS and GPS combination yields improved tropospheric profiles when compared to those inferred from either system alone. A sensitivity test was also performed to investigate how large GPS refractivity error can be and still improve the radiometric retrievals; even tripling the GPS refractivity observation errors yields improved retrievals. Second, the preparations to repeat the same study with real GPS (CHAMP) and ATOVS data are described. Collocations of GPS and CHAMP data were collected and the CHAMP data were evaluated against NWP profiles. CHAMP GPS validation suggested that CHAMP data would be able to improve the ATOVS retrievals as simulations with comparable GPS errors had showed positive retrieval impact. Future work will include application of regression approach for inferring retrievals after a statistically adequate number of collocations have been collected.

Key words: GPS occultations, ATOVS radiances, temperature profile retrieval, statistical regression

1 Introduction

Radiometric techniques using measurements from polar orbiting infrared (IR) and microwave (MW) sounders are used to infer temperature and moisture profiles in the lower and upper troposphere. While the current passive IR/MW remote sensing systems have limited skill for inferring temperature and moisture profiles around the tropopause and in the stratosphere, the GPS radio occultation can provide very accurate upper tropospheric and stratospheric refractivity profiles which are related to temperature and humidity. This study intends to use the RO measurements to get ancillary information around the tropopause region to improve radiometric tropospheric profile retrievals. First, in Section 2 and 3 a simulation study is presented: a statistical regression was used to get temperature and humidity retrievals from the combination of the radiometric and geometric (RO) systems. For validation, the retrievals were compared to those inferred from either system alone. An error sensitivity test was also performed to investigate how big GPS refractivity error can be and still improve the radiometric temperature or humidity retrievals. Second, in Section 4 and 5 we prepare to repeat the same study on real GPS (CHAMP=CHAllenging Minisatellite Payload) and sounder (ATOVS) data: collocations were made and the data was evaluated. The application of regression method used in the simulation study is pending after a statistically adequate number of collocations of occultations, soundings, and radiosonde observations have been collected.

2 Simulation approach

The simulation study uses a one year radiosonde dataset from 1988 (NOAA88), which contains 7547 profiles globally distributed in time and space. The profiles are interpolated into 42 vertical pressure levels between 0.1 and 1050 hPa.

The infrared and microwave brightness temperatures were calculated in a radiative transfer formulation (Huang and Li 1999) to represent the NOAA ATOVS. The ATOVS includes one IR instrument called High Resolution Infrared Sounder (HIRS) with 19 IR channels and two MW instruments called Advanced Microwave Sounder Unit (AMSU) A and B with 20 channels. Nominal instrument noise (Goodrum et al. 2000) plus 0.2 K forward model error is randomly added to the calculated radiances for simulation.

The GPS refractivity profiles were simulated on 16 fixed pressure levels between 5 and 30 km (from 12.6 hPa to 596 hPa) using the NOAA88 radiosonde (RAOB) temperature and humidity profiles. The working equation for GPS relates the atmospheric index of refraction to atmospheric temperature and moisture. The refractivity (N) equation (Smith and Weintraub 1953) is written

$$N = 77.6\frac{P}{T} + 3.73 * 10^5 \frac{P_w}{T^2}, \tag{2.1}$$

where P is the atmospheric pressure (hPa), P_w is the water vapor partial pressure (hPa), T is the temperature of the atmosphere. The magnitude of the noise in the refractivity profile was assumed to be 0.3 % in every level (Kursinski et al. 1997). GPS vertically correlated measurement errors were generated based on the method described in Healy and Eyre (2000).

Ancillary information about the surface conditions was inferred from temperature and humidity values at the lowest level of the radiosonde profiles.

2.1 Retrieval method

Temperature (K) and humidity (mixing ratio g/kg) profiles were retrieved from simulated brightness temperatures, simulated GPS refractivities and simulated surface observations using a statistical regression. Both linear and quadratic terms for all these variables were included in the regression. The regression equation used in this study is the following:

$$x_i = a_{io} + \sum_{j=1}^{M} a_{ij} Tb_j + \sum_{k=1}^{M} b_{ik} Tb_{ik}^2 + \sum_{l=1}^{N} c_{il} N_l + \sum_{m=1}^{N} d_{im} N_{im}^2 + e_i T_{sfc} + f_i T_{sfc}^2 + g_i Rh_{sfc} + h_i Rh_{sfc}^2 , \quad (2.1.1)$$

where x_i is the predictand (variable to be predicted) in the ith pressure level from the 42 pressure levels, Tb is the ATOVS brightness temperature, N is the GPS refractivity, T_{sfc} and Rh_{sfc} is the surface temperature and relative humidity, a, b, c, d, e, f, g , and h are the regression coefficients, M is the number of the ATOVS channels (39) and N is the number of the GPS pressure levels (16). Results for ocean and land are combined in this analysis. Regression coefficients were computed from 90 % of all radiosonde profiles (training dataset) and the remaining 10 % (test dataset) were used for validation. Bias and rms errors were computed between the retrievals and radiosonde profiles in 1 km layers for temperature profiles and 2 km layers for humidity profiles. For humidity, these errors were normalized by the true value of mixing ratio to derive percentage.

3 Results of the simulation study

Retrievals were calculated from the different information sources (e.g. ATOVS, GPS and surface data), and the rms and bias errors were determined between these retrievals and radiosonde profiles. To study the impact of one type of information, the differences of rms and bias errors with and without that information were plotted. Figure 1. shows the improvement (difference of rms errors) of GPS data on temperature (first column) and humidity (second column) retrievals computed from IR brightness temperatures (upper panels) and combination of IR and MW brightness temperatures (lower panels). The figure shows that the combination of radiometric (IR and MW) and geometric (RO) information yields improved tropospheric temperature and moisture profiles when compared to those inferred from the radiometric system alone. The GPS temperature improvement is most significant (max. 2 K) when the HIRS brightness temperatures only were used in the retrievals. Adding AMSU data in thc retrievals the GPS improvement is moderated (max. 0.8 K). For humidity the GPS improvement is less significant than it was for temperature since the refractivity profile was used only above 5 km where the water vapor content is already very low. A study of the impact of various levels of GPS refractivity noise was also performed. The temperature and humidity retrievals were determined from ATOVS brightness temperatures and GPS refractivity using nominal, double, triple GPS observation errors and without GPS data (see Figure 2). Even tripling the GPS refractivity observation errors improved the ATOVS retrievals over retrievals without any GPS information.

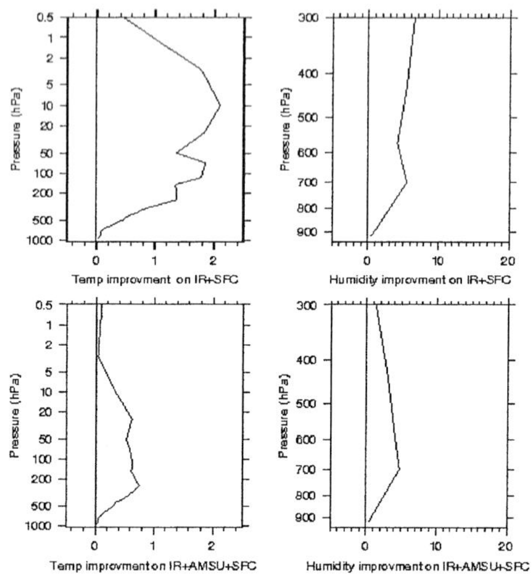

Fig. 1. GPS improvements (difference of rms errors) on temperature (K) (first column) and humidity (%) (second) retrievals derived from infrared (HIRS), microwave (AMSU) and surface (SFC) data. Retrievals without AMSU (upper panels) and with (lower panels) are shown

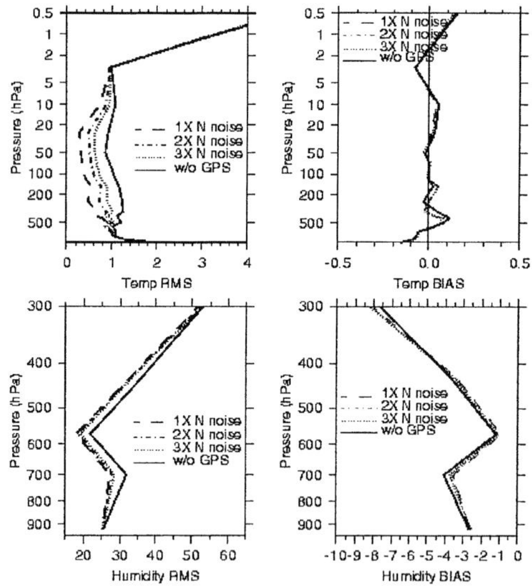

Fig. 2. Study of GPS noise on temperature (K) (upper panels) and humidity (%) (lower panels) retrievals using simulated ATOVS and GPS data. The rms (first column) and bias (second column) errors of retrievals using GPS refractivity profiles with 1X, 2X and 3X noise and without GPS data

4 Preparation of a retrieval study on basis of real data

4.1 Collocations of CHAMP occultations with ATOVS measurements

To start the retrieval study with real data, collocations between ATOVS field of views and CHAMP occultations were collected. 651 CHAMP occultations were available between May 28 and Jun 03, and August 26 and 27, 2001. GPS profiles were used only between 6 and 35 km, with 1 km vertical resolution and interpolated to 19 levels (from 42 levels) between 7 and 400 hPa. We assumed the ATOVS measurements were made at nadir and the occultations are vertical (no horizontal drift of the tangent point). When the drift of the tangent point during an occultation was larger than 300 km between the 6 and 35 km levels, the data were rejected (see the histogram in Figure 3.). The test dataset after filtering contained 583 occultations.

The NOAA16/ATOVS brightness temperatures were calculated at CIMSS/SSEC using the ATOVS and AVHRR Processing Package (AAPP) (Klaes and Schraidt 1999) and the International ATOVS Processing Package (IAPP) (Li et al. 2000). The global data were available with about 60 km horizontal resolution (a box of 3X3 HIRS pixel). Thresholds used for GPS-ATOVS collocations were 3 hours for time and 300 km for geographical distance. In multiple collocations, the selection prefers more clear pixels and closer times of GPS observation. ATOVS brightness temperatures are the average of the clear pixel values (or the average of all 9 cloudy pixel values in overcast conditions). 190 collocations between ATOVS and CHAMP measurements were found which number is very limited for doing statistical regression. So that instead of performing the regression study on real data, a quality check was made to see how realistic our GPS measurements are.

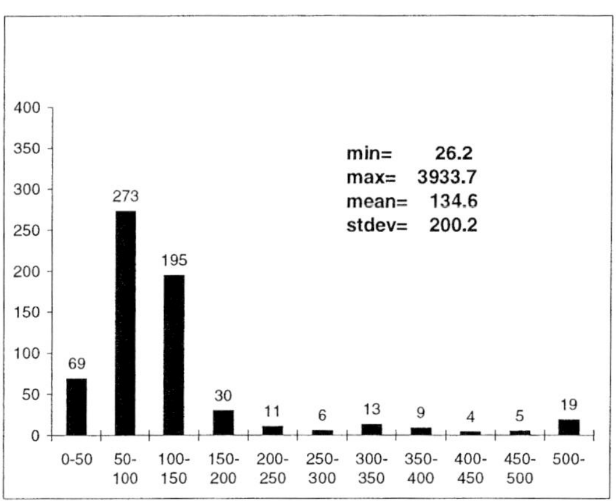

Fig. 3. Histogram of horizontal distances (km) of the occultation tangent points between 6 and 35 km levels

4.2 CHAMP refractivity comparison with NWP calculations

The AVN/NCEP (Aviation Version of the National Center for Environmental Prediction Global Spectral Model) numerical weather prediction (NWP) global model analyses (00, 06, 12, 18 UTC) were used to validate CHAMP refractivity profiles. The NWP temperature and humidity profiles were interpolated in time and space and then refractivity profiles were computed using equation (2.1). The data were horizontally interpolated from the 4 nearest grid points. Vertical interpolation was made from the 42 ATOVS pressure levels. The rms and bias errors of refractivity profiles derived from CHAMP measurements and NWP calculation around the tropopause region are shown on Figure 4. The fact that the rms error is below 0.8 % between 30 hPa and 300 hPa is very promising for retrievals from the combination of radiometric and occultation data, since the simulation GPS noise sensitivity test showed, even with tripling the GPS noise (1.2 %) the radiometric temperature retrievals were improved.

5 Conclusions and future plans

This study uses radio occultation measurements of the tropopause region to improve radiometric (infrared and microwave) tropospheric profile retrievals. First, a simulation study was presented wherein a statistical regression was used to get temperature and humidity retrievals from the combination of the radiometric and

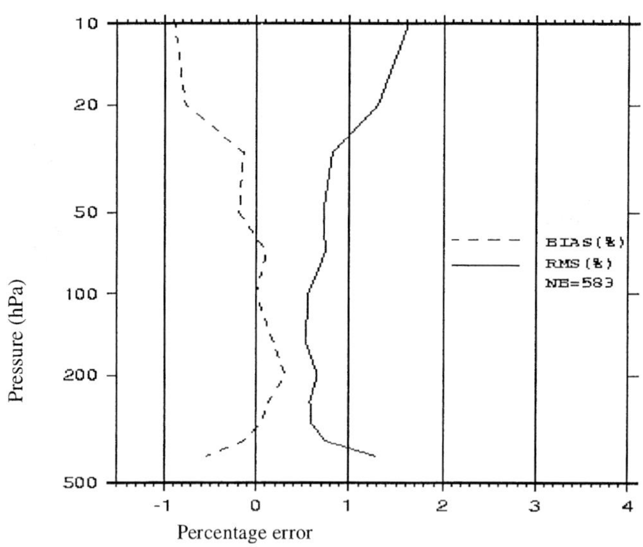

Fig. 4. The rms (solid line) and bias (dashed line) errors (%) of refractivity profiles near the tropopause region between the CHAMP measurements and NWP calculations (583 events)

geometric (RO) systems. The study showed that the combination yields improved tropospheric temperature and moisture profiles when compared to those inferred from either system alone. GPS improved the radiometric temperature profile retrievals around the tropopause level by 0.8 K, and moisture profile retrievals from 570 to 700 by about 4 %. Second, CHAMP data was evaluated against NWP profiles. The CHAMP GPS validation suggests that CHAMP data will be able to improve the radiometric retrievals as simulations with comparable GPS errors showed positive retrieval impact. However the CHAMP dataset needs to be expanded so that the regression computation can be run separately in clear and cloudy cases and also separately over the land and ocean.

Acknowledgements

We thank for GeoForschungsZentrum for providing the CHAMP data. This work was supported in part by NOAA Cooperative Agreement NAO7EC0676.

References

Goodrum G, Kidwell KB, Winston W (2000) NOAA-KLM User's Guide, September 2000 Revision. http://www2.ncdc.noaa.gov/docs/klm/index.htm

Healy SB, Eyre JR (2000) Retrieving temperature, water vapour and surface pressure information from refractivity-index profiles derived by radio occultation: A simulation study. Q J R Meteorol Soc 126: 1661-1683

Huang HL, Li J (1998) Determination of microwave emissivity from Advanced Microwave Sounder Unit measurements. In: Proceedings of SPIE, The International Society for Optical Engineering, 3503: 233– 237

Klaes KD, Schraidt R (1999) The European ATOVS and AVHRR Processing Package (AAPP) development. In: Technical Proceedings of the Tenth International ATOVS Study Conference, Boulder Colorado, pp 288-294

Kursinski ER, Hajj GA, Schofield JT, Linfields RP, Hardy KR (1997) Observing Earth's atmosphere with radio occultation measurement using the Global Positioning System. J Geophys Re 102: 23,429-23,465

Li J, Wolf W, Menzel WP, Zhang W, Huang H-L, Achtor TH (2000) Global Sounding of the Atmosphere from ATOVS Measurements: The Algorithm and Validation. J Appl Meteorol 39: 1248-1268

Smith EK, Weintraub S (1953) The constants in the equation for atmospheric refractive index at radio frequencies. Proc IRE 41: 1035-1037

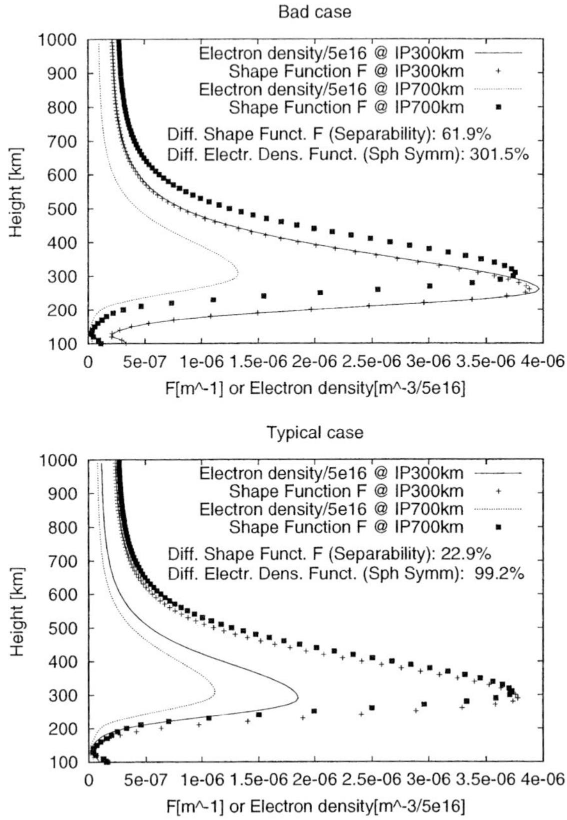

Fig. 4. Electron densities and Shape Functions: Two particular examples with high differences between profiles (top) and a typical case (bottom).

A reformulation of Eq. 1 is shown in Eq. 3, where it expresses the Slant TEC seen by the LEO considering Vertical TEC information.

$$STEC(p_i) = 2l_{ii}VTEC(LT_{ii}, LAT_{ii})F(p_i) + \sum_{j=1}^{j=i-1} l_{ij}\left[VTEC(LT_{ij}, LAT_{ij}) + VTEC(LT'_{ij}, LAT'_{ij})\right] F(p_j) \quad (3)$$

This formula shows how different Vertical TECs are considered in the different parts of the rays, allowing the horizontal gradients of TEC to be taken into account. General results obtained with either spherical symmetry and this separability hypothesis can be seen in [9], where estimations of f_oF2 and f_oE using these two methods and the values provided by a Ionosonde were compared. It was shown that spherical symmetry gives poorer results than the separability assumption. Figure 2 shows the difference between vertical profiles at different locations (A and B). The result of inverting that particular occultation with spherical symmetry gives an averaged profile between the two IRI profiles. If the assumption of separability is applied and the obtained normalised electron density function is multiplied by the corresponding TEC at the location, one is able to obtain the profile provided by IRI. In Figure 2 the Shape Function F has been multiplied by the TEC at the point A, but the IRI profile at location B would be obtained if the TEC at point B was used instead. Note that the IRI profile and the one of separability match closely. In [9] the inversion of synthetic data obtained with IRI using separabiliy and spherical symmetry where compared with the vertical profiles obtained with IRI. The results obtained with the separability assumption were closer to the values of height and electron density provided by IRI.

Two additional examples with real data taking place near the Appleton anomaly can be seen in Figure 5, in this location the VTEC gradients are higher and this leads to significant differences between the two Abel inversion approaches. There is more agreement with the NmF2 value indicated by the Ionosonde when Horizontal TEC information is used, for these epochs and Ionosondes the hmF2 data was not available. Besides, the profiles obtained with the separability hypothesis have lower values in the lower boundary of the Ionosphere compared with spherical symmetry, this is more realistic since at that point electron density must be close to 0.

3 Taking into account the Topside Electron Content on the Abel Inversion

An additional assumption that is made when vertical profiles of electron density are retrieved from radio occultation is that the electron content above the LEO is negligible. This may be realistic to a certain extent for LEOs orbiting at high altitudes such as the GPS/MET or SAC-C (both with orbits above 700km), but this assumption clearly fails for LEOs with low altitude orbit such as the CHAMP (at about 400 km).

In [9] it was assumed an exponential decreasing function for the topside ionosphere, then one may compute the whole vertical profile iteratively in such a way that this assumption helps to converge the normalised electron density profile to the condition formulated in Eq. 4 when TEC based a real data are used.

$$\int F(h) \cdot dh = 1 \tag{4}$$

Figure 6 shows the effect of considering this iterative process with an exponential decreasing topside for two different satellites (GPSMET and

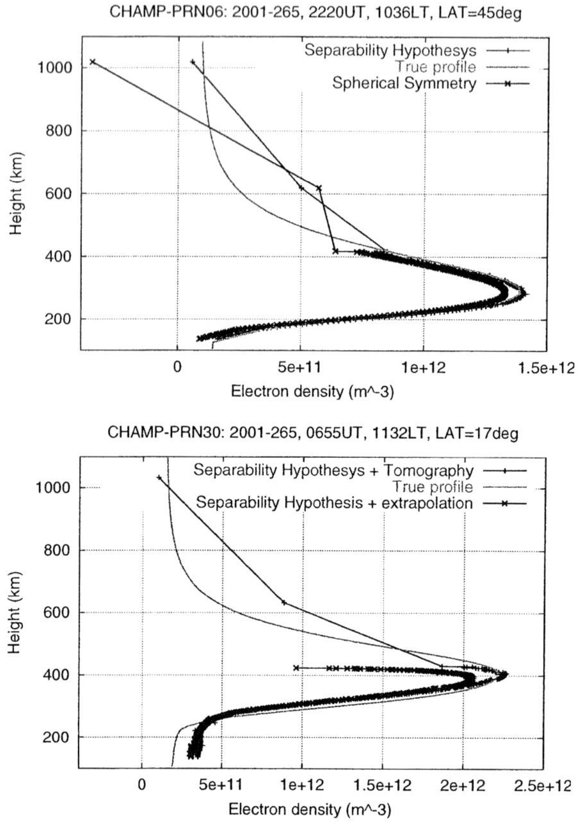

Fig. 7. (Top) Comparison of tomographic technique using spherical symmetry and separability. (Bottom) Comparison of separability assumption using tomographic approach or exponential extrapolation

An additional example corresponds to an occultation taking place over North America, see Figure 8. In the right plot of Figure 8 it is shown that the use of separability and topside voxel modeling provides estimation closer to the values of NmF2 and hmF2 given by Ionosonde.

4 Summary and Conclusions

This study has been focused on the advantages that the Abel transform provides to obtain vertical profiles of electron density: high resolution and low computational load. To exploit this potential, two weak spots are to be

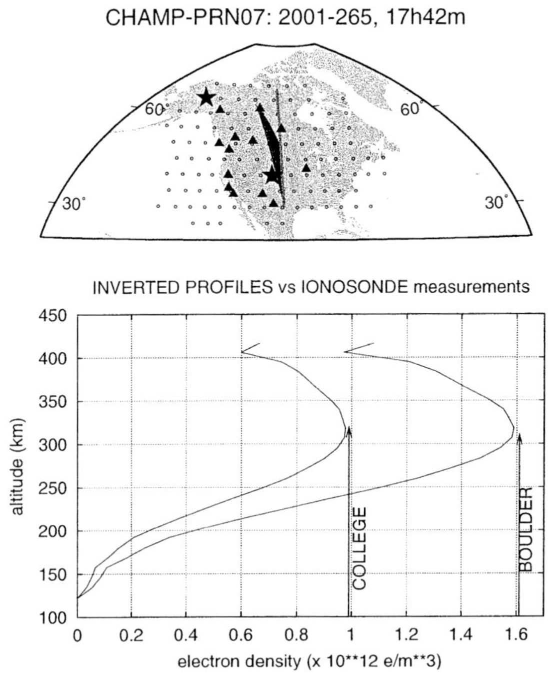

Fig. 8. (Top) Map of occultation taking place over United States. The stars indicated the position of the ionosondes which data will be used for comparison (Boulder, in central North America, and College, in Alaska), while the triangles are the ground GPS stations used to compute the VTEC map that will provide the model with the horizontal VTEC gradients. (Bottom) Comparison of retrieved occultation with ionosonde values, the arrows point to NmF2 and hmF2.

tackled: spherical symmetry and zero value for the topside electron density content. This work proposes two approaches to solve this: horizontal gradients of VTEC have been introduced to the model (separability hypothesis), which can be applied not only on the LI observable but to the bending angle as well) and the topside electron content is estimated simultaneously. This is important to obtain realistic profiles either at mid and low latitudes. Besides topside contribution is computed by means of 2 extra layers above the LEO orbit using positive elevations observations. This is specially important for LEOs with very low orbits such as CHAMP. An overall improvement of up to 40% is obtained in the studied occultations.

5 Acknowledgements

IRI model has been provided by Dr. Bilitza. We are grateful to the IGS, UCAR and the GeoForschungsZentrum for providing with the ground GPS data, the GPS/MET data and the CHAMP observations respectively. The maps have been generated with the software package GMT. This work has been partially supported by the 'Generalitat de Catalunya' under fellowship number 2000FI-00395 and the Spanish projects TIC-2000-0104-P4-03 and TIC2001-2356-C02-02.

References

1. Bilitza D (1990) International Reference Ionosphere 1990. URSI/COSPAR, NSSDC/WDC-A-R&S 90-22.
2. Feltens J, Schaer S. IGS products for the ionosphere. Proceedings of the IGS Analysis Center Workshop, ESA/ESOC Darmstadt, Germany, pp. 225-232.
3. Ganguly S, Brown A (2001) Real-time characterization of the ionosphere using diverse data and models, Radio Science, Volume 36, Number 5, pp 1181-1197, Sept-Oct.
4. García-Fernández M, Hernandez-Pajares M, Juan JM, Sanz J (2001) Combining Ionosonde with ground and LEO GPS data for Electron Density Estimations, (Oral presentation) European Geophysical Society XXVI, Nice (France), March.
5. Hajj GA, Ibañez-Meier R, Kursinski ER and Romans LJ (1994) Imaging the Ionosphere with the Global Positioning System. International Journal of Imaging Systems and Technology, Vol 5, 174-184.
6. Hajj GA, Romans LJ (1998) Ionospheric electron density profiles obtained with the Global Positioning System: Results from the GPS/MET experiment. Radio Science, Vol 33, No 1, 175-190. January-February.
7. Hernandez-Pajares M, Juan JM, Sanz J (1998) Global observation of the ionospheric electronic response to solar events using ground and LEO GPS data, Journal of Geophysical Research (Space Physics), Vol 103 , No A9, 20789-20796.
8. Hernandez-Pajares M, Juan JM, Sanz J (1999) New approaches in global ionospheric determination using ground GPS data. Journal of Atmospheric and Solar Terrestrial Physics. Vol 61, 1237-1247.
9. Hernandez-Pajares M, Juan JM, Sanz J (2000) Improving the Abel inversion by adding ground data LEO radio occultations in the ionospheric sounding. Geophysical Research Letters, Vol 27, No 16, 2743-2746.
10. Schreiner WS, Sokolovskiy SV, Rocken C, Hunt DC (1999) Analysis and validation of GPS/MET radio occultation data in the ionosphere. Radio Science, Vol 34, No 4, 949-966.

Validation of Water Vapour Profiles from GPS Radio Occultations in the Arctic

Michael Gerding and Antje Weisheimer

Alfred Wegener Institute for Polar and Marine Research, Research Division Potsdam, Potsdam, Germany

Summary. The relevance of water vapour in atmospheric physics and climate research contrasts strongly with the availability of humidity data in the Arctic. The most extensive humidity data set is based on approx. 80 radiosonde stations north of 60°N, but suffers from two major problems: Sensor diversity and sensor limitations under Arctic conditions, on the one hand, and lacking radiosonde launch sites in the Arctic Ocean and the Greenland iceshield, on the other hand. Calibration free satellite borne instruments like the GPS receiver onboard CHAMP prevent from both handicaps. Comparisons between radiosonde data and GPS-based humidity profiles are presented for single occultations as well as averaged data from the proof-of-concept experiment GPS/MET and the recent CHAMP satellite mission. The effects of low absolute humidity and uncertain meteorological analyses are examined using additional information from a regional climate model. For the observations of CHAMP in summer 2001, a general dry bias has been found if compared with radiosonde data, apparent both in single and mean profile intercomparisons. In contrast, during February 1997 GPS/MET data show slightly higher humidity in the mid-troposphere, if compared with model data and objective analyses.

Key words: Water vapour, Arctic, Validation, Regional Model, Radiosondes

1 Introduction and data coverage

Water vapour is the most important greenhouse gas in the Earth's atmosphere, inducing about two third of the natural greenhouse effect. For anthropogenic greenhouse effect, General Circulation Models predict the strongest signal in the Arctic region (Walsh 2001). Therefore, the monitoring of the Arctic water vapour distribution is required to seperate anthropogenic effects from natural variability and to understand the radiative feed back from increasing water vapour content due to rising temperatures, waning sea ice extend etc.

GPS radio occultation (RO) soundings provide profiles of atmospheric refractivity with global coverage. Calibration-free measurements and retrievals recommend this technique especially for climate applications (Yuan et al. 1993). In this paper, the reliability and limitations of specific humidity retrieval from RO combined with analyses from European Center for Medium Range Weather Forecast (ECMWF) will be investigated for Arctic conditions. RO humidity data has been provided by GeoForschungsZentrum Potsdam. Figure 1 shows the distribution of 101 individual GPS RO from the GPS/MET prime-time IV period between February 2–16, 1997 and 470 CHAMP RO between May 14–June 10 and August 26-28, 2001. Only occultations within the domain of the regional climate model HIRHAM have been taken into account (see quadrangle in Fig. 1).

66 radiosonde launch sites are situated in the examined area (Fig. 1), most of them performing one to four ascends per day. The meteorological sondes from the Global Telecommunications System (GTS) still provide the most comprehensive humidity data set in the Arctic, although they are of limited reliability (Elliot and Gaffen 1991; Miloshevich et al. 2001). The data set for February 1997 was extended by sondes launched during the "Match campaign" (Schulz et al. 2000) and the FASTEX campaign (http://www.cnrm.meteo.fr/dbfastex/).

Objective analyses from the ECMWF model have been evaluated for February 1997. Additional information is provided by the high resolution climate model HIRHAM (Christensen et al. 1996; Dethloff et al. 1996, 2001).

2 CHAMP observations in summer 2001

61 out of 470 CHAMP radio occultations in the Arctic have been performed simultaneously with radiosonde launches within a spatiotemporal radius of +/-3 h and 300 km, 16 profiles within +/-2 h / 200 km, and only 4 profiles within +/-1 h / 100 km. Fig. 2 compares the specific humidities in 850 and 500 hPa altitude. The correlation between the data sets is moderate (correlation coefficient of 0.51/0.56 in 850/500 hPa). In the lower troposphere CHAMP data has been found about 1 g/kg drier than radiosondes, but about equal in the mid-troposphere. In both altitudes the standard deviation between the data sets is high (1.5/0.45 g/kg in 850/500 hPa). The deviation decreases with decreasing distances of the sounded air parcels. Individual profiles are not compared here because of remaining spatio

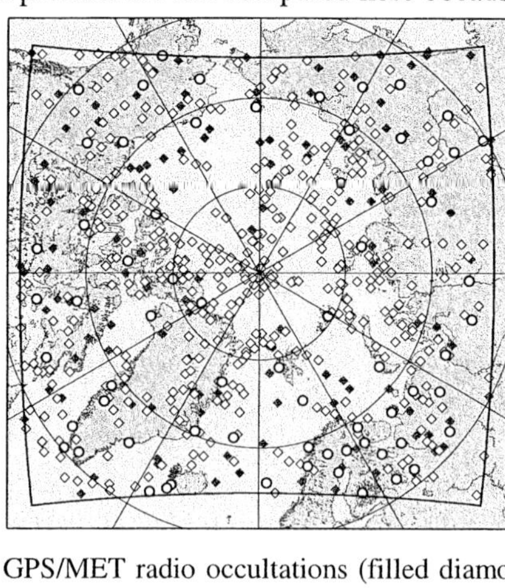

Fig. 1. Distribution of GPS/MET radio occultations (filled diamonds), CHAMP radio occultations (open diamonds), and radiosonde stations (circles). The model domain is marked by the large quadrangle. Periods of the GPS RO data are given in the text.

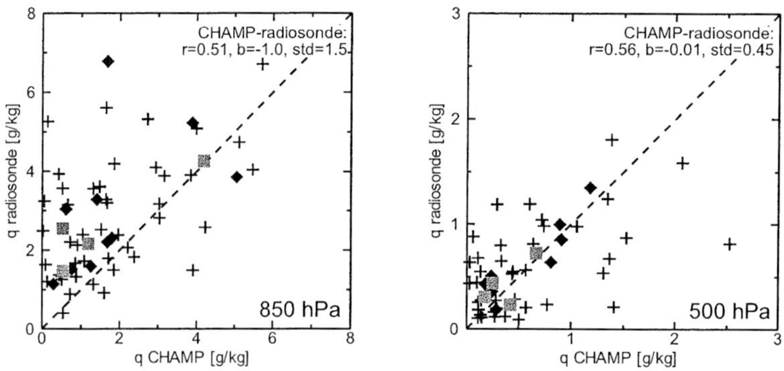

Fig. 2. Specific humidity from radiosonde and CHAMP radio occultations in the Arctic. All radiosondes launched within 3 h and 300 km distance from the occultation point are used. Closer 'matches' are marked (1h, 100km: grey squares; 2h, 200km: diamonds). r: correlation coefficient, b: bias, std: standard deviation

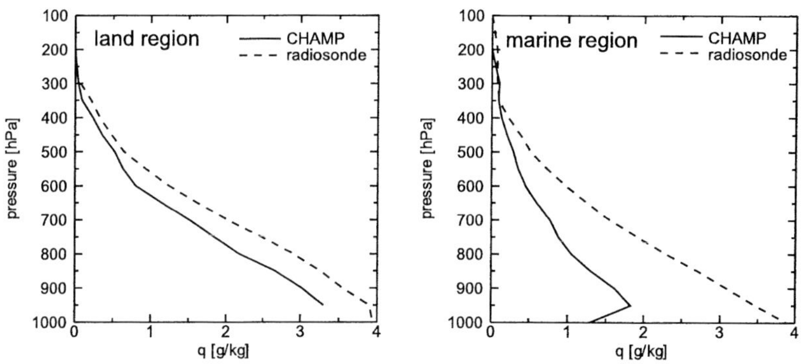

Fig. 3. Mean specific humidity profile for specific Arctic regions for May/June 2001. The data has been analysed on 17 pressure levels (1000 – 100 hPa)

temporal differences. They are of less importance for climatological studies.
Fig. 3 shows mean humidity profiles from all radiosondes and CHAMP radio occultations for the area in Fig. 1, separated by land and coastal/marine regions. Compared with Fig. 2, the number of used single profiles is largely increased, providing the best available data set for climatological investigations in the Arctic region. It is obvious that the CHAMP data is drier by up to 50% compared with radiosonde humidity. Part of this difference may result from the lack of radiosonde data from the dry, central Arctic Ocean and the Greenland iceshield.

3 GPS/MET observations in February 1997

Wintertime GPS RO are available from GPS/MET for February 1997. Comparisons of single profiles with radiosondes suffer from the low number of RO. Therefore, objective analyses from ECMWF have been used for comparison, along with results from the regional model HIRHAM. Specific humidity from RO is compared with model data for the low and mid troposphere (Fig. 4). The model results have been spatiotemporal interpolated to the location of radio occultation. The correlation between the data sets is rather poor in the lower troposphere (0.36/0.35 for comparison with ECMWF/HIRHAM), but higher in 500 hPa (0.71/0.68). In this altitude the GPS/MET data gives slightly higher humidity. For the lower troposphere there are some indications for a small dry-bias of the GPS/MET data, but this difference is not significant because of the high standard deviation.

Comparisons of regional mean profiles (not shown here) reveal that the differences between models and GPS RO occur especially above the Arctic land regions, while above open sea-water all data sets agree well. Radiosonde data available above the Arctic land area confirm the GPS/MET observed profiles and the differences from meteorological analyses. This indicates that in the data sparse Arctic region the analysis results are dominated by the underlying atmospheric model and the assimilation of observational data might have a limited effect only.

4 Discussion and conclusions

Comparison between GPS RO and radiosonde data suffer in general from the limited number of joint soundings and accurate matches of the sounded air parcels. The high standard deviations and only moderate correlation coefficients displayed in Fig. 2 may result from high spatiotemporal variability in the humidity distribution. But still, a significant dry bias of the CHAMP data remains. On the other hand, standard meteorological radiosondes are known to be of limited reliability especially in cold and dry environments (Elliot and Gaffen 1991). However, radiosondes themselves tend to produce dry-biased results if the sounded air gets colder (Miloshevich et al. 2001). Therefore a wet-bias of the radiosondes seems to

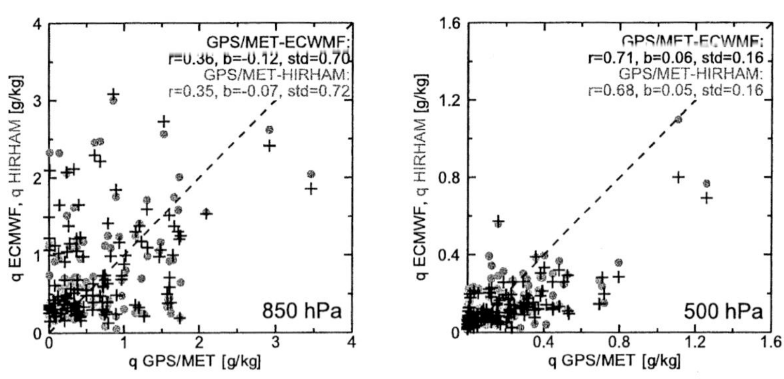

Fig. 4. Specific humidity from ECMWF analyses, HIRHAM, and GPS/MET RO in the Arctic. r: correlation coefficient, b: bias, std: standard deviation

be unlikely. Negative specific humidities are a well-known problem of GPS RO soundings with several possible reasons, like multipath effects, non-spherical symmetry of the Earth's atmosphere, or uncertainties in the ancillary temperature fields used in the retrieval (see, e.g. Rocken et al. 1997; Marquardt et al. 2000). We have neglected negative values for averaging, but cannot exclude that also other profiles are affected by the above mentioned problems.
The humidity inversion found in the marine humidity profile indicates a stable boundary layer above the ice covered part of the ocean. Typically, this area is only poorly masked by radiosonde data, which therefore does not exhibit any inversion. On the other hand, we cannot exclude that these might be erroneous data due to reflections and multipath effects of the GPS signal (Beyerle and Hocke 2001). Further analysis e.g. with atmospheric models is needed to answer this question.

Comparisons between meteorological analyses and GPS/MET RO show large differences in individual profiles, like between CHAMP RO and radiosonde data. The rather good correlation between GPS/MET & ECMWF data and between GPS/MET & HIRHAM data in the middle troposphere reveals the high potential of RO for climate applications. The weaker correlation in the 850-hPa-layer could be explained with the above mentioned reflections and a vanishing GPS signal near the ground (Rocken et al. 1997). Achieving a higher antenna gain, the CHAMP instrument overcame this problem. In contrast to the global findings of Kursinski and Hajj (2001), the Arctic humidity estimates from GPS/MET exhibit slightly wetter conditions than analysed by ECMWF. This coincides with the colder tropospheric air modeled by ECMWF and HIRHAM.

This study suggests that ECMWF objective analyses are of limited accuracy in the data sparse Arctic region and additional observations are needed. Also the humidity retrieval of GPS RO suffers from uncertainties in the analyses. Therefore, new approaches have been discussed in literature using statistical methods or ground-based meteorological data instead (Healy and Eyre 2000; O'Sullivan et al. 2000). Further investigation will show whether these methods can overcome the problems in the remote Arctic region.

The limitations of objective analyses or ancillary observational data will mostly affect the single RO profile data. However, averaged, climatological humidity data from longer periods are still valuable and one of the most promising applications for the GPS radio occultation technique.

Acknowledgement

We would like to thank H. Deckelmann (AWI, Potsdam) for computational support. Radiosonde data has been kindly provided by the British Atmospheric Data Centre, Meteo France, the Danish Meteorological Institute, the Norwegian Institute for Air Research, and the National Institute for Environmental Studies (Japan). We are grateful to GeoForschungsZentrum Potsdam for supplying the radio occultation profiles of specific humidity. The HIRLAM/HIRHAM system was developed by the HIRLAM project group, a cooperative project of the national weather services in Denmark, Finland, Iceland, Ireland, The Netherlands, Norway, Sweden, and Spain. This work was carried out within the HGF "GPS Atmospheric Sounding Project" (BMBF grant No. 01SF9922/2).

References

Beyerle G, Hocke K (2001) Observation and simulation of direct and reflected GPS signals in radio occultation experiments. Geophys Res Lett 28: 1895–1898

Christensen JH, Christensen OB, Lopez P, van Meijgaard E, Botzet M (1996) The HIRHAM4 regional atmospheric climate model. In: DMI Sci Rep 96-4, Dan Meteorol Inst, Copenhagen

Dethloff K, Rinke A, Lehmann R, Christensen JH, Botzet M, Machenhauer B (1996) Regional climate model of the Arctic atmosphere. J Geophys Res 101: 23,401–23,422

Dethloff K, Abegg C, Rinke A, Hebestadt I, Romanov VF (2001) Sensitivity of Arctic climate simulations to different boundary-layer parametrizations in a regional climate model. Tellus 53A: 1–26

Elliot WP, Gaffen DJ (1991) On the utility of radiosonde humidity archives for climate studies. B Am Meteorol Soc 72: 1507–1520

Healy SB, Eyre JR (2000) Retrieving temperature, water vapour and surface pressure information from refractive-index profiles derived by radio occultations: A simulation study. Q J R Meteorol Soc 126: 1661–1683

Kursinski ER, Hajj GA (2001) A comparison of water vapor derived from GPS occultations and global weather analyses. J Geophys Res 106: 1113–1138

Marquardt C, Labitzke K, Reigber C, Schmidt T, Wickert J (2001) An assessment of the quality of GPS/MET radio limb soundings during February 1997. Phys Chem Earth (A) 26: 125–130

Miloshevich LM, Vömel H, Paukkunen A, Heymsfield AJ, Oltmanns SJ (2001) Characterization and correction of relative humidity measurements from Vaisala RS80-A radiosondes at cold temperatures. J Atmos Ocean Tech 18: 135-156

O'Sullivan DB, Herman BM, Feng D, Flittner DE, Ward DM (2000) Retrieval of water vapor profiles from GPS/MET radio occultations. Bull Amer Meteorol Soc 81: 1031–1040

Rocken C, Anthes R, Exner M, Hunt D, Sokolovskiy S, Ware R, Gorbunov M, Schreiner W, Feng D, Herman B, Kuo YH, Zou X (1997) Analysis and validation of GPS/MET data in the neutral atmosphere. J Geophys Res 102: 29,849–29,866

Schulz A et al. (2000) Match observations in the Arctic winter 1996/97: High stratospheric ozone loss rates correlate with low temperatures deep inside the polar vortex. Geophys Res Lett 27: 205–208

Walsh JE (2001) Simulation of the Arctic in global climate models. In: Second Wadati Conference on Global Change and the Polar Climate. Tsukuba Science City, Japan

Yuan L, Anthes RA, Ware RH, Rocken C, Bonner WD, Bevis MG, Businger S (1993) Sensing climate change using the Global Positioning System. J Geophys Res 98: 14,925–14,937

Comparison of DMI Retrieval of CHAMP Occultation Data with ECMWF

Jakob Grove-Rasmussen

Danish Meteorological Institute, Lyngbyvej 100, DK-2100 Copenhagen, Denmark
jgr@dmi.dk

Summary. At DMI a processing chain for GPS radio occultation measurements has been made. It has been used on both GPS/MET and Ørsted data for retrieving both temperature in the stratosphere and troposphere as well as water vapour in the troposphere. For nine days of CHAMP data the Ørsted data processing chain for GPS radio occultation retrieval has been applied. The resulting profiles have been compared to corresponding ECMWF profiles of water vapour and temperature, resulting in a statistical description of the general agreement between the two datasets.

Key words: GPS, occultation, validation, humidity

List of Abbreviations and Symbols

- CHAMP CHAllenging Microsatellite Payload
- DMI Danish Meteorological Institute
- ECMWF European Centre for Mediumrange Weather Forecast
- GPS Global Positioning System
- LEO Low Earth Orbit
- SNR Signal-to-noise ratio
- N Refractivity ($N = 10^6(n-1)$)
- n Index of refraction
- P Dry pressure (hPa)
- e Water vapour partial pressure (hPa)
- T Temperature (K)

1 Radio occultation methodology

Profiles of the atmospheric parameters pressure, temperature and humidity can be derived from measurements of changes in the signals from GPS satellites setting behind the limb of the Earth, as observed from a receiver on-board a LEO satellite [4]. The observed parameter is the complex microwave signal, which contains information on the changes with height of the refractive index of the atmosphere. Knowing the geometry of the occultation, i.e. position and velocity of both satellites, the bending angle of each

3 Water vapour statistics

The same set of occultation profiles has been inverted to water vapour profiles through the iterative method [2] at altitudes below 15 km. This method does not take into account the total error budget but gives a reliable calculation of water vapour pressure. The profiles have been compared to ECMWF humidity profiles (available up to 10 hPa) and averaged in 3 latitude intervals (high-, mid- and low-latitude). This has been done to separate the effect of huge changes in atmospheric structure with latitude, having the warm humid atmosphere at the equator the cold atmosphere at the poles.

Statistical comparison of the derived water vapour profiles to the humidity profiles from ECMWF reveals (fig. 3) a significant average difference below 5 km, especially at low latitudes. Here also a large standard deviation is calculated which is of comparable size to the average difference, which implies a significant bias. At high latitudes the average difference is below 1 hPa, shifting from negative to positive around 2 km. The mid-latitudes show an increasing average difference, ending around 1 hPa at the surface.

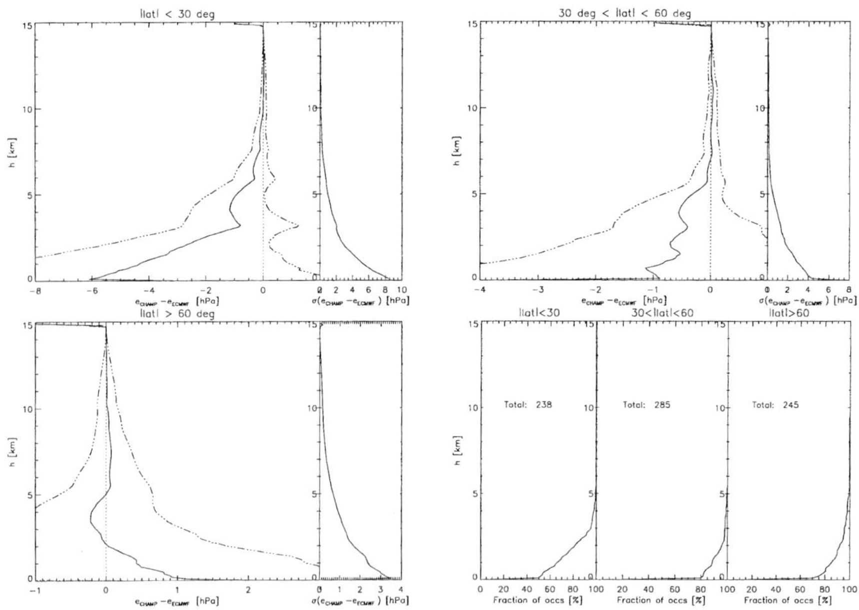

Fig. 3. Upper panels and lower left: Mean water vapour pressure difference as compared to ECMWF in the 3 latitude bands. Line styles as in fig. 2. Lower right panel: penetration depth in the 3 latitude bands.

4 Discussion and conclusions

The average temperature difference is close to or below 1 K throughout the lower part of the stratosphere, only increasing in case of water vapour or problems in the upper boundary initialisation. This result is comparable to corresponding procedures performed on GPS/MET observations [6][plate 11].

Variations are found with latitude in the water vapour statistics, the low latitude occultations having the largest average difference and standard deviation, −6 and 9 hPa respectively near the surface. This relatively large error needs further investigation but could arise from measuring in a very moist region where the vater vapour partial pressure can be high, and atmospheric multipath phenomena can occur and cause interpretation problems of the measured excess phase.

The penetration depth of the occultations looks promising but is partly the result of the use of fly-wheeling in the receiver in case of loss of phase lock. In this situation the receiver continues to give an output based on extrapolation from the last sample with phase lock. In case of fly-wheeling the SNR is very low, which indicates that SNR-wheighting should be used in the inversion. In the equatorial region the tracking is, as expected due to expected large humidity gradients, more often lost before the ground is reached (see fig. 3 lower right panel).

5 Acknowledgement

This work was funded by the Danish National Science Foundation under the ESA follow-up research programme. The author wish to thank B. Amstrup for the ECMWF data preparation and G.B. Larsen for rewarding comments.

References

1. Bevis M, Businger S and Chiswell S (1994) GPS Meteorology: Mapping Zenith Wet Delay onto Precipitable Water. *Journal of Applied Meteorology*, 33:379–386.
2. Gorbunov M E and Sokolovskiy S V (1993) Remote sensing of refractivity from space for global observations of atmospheric parameters. Technical Report 119, Max-Planck-Institut für Meteorologie.
3. Høeg P, Larsen G B, Benzon H-H, Grove-Rasmussen J, Syndergaard S, Mortensen M D, Christensen J, and Schultz K (1998) GPS Atmosphere Profiling Methods and Error Assessments. Technical Report 98-7, DMI. ISBN: 87-7478-374-2.
4. Melbourne W G, Davis E S, Duncan C B, Hajj G A, Hardy K R, Kursinski E R, Meehan T K, Young L E, and Yunck T P (1994) The Application of Spaceborne GPS to atmosphere Limb Sounding and Global Change Monitoring. Technical Report 94-18, JPL, NASA.

systems at many NWP centres (e.g. NCEP, ECMWF, Météo France, Met Office). VAR provides a clear framework for effective and consistent treatment of meteorological information of widely different types and enables limitations apparent in early assimilation schemes to be overcome. In particular, a key advantage of VAR is that it allows us to consider the assimilation of observations close to their raw form, reducing the need for various forms of pre-processing or retrieval operation. This tends to lead to simpler observation error characteristics, which can be described more easily. In addition, more specifically in the context of RO measurements, this approach also circumvents the well known "water vapour ambiguity".

In section 2 the variational approach to data assimilation is outlined. This is followed by a discussion of the assimilation of RO measurements in section 3. Recent progress at the Met Office is outlined in section 4 and then the conclusions are presented in section 5.

2 Variational Data assimilation

The VAR approach for assimilating data into an NWP system has been described in detail (e.g., [8, 9]) and review of the assimilation of remotely sensed observations has been presented by Eyre [10]. A schematic representation of a VAR system is shown in Figure 1. The aim is to find the most probable atmospheric state given the observations and the NWP forecast. This is equivalent to minimizing a penalty function $J(\mathbf{x})$ with respect to $\mathbf{x}$, the vector representing the atmospheric state, where $J(\mathbf{x})$ measures the degree of fit to the observations, NWP forecast (background) and possibly other physical and dynamical constraints. If the observation and background information have unbiased, Gaussian errors, then the maximum likelihood penalty function is,

$$J(\mathbf{x}) = \frac{1}{2}(\mathbf{x}-\mathbf{x}^{\mathrm{b}})^T\mathbf{B}^{-1}(\mathbf{x}-\mathbf{x}^{\mathrm{b}}) + \frac{1}{2}(\mathbf{y}^{\mathrm{o}} - H(\mathbf{x}))^T(\mathbf{E}+\mathbf{F})^{-1}(\mathbf{y}^{\mathrm{o}} - H(\mathbf{x})) + J_c \quad (1)$$

where $\mathbf{y}^{\mathrm{o}}$ is the vector of observations, $H(\mathbf{x})$ is the equivalent values corresponding to state $\mathbf{x}$, $\mathbf{E}$ is the expected error covariance of the observations, $\mathbf{F}$ the expected error covariance of the "observation operator" or "forward model", H, $\mathbf{x}^{\mathrm{b}}$ is the background state (NWP forecast) and $\mathbf{B}$ is its expected error covariance, and J_c represents additional constraints. The superscripts T and -1 denote matrix transpose and inverse respectively.

The relative weight given to the background and observation vectors depends on the covariance matrices. Background NWP error covariance matrices are often derived from the statistics of differences between 24 and 48 hour forecasts valid at the same time (e.g., [12]). Observation errors are usually based on the specifications of the measurement system and simulation exercises to estimate the errors arising in the "pre-processing" or retrieval code. The forward model errors can be divided into two classes. The first class arise

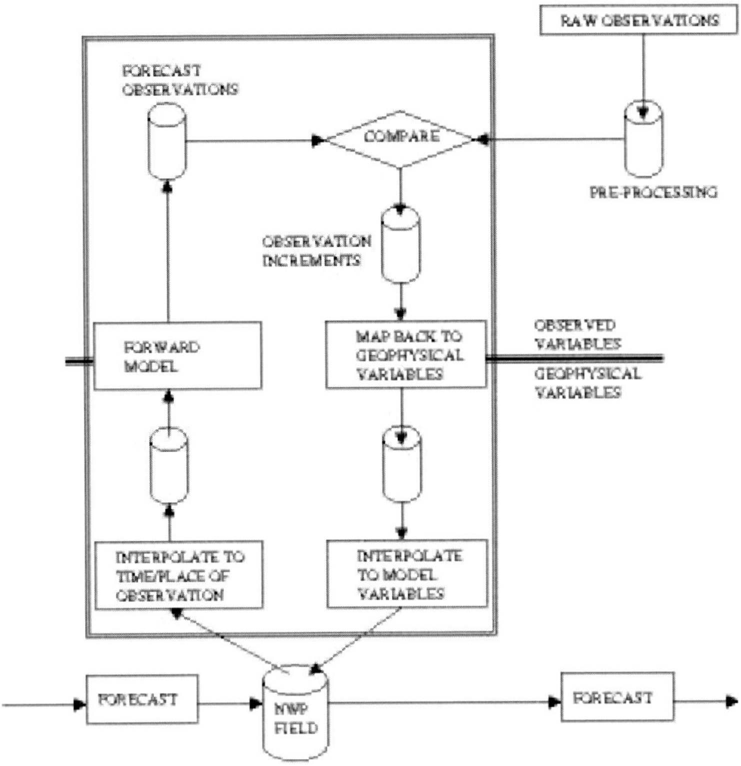

Fig. 1. A schematic of a variational assimilation system

because H often contains simplified physics and approximations to speed up the calculation so that it can be used in an operational system. The second class arises because the finite NWP representation of the atmosphere may smooth out real structure that the observation is sensitive to. This signal has to be treated as noise in the assimilation system, and it is known as representation error. For example, it was estimated that the bending angle errors caused by fine, vertical structure not represented by the 19-level Met Office global forecast model used in 1995 at the time of GPS/MET, were around $\sim 2\%$ [14].

3 Assimilation of RO measurements

The options for assimilating RO data have been discussed in detail by Eyre [10] and Kuo et al [13].

3.1 The assimilation of bending angle

The assimilation of bending angles is advantageous because it is closer to the "raw" measurement and therefore the error characteristics are likely to be simpler. The approach requires the development of "fast" models for calculating bending angles as a function of impact parameter, which are analogous to "fast" radiative transfer models used routinely in the assimilation of passive sounding measurements. The aim of a "fast" model is to implement physically reasonable approximations that speed up the bending angle calculation, without introducing additional errors that degrade the quality of the meteorological information that can be derived from the measurement. These simplifications and approximations are necessary because of the strict time constraints on the operational assimilation of measurements. A fast bending angle model for RO may be quite nonlinear, although clearly the degree of nonlinearity will depend on the precise formulation of the model. For example, if the impact parameter is used to derive the tangent point height [10, 14], the tangent point location can jump from one model vertical grid box to the next during the iterations of the minimization routine. This may slow down convergence.

Assimilating bending angles does not remove errors associated with horizontal gradients because the assumption of spherical symmetry is used in order to derive the bending angle and impact parameter from the Doppler shift and the errors can be large in frontal regions [15]. Zou et al [16] presented a forward model that formally minimized the horizontal gradient errors by forward modelling Doppler shift and then deriving the bending angle and impact parameter values assuming spherical symmetry, thereby reproducing the processing of the measurement. However, if the Doppler shifts can be forward modelled accurately and efficiently, then these should be assimilated. In fact, the Doppler shift modelled by Zou et al is approximate because the satellites are effectively moved to intersect the ray path, rather than iteratively finding the ray path that intersects the satellites. This approximation may be reasonable but the forward model error associated with it needs to be quantified. In addition, it is also necessary to consider how well the NWP model can represent horizontal gradients in the real atmosphere. Clearly, this form of representation error will depend on the horizontal resolution of the NWP model.

Kuo et al [13] noted that the assimilation of bending angle does not have to be performed as a function of impact parameter, and could be done as function of time, given the time-dependent position and velocities of both the transmitter and receiver. This requires finding the ray-path that intersects the satellites as a function of time. Whilst this is a reasonable approach, in effect it is the same as assimilating the Doppler shift and is likely to be more expensive.

3.2 The assimilation of refractivity

The assimilation of refractivity profiles involves an additional layer preprocessing with the Abel transform. Therefore, the observation errors will be more complicated and the integration contained in the Abel transform [17] produces vertically correlated errors even if the bending angle errors are uncorrelated. In addition, although the transform is linear and well conditioned, additional errors can arise, for example when extrapolating the observed bending angle profile above the top of the measurement [18]. The extrapolation will be based on some *a priori* assumptions regarding the shape of the bending angle profile, and these assumptions may introduce errors. However, the approach is worth considering because the forward model is relatively simple to develop and not computationally expensive. Extrapolation of the NWP data above the model top is not required and the forward model will be less nonlinear than the bending angle case. Clearly, errors will arise as a result of horizontal gradients, but these may be mitigated by interpreting the output of the Abel transform as plane averaged refractivity, given a suitable horizontal weighting function. A Gaussian weighting function, with a width proportional to the square-root of the refractivity scale-height [19], appears to be a reasonable approach. Error simulations within the domain of the Met Office mesoscale model have indicated that the refractivity errors 1km above the surface caused by horizontal gradients can be reduced from 3.5 refractivity units to 2.5 refractivity units as a result of the plane averaging (Figure 2).

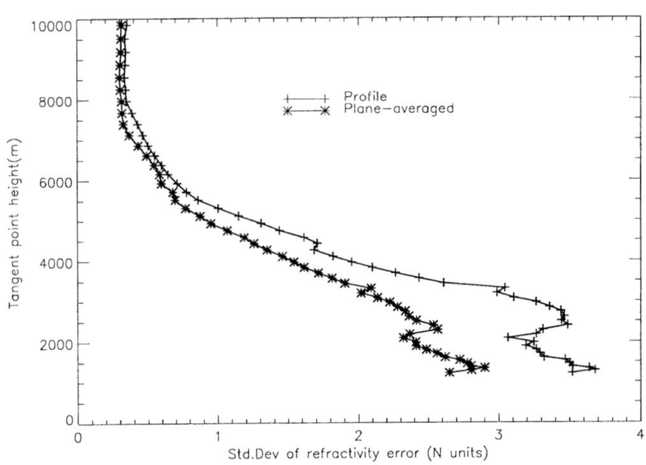

Fig. 2. The refractivity errors caused by horizontal gradients when the results are interpreted as profile and plane-averaged quantities

3.3 Alternative approaches

It is unlikely that temperature and humidity profiles derived from RO measurements will be assimilated directly because they will have been derived using a priori information meaning the observation errors will be complicated and difficult to model. However, there are other potential approaches. For example, the assimilation of Doppler shift would remove the assumption of spherical symmetry completely, but it would probably require an iterative, ray tracing forward model based on a shooting method to find the ray path that intersects the satellites, and this may be computationally expensive. Alternatively, phase and amplitude may be assimilated directly if fast wave optics forward models can be developed [20].

4 Progress at the Met Office

Forecast impact trials with CHAMP data will be performed during 2002, using the Met Office 3DVAR assimilation system [21]. Refractivity profiles will be assimilated initially. However, before the assimilation trials begin further work is required in order to understand and hopefully remove the refractivity bias seen in the CHAMP data, because the assimilation system assumes that the observation errors are unbiased. The forward model we will use will differ from that outlined in Healy and Eyre [22] because the Met Office is moving to a new NWP forecast model which has a height based vertical co-ordinate system. Planar averaging may be incorporated into the forward model at a later date.

In the longer term, bending angles may be assimilated. A "fast" forward model based on the model outlined by Eyre [10] has been tested against a 3D raytracer for simulated occultations in the domain of the mesoscale model [23]. The principal approximation made in the fast model is that the calculation is started at the assumed tangent point and the bending angle found by integrating towards the LEO and GPS satellites. However, we do not attempt to find the path that intersects the satellites. The horizontal location of the tangent point can be approximated with a "representative" value and the vertical height is derived from the impact parameter. Since the impact parameter is itself derived assuming spherical symmetry, horizontal gradients cause forward modelled bending angle errors. The comparisons against the 3D ray tracer have enabled us to investigate the forward model errors associated with the various approximations used in the fast model, such as not intersecting the satellites, the use of a single, representative tangent point for the occultation and horizontal gradient errors. Figure 3 summarises the fast model errors when all of the above approximations are implemented. We have found that using the impact parameter to derive the tangent point height is the largest single component of the error. However, the magnitude of these errors need to be placed in some context. NWP forecast errors mapped into

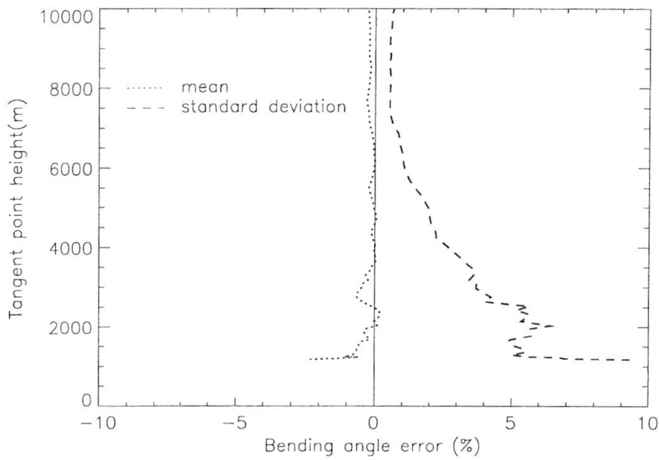

Fig. 3. The "fast" forward model bending angle errors when compared with a 3D ray tracer

bending angle space are ~ 20% near the surface. In addition, it may be possible to reduce the horizontal gradient errors with a correction strategy using analytical expressions for the bending angle and impact parameter errors derived previously [15].

5 Conclusions

GPS RO data has the potential to play an important role in operational NWP in the future, but significant work is required in order to realise this potential. It is crucial for NWP users to understand the steps between the "raw" and "pre-processed" measurements that are presented to the NWP system in order to obtain realistic observation covariance matrices for use in the variational assimilation scheme. This will require close collaboration between the data providers and NWP users. The direct assimilation of bending angle or refractivity currently seem to be the most reasonable approaches. The Met Office intends to perform forecast impact trials with CHAMP data during 2002.

References

1. Kursinski ER, Hajj GA, Bertiger WI, Leroy SS, Meehan TK, Romans LJ, Schofield JT, McCleese DC, Melbourne WG, Thornton CL, Yunck TP, Eyre JR (1996) Initial results of radio occultation of the Earth's atmosphere using the Global Positioning System. Science, **271**, 1107-1110

2. Rocken C, Anthes R, Exner M, Hunt D, Sokolovskiy S, Ware R, Gorbunov M, Schreiner W, Feng D, Herman B, Kuo YH, Zou X (1997) Analysis and validation of GPS/MET data in the neutral atmosphere. J Geophys Res, **102**, 29,849-29,866
3. Wickert J, Reigber C, Beyerle G, Konig G, Marquardt C, Schmidt T, Grunwald L, Galas R, Meehan TK, Melbourne WG, Hocke KA (2001) Atmosphere sounding by GPS radio occultation: First results from CHAMP. Geophys Res Lett **28** 28, 3263-3266
4. Anthes RA, Rocken C, Kuo YH (2000) Applications of COSMIC to meteorology and climatology. Terrestrial Atmospheric and oceanic Sciences **11**, 115-156
5. Talagrand O (1997) Assimilation of observations, an introduction. J Meteorol Soc Japan, **75(1B)**, 191-209.
6. Eyre JR, Lorenc AC (1989) Direct use of satellite sounding radiances in numerical weather prediction. Meteorological Magazine, **118** 13-16.
7. Eyre JR, Kelly GA, McNally AP, Andersson E, Persson A (1993) Assimilation of TOVS radiance information through one-dimensional variational analysis. Q J R Meteorol Soc **119**, 1427-1463.
8. Lorenc AC (1986) Analysis methods for numerical weather prediction, Q J R Meteorol Soc, **112**, 1177-1194
9. Le Dimet FX, Talagrand O (1986) Variational algorithms for analysis and assimilation of meteorological observations: Theoretical aspects. Tellus, **38A**, 97-110
10. Eyre JR (1997) Variational assimilation of remotely-sensed observations of the atmosphere. J Meteorol Soc Japan, **75(1B)**, 331-338
11. Eyre JR (1994) Assimilation of radio occultation measurements into a numerical weather prediction system, ECMWF Tech Memo, **199**
12. Ingleby NB (2001) The statistical structure of forecast errors and its representation in the Met Office 3-Dimensional variational data assimilation scheme. Q J R Meteorol Soc, **127**, 209-231
13. Kuo YH, Sokolovskiy SV, Anthes RA, and Vandenberge F (2000) Assimilation of radio occultation data for numerical weather prediction. Terrestrial, Atmospheric and Oceanic Sciences, **11**, 157-186
14. Palmer PI, Barnett JJ, Eyre JR, Healy SB (2000) A non-linear optimal estimation method for radio occultation measurements of temperature, humidity and surface pressure. J Geophys Res, **105**, 17,513-17,526
15. Healy SB (2001) Radio occultation bending angle and impact parameter errors caused by horizontal refractive index gradients in the troposphere: A simulation study. J Geophys Res, **106**, 11875-11889
16. Zou X, Vandenberghe F, Wang B, Gorbunov ME, Kuo YH, Sokolovsky S, Chang JC, Sela JG, Anthes RA (1999) A ray-tracing operator and its adjoint for the use of GPS/MET refraction angle measurements. J Geophys Res, **104**, 22,301-22,318
17. Fjeldbo GA, Kliore AJ, Eshleman VR (1971) The neutral atmosphere of Venus studied with the Mariner V radio occultation experiments. Astron J, **76**, 123-140
18. Kursinski ER, Hajj GA, Hardy KR, Schofield JT, Linfield R (1997) Observing Earth's atmosphere with radio occultation measurements using GPS. J Geophys Res, **102**, D19, 23429-23465,1997.
19. Melbourne WG, Davis ES, Duncan CB, Hajj GA, Hardy KR, Kursinski ER, Meehan TK, Young LE, Yunck TP (1994) The application of GPS to atmospheric limb sounding and global change monitoring. JPL Publication **94-18**

20. Levy MF, Anderson KD (2000) Modelling refractive effects on earth-space paths with hybrid parabolic equation techniques. Presented at AP2000, 9-14 April 2000, Davos Switzerland
21. Lorenc AC, Ballard SP, Bell RS, Ingleby NB, Andrews PLF, Barker DM, Bray JR, Clayton AM, Dalby TD, Li D, Payne TJ, Saunders FW (2000) The Met Office global three dimensional variational assimilation scheme. Q J R Meteorol Soc, **126**, 2,991-3,012
22. Healy SB, Eyre JR (2000) Retrieving temperature, water vapor and surface pressure information from refractive index profiles derived by radio occultation: A simulation study. Q J R Meteorol Soc, **126**, 1661-1683
23. Healy SB, Eyre JR (2002) A fast forward model for the direct assimilation of radio occultation bending angle measurements. In preparation.

5 Validation results

Before any retrieved electron density profiles can be used in ionospheric physics, the accuracy and reliability of the retrievals have to be estimated. This can be done by comparison with measurements obtained by quite different techniques such as vertical sounding, incoherent scatter radar or in situ measurements. Principally, it has to be stated that the IRO derived electron density profiles provide a unique measure. The limb sounding technique provides profiles that describe the mean vertical electron density distribution over a fairly large area with a diameter of about 2000 km. Thus, a comparison with precisely located vertical sounding data is expected to yield differences because they are are strongly influenced by the local ionospheric structure.

In Fig. 2 IRO retrieved electron density profiles are compared with corresponding vertical sounding (digisonde) data obtained in Juliusruh (54.6°N; 13.4°E) and Athens (38.0°N; 23.5°E) around noon times on 27 and 29 April 2001, respectively. The plots indicate the capabilities for an estimation of the systematic differences between both techniques. Related studies are on-going. In this paper we confine our attention to the profile key parameters such as the peak density NmF2 or peak plasma frequency f0F2 and the peak height hmF2. The deviations from vertical sounding data (coincidence radius: 8° , time window: 20 minutes) are estimated by using globally distributed stations that were available in the SPIDR data basis [6]. The computed absolute and percentage deviations are shown in Fig. 3. The distribution of the deviations is based on 267 and 119 values for f0F2 and hmF2, respectively. Both the shift of the absolute distribution function (0.07 MHz for f0F2 and -1.1 km for hmF2) as well as that of the percentage distribution (2.17 % for f0F2 0.6 % for hmF2) are rather small indicating a nearly symmetrical scatter. The width of the distribution corresponds to a RMS deviation of 16.6% and 12.7% for f0F2 and hmF2, respectively, agreeing with former estimates [8], [4]. It is

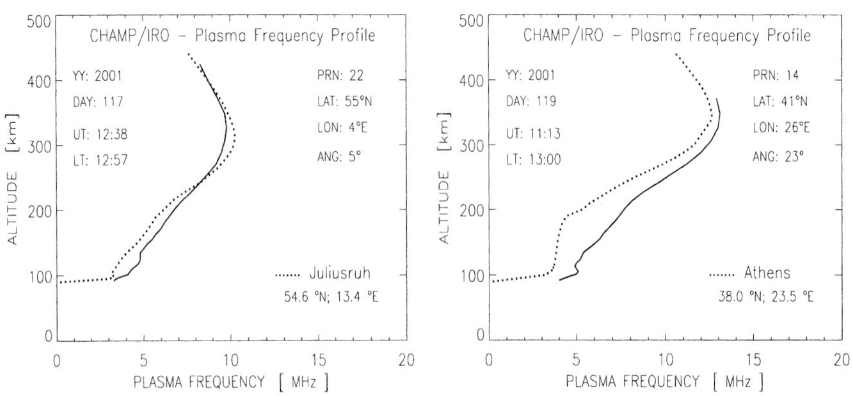

Fig. 2. Comparison of IRO derived plasma frequency profiles with corresponding vertical sounding profiles obtained at the stations Julisruh (left) and Athens (right).

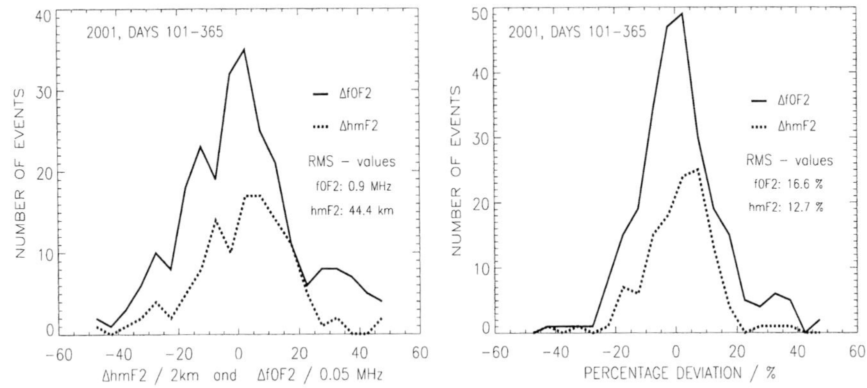

Fig. 3. Distribution of absolute (left) and percentage (right) differences between IRO derived f0F2 (N=267) and hmF2 (N=119) values and corresponding vertical sounding observations taken as reference. The center of the difference distribution functions are shifted by +0.07 MHz/2.17 % for f0F2 and by -1.1 km/0.6 % for hmF2.

assumed that a better agreement between IRO and VS derived key parameters can only be achieved in case of a rather 'smooth' ionosphere. Further data will be collected and analyzed to improve the statistics.

A very effective way of checking the topside electron density profiles is a comparison with the Planar Langmuir Probe (PLP) data obtained along the satellite track. The PLP is part of the Digital Ion Drift Meter and was provided by the US Air Force Research Laboratory, Hanscom, MA. Fig. 4 shows

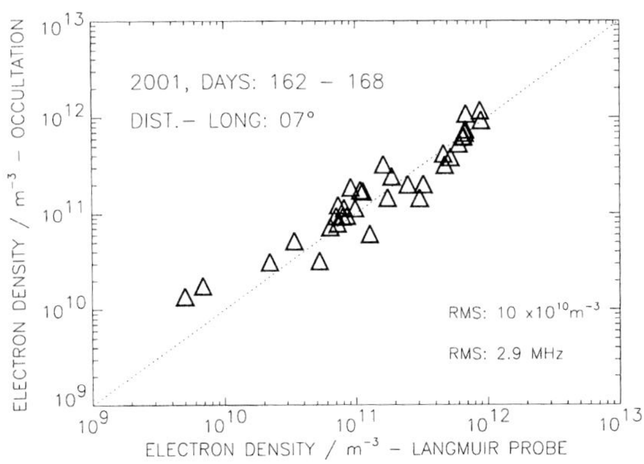

Fig. 4. Scatter plot of in situ measurements of the electron density obtained from the Langmuir probe versus IRO derived electron densities in CHAMP orbit height. The occultation measurements are separated from the CHAMP orbit plane by less than 7° (< 800km).

havior of this plasma scale height in the transition region between model and measurements indicates a stable retrieval procedure. In both figures the summer hemisphere indicates higher plasma scale heights that can be explained by enhanced plasma temperatures. In fact, this effect is emphasized in the computed equivalent slab thickness $\tau_b = TEC(hmF2)/NmF2$ of the bottom side electron density profile (Fig. 8). This slab thickness parameter seems to

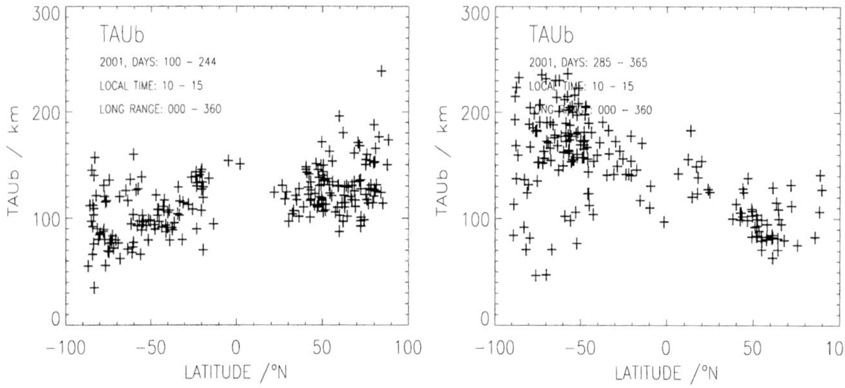

Fig. 8. Latitudinal distribution of the IRO derived bottomside slab thickness during day-times (10-15 LT) on the days 100-244 (left) and 285-365 (right) at all longitude zones.

be a very sensitive parameter related to the plasma temperature. In both figures strong gradients between the summer and winter hemisphere exist, indicating that in particular the bottomside ionosphere is very sensitive to solar radiation variations.

7 Conclusions

We have reported the status of the generation of ionospheric data products computed from GPS radio occultations on board CHAMP since 11 April 2001. Although the validation work is not yet finished, ionospheric data products are available via the ISDC. Potential users should be aware of the principal averaging character of IRO derived electron density profiles. If compared with localized vertical sounding measurements, the F2 layer peak electron density f0F2 and the corresponding height hmF2 agree within a spread of 17 % and 13 % , respectively. This agreement might be improved, if additional information, e.g. on horizontal gradients or local densities, is included in the retrieval procedure. Horizontal gradients can be deduced from horizontal TEC maps as they are produced from ground based GPS measurements in DLR/IKN since 1995 (e.g. [2], http://www.kn.nz.dlr.de). Furthermore, the in situ electron density data of the PLP may be used as an additional input to improve the retrieval results in comparison with ionosonde data, at least in post-processing. An essential improvement is expected, if IRO retrievals and tomographic reconstructions of the topside electron density distribution (see Heise et al. in this issue) are combined to get a comprehensive view on the entire vertical electron density structure of the ionosphere from the bottom-side up to GPS orbit heights. The discussion of the early results has shown that the IRO retrieved electron density profiles and deduced parameters provide a consistent description of the general ionospheric behavior. This fact indicates that IRO data should have a big potential for studying a number of ionospheric phenomena on global scale and for improving ionospheric models. Since IRO derived ionospheric electron density profiles may be retrieved operationally, satellite missions carrying a GPS receiver on board such as CHAMP are very promising candidates for space weather monitoring tasks.

Acknowledgements

The authors are very grateful to all the colleagues from the international CHAMP team. We thank Jens Mielich (IAP, Germany) and Ioanna Tsagouri (NOA, Athens) for providing vertical sounding data for comparison. This study was carried out under grant number 01SF9922/2 of the German Federal Ministry of Education and Research (BMBF).

References

1. Hajj GA and Romans LJ (1998) Ionospheric electron density profiles obtained with the Global Positioning System: Results from the GPS/MET experiment. Radio Sci, **33**, 175-190
2. Jakowski N (1996) TEC Monitoring by Using Satellite Positioning Systems. In: Kohl H, Rüster R, Schlegel K (Eds) Modern Ionospheric Science. EGS, Katlenburg-Lindau, ProduServ GmbH Verlagsservice, Berlin, pp 371–390
3. Jakowski N (1999) Capabilities of radio occultation measurements onboard LEO satellites for ionospheric monitoring and research. In: Vernon A (Ed) Proc. 4th COST 251 Workshop 'The Impact of the Upper Atmosphere on Terrestrial and Earth-Space Communications', 22-25 March, Funchal, Madeira, Portugal, pp116-121
4. Jakowski N, Wehrenpfennig A, Heise S, Reigber Ch, Lühr H, Grunwaldt L, Meehan T (2002) GPS Radio Occultation Measurements of the Ionosphere from CHAMP: Early Results. Geophys Res Lett, accepted
5. Kursinski ER, Hajj GA, Schofield JT, Linfield RP, Hardy KR (1997) Observing Earth's atmosphere with radio occultation measurements using the Global Positioning System. J Geophys Res, **102**, D19, 23429-23465
6. O'Loughlin KF (1997) SPIDR on the Web: Space Physics Interactive Data Resource on-line analysis tool. Radio Sci, **32**, 2021-2026

7. Reigber Ch, Lühr H, and Schwintzer P (2000) CHAMP mission status and perspectives. Suppl to EOS, Transactions, AGU, **81**, 48, F307
8. Schreiner WS, Sokolovskiy SV, Rocken C (1999) Analysis and validation of GPS/MET radio occultation data in the ionosphere. Radio Sci, **34**, 949-966
9. Wehrenpfennig A, Jakowski N, Wickert J (2001) A Dynamically Configurable System for Operational Processing of Space Weather Data. Phys Chem Earth, **26**, 601-604
10. Wickert J, Reigber Ch, Beyerle G, König R, Marquardt Ch, Schmidt T, Grunwaldt L, Galas R, Meehan T, Melbourne WG, Hocke K (2001) Atmosphere sounding by GPS radio occultation: First results from CHAMP. Geophys Res Lett, **28**, 3263-3266

NWP Model Specific Humidities Compared with CHAMP/GPS and TERRA/MODIS Data

Klaus-Peter Johnsen

Institute for Coastal Research, GKSS Research Centre Geesthacht GmbH, Geesthacht, Germany

Summary. Limb sounding observations of the Earth's atmosphere by the GPS receiver onboard CHAMP allow to determine vertically resolved profiles of the specific humidity. The specific humidity shows a high variability and is a critical parameter for short term numerical weather prediction (NWP). Specific humidity profiles calculated with the hydrostatic NWP model HRM (High resolution Regional Model) of the Deutscher Wetterdienst over the BALTEX (Baltic Sea Experiment) modelling area are compared with profiles from CHAMP/GPS radio occultation data (14th of May to 10th of June 2001) and from the MODIS detector onboard the TERRA spacecraft (dataset from 1st of December 2001, 1100 UTC). Radiosonde data from the station Lindenberg/Germany (1st of May to 10th of June 2001) and ECMWF data (14th of May to 10th of June 2001) are used to evaluate the results. The HRM model used as boundary conditions 6-hourly analyses of the European Centre for Medium-Range Weather Forecast (ECMWF). Consecutive 30 h forecast starting each day at 0 UTC were performed. The mean difference between the HRM and the radiosonde data or the ECMWF data is below 0.4 g/kg while the biases between HRM and CHAMP/GPS are up to 1.7 g/kg and the biases between HRM and TERRA/MODIS are around 0.5 g/kg. The standard deviations of the specific humidity differences between HRM and CHAMP/GPS, TERRA/MODIS, ECMWF or the radiosonde data from the station Lindenberg/Germany are about 1 to 2 g/kg in the lower troposphere (below about 4 km) and decrease with increasing height.

Key words: HRM model, BALTEX, CHAMP/GPS, TERRA/MODIS, radiosonde data, ECWMF data

1 Introduction

With different networks of groundbased receivers it was shown that the US Global Positioning System (GPS) is an accurate technique to determine the vertically integrated water vapor (IWV) within the atmosphere [1],[2],[3],[4]. Comparisons with the hydrostatic **H**igh resolution **R**egional weather forecast Model **HRM** of the Deutscher Wetterdienst have *e.g.* shown a mean bias of 0.08 kg/m^2 and standard deviations of IWV differences of HRM and GPS of about 2.5 kg/m^2 over Germany. Similar results were found for mean IWV differences between the HRM model and data of the Advanced Microwave Sounding Unit A (AMSU-A), [4].

Between 1995 and 1998 GPS/MET onboard the MicroLab 1 successfully demonstrated the radio occultation technique to probe the Earth's atmosphere through its refraction effects on the signals transmitted by the GPS satellites [5]. Thus also specific humidity profiles could be obtained.

Here specific humidities derived from HRM and compared with CHAMP/GPS and TERRA/MODIS data are presented. Radiosonde and ECMWF data are used to evaluate the results. The objective is to investigate whether specific humidity profiles from CHAMP/GPS and/or from TERRA/MODIS data can be assimilated into NWP models.

2 Comparison of HRM with ECMWF and radiosonde data

The hydrostatic weather forecast model HRM used as boundary conditions 6-hourly analyses of the ECMWF (European Centre for Medium range Weather Forecast) and was initialised by interpolation of the 0 h analysis data. It calculates the water vapor with a horizontal resolution of 0.125° (about 14 km) in 30 vertical layers on a rotated latitude/longitude grid. The time step was chosen to 90 s. Consecutive 30 h forecasts starting each day at 0 UTC including a 6 h spin-up time of the model were performed. The BALTEX (Baltic Sea Experiment, [6]) modelling area shown in Figure 1 is used. Figure 2 (left) shows mean specific humidity differences and their standard deviations of HRM and the ECMWF data between 14th of May and 10th of June 2001. 123 profiles of HRM and ECMWF colocated with the CHAMP/GPS data and interpolated to the GPS levels and times were used. The standard deviations are below 1 g/kg. The HRM model shows slightly (around 0.1 g/kg) smaller specific humidities for most of the heights below about 3 km and

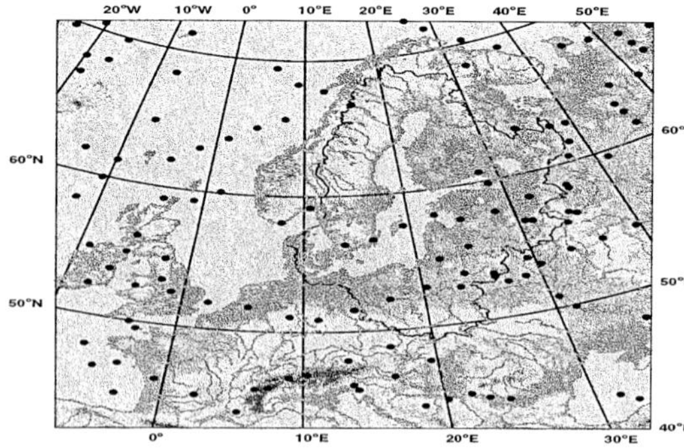

Fig. 1. BALTEX modelling area with 123 CHAMP radio occultations between 14th of May and 10th of June 2001.

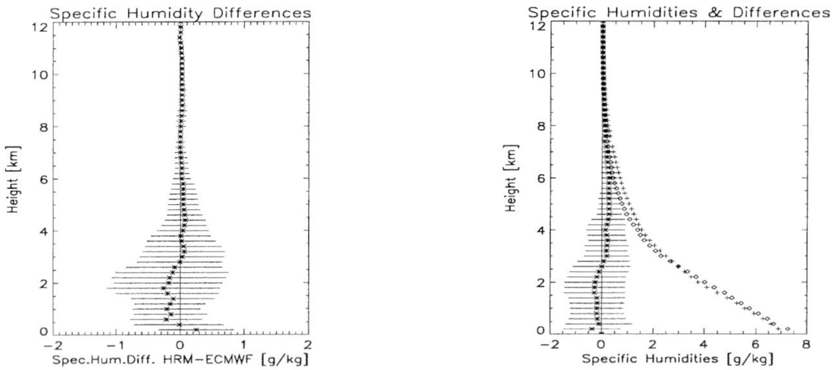

Fig. 2. Left: Mean specific humidity differences between HRM (diamonds) and ECMWF data (diamonds with line) and their standard deviations. **Right:** Mean differences and standard deviations of the specific humidities derived from the HRM model (diamonds) and from 145 radiosonde profiles in Lindenberg/Germany (diamonds with line) between the 1st of May 2000 and the 6th of June 2000.

slightly larger specific humidities above 3 km. Similar results were obtained by comparing the HRM model results with 145 radiosonde ascents from Lindenberg/Germany between the 1st of May and the 6th of June 2000 (Figure 2, right). The standard deviations are below 1.5 g/kg. The mean absolute differences are below 0.4 g/kg.

3 Comparison of HRM with CHAMP/GPS

To derive the specific humidity from CHAMP refractivity (N) data tropospheric temperature information from ECMWF data on 11 standard pressure levels between the surface and 100 hPa is used. To account for the limited vertical resolution of the temperature data the refractivity profiles have been smoothed in the vertical by applying a 1 km boxcar filter of log(N) prior to the derivation of humidity.

Figure 3 (left) shows the mean specific humidities as derived from CHAMP/GPS and calculated with the HRM model at the points shown in Figure 1. The specific humidity differences from the HRM and CHAMP/ GPS are also represented together with their standard deviations. CHAMP/GPS shows lower specific humidities up to about 1.7 g/kg. The difference increases with decreasing height, maybe due to multipath effects within the lower troposphere [7]. However, recent studies show [8] that the bias problem of the CHAMP refractivities does not seem to be caused solely by multipath effects but corrections e.g. by the canonical transform method show an improvement of the data quality. Effects influencing the GPS phase tracking process causing cycle slips could maybe also explain the bias. For further details see e.g. [11].

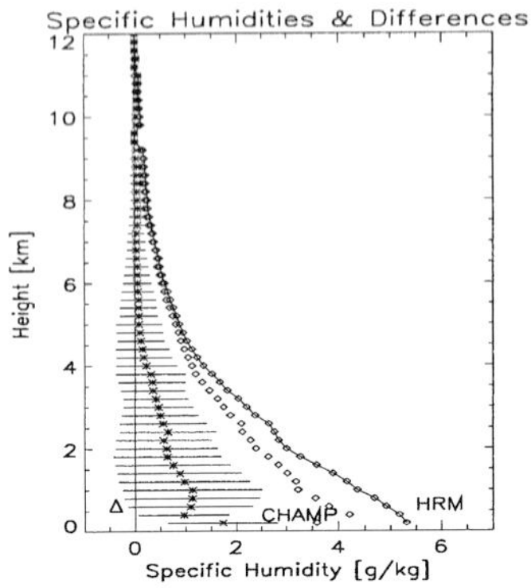

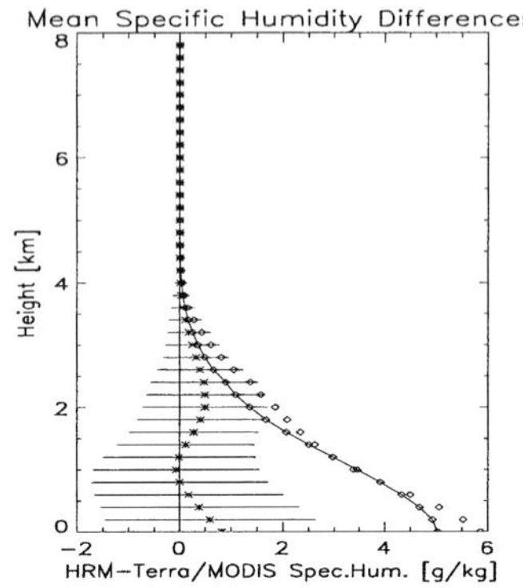

Fig. 3. Left: Mean specific humidities derived from HRM (diamonds) and from CHAMP/GPS (diamonds with line) and their differences Δ calculated with CHAMP radio occultations within the BALTEX modelling area (Figure 1) between 14th of May and 10th of June 2001. **Right:** Mean specific humidities from HRM (diamonds) and TERRA/MODIS (diamonds with line) and its differences and standard deviations for 18142 pixels of the TERRA/MODIS dataset of 1st of December 2001, 1100 UTC.

Slightly lower standard deviations and smaller biases were obtained for the specific humidity differences of HRM and temporally interpolated ECMWF data (Figure 2, left) or radiosonde ascents from Lindenberg (Figure 2, right).

4 Comparison of HRM with TERRA/MODIS

MODIS is a scanning spectroradiometer onboard the TERRA spacecraft with 36 spectral bands between 0.645 and 14.235 μm [9]. Following the approach of Smith et al. [10] clear sky temperature and moisture profiles (MOD-07 product) together with the surface temperatures were calculated simultaneously with the help of a radiative transfer equation. Cloud filtering is achieved with the aid of the cloud mask product (MOD-06). The horizontal spatial resolution of one vertical profile is $5{\times}5$ km^2.

18142 vertically interpolated specific humidity profiles of TERRA/MODIS were compared with colocated profiles derived with the HRM model. The comparison for this dataset of the 1st of December 2001, 1100 UTC is shown in Figure 3, right: Similar like the CHAMP/GPS dataset this dataset contains a large range of specific humidities (e.g. between 0.5 and 7.5 g/kg in a height of 1 km). It covers nearly the half of the BALTEX modelling area (MODIS data between 36.7° and 58.4° N and between −14.7° and 22.7° E). The HRM model shows slightly (around 0.5 g/kg) larger mean specific humidities for most of the heights between the surface and about 4 km. The standard deviation at the surface between both datasets ($\approx$ 2 g/kg) is a little bit larger compared with CHAMP/GPS. It decreases to less than 0.01 g/kg at 4 km height.

5 Conclusions

In this study a comparison between the specific humidity calculated with the hydrostatic NWP model HRM of the Deutscher Wetterdienst and CHAMP/GPS radio occultation data is presented. Between the 14th of May and the 10th of June 2001 123 profiles were obtained from the GPS receiver onboard CHAMP over the BALTEX modelling area. The comparison has shown that CHAMP/GPS measures significantly lower mean specific humidities below about 4 km. This is supported by comparisons between the HRM model and the ECMWF analysis data as well as between the HRM model and radiosonde ascents at Lindenberg/Germany which have shown lower mean absolute differences of about 0.2 g/kg. The standard deviations of the difference between HRM and CHAMP/GPS are similar to the standard deviations between HRM and the radiosonde ascents. The TERRA/MODIS dataset of the 1st of December 2001 1100 UTC shows slightly larger standard deviations of up to 2 g/kg but smaller mean specific humidity differences to the HRM model by comparing 18142 profiles with similar large ranges of specific humidities like from CHAMP/GPS.

If the bias due to the multipath problem or due to noise contributions influencing the GPS phase tracking process can significantly be reduced, the CHAMP/GPS data are a valuable source of specific humidity data and should consequently be assimilated into NWP models for Europe. The slightly larger standard deviations of TERRA/MODIS compared with the HRM could be a problem by using the TERRA/MODIS data for assimilation into NWP models which maybe solved by taking mean values before assimilation. For further details about the assimilation techniques see e.g. [12]. However, due to the good coverage, the high horizontal resolution of the TERRA/MODIS data of about 5×5 km^2 and the acceptable low mean specific humidity difference between TERRA/MODIS and HRM the MOD-07 product could also supplement the radiosonde data quite well, especially in data sparse regions, and should be assimilated into NWP models.

6 Acknowledgements

The provision of data by various institutions is gratefully acknowledged: GeoForschungsZentrum, Potsdam, Germany (CHAMP/GPS data), Deutscher Wetterdienst (radiosonde data), European Centre for Medium Range Weather Forecast (ECMWF analysis data), NASA/GSFC (TERRA/MODIS data).

References

1. Emardson TR, Johansson J, Elgered G (2000) The Systematic Behavior of Water Vapor Estimates Using Four Years of GPS Observations. Transactions on Geoscience and Remote Sensing, **38** (1), 324–329
2. Ohtani R, Naito I (2000) Comparisons of GPS-derived precipitable water vapors with radiosonde observations in Japan. Journal of Geophysical Research, **105** (D22), 26917–26929
3. Cucurull L, Navascues B, Ruffini G, Elosegui P, Rius A, Vila J (2000) The Use of GPS to Validate NWP Systems: The HIRLAM Model. Journal of Atmospheric and Oceanic Technology, **17**,773–787
4. Johnsen K-P, Kidder SQ (2002) Water vapor over Europe obtained from remote sensors and compared with a hydrostatic NWP model. Physics and Chemistry of the Earth, **27**, 371–375
5. Rocken C, Anthes R, Exner M, Hunt D, Sokolovskiy S, Ware R, Gorbunov M, Schreiner W, Feng D, Herman B, Kuo Y-H, Zuo X (1997) Analysis and validation of GPS/MET data in the neutral atmosphere. Journal of Geophysical Research, **102** (D25), 29849–29866
6. Raschke E, Meywerk J, Warrach K, Andrea U, Bergstroem S, Beyrich F, Bosveld F, Bumke K, Fortelius C, Graham L-P, Gryning S-E, Halldin S, Hasse L, Heikinheimo M, Isemer H-J, Jacob D, Jauja I, Karlsson K-G, Keevallik S, Koistinen J, van Lammeren A, Lass U, Launianen J, Lehmann A, Liljebladh B, Lobmeyr M, Matthaeus W, Mengelkamp T, Michelson DB, Napiorkowski J, Omstedt A, Piechura J, Rockel B, Rubel F, Ruprecht E, Smedman A-S, Stigebrandt A (2001) The Baltic Sea Experiment (BALTEX): A European Contribution to the Investigation of the Energy and Water Cycle over a Large Drainage Basin. Bulletin of the American Meteorological Society, **82** (11), 2389–2413
7. Gorbunov ME, Gurvich AS (1998) Microlab-1 experiment: Multipath effects in the lower troposphere. Journal of Geophysical Research, **103** (D12), 13819–13826
8. Beyerle G, Gorbunov ME, Marquardt C, Reigber C, Schmidt T, Wickert J (2002) Inverting GPS Radio Occultation Data using the Canonical Transform Method: Results from Simulation Studies and CHAMP Observations. Poster, European Geophysical Society assembly, Nice, France
9. Menzel WP, Gumley LE (1998) MODIS Atmospheric Profile Retrieval Algorithm Theoretical Basis Document. Technical Report Version 4, NOAA/NESDIS, University of Wisconsin-Madison, Madison, WI, USA
10. Smith WL, Woolf H-M, Schriener AJ (1985) Simultaneous retrieval of surface and atmospheric parameters: a physical and analytically direct approach. In: Deepak A, Fleming HE, Chahine MT (ed.), Advances in Remote Sensing, ISBN 0-937194-07-7, 221–232
11. Hajj GA, Kursinski ER, Romans LJ, Bertiger WI, Leroy SS (2002) A technical description of atmospheric sounding by GPS occultation. Journal of Atmospheric and Solar-Terrestrial Physics, **64**, 451-469
12. Kuo Y-H, Guo Y-R, Westwater ER (1993) Assimilation of Precipitable Water Measurements into a Mesoscale Numerical Model. Monthly Weather Review, **121**, 1215–1238

Analysis of Gravity Waves from Radio Occultation Measurements

Martin Lange and Christoph Jacobi

Institute for Meteorology, Stephanstr. 3, 04103 Leipzig
mlange@uni-leipzig.de, jacobi@uni-leipzig.de

Summary. In the height range 10-30 km atmospheric gravity waves lead to periodic perturbations of the background temperature field in the order of 2-3 K, which can be resolved in temperature profiles derived from radio occultation measurements. Due to the spherical symmetry assumption in the retrieval algorithm and the low horizontal resolution of the measurement weakening in the amplitude and phase shift of the waves occurs. The influence of the geometric wave parameters and the measurement geometry on a homogeneous spectrum of plane gravity waves in the range 100-1000 km horizontal and 1-10 km vertical wavelength is investigated with a 2D-model ranging $\pm$1000 km around the tangent point and 10-50 km in height. The investigation shows that with radio occultation measurements more than 90% of the simulated wave spectrum can be resolved with relative amplitudes above the 1/e level. Considering the total variance, about 80% is retrieved in the worst case when the GPS receiver scans perpendicular to the wave crests and about 88% as mean value when the view angle of the receiver through the gravity waves is arbitrarily oriented. More realistic wave spectra lead to slightly smaller values.

Key words: Gravity waves, gravity wave activity, radio occultations, satellite measurements, model simulations

1 Introduction

Gravity waves (GW's) play a main role in the circulation of the middle atmosphere. GPS radio occultation (RO) measurements provide a new tool to estimate GW activity on a global scale (Tsuda et al., 2000). Unfortunately, due to the low horizontal resolution of about 200-400 km (Kursinski et al., 1997) and the spherical symmetry assumption in the retrieval algorithm for atmospheric parameters, weaker amplitudes and a phase shift is derived in the temperature profiles depending on the gravity wave parameters. Short waves with horizontal wavelengths smaller than 50-100 km are filtered out due to the averaging over positive and negative phase variations of the GPS signal, whereas waves with larger horizontal wavelengths than 1000 km match the spherical symmetry assumption (Belloul and Hauchecorne, 1997). Therefore the vertical profiles of small gravity waves cannot be derived directly, and the potential energy derived from the temperature variance used to describe climatologies of gravity wave activity is estimated too low.

3 Results

3.1 Visibility of plane gravity waves

Fig. 2 shows a general picture of the visibility and the vertical phase shift of gravity waves in dependence on the horizontal and vertical wavelength. Following an approach similar to that of Preusse et al. (2001), who derived gravity waves from thin layer radiation of simulated GW temperature perturbations, visibility is defined here as the ratio between the derived temperature amplitude and the original amplitude. The dashed lines in the left of Fig. 2a) indicate visibility of 1/e and 1/e^2. The 90% isoline apears above the diagonal of the plot. Thus more than 50% of the waves are retrieved with an amplitude above that level. Calculating the total temperature variance ΔT^2 for all waves (not explicitly shown), about 80% of the original variance is derived. Nevertheless the phase shift with respect to height, that occurs from the retrieval algorithm (panel b), is quite remarkable for a major part of the waves. Although it decreases with increasing visibility, about 40% of the waves undergo a phase shift of more than 45°. At combinations of short horizontal and long vertical wavelengths, see the left side of panel b), where the phase shift exceeds 60°, the contour lines become irregular. For example at a horizontal wavelength of 200 km the phase shift maximizes at a vertical wavelength of 6 km with values above 160° and decreases again at larger wavelengths.

All the results hitherto described are obtained under the assumption, that the LEO satellite views perpendicular to the wave crests. The same temperature amplitudes may be derived for even shorter horizontal wavelengths, if the view angle between the line of sight and the horizontal wave vector is larger. Therefore the worst case of geometry has been considered for plane gravity waves. To estimate this effect, the visibility in dependence on the

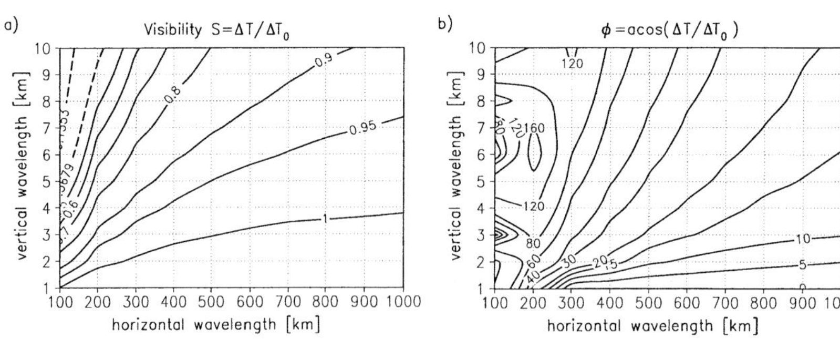

Fig. 2. Visibility of plane gravity waves (a) defined as amplitude ratio between the retrieved and the original temperature perturbation and phase delay ϕ of the derived wave perturbation (b).

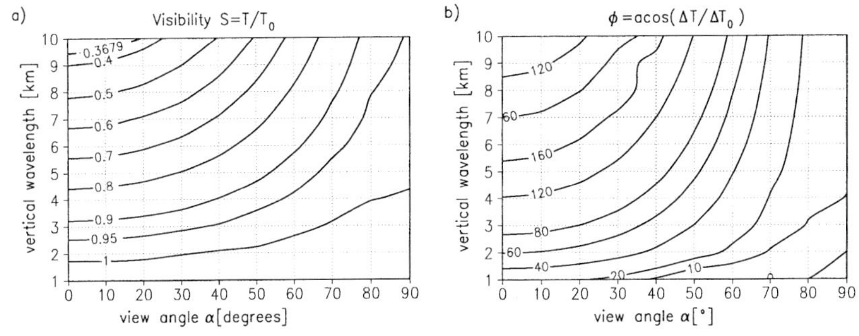

Fig. 3. Visibility of plane gravity waves (a) and phase difference ϕ (b) between the derived and the original wave dependend on the horizontal view angle between the line of sight and the horizontal k-vector of the wave. The respective horizontal wavelength is 200 km.

vertical wavelength and the horizontal view angle is considered.

Figure 3 shows the visibility (a) and the phase difference (b) between the derived and the original gravity wave perturbation dependent on the view angle α between the line of sight and the horizontal wave vector for a horizontal wavelength of 200 km. Since the horizontal wavelength seen by the GPS receiver increases with sec(α), the visibility approaches unity while turning the occultation ray in the direction parallel to the wave crests. Also the phase delay decreases with increasing view angle as expected from Fig. 2 for longer horizontal wavelength. At vertical wavelengths above 7 km the phase shift cycles through a maximum with increasing view angle reflecting the retrieval error when the wave vector is tilted against the vertical and spherical symmetry is disturbed. Considering a homogeneous distribution of the view angles through the gravity waves for the spectrum considered in Fig. 2 the total derived mean variance increases from 80% for $\alpha = 0°$ to 88%. Assuming a more realistic vertical wavelength spectrum with a slope of $k^{-5/3}$ in power spectral density (not shown) the mean observed variance reduces additionally by 2-3% depending on the wave parameters. The larger contribution from long vertical wavelength accounts for that.

4 Conclusions

The signal of temperature perturbations of plane gravity waves, derived from radio occultation measurements, depends on the geometrical wave parameters and the horizontal view angle between the occultation ray and the wave. Incomplete resolution of the amplitudes and vertical phase shift are introduced during the retrieval increasing the effect with shorter horizontal and larger vertical wavelength. For the simulated gravity wave spectrum with constant amplitude in the range 100-1000 km horizontal and 1-10 km vertical wavelength about 80% of the total variance of the spectrum is derived in the worst case, when the view angle of the GPS receiver is perpendicular through the wave crests. For isotropic distributed view angles, as can be assumed for large numbers of measurements used to derive climatologies of gravity wave activity, a mean variance of 88% is derived. Considering a more realistic vertical wavenumber spectrum with a slope of $k^{-5/3}$ in power spectral density the mean observed variance reduces additionally by 2-3% due to the larger contribution of the long vertical wavelengths. Phase delay and amplitude damping increase simultaneously, therefore the part of the spectrum, that is difficult to identify is small compared to the rest. Additional information on gravity wave parameters will be provided by a more dense network of radio occultation measurements in future that allows for multiple measurements near the same tangent point at time scales shorter than one wave period. At horizontal wavelength below 200-300 km the resolution of gravity waves is enhanced when fresnel diffraction theory is taken into account in the retrieval algorithm.

Acknowledgements

This study was supported by the Deutsche Forschungsgesellschaft under grant JA 836-4/1 and by INTAS under grant 991-1186. The project is imbedded in the CHAMP mission lead by the GFZ Potsdam.

References

Belloul MB, and Hauchecorne A, (1997) Effect of periodic horizontal gradients on the retrieval of atmospheric profiles from occultation measurements. Radio Sci, **32**, 469–478.

Fjeldbo G and Eshleman VR (1965) The bistatic radar occultation method for the study of planetary atmospheres. J. Geophys Res, **13**, 3217–3225.

Hocke K (1997) Inversion of GPS meteorology data. Ann Geophys, **15**, 443-450.

Kursinski ER, Hajj GA, Schofield JT, Linfield RP, and Hardy KR (1997) Observing earth's atmosphere with radio occultation measurements using the global positioning system. J Geophys Res **102**, 23,429–23,465.

Preusse P, Dörnbrack A, Eckermann SD, Riese M, Schaeler B, Bacmeister JT, Broutman D, and Grossmann KU (2001 Space based measurements of stratospheric mountain waves by CRISTA. 1. Sensitivity, analysis method and a case study. J Geophys Res, accepted.

Tsuda T, Nishida M, Rocken C and Ware RH (2000) A global morphology of gravity wave activity in the stratosphere revealed by the GPS occultation data (GPS/MET). J Geophys Res, **105**, 7257–7273.

GPS Atmosphere and Ionosphere Methods used on Ørsted Data and Initial Application on CHAMP Data

Georg Bergeton Larsen, Per Høeg, Jakob Grove-Rasmussen, Martin B. Sørensen

Danish Meteorological Institute, Lyngbyvej 100, DK-2100, Copenhagen, Denmark
keywords: GPS, occultation, atmosphere, ionosphere

Summary. The launch of the Ørsted satellite started a range of activities to verify GPS occultation measurements of the Earth atmosphere and ionosphere based on observations of GPS signal delay from a low Earth orbiting satellite. The Ørsted satellite followed the GPS/MET satellite experiment as the second mission in space to perform these types of measurements. Now the CHAMP satellite and also the SAC-C satellite are producing GPS occultation measurements in large numbers in long time-series and the data are just available since November 2001. The aim of this paper is to present an overview of the retrieval techniques used on the observations with the TurboRogue receiver on Ørsted and the results obtained both for the atmosphere and ionosphere. The results of CHAMP occultations using Ørsted processing tools will be presented and compared with Ørsted.

Key words: GPS, occultation, ionosphere, Ørsted, atmosphere

1 Introduction

The Danish Ørsted satellite was launched on February 23, 1999 with the primary mission to measure the Earth magnetic field. One of the other scientific instruments onboard is the TurboRogue GPS (Global Positioning System) receiver developed by JPL. This made Ørsted the second satellite capable of measuring GPS occultations following the very successful GPS/MET limb sounding experiment [1]. Today two other missions, CHAMP (CHAllenging Minisatellite Payload) [2] and SAC-C, are providing very accurate GPS limb sounding measurements.

The first Ørsted measurements of GPS occultations were performed on April 17, 1999, and a dedicated measurement campaign was successfully completed in the period from February 3-22, 2000. In this period the TurboRogue was measuring continuously in a 24 hrs mode resulting in an average number of 50 daily profiles.

The technical constraints on the Ørsted GPS occultation measurements arise from a combination of low antenna gain and platform dependant parameters such as attitude control made data processing a difficult task. The antenna was pointed perpendicular to the velocity direction for most of the observations resulting in less observed occultations and only limited measurements below 10 km in tangent height. Data gaps every 10 second were observed in most profiles, this was repaired and was only a very small limit on the accuracy. Measurements of the L2 frequency were not reliable resulting in the dominant limit on retrieval accuracy and an alternative approach described below was necessary to perform corrections for the ionosphere term. Despite these technical difficulties we have succeed to derive profiles of both temperature in the lower atmosphere and electron density in the ionosphere.

2 Atmosphere profiles

A single frequency method was developed and used on the measured occultations to perform the ionosphere correction. The contribution from the ionosphere can be expressed in terms of the total electron content (TEC). Usually TEC is derived using both GPS frequencies, however accepting less accuracy then TEC can be derived as shown in equation (2) using the pseudo range measurements obtained at 1Hz. The Ørsted processing procedure can be summarised as follows [3],

□ Ørsted input data: L1, C1. Fiducial station input: L1, C1, L2
    Cycle slip detection and correction
□ Data filtering through a regularisation method
    Regularisation of all ingoing signals L1, L2*(derived based on L1-C1)
    Data gap filling
□ Orbit arc construction (initial orbits; 10 Hz interpolation)
□ Double differencing
    Zero-order troposphere correction
    First-order relativistic correction
    Second-order ionosphere correction
□ Ionosphere correction on phase (L1-L2*)
□ Bending angle construction
    Bending angle bias correction (Based on the MSIS90 model)
□ Abel transform
□ Retrieval of atmosphere parameters

All the profiles measured during the Ørsted GPS campaign in February 2000 have been validated against data from ECMWF (European Centre for Medium-Range Weather Forecasts) as part of the CLIMAP project [4]. The results are presented in fig. 1. as temperature differences compared to the ECMWF analysis. A total of 921 occultations split in polar and equatorial regions are used in the statistics. The standard deviation grow large close to the surface caused by the limited number of measurements below 10 km in tangent height. The best accuracy is achieved close to the tropopause and can be roughly summarised to be less than 1 K in mean difference and with about 3 K standard deviation.

Using the Ørsted processing algorithms on CHAMP measurements taking as input the RINEX files, and using initial orbits computed with the GIPSY/OASIS tool [5] we have computed a limited number of temperature profiles from Novem-

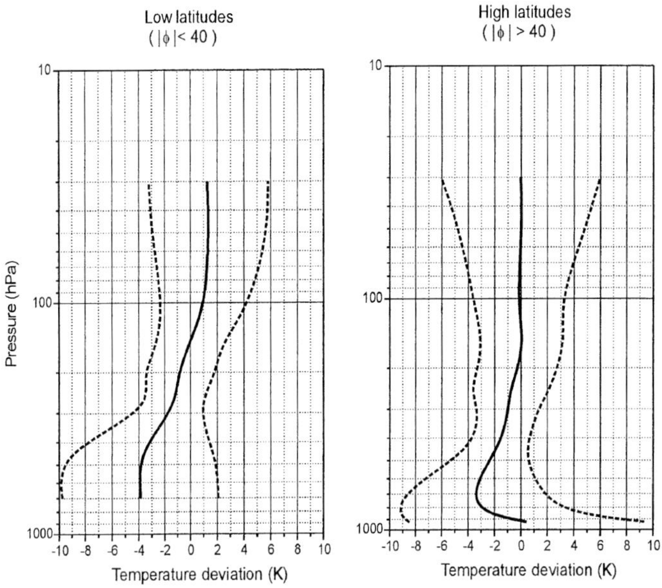

Fig. 1. Statistics comparing Ørsted temperature profiles with ECMWF. Left side represents the equatorial region and is based on 432 profiles. Right hand side contains 489 profiles for the mid/polar regions. The solid curve is the mean difference and the dashed curves represent one standard deviation.

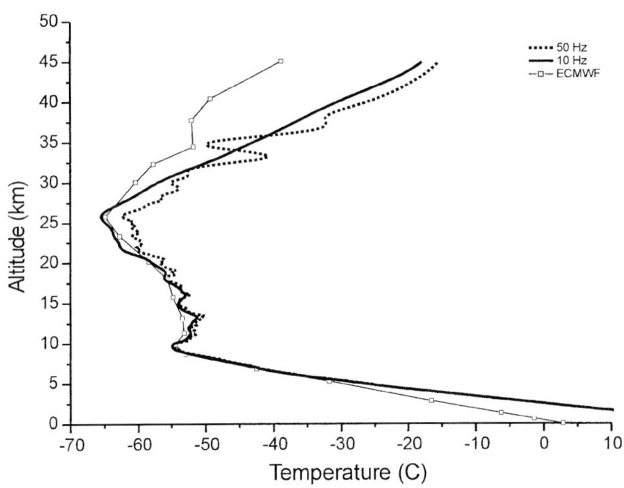

Fig. 2. Temperature profiles derived using CHAMP data. Two different processing methods were applied in this example, one using the 10 Hz method which the Ørsted method was designed at, and one using the full 50 Hz sampling of the CHAMP data.

ber 5, 2001. In fig. 2 we show an example of a CHAMP temperature profile and compare it with data from ECMWF. A close match is observed between 5-20 km. The Ørsted sampling rate was only 10 Hz and the filtering method used on the phase delay is a regularisation method with parameters tuned for 10 Hz sampling. The resulting bending angles are subsequently bias corrected using a model fit to bending angles derived based on the MSIS90 model. The model bending angle is based on a zonal selection. In fig. 2 above 20 km the resulting temperature profiles show clear differences. This is caused by the difference on the filtering parameters between 10 and 50 Hz and the selected bending angle model, which is different in the two cases.

3 Ionosphere profiles

The Satellite-to-Satellite Total Electron Content (SS-TEC) is derived using only the L1 signal. This is possible by combining the measured carrier phase, L_1, and the C/A pseudo range, C_1. The effect on the signal caused by the ionosphere has opposite sign for phase and pseudo range:

$$L_1 = \int_1 (1 - C\frac{N_e}{f_1^2})ds \qquad \text{and} \qquad C_1 = \int_1 (1 + C\frac{N_e}{f_1^2})ds \tag{1}$$

where N_e is the electron density, $C = 40.3\ \mathrm{m^3 s^{-2}}$ and f_1=1.57542 GHz. Combining these two equations we obtain the following expression for the SS-TEC:

$$I = \frac{f_1^2}{2C}[C_1 - L_1] \tag{2}$$

where I is the SS-TEC. The ionosphere occultation measurements are performed at a sampling rate of 0.1 Hz, corresponding to a vertical resolution of about 30 km.

The ray path of a radio wave traveling through the ionosphere suffers a small bending, α, caused by the gradients in the background electron density. Expansion of the appropriate equations to first order results in the following simple and sufficiently accurate expression for the (approximate) bending angle [6], since the bending angle α is very small in the ionosphere, of the order 10^{-2} degree,

$$\alpha \approx \frac{C}{f_1^2}\frac{dI}{da_0} \tag{3}$$

Here a_0 is the impact parameter of the straight line connecting the Ørsted and the GPS satellite computed to first order.

Under the assumption of local spherical symmetry, the index of refraction and subsequently the electron density is then derived from the bending angle [7], [8] through the use of the Abel transform. The bending angle is extrapolated above Ørsted orbit height using an exponential [9].

Fig. 3 shows a recently measured electron density profile, which is located close to a digisonde station. This measurement is part of a campaign from February 16-22, 2002 to measure the ionosphere cusp region using both satellite data from CHAMP, SAC-C and Ørsted and with radar measurements from ground at Svalbard and Tromsø. A close agreement between the Ørsted profile and the digisonde is observed.
The geometry of this particular occultation is favorable, with an azimuth angle less then 5 degrees, but still a small negative bias caused by the assumption of spherical symmetry is seen for heights below 200 km. To overcome the assumption of local spherical symmetry the Ørsted TEC measurements of both occultation links and GPS satellites tracked above orbit height have been combined in a tomographic solution of electron density to produce two-dimensional maps of electron density [10].

4 Summary

Validation of GPS occultations measured by the Ørsted satellite has shown that despite many technical difficulties useful profiles of both the neutral atmosphere and the ionosphere have been obtained. The accuracy of temperature profiles is reduced by roughly a factor of two compared to the results from GPS/MET [1] and a similar factor is expected for CHAMP. This reduced accuracy on Ørsted

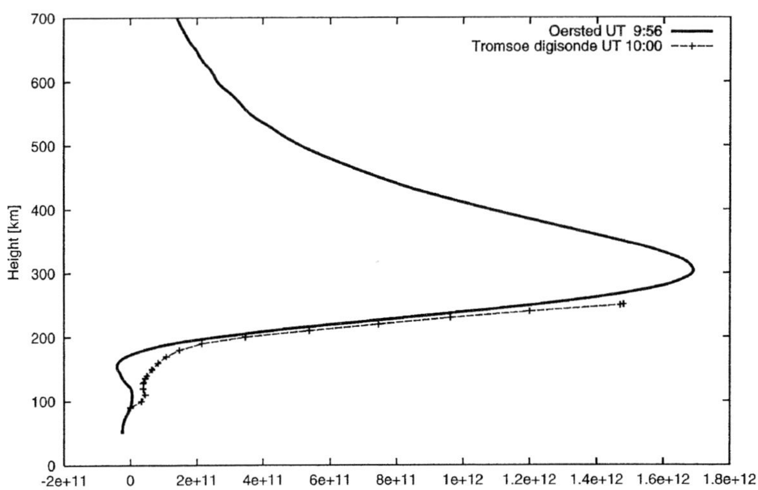

Fig. 3. Ionosphere occultation from February 20, 2002 compared against a digisonde measurement. Distance between the tangent point of the Ørsted profile and the digisonde is about 300 km.

profiles is caused primarily by the lack of L2 measurements. The ionosphere profiles also have reduced accuracy; the TEC obtained has a standard deviation of about one TEC unit. This limits in particular nighttime retrievals when the electron density is low.

In the recent February 2002 campaign the Ørsted satellite, now 3 years since launch, has provided the best quality GPS data yet observed. The attitude control of the satellite is very stable and the occultation antenna is pointing directly in the anti-velocity direction. More Ørsted GPS measurements are planned during the next couple of months to compare and combine with CHAMP and SAC-C occultations.

References

1 Rocken C et al (1997) Analysis and validation of GPS/MET data in the neutral atmosphere. J Geophys Res, 102 (D25), 29849-29866

2 Wickert J et al (2001) Atmosphere sounding by GPS radio occultation: First results from CHAMP. Geophys Res Lett, 28 (17), 3263-3266

3 Høeg P et al (2000) GPS Atmosphere and Ionosphere Profiling Results from the Ørsted Mission. In: Neubert T (ed) Ørsted International Science Team meeting. Grasse, France

4 Hass F et al (2000) Climate and environment Monitoring with GPS Atmospheric Profiling (CLIMAP) Final Report. PL 970695, Copenhagen

5 Webb FH, Zumberge JF (1997) An Introduction to GIPSY/OASIS. JPL D-11088, Jet Propulsion Laboratory, Pasadena, California

6 Høeg P, Larsen GB, Benzon H-H, Grove-Rasmussen J, Syndergaard S, Mortensen MD, Christensen J, Schultz K (1998) GPS Atmosphere Profiling Methods and Error Assessments. Scientific Report 98-7, DMI, Copenhagen

7 Hajj GA, Romans LJ (1998) Ionospheric electron density profiles obtained with the Global Positioning System: Results from the GPS/MET experiment. Radio Science, 33 (1),175-190

8 Schreiner WS, Sokolovskiy SV, Rocken C, Hunt DC (1999) Analysis and validation of GPS/MET data in the ionosphere. Radio Science, 34 (4), 949-966

9 Larsen GB et al (2000) Comparison of Electron Density Profiles from Ørsted GPS Occultation Data and Ground-Based Radar Observations. In: Neubert T (ed) Ørsted International Science Team meeting. Grasse, France

10 Escudero A, Schlesier A, Rius A, Flores A, Rubek F, Larsen GB, Syndergaard S, Høeg P (2001) Ionospheric Tomography Using Ørsted GPS measurements. Phys Chem Earth, 25 (2), 123-126

Combining Radio Occultation Measurements with Other Instruments to Map the Ionospheric Electron Concentration.

Cathryn N Mitchell

Department of Electronic and Electrical Engineering, University of Bath, Bath, UK.

Summary. The Multi-Instrument Data Analysis System (MIDAS) algorithm is a spatial and temporal imaging program for the atmosphere and ionosphere. This analysis algorithm can routinely use dual-frequency ground-based observations from the GPS satellites to produce four-dimensional images of electron concentration over very large geographical regions (potentially globally). MIDAS also has the facility to incorporate other ionospheric measurements, such as electron-density profiles from inverted ionograms or *in-situ* measurements of ionization concentration from satellites in low-earth-orbit.

The main limitation to the spatial resolution in ionospheric tomography is due to the limited geometrical coverage provided by the satellite to ground ray paths. This problem, known as the missing horizontal rays, can be overcome by the data collected by the new radio-occultation satellites. The simultaneous inversion of both satellite-to-ground and satellite-to-satellite GPS data in a single algorithm (MIDAS) provides a greatly improved geometry in comparison to that found when using either data set independently. Experimental results demonstrate the inversion imaging when radio-occultation data are included in the inversion. The potential benefit in using CHAMP data in such as system is discussed.

Key words: Ionosphere, tomography, imaging, GPS/MET

1 Introduction

The tomographic technique, so successful in the medical field, has been applied relatively recently to producing images of the electron concentration in the ionosphere. In conventional ionospheric tomography dual-frequency radio signals, transmitted from a polar-orbiting satellite are received by a chain of ground-based receivers. The refractive effect of the ionosphere on the two trans-ionospheric radio waves results in a phase difference between them. This phase difference can then be related to the integrated amount of ionization, or electron concentration, present along the propagation path. Obviously the individual TEC measurements contain no information about the spatial variation of the electron concentration along the ray path. Hence the task of finding the spatial distribution of electron concentration requires a tomographic solution.

Many ionospheric phenomena have already been imaged using conventional tomography. Images showing travelling ionospheric disturbances (TIDs) (*Pryse et al.* [1995]) have been presented. These waves-like structures are the manifestation of internal atmospheric gravity waves in the ionosphere. *Mitchell et al.* [1995] presented tomographic images of magnetic-field-aligned irregularities and E-region enhancements in the auroral region above northern Scandinavia. *Mitchell et al.* [1995] demonstrated that the technique could be used to make images of large-scale ionization depletions known as troughs, generally found on the night-side auroral mid-latitude boundary. Results from the polar regions have indicated ionospheric signatures of processes occurring further out in space, such as magnetic reconnection events *Walker et al.* [1998]. A review of the subject is given by *Bernhardt et al.* [1998].

This paper shows the extensional of tomographic (two dimensions) to multi-dimensional imaging. Sections 2 and 3 contain a description of the concepts behind this type of imaging and show how the two-dimensional inversion can easily be extended to encompass 3-D space and time. Example results from winter 1997 using ground-based GPS and GPS/MET data into a series of one-hour inversions are displayed in Section 4.

2 Tomographic Imaging

Tomography involves imaging a parameter in two dimensions using measurements taken along a large number of intersecting paths. Each separate view of the object along the different paths provides information in a specific direction and is known as a projection. These one-dimensional projections are combined using a reconstruction algorithm to obtain a tomographic image of the parameter. Initial tomographic measurements may be from diffracting or non-diffracting sources. If the projections are from a non-diffracting source the imaging is called ray tomography. In this case the projections can be considered to be line integrals of some parameter along paths through the object.

To facilitate an understanding of tomography it is instructive to consider the following simple problem. A square grid contains four unknown numbers (Figure 1a). Each element of the grid has sides of unit length and is known as a pixel. The summation of the numbers along particular lines (projections or ray paths) passing through the pixels are measured. These projections constitute a set of tomographic observations.

Fig. 1. a) Two-by-two grid providing a simple tomographic illustration, b) Two-by-two grid with indexed projections, i, and pixels, j.

The problem is to find the numbers in the grid using only the information gained from the projections. For example, if each of the numbers within the grid were equal to one, then the summation directly along any row or column would be two. For ray paths which cross the grid at an oblique angle it becomes necessary to utilize the amount by which each ray path crosses each grid element. Hence the summation along the diagonal of the grid of Figure 1 would be $\sqrt{2}+\sqrt{2}$. By simple geometry the summation of numbers along any arbitrary ray path can be calculated (for example for projection entering the grid at an angle of 22.5° to the horizontal = 2.4). From knowledge of the locations of the start and end point of each projection (from the locations of the transmitter and receiver), and the tomographic measurements the problem can therefore be formulated as follows.

Each projections is given an index, *i* and each pixel an index *j* (Figure 1b). The measurements are given the symbol y_i and the known length of each projection (*i*) through each pixel (*j*) the symbol Δ_{ij}. The unknown number in each pixel is designated x_j. Hence the information provided in the form of measurements a, b and c can be re-written in a manner that attributes the total measurement to the pixels in accordance with the path lengths. For the first ray path this would be written as,

$$y_1 = \Delta_{11}x_1 + \Delta_{12}x_2 \quad [1]$$

The task is to find the value of each unknown, x_j, given the set of measurements, y_i.

Two further points can be drawn from this illustration. Firstly, for a unique mathematical solution sufficient measurements must be made. Clearly if only one measurement was taken, Equation 1 could be satisfied with many different solutions for x_1 and x_2. Secondly, the measurement set should cross each of the pixels and project over the full range of viewing angles. For the illustration an example of an ideal projection set would be two horizontal measurements and two vertical measurements, thus ray paths would cross each pixel twice. In many practical applications of tomography it is not possible to satisfy these conditions and it becomes necessary to constrain the solution with other information, such as physical models or a-priori knowledge.

3 Multi-Dimensional Inversion

Conventional ionospheric tomography implements a two-dimensional inversion from a single polar-orbiting satellite. During the over-fly of this satellite (typically taking 20 minutes) the ionisation distribution is assumed to be stationary. At the start of GPS being used for ionospheric research it had been thought that these new navigation satellites could not be used for ionospheric imaging because the satellite orbits are high and so the transit time is many hours – too long to image a moving ionosphere. However, multiple GPS satellites are in view at any time MIDAS (Multi-Instrument Data Analysis System) exploits this fact to allow a fully four-dimensional determination of the ionisation distribution so that the time evolution of three-dimensional structures can be studied. The inversion thus results in a 3-D 'movie' rather than the static 2D slice produced by a tomographic inversion. This obviously has great potential as a scientific tool.

Dual frequency radio signals that propagate through the ionosphere are subject to a differential phase change due to the dispersive nature of the medium. Provided lock is maintained on both frequencies the change in the differential phase shift is directly proportional to the change in integrated total electron content (TEC) between the transmitter and receiver. In the case of GPS, since the actual satellite-to-receiver ionospheric path is constantly changing, this change in TEC is a composite of spatial and temporal effects. In addition to the relative phase differences the GPS satellites also provide P-code observations (time delay) enabling an estimate of the absolute TEC along the ray path. However, these time-delay estimates are subject to transmitter and receiver clock biases and to the effects of multi-path. The current implementation of MIDAS allows for both types of observations (differential phase and differential time) to be included in the inversion.

One of the main limitations to the accuracy in tomographic imaging with ground-based instrumentation is the geometrical bias imposed by the satellite-to-ground geometry. *Hajj et al.* [1994] proposed the use of GPS-to-LEO data to overcome this problem. Figure 2 shows an impression of the ray path geometry that is possible with ground-based receivers and space-based receivers. It is obvious that the problem is unlikely to be solvable as a two-dimensional inversion. Moreover, considering the time-differences involved when coordinating many different instruments, and the non-stationary nature of the ionization it becomes clear that such a measurement set can be solved optimally in a four-dimensional imaging system.

The first stage in the inversion problem is to set up a grid of three-dimensional voxels in a spherical volume and compute the elemental contributions from each of the satellite-to-receiver ray path integrals. Defining the electron concentration within each voxel as x, the problem may be expressed as,

$$Ax = b \qquad [2]$$

where A is a sparse matrix of the path lengths within each voxel and b are the observed line integrals (TEC). The matrix A is highly singular and incorporates no prior information as to the likely solution. To overcome this difficulty a mapping matrix, M, is used to transform the problem to one for which the unknowns are a set of appropriately selected ortho-normal basis functions. This is expressed mathematically as,

$$AMX = b \qquad [3]$$

where the matrix M defines the mapping from a voxel based representation to an ortho-normal representation of the reconstruction volume. Since the ray-paths integrals derived from phase are subject to an unknown cycle offset adjacent rows of the matrices AM and b are differenced so as to negate the effect of this on the solution. The unknowns, X, now represent the relative contribution of the basis lution. The unknowns, X, now represent the relative contribution of the basis functions where,

$$X = (AM)^{-1} b \qquad [4]$$

which is solved for the least-squares estimate using Singular Value Decomposition. Finally, the electron densities within each voxel are recovered using,

$$x = MX \qquad [5]$$

4 Example Results

In this Section results are presented from a case study of the ionosphere over the USA during February 1997. The geomagnetic conditions were fairly quiet, with Kp values between 0 and 4. This time period was chosen due to the availability of GPS/MET data towards the end of the satellite's life, when the ground-based IGS data were becoming more extensive. Figure 3 shows the network of ground-based receivers that were used for the inversions. Each inversion used data collected over a one-hour period and resulted in a movie (with a frame every thirty seconds) of the 3-D electron concentration distribution. Figure 4 shows a single frame from the 1600 to 1700 UT movie. The image shows the vertical TEC obtained by vertical integration through the 3-D grid frame at 1630 UT.

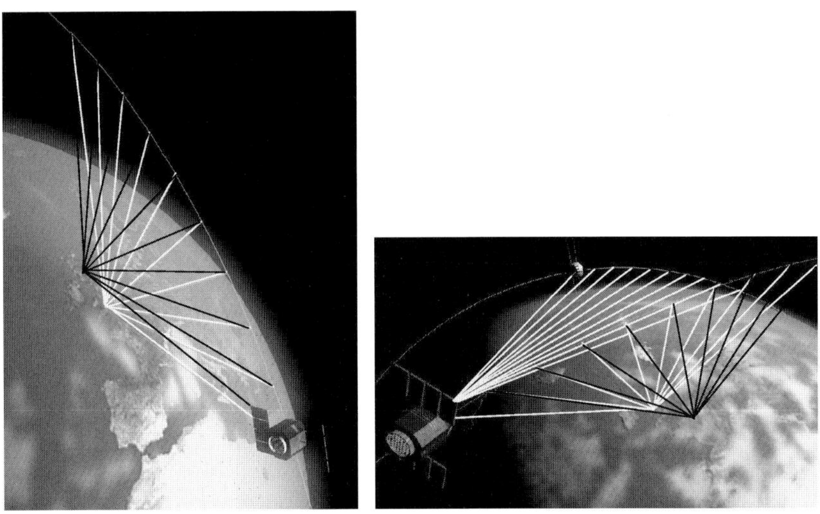

Fig. 2. Diagrams illustrating the geometrical biases involved in the special case of conventional TRANSIT-to-ground (left) and the more general GPS to LEO and LEO-to-ground (right). The orientations of the projections from radio occultation complement those from satellite-to-ground, but do not provide a perfect geometrical coverage for a fully determined tomographic inversion.

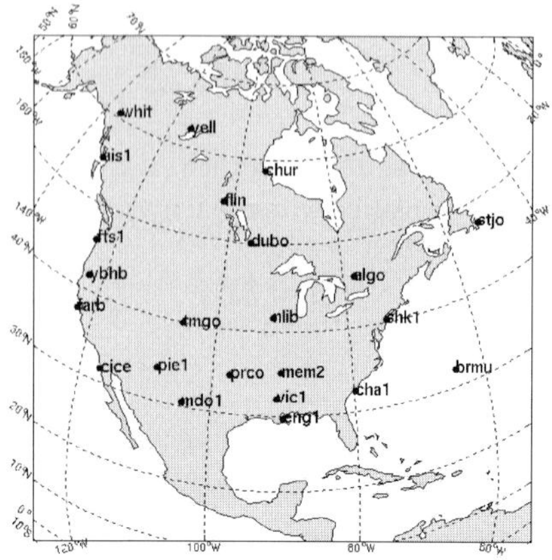

Fig. 3. Plot of IGS receivers used for the 1997 case study.

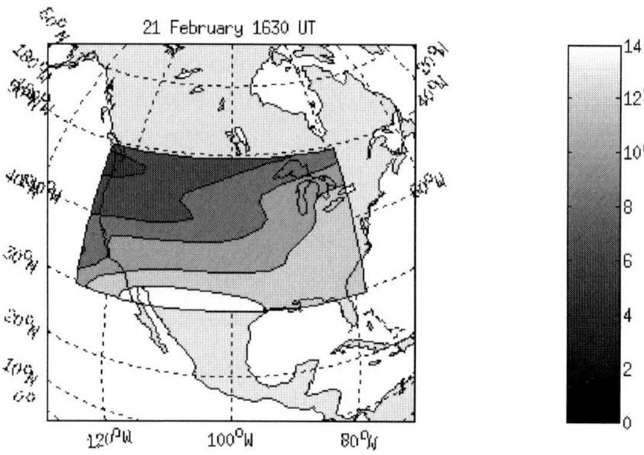

Fig. 4. Vertical TEC at 1630 UT on 21 February 1997 obtained from integration through the 3-D image structure at the mid point time of the one-hour immersion

Figure 5 shows a series of four cross sections of electron concentration taken from the 3-D movies. Each cross section is for a longitude of 98°W and extends from 30 to 58°N. The daytime peak height is typically 270 km, whereas the night-time one is higher between 300 km and 400 km. The scale-heights are also larger at night. Complete results from this study have been submitted to the journal *Radio Science.* In this more extensive paper quantitative results confirm the improvements to the imaging with the inclusion of the GPS/MET data.

Summary

The technique of tomographic imaging has been introduced and the extension of conventional ionospheric tomography into four-dimensional imaging has been

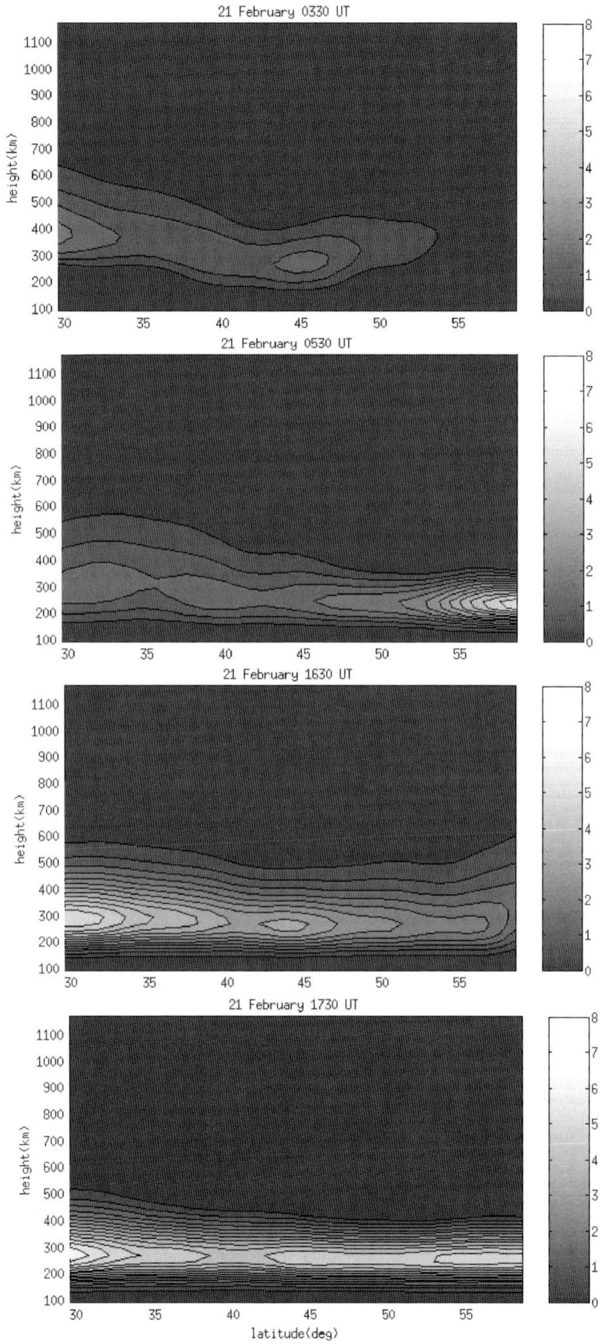

Fig. 5. Cross-sections of electron concentration $(x10^{11}m^{-3})$ for 21 February 1997. The diurnal variation in peak height and scale height can be clearly seen.

Measurement	Advantages
Satellite-to- ground	Data easily available Horizontal resolution
Satellite-to- satellite	Vertical resolution Global coverage including oceans and poles Topside and plasmasphere
Sea-reflecting radar	Oceans covered Horizontal gradients
Ionosonde	Vertical resolution
***In situ* measurement**	Non-integrated (anchor-point for the inversion)

Table 1. Types of ionospheric measurements that can be put into the MIDAS inversion software and their strengths.

shown. Conventional techniques have involved the independent analysis of either (i) satellite-to-satellite radio-occultation observations or (ii) satellite-to-ground GPS. However satellite-to-ground GPS observations demonstrate a good horizontal resolution capable of imaging the main trough but have limited vertical resolution. Satellite-to-satellite radio-occultation observations demonstrate good vertical resolution but have poor horizontal resolution. Consequently, the simultaneous inversion of both satellite-to-ground and satellite-to-satellite GPS data in a single algorithm provides a greatly improved geometry in comparison to that found when using either data set independently (Mitchell and Spencer, 2002). Table 1 summarizes the different types of measurements that can be put into the MIDAS inversion software and details their strengths. It can be seen that the satellite-to-satellite measurements, as provided by the GPS to CHAMP radio occultation and navigation data are extremely valuable. The planned launch of many new occultation satellites over the next decade offers great promise for imaging the electron concentration in the near-Earth space for practical application and scientific study.

References

Bernhardt PA, McCoy RP, Dymond KF, Picone JM, Meier RR, Kamalabadi F, Cotton DM, Charkrabarti S, Cook TA, Vickers JS, Stephan AW, Kersley L, Pryse SE, Walker IK, Mitchell CN, Straus PR, Na H, Biswas C, Bust GS, Kronschnabl GR, and Raymund TD (1998) Two-dimensional mapping of the plasma density in the upper atmosphere with computerized ionospheric tomography (CIT), *Physics of Plasmas*, *5* (5), 2010-2021.

Hajj GA, Ibanezmeier R, Kursinski ER, and Romans LJ (1994) Imaging the Ionosphere With the Global Positioning System, *International Journal of Imaging Systems and Technology*, *5* (2), 174-&.

Mitchell CN, Jones DG, Kersley L, Pryse SE, and Walker IK (1995) Imaging of field-aligned structures in the auroral ionosphere, *Annales Geophysicae-Atmospheres Hydrospheres and Space Sciences*, *13* (12), 1311-1319.

Mitchell CN and Spencer PSJ (2002) Development of tomographic techniques for large-scale ionospheric imaging, Proc Ionospheric Effects Symposium.

Pryse SE, Mitchell CN, Heaton JAT, and Kersley L (1995) Travelling ionospheric disturbances imaged by tomographic techniques, *Annales Geophysicae-Atmospheres Hydrospheres and Space Sciences*, *13* (12), 1325-1330.

Walker IK, Moen J, Mitchell CN, Kersley L, and Sandholt PE (1998) Ionospheric effects of magnetopause reconnection observed using ionospheric tomography, *Geophysical Research Letters*, *25* (3), 293-296.

Acknowledgements

The MIDAS inversion program (© University of Bath) was developed under a grant awarded by the UK Engineering and Physical Sciences Research Council.
The evaluation of the MIDAS inversion over USA was supported by the USAF EOARD.
The data were provided by:

- IGS Service ground-based GPS data.
- UCAR, USA GPS/MET LEO radio occultation data.

termine parameters of focused synthetic aperture and magnitude of spatial resolution.

Sliding aperture method described by *Gorbunov, 2001, section 1.3,* is equivalent to Doppler selection. In this case spatial resolution is of about 0.5-1 km and is limited by uncertainty relation (*Gorbunov, et al., 2000*). *Igarashi et al., 2000, 2001* used exponential model of the refractivity in the atmosphere and IRI-95 model for two parts of the ionosphere in the RO region to calculate temporal dependence of $\psi_m(t)$ and obtained spatial resolution of about 20-70 m. They applied Fourier transform to product of RO and reference signal to obtain the angular spectrum $A(\beta(\omega),p(\omega))$ of RO signal:

$$A(\beta(\omega),p(\omega))=\int_{-T/2}^{T/2} dt E(t)E_m(t)\exp(-\omega t);\sin\beta=\omega/(kv)+\sin\beta_m;p=\omega R_1/(kv)+p_m; \quad (2)$$

where β is the angle between directions PO and PG at the point P (Fig. 1), β_m, p_m are the angle β and the impact parameters p corresponding to trajectory of reference signal at the point P (Fig. 1), T is the time of coherent data handling, $k=2\pi/\lambda;\lambda$ is the wavelength, $v=R_1 d\theta/dt$. Equations (2) give presentation of radio field in ray coordinates β, p. The second and third equations (2) allow independent measuring the functions p and $\xi(p)$ corresponding to the main ray of RO signal and to each ray in the angular spectrum $A(\beta,p)$ (*Hocke et al., 1999, Igarashi et al., 2000, Pavelyev et al., 2002*). Vertical resolution of the focused synthetic aperture method $\Delta h\approx\lambda R_1/(2vT)$ (typically T ~ 2-3 sec) is essentially higher than vertical resolution corresponding to the sliding aperture method. Note that canonical transform introduced by *Gorbunov, 2001,* as indicated his analysis, gives also a possibility of measurements of the refraction angle in multibeam areas at a high resolution. Usually in the lower ionosphere the electron density is not high and the single ray propagation prevails. In the case of single ray propagation, the amplitude and phase of RO signal may be considered as two independent information channels of radioholograms at frequencies F1 and F2. The phase channel is usually used for estimation of the vertical profiles of the refractivity on Doppler frequency displacement (e.g. *Gorbunov et al., 1996, Feng and German, 1999, Liou and Huang, 2000*). *Pavelyev et al., (1986)* found exact solution of inverse problem for amplitude part of RO signal. Simplified form of this solution is described by:

$$p(t)-p(t_o)=\int_{t_o}^{t} dt\, X[t(p_s)]dp_s/dt\ ;\ p_m(t)-p_m(t_o)=\int_{t_o}^{t} dt\, X_m[t(p_s)]dp_s/dt, \quad (3)$$

where X_m, X are the power attenuation relative to free space owing to refraction effect for model and experimental data. Equation (3) allows finding the temporal dependence of impact parameter p(t) from amplitude data if initial condition is given. The power attenuation X(t)depends on the derivative $d\xi/dp$:

$$d\xi/dp=[1-1/X(t)]\,B(p_s);\ B(p)=(R_1^2-p^2)^{-1/2}+(R_2^2-p^2)^{-1/2}. \quad (4)$$

Thus the amplitude data can be used for restoration of the functions p(t), dξ/dp(t), and the refractivity gradient $N'_h(h)$

$$J(p)=1/\pi\int_p^{\infty} d\xi'_x(x)/dx(x^2-p^2)^{1/2}dx; \quad N'_h(h)=-n^2(h)J(p)/\{p[1+J(p)]. \qquad (5)$$

The amplitude information may be used for retrieving the vertical gradient of the temperature. As shown by *Pavelyev et al., 2001,* for the case of wet atmosphere a connection exists between the vertical temperature and refractivity gradients:

$$[dT_w(h)/dh]/T_w(h)=-[N(h)]^{-1}dN(h)/dh-T_x/T_a(h), \qquad (6)$$

where T(h) is the temperature of the atmosphere expressed in Kelvin's degrees. At the height above 10 km (6) may be used for finding vertical gradient of temperature if the refractivity gradient is known. It may be noted that the way of restoration of the vertical temperature and electron density profiles from amplitude data, which in some details is similar to our approach, has been considered also by *Vorob'ev et al., 1999,* and *Sokolovskiy, 2000.* In distinction we use model for control of multibeam and diffraction effects and apply directly our exact solution of inverse problem.

3 Vertical Gradients in the Mesosphere and Atmosphere

Derived in section 2 approach has been used for analysis of amplitude RO data. Experimental data are shown by the curves A1, A2 in Fig. 2 (left panel) for GPS/MET occultation event 0393, February 14, 1997. Amplitudes corresponding to reference signal at two frequencies F1, F2 are described by smooth curves M1 and M2 in Fig. 2. Amplitudes M1, M2 has been found by numerical ray tracing from model of electron density $N_m(h)$ and its gradient $dN_m(h)/dh$ shown in Fig. 2 (right panel) by curves 2 and 4. Curves A2, M2 are displaced for comparison. There is satisfactory correspondence between amplitudes of reference signal and experimental data. Features at levels 92 and 105 km corresponding to sporadic E-layers are clearly seen in the experimental and computed data. Retrieved by method described in section 2 vertical gradient $dN_e(h)/dh$ (curve 5), electron density $N_e(h)$ (curve 1) and difference $dN_e(h)/dh-dN_m(h)/dh$ (curve 3) are shown in Fig. 2 (right panel). The highest positive maximums of $dN_e(h)dh$ $48\cdot10^9$ and $43\cdot10^9$ $[m^{-3}km^{-1}]$ are seen at heights 92 and 105 km. Corresponding two maximums in the electron density are of about 25-20$\cdot10^9$ $[1/m^3]$ at levels 94 and 106 km. Also secondary maximums are seen in the Fig. 2 (right panel). Results shown in Fig. 2 illustrate sensitivity of amplitude channels of RO signal to variations of the vertical gradients of the electron density in sporadic E-layer of the ionosphere. Another example of amplitude variations that corresponds to GPS/MET occultation event June 19, 1995, No. 0583 is shown in Fig. 3 (curves A1, A2, left panel).

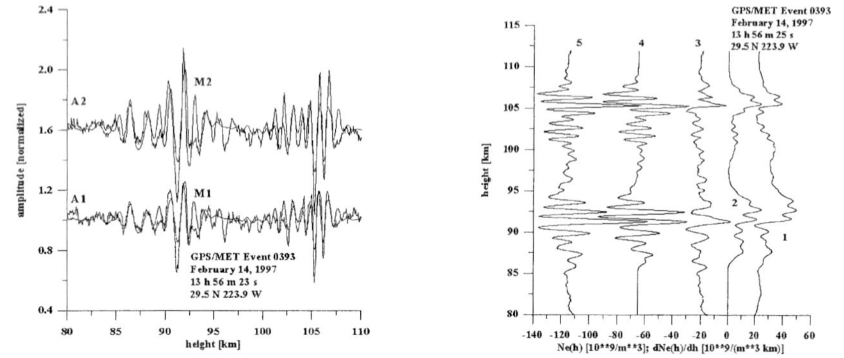

Fig. 2. Left panel. Comparison of amplitude variations A1, A2 at two frequencies with amplitudes of reference signal (smooth curves M2, M1). Right panel. Results of restoration of vertical distribution of the electron density $N_e(h)$ and its gradient $dN_e(h)/dh$. Curves 1, 3, 4, 5 are displaced by 20 $[10^9/m^3]$; –20; –65; –115 $[10^9/m^3 km^{-1}]$, respectively.

Smooth curves M in Fig. 3 show amplitude of reference signals calculated by ray tracing from models of $N_m(h)$, $dN_m(h)/dh$ (curves 2, 4, right panel). Amplitude variations A1, A2 are strongly correlated, their level is proportional to the ratio f_2^2/f_1^2 demonstrating their ionospheric origin. In the amplitude data influence of sporadic E-layer structure can be seen at the heights 90-94 km and 98-100 km. Retrieved vertical profiles $N_e(H)$ and its gradient $dN_e(H)/dH$ are shown in Fig. 3, right panel, by curves 1, 4, respectively. The maximum value of the negative gradient of about $-46{\cdot}10^9$ $[m^{-3}km^{-1}]$ is located at height 94.5 km (curve 4 in Fig. 3). The vertical profile $N_e(h)$ (curve 1 in Fig. 3) coincides with its gradient $dN_e(h)/dh$ (curve 4) showing feature at 94 km with maximum of about $50{\cdot}10^9$ $[m^{-3}]$.

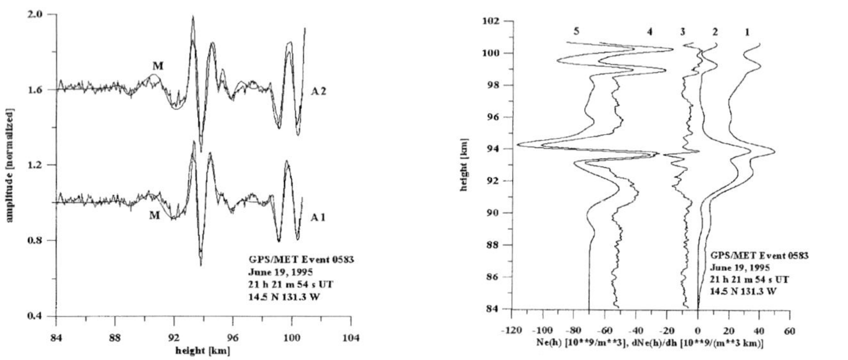

Fig. 3. Left panel. Comparison of amplitude variations A1, A2 at two frequencies with amplitudes of reference signal M (smooth curves M). Right panel. Results of restoration of vertical distribution of the electron density $N_{em}(h)$ and its gradient $dN_e(h)/dh$. Curves 3, 4, 5 are displaced by –10; –55; –70 $[10^9/m^3 km^{-1}]$ respectively. Curve 3 represents difference $dN_e/dh - dN_m/dh$.

Analysis of information containing in the amplitude channels of RO signal gives detailed picture of vertical gradient distribution in the lower ionosphere. This is important because Earth-based tools usually give vertical profiles of the electron density distribution below their maximum. Vertical gradients of refractivity may be studied in the atmosphere also using amplitude channels of RO signal. For example amplitude variations at frequency F1 are shown in Fig. 4 (curve A) for CHAMP RO event 09 (April 06, 2001). Normalized amplitude variations shown in Fig. 4 are scaled by 20 and displaced by 10 for comparison with retrieved vertical gradient of the temperature dT/dh (curve B) and refractivity dN/dh (curves C, D). Smooth curve A in Fig. 4 corresponds to amplitude changes owing to standard vertical distribution of the refractivity N(h)=332exp(-h/8) [N-unit]. Curve A accounts for the absorption in oxygen at small height also. As seen from Fig. 4 the regular refraction attenuation (with absorption) changes from 0.3 (at height 3 km) up to 1 (at height 40 km). Curve A, in average, gives good approximation to experimental amplitude variations. Measured amplitude variations are intense in the height interval 3-18 km. Quasi-periodical structures are seen in amplitude variations between levels 10 and 25 km. Above 25 km the amplitude variations are smaller. Curve D in Fig. 4 shows variations in the vertical gradients in the height interval 3-37 km. Curve C is scaled version of the curve D. Curve C shows small changes of the vertical gradient of refractivity at the heights 12 to 37 km. At the

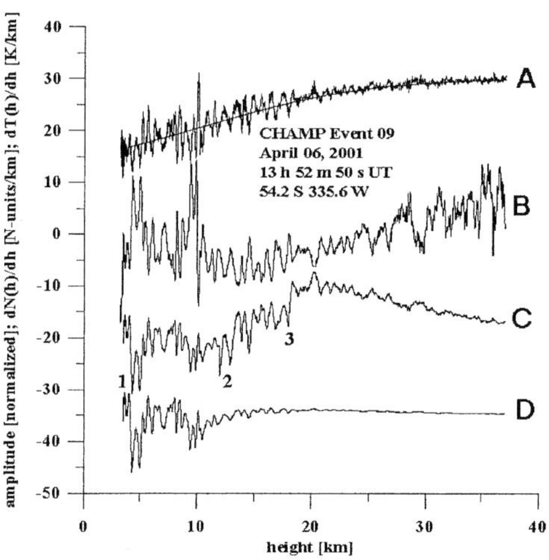

Fig. 4. Comparing amplitude variations at the first CHAMP frequency (curve A), retrieved vertical gradient of refractivity dN(h)/dh $-G_s(h)-20$; dN(h)/dh $-G_s(h)-35$ [N-units/km] (curves C, D) and temperature dT(h)/dh (curve B). Curve C is scaled by 5 times in interval 2-3 (heights between 12 and 18 km), and by 10 times at heights above 18 km. Scaling is absent in the interval 1-2 (heights between 3 and 12 km). Vertical gradient $G_s(h)$ corresponds to model A used for calculation of refraction attenuation: $G_s(h)=-42\ \exp(-h/8)$ [N-units/km].

height levels between 3-11 km variations of the vertical gradient of refractivity are about of ±5 N-units/km. In the tropopause area (heights between 12-19 km) the variations of vertical gradient dN/dh are about ±2 N-units/km. Vertical distribution of temperature gradient between level 3-37 km (curve B) reveals features at height 4-6, 9-10 km with positive values of about 6-9 ^{o}K/km. Vertical distribution of temperature gradient reveals waves structures with spatial period 0.8-4 km. These variations may be connected with propagation of internal waves in the atmosphere.

4 Conclusion

It is shown that amplitude channel of RO signals may be used independently from phase channel for detailed retrieving the vertical profiles of the refractivity gradient in the atmosphere and ionosphere. Amplitude channel is less sensitive to upper ionosphere contribution than phase channel. The first results of measurements of refractivity gradients in the equatorial atmosphere revealed layered structure below tropopause in the troposphere. Fine structures in the vertical gradient of electron density have been retrieved from RO amplitude channels in the sporadic E-layers (heights interval 92-105 km). This demonstrates that amplitude method may be used for high-precision measurements of atmospheric and ionospheric parameters in existing CHAMP and future COSMIC RO mission.

Acknowledgements. We are grateful to UCAR for access to the GPS/MET data. We are grateful to National Science Council of Taiwan, R.O.C., for financial support under the grants NSC 91-2811-M008-001, NSC 90-2111-M008-047-AGC, and Office of Naval Research (ONR) of USA under grant N00014-00-0528. Work has been partly supported by Russian Fund of Basic Research, grant No. 01-02-17649.

References

Feng DD and Herman BM (1999) Remotely sensing the Earth's atmosphere using the Global Positioning System (GPS) - the GPS/MET data analysis. *Journal of Atmospheric and Ocean Technology* **16**, 990-1002.

Gorbunov ME, Gurvich AS, Bengtsson L (1996) Advanced algorithms of inversion of GPS=MET satellite data and their application to reconstruction of temperature and humidity. Report 211, Max-Planck-Institute for Meteorology ISSN0937-1060.

Gorbunov ME (2001) Radio holographic methods for processing radio occultation data in multipath regions. *Scientific report 01-02*. Danish Meteorological Institute (DMI), Copenhagen.

Gorbunov ME, Gurvich AS, and Shmakov AV (2002) Back-propagation and radio-holographic methods for investigation of sporadic ionospheric E-layers from Microlab-1 data. *Int J Remote Sensing,* **23**, No 4, 675-685.

Gorbunov ME, Gurvich AS, and Kornblueh L (2000) Comparative analysis of radio-holographic methods of processing radio occultation data, *Radio science,*.**35**, No 4, 1025-1034.

Hocke K, Pavelyev A, Yakovlev O, Barthes L and Jakowski N (1999) Radio occultation data analysis by radio holographic method. *JASTP,* **61**, P 1169-1177.

Igarashi K, Pavelyev A, Hocke K, Pavelyev D, and Wickert J (2001) Observation of wave structures in the upper atmosphere by means of radio holographic analysis of the radio occultation data. *Advances in Space Research*, **27**, No 6-7, 1321-1327.

Igarashi K, Pavelyev A, Wickert J, Hocke K, Pavelyev D (2002) Application of radio holographic method for observation of altitude variations of the electron density in the mesosphere/lower thermosphere using GPS/MET radio occultation data. *JASTP,* **64**, in press.

Igarashi K, Pavelyev A, Hocke K, Pavelyev D, Kucherjavenkov I, Matugov S, Zakharov A, and Yakovlev O (2000) Radio holographic principle for observing natural processes in the atmosphere and retrieving meteorological parameters from radio occultation data. *Earth Planets Space*, **52**, 968-875.

Karayel ET, and Hinson DP (1997) Sub-Fresnel-scale vertical resolution in atmospheric profiles from radio occultation. *Radio science,*.**32**, No 2, 411-428.

Marouf EA, and Tyler GL (1982) Microwave edge diffraction by features in Saturn's rings: Observations with Voyager 1. *Science* **217**, 243-245.

Lindal GF., Lyons JR, Sweetnam DN, Eshleman VR, Hinson DP, and Tyler GL (1987) The atmosphere of Uranus: Results of radio occultation measurements with Voyager 2, *J Geophys Res,* **92** (A13), 14,987--15,001.

Pavelyev A, Yakovlev OI, Matyugov SS, Pavelyev D, Anufriev VA (2001) Advanced algorithms of inversion of GPS radio limb sounding data. Scientific Technical Report STR01/06. GeoForschungZentrum Potsdam.*and Electronics*, **43**, No. 8, 126-131, 1998.

Pavelyev A, Matyugov S, Kalashnikov I, and Yakovlev O (1986) Analysis of the features of radio occultation method for the Earth's atmosphere study. In the book *Electromagnetic waves in the atmosphere and space*, pp 208-218. "Nauka" Ed Moscow, 1986 (in Russian).

Pavelyev A, Igarashi K, Reigber Ch, Hocke K, Wickert J, Beyerle G, Matugov S, Kucherjavenkov S, Pavelyev D, and Yakovlev OI (2002) First application of the radio holographic method to wave observation in the upper atmosphere. *Radio science,* **37**, in press.

Sokolovskiy SV (2000) Inversion of radio occultation amplitude data, *Radio Science*, **35**(1), 97-105.

Sokolovskiy SV, Schreiner W, Rocken C, Hunt D (2002) Detection of high-altitude ionospheric irregularities with GPS/MET, *Geophys Res Letters*, **29**(3), 621-625.

Vorob'ev VV, Gurvich AS, Kan V, Sokolovskiy SV, Fedorova OV, Shmakov AV (1999) Structure of the ionosphere from the radio-occultation GPS-"Microlab-1" satellite data: preliminary results. *Earth Observations and Remote Sensing* **15**, 609–622.

Yuei-An Liou, and Cheng-Yung Huang (2000) Active Limb Sounding of Atmospheric Refractivity, Pressure, and Temperature Profiles from GPS Occultation by 3-D Vector Analysis. *Proceedings of COSMIC International Workshop. September 27-29, 2000,* pp 60-69. Taipei, Taiwan.

Observation of Reflected Signals in MIR/GEO and GPS/MET Radio Occultation Missions

Dmitrii Pavelyev[1], Jens Wickert[2], Alexander Pavelyev[1(✉)], Igor Kucherjavenkov[1], Stanislav Matyugov[1], Klemens Hocke[3], Georg Beyerle[2], Oleg Yakovlev[1], Anatolii Kucherjavenkov[1]

[1]Institute of Radio Engineering and Electronics of the Russian Academy of Sciences, (IRE RAS), Fryazino, Vvedenskogo sq. 1, 141191 Moscow region, Russia, *agp117@ire216.msk.su*
[2]GeoForschungsZentrum Potsdam (GFZ-Potsdam), Telegrafenberg, 14473 Potsdam Germany, *wickert@gfz-potsdam.de*
[3]Communication Research Laboratory, Independent Administrative Institute, 4-2-1, Nukui-Kita Machi, Koganei-shi, Tokyo 184-8795 Japan, *hocke@crl.go.jp*

Summary. Recovering reflected signals in MIR/GEO (wavelength 2 and 32 cm), GPS/MET and CHAMP (wavelength 19 and 24 cm) radio occultation experiments opened new perspectives for bistatic monitoring of the Earth at small elevation angles. Experimental observation of the propagation effects at low elevation angles has also principal importance for fundamental theory of radio waves propagation along the Earth' surface. The aim of this contribution consists of a short review of parameters of reflected signals, observed in MIR/GEO and GPS/MET experiments at wavelength 2, 32, 19 and 24 cm and in consideration of effect of boundary layer on parameters of reflected signal at low elevation angles.

Key words: Reflection, radio occultation, radioholography, atmosphere, refraction

1 Introduction

Successful observation of reflected signal in MIR/GEO, GPS/MET and CHAMP radio occultation (RO) data opens new direction in monitoring terrestrial surface from space. The first observations of signals reflected from terrestrial surface have been provided in MIR/GEO RO experiments at wavelength 32 cm [Rubashkin et al., 1993, Pavelyev et al. 1996, 1997]. Recently Beyerle and Hocke, 2001, Igarashi et al., 2001, demonstrated evidences of existence of reflected signals in GPS/MET RO data. Beyerle et al., 2002, provided analysis of reflected signals in CHAMP RO data. Reflected signals in GPS/MET and CHAMP RO data have been discovered by radio holographic method elaborated for analysis of RO signals by Pavelyev, 1998, Hocke et al., 1999, Igarashi et al., 2000. In future, combination of GPS RO and new area of GPS reflection may provide a simple measurement tool for global observation of the interaction of ocean-atmosphere-ionosphere [Pavelyev et al., 1996, Komjathy et al., 1999; Beyerle and Hocke, 2001, Beyerle et al., 2002] because CHAMP can perform continuously more than 200 occultations daily [Wickert et al., 2001].

In this contribution results of observation of reflected signals at wavelength 2, 32 cm (MIR/GEO mission) and 19, 24 cm (GPS/MET mission) will be presented and influence of the atmospheric boundary layer on parameters of reflected signal would be described.

2 Reflected Signals in MIR/GEO Radio Occultation Experiments

Scheme of RO experiments is shown in Fig.1 (left panel). MIR/GEO team used a transmitter installed on the orbital station "MIR" (point T) and receivers installed on GEO satellites (point G) at λ=32 and 2 cm (monochromatic signals at both frequencies) [*Yakovlev et al., 1995*]. RO signals have been transmitted from GEO satellites to Earth-based station for analysis. Open-loop phase tracking method was used for removing Doppler effect due to orbital motion of the satellite and expected atmospheric influence. Then spectral method was applied for revealing reflected signal in the RO data. The Doppler spectra of RO signal reflected from the surface of Indian Ocean are shown in Fig. 1 (right panel, wavelength λ= 32 cm). The Doppler frequency (expressed in Hz) is shown at horizontal axis, the spectral intensity of reflected signal (normalized to the spectral intensity of direct signal) is given on the vertical axis. The data are shown in Fig. 1 (right panel) in succession with time interval between neighboring spectra equal to 1/3 sec. As follows from Fig. 1 the distribution of spectral power density of reflected signal depends on time and may be connected with features in long-scale slope variations of the sea surface and physical properties of boundary layer of the atmosphere. Reflected signal has been observed also at wavelength 2 cm during MIR/GEO experiment (RO event 28 May, 1998). Time-Doppler frequency story of signal reflected from

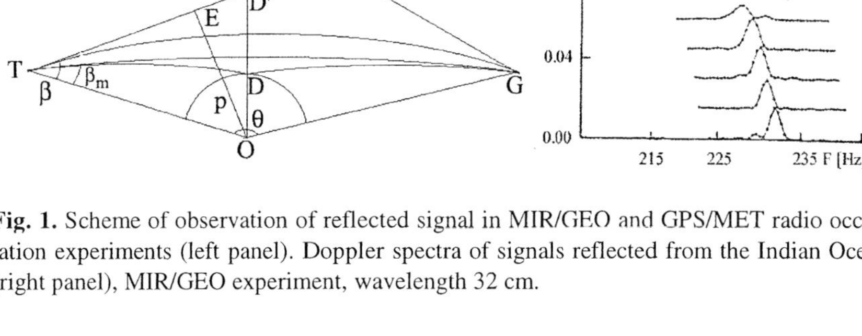

Fig. 1. Scheme of observation of reflected signal in MIR/GEO and GPS/MET radio occultation experiments (left panel). Doppler spectra of signals reflected from the Indian Ocean (right panel), MIR/GEO experiment, wavelength 32 cm.

the Aral Sea (MIR/GEO data, wavelength 2 cm) is shown in Fig. 2. Weak reflected signal (sharp strip above, near frequency 20 KHz) is observed together with more intense direct signal (broad bright strip below near frequency 25 KHz). Difference of Doppler frequencies of reflected and direct signal gradually diminishes as time increases. Also variations in the amplitude of reflected signal are seen in Fig. 2. These variations may be connected with changes of the surface roughness and horizontal inhomogeneities of the atmosphere. Reflected signal disappeared after departure of specular point from Aral Sea because of the roughness of the land surface. Time–Doppler frequency story and temporal amplitude dependence of reflected signal contain important information on the structure of boundary layer of the atmosphere.

3 Influence of Boundary Layer on Parameters of reflected Signal in GPS/MET and MIR/GEO Radio Occultation Experiments

In Fig. 3 (right panel) the dependence of Doppler frequencies difference Δf(Hz) of reflected and direct signal is shown as function of time t for RO event December 25, 1990. The RO region located above the Indian Ocean at 31.2° South latitude and 68.2° East longitude. Curve 1 describes experimental data, curves 2,3 correspond to two models of Δf(t). The first model describes the function Δf(t) for the

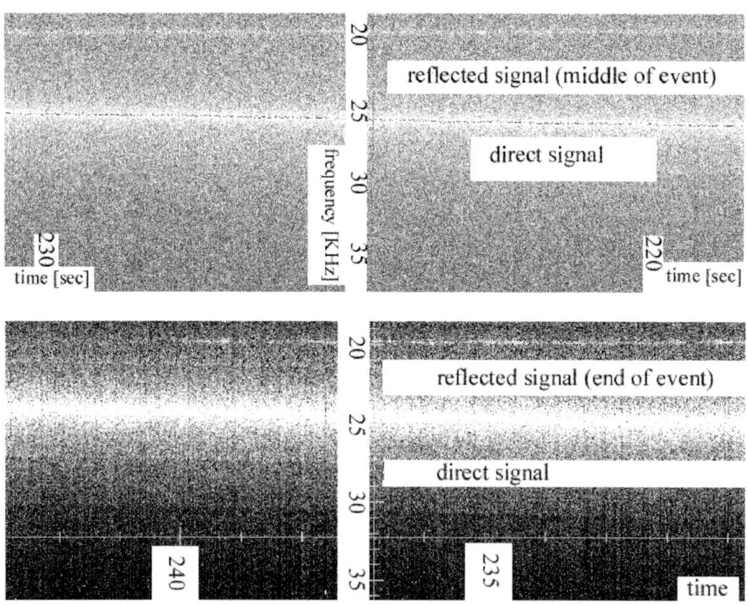

Fig. 2. Time-Doppler frequency story of signal reflected from the Aral Sea (MIR/GEO data,wavelength 2 cm).

case of free space propagation (curve 2). The second model shows influence of real atmosphere with parameters of boundary layer equal to N_o = 350 N-units, dN/dh=–36.8 (N-units/km) (curve 3). As follows from Fig. 3 (right panel) the atmosphere influence is essential near zone of radio shadow. Intensity of reflected signal has sharp changes near radio shadow region (Fig. 3, left panel). The reflection coefficient $\eta^2(t)$ varies from nominal level 0.02…0.04 up to 0.08…0.14 in the time interval 250…270 sec and then indicates sharp changes from 0.03 up to 0.5. These variations reveal horizontal inhomogeneity of the atmosphere along the trajectory of motion of specular reflection point. As follows from the model given by Pavelyev et al., 1997, 2001, reflection coefficient depends on the vertical gradient of refractivity in the boundary layer. Theoretical dependence of power of reflected signal on vertical gradient of refractivity is given in Fig. 4. The central angle θ (expressed in degrees) (Fig. 1) is shown at horizontal axis. Power attenuation of the reflected signal normalized relative to free space level is shown on the vertical axis in Fig. 4. The curve 1 describes the case of absence of the atmosphere. The curve 2 corresponds to value of vertical gradient –35.4 N-units km^{-1}. The curves 3,4,5 correspond to vertical gradient equal to G_1=–43.0; G_2=–50.0 and G_3=–57.0 N-units km^{-1}. The angular position of radio shadow is different for indicated values of vertical gradient G and changes from 102.3° (absence of the atmosphere, G=0) to 103.5° (G_o=–35.0), 103.75° (G_1=–43.0), 103.9° (G_2=–50.0), 104.1° (G_1=–57.0). As follows from Fig. 4 the power of reflected signal near boundary of radio shadow is sharp function of the central angle θ and vertical gradient dN(h)/dh. Thus amplitude of reflected signal near radio shadow is indicator of vertical gradient of refractivity in the boundary layer of the atmosphere. It is useful to consider radio image of the boundary layer for more careful analysis of the connections of the parameters of reflected signal with structure of boundary layer. Radio image

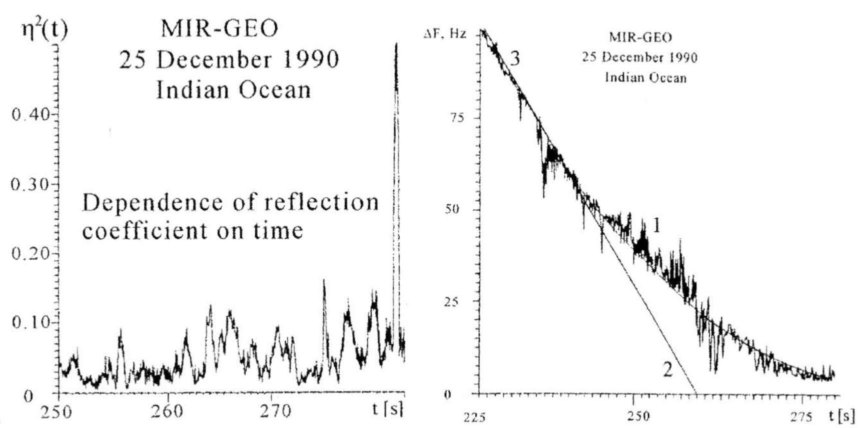

Fig. 3. Dependence of reflection coefficient and Doppler frequency of reflected signal on time (MIR/GEO experiment, wavelength 32 cm).

References

Beyerle G, Hocke K (2001) Observation and simulation of direct and reflected GPS signals in radio occultation experiments, *Geophys Res Lett,* **28**, No 9

Beyerle G., Hocke K, Wickert J, Schmidt T, Marquardt C and Reigber Ch (2002) GPS radio occultations with CHAMP: A radio holographic analysis of GPS signal propagation in the troposphere and surface reflections, *J Geophys Res,* in press.

Gorbunov ME, Sokolovskii SV and Bengtsson L (1996) Space Refractive Tomography of the Atmosphere: Modelling of Direct and Inverse Problems *Paper 210* Max-Planck-Institute for Meteorology ISSN 0937-1060

Gorbunov ME and Gurvich AS (1998) Microlab-1 experiment: Multipath effects in the lower troposphere. *J Geophys Res* **103**, 13,819-13,826

Hocke K, Pavelyev A, Yakovlev O, Barthes L and Jakowski N (1999) RO data analysis by radio holographic method. *JASTP,* **61**, 1169-1177

Igarashi K, Pavelyev A, Hocke K, Pavelyev D and Wickert J (2001) Observation of wave structures in the upper atmosphere by means of radio holographic analysis of the RO data. Advances in Space Research **27**, No 6-7, 1321-1327

Igarashi K, Pavelyev A, Hocke K, Pavelyev D, Kucherjavenkov IA, Matugov S, Zakharov A and Yakovlev O (2000) Radio holographic principle for observing natural processes in the atmosphere and retrieving meteorological parameters from RO data. *Earth Planets Space*, **52**, 968-875

Komjathy A, Garrison JL and Zavorotny V (1999) GPS: Anew tool for ocean science, *GPS World*, April, 50-56

Pavelyev A (1998) On the possibility of radio holographic investigation on communication link satellite-to-satellite. *Radioteknika i elektronika* **43,** No 8, 939-944 (in Russian).

Pavelyev A, Zakharov AI, Kucherjavenkov AI, Molotov EP, Sidorenko AI, Kucherjavenkova IL and Pavelyev DA (1997) Propagation of Radio Waves Reflected from Earth's Surface at Grazing Angles between a Low-Orbit Space Station and a Geostationary Satellite. *Journal of Communication Technology and Electronics.* **42,** No 1. pp 51-57

Pavelyev DA and Pavelyev AG (1997) Model of propagation of reflected from the Earth's surface signals near boundary of radio shadow on telecommunication line geo-*Radiotekhnika* (Moscow), No 11, 21-27. stationary – LEO satellites.

Pavelyev A, Yakovlev OI, Matyugov SS, Pavelyev D, Anufriev VA (2001) Advanced algorithms of inversion of GPS radio limb sounding data. Scientific Technical Report STR01/06. Part 4. Ionospheric correction and revealing ionospheric features. pp 52-72. GeoForschungZentrum Potsdam

Rubashkin SG, Pavelyev AG, Yakovlev OI, Kucherjavenkov AI, Sidorenko AI and Zakharov AI (1993) Observations of reflected signals on telecommunication line MIR-geostationary satellites. *Radiotekh. Elektron.* (Moscow), **38**, No 3, pp 447-453

Yakovlev OI, Matyugov SS and Vilkov IA (1995) Attenuation and scintillation of radio waves in the Earth's atmosphere from RO experiments on satellite-to-satellite links, *Radio Science* **30**, No 3, 591-602

Wickert J, Reigber Ch, Beyerle G, Koenig R, Marquardt C, Schmidt T, Grunwaldt L, Galas R, Meehan TK, Melbourne WG and Hocke K (2001): Atmosphere sounding by GPS radio occultation: First results from CHAMP, Geophys Res Lett, **28**, 3263-3266

Assimilation Experiments of One-dimensional Variational Analyses with GPS/MET Refractivity

Paul Poli[1,3] and Joanna Joiner[2]

[1] Joint Center for Earth Systems Technology (JCET), University of Maryland Baltimore County, Baltimore, MD 21250 USA *ppoli@dao.gsfc.nasa.gov*
[2] NASA Laboratory for Atmospheres, Goddard Space Flight Center, Greenbelt, MD 20771 USA *jjoiner@dao.gsfc.nasa.gov*
[3] also at Météo France Centre National de Recherches Météorologiques, France

Summary. The GPS/MET 1995 experiment demonstrated the feasability of remote soundings of the Earth's atmosphere using the GPS radio occultation technique [1]. Subsequent studies assessed the information content of such measurements (e.g. [2]). Several groups have attempted to assimilate the GPS radio occultation measurements into a global Data Assimilation System (DAS). These attempts followed in general the methodology proposed by *Eyre* [3], and range from direct four-dimensional variational assimilation of bending angles to assimilation of inverted profiles of temperature and/or humidity. We present here an hybrid approach. We attempted to take advantage of the accuracy of current weather models to constrain one-dimensional variational analyses using GPS refractivity. Retrieved profiles of atmospheric parameters were then assimilated like other types of observations in a global DAS for issuing the next forecast period. We compare analyses (as well as forecasts) performed with and without assimilating GPS radio occultation observations.

Key words: GPS, Radio Occultation, Refractivity, Variational Assimilation

1 One-dimensional Variational (1DVAR) Analyses with Refractivity

The application of one-dimensional variational analysis (1DVAR) to the retrieval of geophysical parameters from GPS radio occultation has already been discussed by various authors (e.g. [2], [4]). We briefly describe the main ideas. Technical details can be found in [5].

1.1 1DVAR Analysis Theory

Air refractivity N for GPS signals frequencies (L-band) can be written (neglecting scattering and ionospheric contribution)

$$N = b_1 \frac{P}{T} + b_2 \frac{P_w}{T^2} \tag{1}$$

[6], where P is the air pressure in hPa, T the temperature in K, P_w the water vapor partial pressure in hPa, $b_1 = 77.6$ N-unit·K·hPa^{-1}, and $b_2 = 3.73\times10^5$ N-unit·K^2·hPa^{-1}. It is obvious from equation (1) that the direct inversion of refractivity measurement into temperature, humidity and pressure is not possible without assumptions. One way to address the temperature/humidity ambiguity is to use physical constraints *via* a 1DVAR approach. *A priori* information (or background) is combined with GPS refractivity observations in a statistically optimal way to yield analyzed profiles of temperature and humidity. An important hypothesis is that all errors are supposed to be non-biased and normally distributed.

1.2 GPS/MET Observations

We used 797 occultations collected during 'Prime Time 2' of the GPS/MET 1995 experiment, when the GPS Anti-Spoofing encryption had been turned off from June 21st until July 4th, 1995. This, along with changes in the software of the receiver enabled tracking of the GPS occultations deep in the troposphere. In order to use these observations in the 1DVAR scheme we assumed the following standard deviation error in refractivity: 1% below 5 km altitude, 0.2% between 5 km and 30 km altitude. We discarded the refractivity observations above 30 km altitude.

1.3 Background Information

The Data Assimilation Office (DAO) Finite Volume Data Assimilation System (FVDAS) [7] provided the 6-hour forecasts used as a background in the 1DVAR implementation. The forecasts were bilinearly interpolated from 2×2.5 degrees grid to the location of the GPS occultation. No time interpolation was performed. The forecast fields (from the surface to the 0.01 hPa level) were interpolated to 46 levels for the 1DVAR analysis.

1.4 GPS 1DVAR Results

Due to a positive bias in the background minus observation refractivity the analyzed 1DVAR humidity profiles presented systematic departure from the original 6-hour forecast used to initiate the retrieval. These departures were inconsistent with humidity measurements reported by nearby radiosondes [5]. We hence decided to investigate this subject further before assimilating these analyzed humidity profiles into the FVDAS.

Figure 1 shows temperature comparisons with nearby RS (less than 280 km, +/- 3 hours). The 1DVAR analysis has smaller bias than the background above the 200 hPa level. The standard deviation of the background is larger than the 1DVAR analysis between 850 hPa and 20 hPa. This suggests that the 1DVAR analyzed temperature profiles may provide useful information to a global Data Assimilation System (DAS).

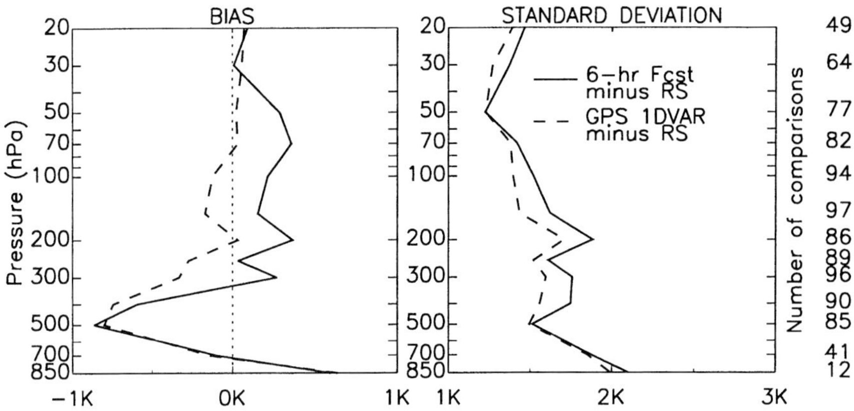

Fig. 1. Temperature comparisons of 6-hour forecasts and GPS 1DVAR with nearby radiosondes; number of comparisons on the right-hand side

2 Assimilation of GPS 1DVAR Profiles in the Finite Volume Data Assimilation System (FVDAS)

2.1 Description of the FVDAS and Link with GPS 1DVAR Analysis

The FVDAS relies on a NASA/NCAR (National Aeronautics and Space Administration/National Center for Atmospheric Research) Finite Volume General Circulation Model (GCM). The GCM dynamical core uses a Lagrangian vertical coordinate system [8]. The assimilation part of the FVDAS is a Physical-space Statistical Analysis System (PSAS) [7]. The FVDAS assimilates conventional observations such as radiosondes (RS) as well as 1DVAR cloud-cleared retrieved profiles of temperature and humidity derived from TIROS Operational Vertical Sounder (TOVS) level-1b brightness temperatures [9].

When running the FVDAS from June 21st, 1995 to July 4th, 1995, the latest available forecast was used to perform the GPS 1DVAR analysis. Then, the GPS 1DVAR analyzed temperature profiles were assimilated in the FVDAS using PSAS. They were assigned the same error characteristics as radiosondes. Due to significant refractivity differences between 6-hour forecast background and observation [5], we chose to restrict the assimilation of 1DVAR retrievals to the region 500 hPa–30 hPa in the [30°S,30°N] latitude band. At other latitudes, 1DVAR retrievals were assimilated between the surface and the 30 hPa level. We refer to this experiment as 'GPS'.

In order to evaluate the impact of the GPS 1DVAR analyses on the FVDAS we also performed a 'CONTROL' experiment with no GPS data.

2.2 Impact on Analyses

Without any other observation other than those assimilated in the two experiments, it is difficult to determine which analysis is the closest to the atmospheric "truth". Figure 2 shows the difference between an analysis with GPS data minus the corresponding 'CONTROL' analysis for 1995-06-29 at 00GMT. In this example we used the same first guess for perform both analyses. Consequently, the differences we see are solely due to the GPS temperature observations assimilated at 00GMT on 1995-06-29. The differences in the analyzed winds enables to assess how the GPS temperature observations changed the atmospheric circulation through the mass-wind balance. For example, negative temperature increments corresponding to negative geopotential heights increments induced cyclonic-type increments in winds, i.e. horizontal circulation differences oriented clockwise in the Southern hemisphere.

2.3 Impact on 6-hour Forecasts

We compared the 6-hour forecasts from the two experiments with radiosonde observations. These observations may be considered as an independent validation. Figure 3 shows zonal averages of the standard deviation of the observation minus forecast for the meridional wind. The 'CONTROL' has larger errors in the Southern hemisphere than in the Northern hemisphere where it

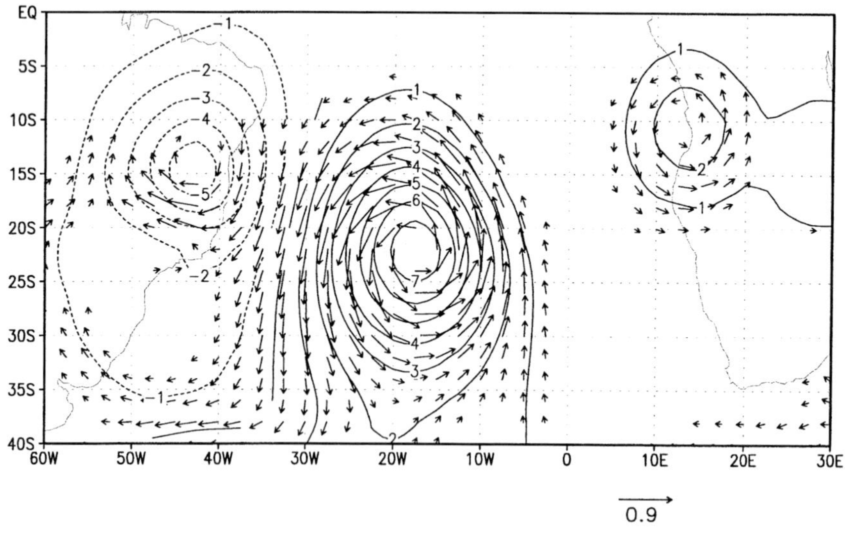

Fig. 2. Difference 'GPS' analysis minus 'CONTROL' at 250 hPa. Contours designate geopotential heights, arrows represent wind.

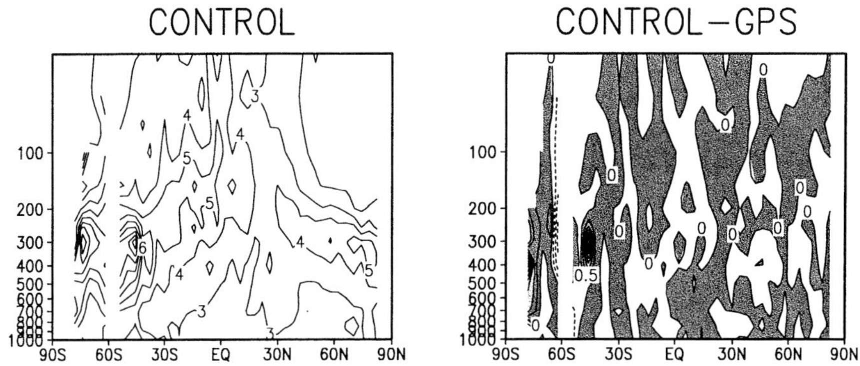

Fig. 3. Radiosonde observations minus 6-hour forecast standard deviation error for meridional wind. Negative contours (dotted lines) in the 'CONTROL' minus 'GPS' suggest a negative impact of the GPS data; Shaded areas indicate a slight positive impact.

is more constrained by *in situ* observations. The shaded areas in the 'GPS' minus 'CONTROL' indicate reduced errors in the 'GPS' 6-hour forecast as compared to the 'CONTROL'. Differences are more apparent in the Southern hemisphere.

2.4 Impact on 5-day Forecasts

We ran a series of 5-day forecasts from daily 00GMT analyses from June 21st until July 4th, 1995. Forecast scores were computed in the form of Root Mean Square (RMS) and anomaly correlation for sea level pressure and 500hPa geopotential height. We used three different verifying analyses: (a) own analysis for each experiment, (b) 'CONTROL' analyses, and (c) European Centre for Medium-range Weather Forecasts (ECMWF) operational 1995 analyses. The GPS data induced slight differences on forecast scores of individual forecasts. The average of the 14 forecasts showed very little difference between 'CONTROL' and 'GPS'. The small number of forecasts issued and the limited amount of GPS data assimilated does not enable us to fully assess the impact of GPS refractivity data in a global data assimilation system.

3 Conclusions and Future Work

We have implemented a 1DVAR scheme to analyze temperature and humidity profiles using GPS radio occultation refractivity. This scheme was incorporated in the assimilation cycle of the FVDAS and we performed assimilation experiments using GPS/MET refractivity. We found mixed results in terms

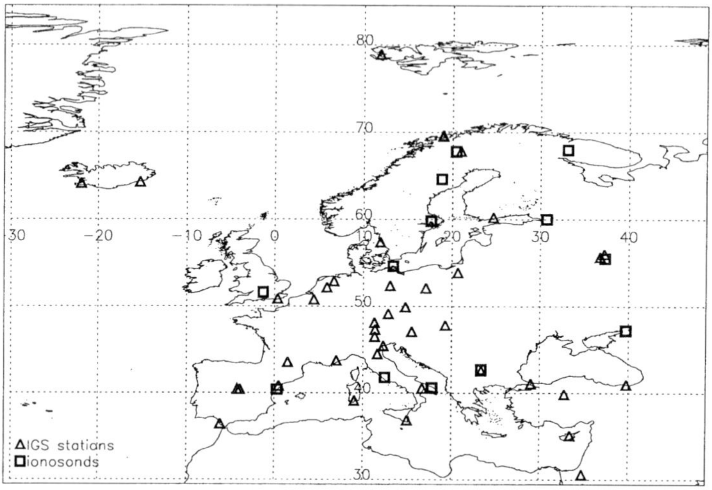

Fig. 1. Distribution of European IGS stations and ionosondes

and radio occultation measurements, which improve the vertical resolution but are not permanently available for the selected region. A description of the data processing and first results will be given in the following sections.

2 Data Processing

The monitored region selected for this approach extends from -20E to 40W and from 32.5N to 70N. GPS data from 47 selected European IGS stations as shown in Fig. 1 form the database for the tomographic reconstrution of the ionosphere. In a first step the collected GPS code and carrier phase data are preprocessed to estimate the TEC for every single ray path [6]. In order to remove the plasmaspheric contribution of the given GPS measured TEC, a plasmaspheric correction factor is applied (In principle it possible to include the plasmaspheric region in the tomography, but for this approach we concentrate on the ionosphere). The factor describes the proportion of the TEC fraction above and below 1000km altitude for each ray path. It is computed using a combination of the ionospheric model IRI and the plasmaspheric Gallagher model [2]. As signal bending effects can be ignored in ionospheric tomography, the ray paths are assumed as straight lines between satellite and receiver.

For tomographic reconstruction the ionosphere in the selected region is represented as a three-dimensional array of voxels. Here the voxels are defined at a height range from 80 km up to 1000 km, with a resolution of 5° in longitude, 2.5° in latitude and 10 km in height. The inversion problem can be expressed as a set of linear equations:

$$\mathbf{A}\mathbf{x} = \mathbf{g}, \tag{1}$$

where $\mathbf{g}$ represents the vector of the given TEC measurements, $\mathbf{x}$ the unknown electron density in the voxels and $\mathbf{A}$ is a large sparse matrix containing the voxel path lengths for each ray.

To solve this illposed problem (16560 unkowns and only about 400 measurements) mulitiplicative algebraic reconstruction technique (MART) is used. In MART, each measurement is iteratively compared to a predicted (forward projection) value computed from the current reconstructed data estimate. A multiplicative correction factor is then applied to each voxel covered by the measurement set. This causes the forward projection to approach the measurement. The algorithm is given by

$$\mathbf{x}_\mathbf{j}^{\mathbf{new}} = \mathbf{x}_\mathbf{j}^{\mathbf{old}} \left(\frac{g_i}{(\mathbf{x}^{\mathbf{old}}, \mathbf{a}_\mathbf{i})} \right)^{\lambda_k a_{ij} / \mathbf{a}_\mathbf{i}}, \tag{2}$$

where a_{ij} are the elements of the matrix $\mathbf{A}$ and a_i is the row vector of the matrix $\mathbf{A}$. As intitial guess for the iteration process the IRI model is used. After all measured TEC values have been used during the iteration, NmF2 data from selected ionosonde measurements (here Rome and Juliusruh) are compared with the tomographic result for these particular locations. The iteration is repeated until no essential improvements are obtained by the algorithm.

In order to fill the data gaps in the voxels which are not intersected by rays and also to stabilize the solution a simple three-dimensional distance weighted gaussian-like boxcar average of the specific width w is applied after the iteration. Therefore we rewrite the vector $\mathbf{x}$ to a three-dimensional matrix $\mathbf{B}$ with the dimension L, M, N and apply

$$R_{l,m,n} = \begin{cases} \dfrac{\sum\limits_{i=0}^{w-1} \sum\limits_{j=0}^{w-1} \sum\limits_{k=0}^{w-1} \omega_{i,j,k} B_{l+i-w/2, m+i-w/2, n+i-w/2}}{\sum\limits_{i=0}^{w-1} \sum\limits_{j=0}^{w-1} \sum\limits_{k=0}^{w-1} \omega_{i,j,k}} \\ \\ B_{l,m,n}, \quad \text{otherwise} \end{cases} \tag{3}$$

where

$$\begin{aligned} l &= w/2, \cdots, L - w \\ m &= w/2, \cdots, M - w \\ n &= w/2, \cdots, N - w \end{aligned} \tag{4}$$

and ω represents a weighting factor obtaind by

$$\omega_{i,j,k} = e^{\frac{(i-w/2)^2}{c_1^2} + \frac{(j-w/2)^2}{c_2^2} + \frac{(k-w/2)^2}{c_3^2}}. \tag{5}$$

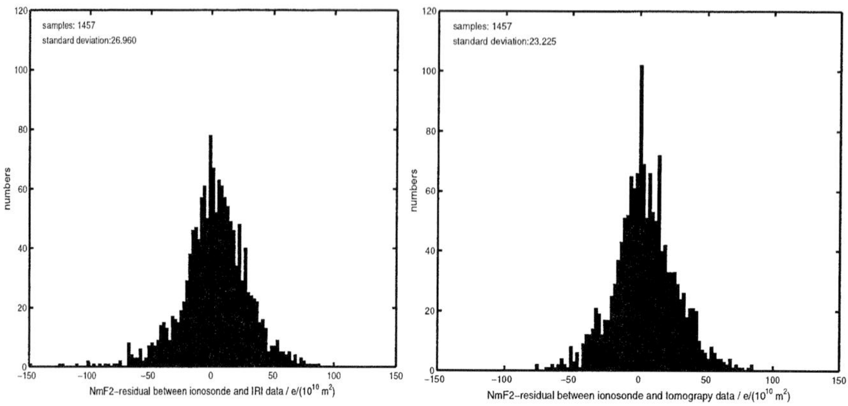

Fig. 2. Residual between NmF2 data retrieved from selected ionosonde and the IRI model(left panel) and tomography (right panel)

The values of the coefficients c_1, c_2, c_3 and w, have been estimated by serveral runs, cross-checking the tomographic result with ionosonde and TEC data from the Neustrelitz maps. After the smoothing has been applied, $\mathbf{B}$ is rewritten to $\mathbf{Ax}$ and the iteration is restarted. The whole process is repeated until no significant changes appear in the iteration process.

3 Evaluation

An important task in ionospheric tomography is the evaluation of the results. In principle there are no sources of continuous data covering the whole European region and giving vertical information about the internal structure of the ionosphere up to 1000 km. Therefore we have used NmF2 and hmF2, obtained by ionosonde measurements, as a key parameter for the evaluation of vertical resolution of the tomographic reconstruction. In particular, 3-hourly NmF2 and hmF2 estimates from April 1st to April 21th, 2000 of 11 European ionosondes (see Fig. 1) were used for the comparison (the ionosonde stations Juliusruh and Rome where not used in the comparison, because their data have been taken for the tomogram). This period was chosen, because a geomagnetic storm event occured at April 6th, 2000. Such events, causing effects which are not reproduced by the inital model, give good opportunity to prove the potential of this very model dependent approach. The results for NmF2 and hmF2 residuals between the ionosonde data and the data obtained by tomography are given in Fig. 2 and 3. For comparison residuals under the same conditions but using the initial model instead of tomography are computed (see also 2 and 3).

The results show that the tomographic approach has led to an overall improvement of the NmF2 model estimate of about 14% (comparing the

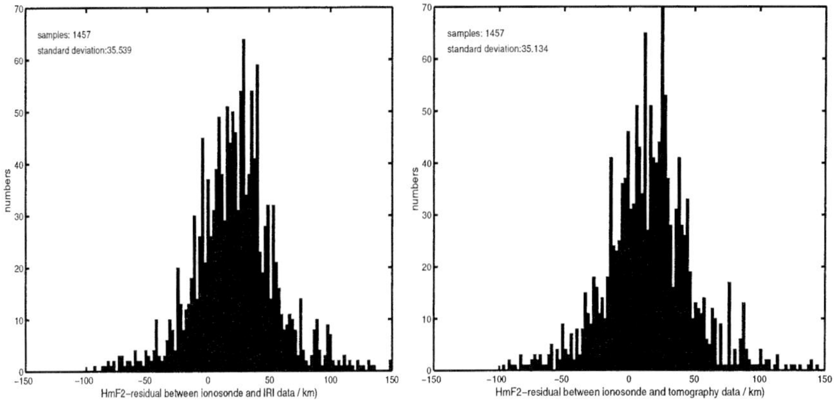

Fig. 3. Residual between hmF2 data retrieved from selected ionosondes and the IRI model (left panel) and tomography (right panel)

dispersion and the bias to the ionosondes in both statistics), whereas no significant improvement can be found in the hmF2 data.

4 Including Radio Occultation Data

As shown, pure ground-based GPS measurements contain hardly any information on the vertical ionospheric profile. An opportunity to improve the vertical resolution in tomography is the use of GPS radio-occultation data as those obtained by the CHAMP satellite. To prove the potential improvement obtained by such data, a set of calibrated horizontal TEC data from the CHAMP satellite [4] where included in the tomography and compared with ionosonde data of the Juliusruh station, which was close to the ray path of the occultation links. As an example Fig. 4 shows vertical profiles over Juliusruh, obtained by the IRI model, ionosonde data (with a model extension above the electron density peak) and ground based GPS tomography. Fig. 4 shows the corresponding results with CHAMP data included in the tomographic algorithm. The improvement of the model estimate obtained by the use of ground-based GPS data is shown in Fig. 4, as it has already been indicated by the statistical investigations. Model correction by real data is significantly better by combining ground-based GPS and radio occultation data as demonstrated in Fig.4.

5 Conclusion

So far GPS based radio tomography is still in a developing state. However, it was shown that the use of GPS networks as that of IGS can be the base for a

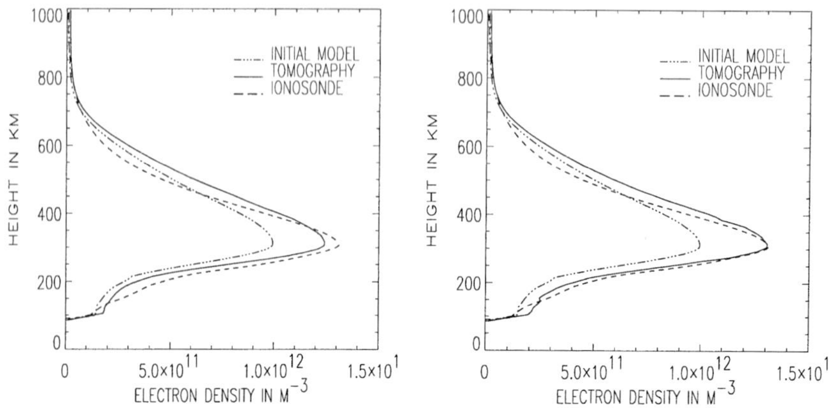

Fig. 4. Height profile of electron density at Juliusruh (Germany), obtain by IRI, ionosonde (the upper part is extended by a model) and GPS-tomography using IGS data (left panel) and using combined IGS and CHAMP data (right panel).

continous three-dimensional ionospheric monitoring system. As exemplarily shown, tomographic combination of ground-based GPS and radio-occultation data can lead to promising results.

Regardless of the inversion method, to obtain tomographic results with a resolution that allows the investigation of even smaller structures and phenomena in the ionosphere, more and especially horizontal measurements are needed. Future navigation systems as the European Galileo and further radio-occultation missions may improve such data base.

Nevertheless, even under better data conditions, satellite radiotomography has to include additional information. In our case, using the recursive MART algorithm, data of the IRI model were used as a priori information and ionosonde data to obtain a stop criteria for the iteration process.

Acknowledgements

The authors are grateful to the CHAMP team, co-ordinated by the GFZ Potsdam, for the given support and data. We also thank all members of the IGS community for providing their data service. Ionosonde data have kindly been provided by the IAP Kühlungsborn.

References

1. Dick G, Gent G, Reigber C (2001) First experience with near real-time water vapor estimation in a German network. J of Atmos and Solar-Terr Phys, **63**, 1295–1304

2. Gallagher DL, Craven PD, Comfort RH (2000) Global core plasma model. J Geophys Res,**105(A8)**,18819–18833
3. Jakowski N, Sárdon E, Engler E, Jungstand A, Klähn D (1996) Relationships between GPS signal propagation errors and EISCAT observations. Annal Geophysicae, **14**, 1429–1436
4. Jakowski N, Wehrenpfennig A, Heise S, Reigber Ch, Lühr H, Grunwaldt L, Meehan TK (2002) GPS Radio Occultation Measurements of the Ionosphere from CHAMP: Early results. Geophys Res Lett, (in press).
5. Jakowski N, Schlüter S (1999) Ionospheric storms detected by GPS based TEC monitoring., In: Hanbaba R, de la Morena BA (eds) Proc 3rd COST251 Workshop, 103–107
6. Sardón E, Rius A, Zarraoa N (1994) Estimation of the transmitter and receiver differential biases and the ionospheric total electron content from Global Positioning System observations. Radio Sci, **29(3)**, 577–586

Comparison of Three Different Meteorological Datasets (ECMWF, Met Office and NCEP)

Katrin Schöllhammer[1], Christian Marquardt[2], Karin Labitzke[1].

[1]Freie Universität Berlin, Institut für Meteorologie, Carl-Heinr.-Becker Weg 6-10, D-12165 Berlin, Germany.
[2]GeoForschungsZentrum Potsdam (GFZ), Germany. Current Address: Met Office, London Road, Bracknell, Berkshire, RG12 2SZ, UK.

Summary. Analyses of the global stratospheric circulation are routinely provided by weather services as part of their regular data assimilation and forecasting practice. Nevertheless, the stratosphere, compared to the troposphere, remains a comparatively data poor region of the Earth's atmosphere. Thus, accurate temperature measurements from GPS based radio occultation soundings like those provided by CHAMP are likely to provide important additional information to our understanding of the stratospheric circulation. In this study, we examine the differences between present day analyses of the stratosphere in order to identify those regions and seasons where radio occultation soundings can be expected to have the largest positive impact. We show that the largest deviations and systematic biases between present day analyses occur during Northern Hemisphere wintertime in the mid (10 hPa) stratosphere. Thus, CHAMP observations could provide the largest impact on meteorological analyses in the mid stratosphere if radio occultation measurements at these altitudes do indeed give independent temperature information. In the lower stratosphere, deviations between analyses and uncertainties of CHAMP temperature soundings are of the same order of magnitude.

Key words: stratosphere, temperature (analyses), comparisons

1 Introduction

Nearly all present day global meteorological analyses cover the stratospheric circulation at least up to 10 hPa (about 30 km). The stratospheric part of these analyses is mainly based on the small amount of data provided by radiosondes ascending up into the stratosphere. As not all balloons reach these altitudes, radiosonde data coverage is significantly smaller in the stratosphere than in the troposphere. For example, during December 2001, a total of 33315 radiosonde observations were available at the 500 hPa level (averaging to 1075 radiosonde launches per day), but only 9491 (306 per day) were reported for the same period at the 10 hPa level. Data assimilation systems like those implemented at the European Centre for Medium-Range Weather Forecasts (ECMWF), the National Centers for Environmental Prediction (NCEP), or the UK Met Office (MetO) use additional data stemming from nadir sounding satellite instruments like ATOVS (Advanced Tiros Operational Vertical Sounder) and HIRS (High resolution Infrared Radiation

Sounder) onboard the operational polar orbiting NOAA (National Oceanic and Atmospheric Administration) satellites. These instruments exhibit broad vertical weighting functions in the stratosphere, providing a low vertical resolution of the order of several km only. Under strongly perturbed conditions (e.g. during minor and major stratospheric warmings in Northern Hemisphere (NH) winters), they cannot provide detailed information on the vertical temperature structure. Thus, stratospheric analyses may have larger errors, and deviations between different analyses may be larger during these periods than at other times.

Due to their unique characteristics, radio occultation (RO) measurements like those from CHAMP (Reigber et al. 2000; Wickert et al. 2001) could provide a significant positive impact on stratospheric analyses: RO measurements exhibit their highest accuracy of stratospheric temperature soundings in the lower and mid stratosphere. Furthermore, they combine global data coverage with a high vertical resolution (see, e.g., Kursinsky et al. 1995; Rocken et al. 1997). While a single satellite like the German CHAMP already provides about 200 daily vertical temperature profiles, future upcoming satellite missions like the US constellation COSMIC will provide thousands of daily temperature profiles.

In this paper, we analyse deviations between three state-of-the-art operational stratospheric analyses (provided by ECMWF, NCEP, and the Met Office) during July and December 2001. We use deviations between different analyses as proxy for their uncertainty. We will identify those regions and seasons where uncertainties of stratospheric analyses are largest, and where radio occultation data are most likely to improve them.

2 Data

We have used gridded analysis data on standard pressure levels as provided be the various weather centres. ECMWF analyses are produced by a 4D Var data assimilation system (Rabier et al. 2000; Mahfouf et al. 2000; Klinker et al. 2000), using a T511 spectral model (about 0.4° resolution); the 60 hybrid model levels cover the atmosphere from the ground to 0.1 hPa. The data used in this study has been obtained on standard pressure levels, and interpolated onto a regular 2.5° × 2.5° longitude-latitude grid in order to match the resolution available for the NCEP analyses.

Operational NCEP analyses are based on a 3D Var system ("Spectral Statistical Interpolation", SSI; Derber et al. 1991; Parrish and Derber 1992; Derber and Wu 1998); the underlying spectral model is run at a T170 resolution (about 1°) and exhibits 42 sigma levels between the ground and 2 hPa. Similar to the ECMWF analyses, satellite radiances are used in the data assimilation rather than retrieved temperature profiles (McNally et al., 2000). We have used data on standard pressure levels and with standard resolution (2.5° × 2.5°) as distributed by NCEP.

Met Office stratospheric analyses (Swinbank et al. 1994) also use a 3D Var data assimilation system assimilating satellite radiances (Lorenc et al. 2000; Ingleby 2001). The underlying 2.5° (latitude) by 3.75° (longitude) grid point model exhibits 40 pressure levels between the ground and 0.3 hPa; the data is available on 22 standard pressure levels. As two of these are identical or very close to 10 and 30 hPa we have used these pressure levels for our study. We have also regridded the Met Office data to the same 2.5° × 2.5° longitude-latitude grid used by the other analyses, again in order to make the calculation of gridded statistics feasible.

3 Monthly Statistics

Table 1 lists the maximum and minimum values of the monthly mean temperature deviations between ECMWF and the other two analyses at 30 and 10 hPa during July and December 2001. Also shown are the maximum values of the monthly mean root mean square (rms) temperature deviations for the same periods.

	July			December		
	min	max	RMS	min	max	RMS
			ECMWF-MetO			
10 hPa	-4,0	0,1	4,1	-5,0	0,4	6,1
30 hPa	-1,3	2,1	2,2	-1,7	1,8	2,2
			ECMWF-NCEP operational analysis			
10 hPa	-1,9	1,8	2,5	-2,4	3,2	4,3
30 hPa	-3,0	1,8	3,0	-3,0	0,4	3,1

Table 1. Minimum and maximum monthly mean and maximum monthly rms deviations of temperature between ECMWF, MetO and NCEP during July and December 2001.

We note that the maxima of the monthly mean deviations between ECMWF and both MetO and NCEP at 10 hPa are larger during NH winter (i.e., in December) than summer. Fig. 1, showing the global distribution of the December 2001 rms deviations between ECMWF and MetO, illustrates that the largest rms deviations at 10 hPa occur in NH mid and high latitudes, with maximum rms deviations of up to 6 K over Siberia. Further local maxima are located over the mid-latitude Atlantic, Central Asia (above 5 K) and the NH subtropical Pacific (above 4 K). Areas of large rms deviations are partially linked to areas where no or only few conventional meteorological data is available (e.g., over the oceans and much of Asia). However, even data rich regions like Europe exhibit rms deviations above 4 K. At the same time, rms deviations in the Southern Hemisphere (SH) are smaller, typically ranging between 2 and 4 K, obwohl SH is less well covered with regular meteorological observations. It must be kept in mind that NH wintertime dynamics can be grossly perturbed. In December 2001, the stratospheric circulation in the NH was characterised by a strong major warming [Naujokat et al. 2002]. Due to the complex dynamics occuring during such events, numerical models used in data assimilation systems may have difficulties in following the rapid synoptical development. They therefore may not represent the 'true' state of

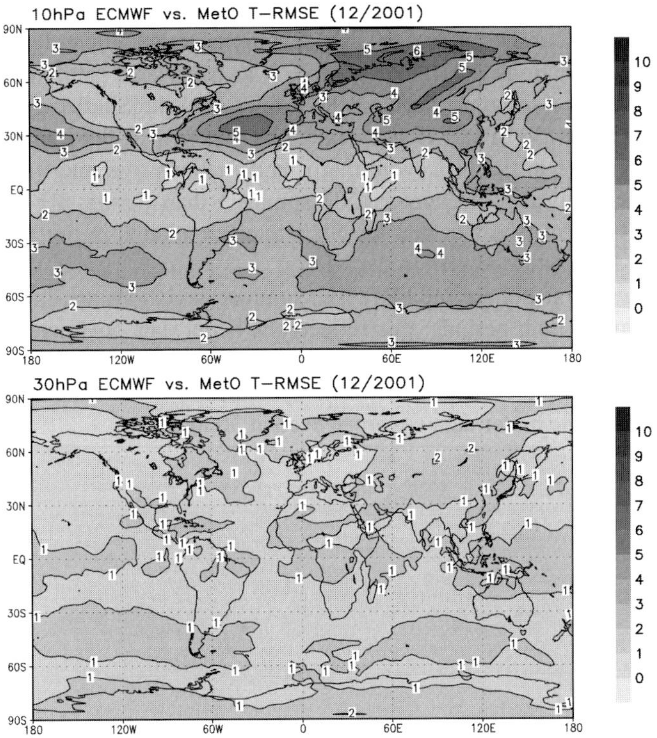

Fig. 1. Rms temperature deviations between ECMWF and MetO in December 2001 at 10 hPa (top) and 30 hPa (bottom).

the atmosphere. This may yield to large deviations between different stratospheric analyses. The large rms deviations between ECMWF and MetO data observed over Europe are an example of this: during most of December, the polar vortex edge was located above Europe, but moved significantly as the warming progressed over time. Due to temperature gradients associated with the vortex edge, small dislocations of the polar vortex in two analyses have caused increased rms deviations in the monthly statistics.

Both Table 1 and Fig 1 indicate that rms deviations between different analyses are larger at 10 hPa than at 30 hPa during December 2001: rms deviations at 30 hPa are below 2 K everywhere, with maximum values found over the data poor regions of Russia and the South Pole. Rms deviations above 1 K are also found in the tropics and the mid-latitude SH oceans. Data rich regions like the US and Europe, but also large regions over the world ocean, exhibit rms deviation below 1 K. Despite the better data coverage in the NH, regions of rms deviations above

Thus, large systematic differences of the order of several K can be found between the analyses at 10 hPa over large regions. Note that these large deviations occur mainly at the 10 hPa level; at 30 hPa mean deviations are typically smaller (not shown, but see Table 1).

The spatial patterns of NH rms deviations against ECMWF are very similar for both data sets, again suggesting that non-systematic differences between analyses are linked to strong perturbations in stratospheric dynamics. Rms deviations do, however, deviate in magnitude. The largest rms deviations of up to 6 K (ECMWF vs. MetO) are found in Asia over China and Siberia and over the mid latitude Atlantic. Between ECMWF and NCEP, values of up to 4 K are reached over Asia and the mid latitude Atlantic.

5 Discussion

We have shown that deviations between present day meteorological analyses of the stratosphere agree within 1-2 K (bias) and 2-3 K (rms) in the lower stratosphere (i.e., 30 hPa) during the periods discussed. The agreement of CHAMP stratospheric temperature retrievals with both meteorological analyses and radiosonde data (see Marquardt et al., this issue) is of the same order of magnitude. Thus, meteorological analyses of the lower stratosphere will almost certainly benefit from the assimilation of this new type of remote sensing measurements. On the other hand, the uncertainty between various analyses is of the same order of magnitude as the uncertainty of the CHAMP measurements. Thus, CHAMP radio occultation measurements are not able to distinguish between 'correct' and 'erroneous' analyses in the lower stratosphere.

In the mid stratosphere (i.e., 10 hPa), both mean and rms deviations between analyses are significantly increased during NH winter. The largest deviations occur in the Northern Hemisphere and are related to the complex and rapid dynamical development that can take place during minor or (in the case of December 2001) major warming events. Thus, the complicated stratospheric winter time dynamics more than compensates for the better data coverage of the NH. During December 2001, systematic deviations between ECMWF and MetO stratospheric analyses were as large as -5 K (with ECMWF being systematically colder), or +/- 2 K between ECMWF and NCEP. Rms differences peak at 6 and 3 K respectively. Day-to-day differences locally exceeding 10 K (not shown) have been found.

It is obvious that the largest benefit of RO soundings to stratospheric analyses can be expected where the analyses are weakest - i.e., in the mid stratosphere during Northern Hemisphere winter. Especially the identification of large scale systematic biases that apparently exist for some of the meteorological analyses at 10 hPa seems important. However, this requires unbiased and sufficiently accurate temperature estimates from radio occultation measurements at this altitude. Unfortunately, details of the smoothing algorithms and the use of climatological or operational weather data within the retrieval of temperatures from RO soundings can affect temperature retrievals at 10 hPa significantly (see, e.g., Marquardt et al., this issue). To our understanding, the influence of smoothing and the use of *a priori* data in the RO processing chain has not yet been addressed properly in the literature. Error estimates for mid and upper stratospheric RO temperature retrievals that take these issues into account should be provided before RO data can be used to infer reliably about shortcomings of analyses in the stratosphere.

References

Derber JC, Parrish DF, Lord SJ (1991) The new global operational analysis system at the National Meteorological Center. Wea. and Forecasting. 6:538-547

Derber JC, Wu W-S (1998) The use of TOVS cloud-cleared radiances in the NCEP SSI analysis system. Mon Wea Rev 126:2287-2299

Ingleby NB (2001) The Statistical Structure of Forecast Errors and its Representation in The Met. Office Global 3D-Var. Quart J Roy Meteorol Soc 127:209-231

Klinker E, Rabier F, Kelly G, Mahfouf J-J (2000) The ECMWF operational implementation of four dimensional variational assimilation. Part III: Experimental results and diagnostics with operational configuration, Quart. J. Roy. Meteorol. Soc., 126:1191-1218

Kursinski ER; Hajj GA, Romans LJ (1995) Observing tropospheric water vapor by radio occultation using the global positioning. Geophys Res Lett 17:2365-2368

Lorenc AC, Ballard SP, Bell RS, Ingleby NB, Andrews PLF, Barker DM, Bray JR, Clayton AM, Dalby T, Li D, Payne TJ, Saunders FW (2000) The Met. Office Global 3-Dimensional Variational Data Assimilation Scheme. Quart. J. Roy. Meteorol. Soc., 126, 2991-3012

Mahfouf J-F, Rabier F (2000) The ECMWF operational implementation of four dimensional variational assimilation. Part II: Experimental results with improved physics. Quart J Roy Meteorol Soc 126:1171-1190

Marquardt C, Schöllhammer K, Beyerle G, Schmidt T, Wickert J, Reigber Ch (2002) Validation and Data Quality of CHAMP Radio Occultation Data. this issue

McNally AP, Derber JC, Wu W-S, Katz BB (2000) The use of TOVS level-1B radiances in the NCEP SSI analysis system. Quart J Roy Meteorol Soc 126:689-724

Naujokat B, Krüger K, Matthes K, Hoffmann J, Kunze M, Labitzke K (2002) The early major warming in December 2001 – exceptional? Geophys Res Lett in press

Parrish DF, Derber JC (1992) The National Meteorological Center's spectral statistical interpolation analysis system. Mon Wea Rev 120:1747-1763

Rabier F, Järvinen H, Klinker E, Mahfouf J-F, Simmons A (2000) The ECMWF operational implementation of four dimensional variational assimilation. Part I: Experimental results with simplified physics. Quart J Roy Meteorol Soc 126:1143-1170

Reigber Ch, Lühr H, Schwintzer P, (2000) CHAMP mission status and perspectives. Suppl. to EOS, Transactions, AGU, 81, 48, F307, 2000.v

Rocken Ch, van Hove T, Ware R (1997) Near real-time GPS sensing of atmospheric water vapor. Geophys Res Lett 24:3221-3224

Swinbank R, O'Neill A (1994) A stratosphere-troposphere data assimilation system, Monthly Wea Rev 4:686-702

Wickert J, Reigber Ch, Beyerle G, König R, Marquardt C, Schmidt T, Grunwaldt L, Galas R, Meehan Th K, Melbourne W G, Hocke K (2001) Atmosphere sounding by GPS radio occultation: first results from CHAMP. Geophys Res Lett 17:3263-3266

Radio Occultation Data Processing at the COSMIC Data Analysis and Archival Center (CDAAC)

Bill Schreiner, Doug Hunt, Chris Rocken, Sergey Sokolovskiy

University Corporation for Atmospheric Research (UCAR), Boulder, CO, USA

Summary. This paper gives an overview of the COSMIC Data Analysis and Archival Center (CDAAC) and presents initial results that were obtained from neutral atmospheric radio occultation data processing of the GPS/MET, CHAMP, and SAC-C datasets. Approximately 45 % of CHAMP and SAC-C retrieved radio occultation profiles reach below 1 km altitude, compared to only 35 % for GPS/MET. All missions exhibit a negative refractivity bias in the lower troposphere of nearly 1 % compared to NWP models. When constrained to the tropics, the CHAMP data show a negative refractivity bias of approximately 1.7 %, and only 20 % of occultation profiles reach 1 km. Current CDAAC LEO orbit determination for CHAMP agrees with JPL orbits at the 30 cm 3D root mean square level, which can result in temperature errors of about 0.3 degrees K at 30 km altitude.

Key words: Global Positioning System (GPS), Radio occultation, Remote sensing, COSMIC, SAC-C

1 Introduction

In 1995, the University Corporation for Atmospheric Research (UCAR) in collaboration with the Jet Propulsion Laboratory (JPL), the University of Arizona, and Orbital Sciences Corporation demonstrated the concept of GPS active limb sounding with the launch and successful two-year operation of the GPS/MET receiver payload. This success has led to a joint U.S.-Taiwan six spacecraft follow-on mission called the Constellation Observing System for Meteorology, Ionosphere and Climate (COSMIC), which is now funded and due to launch near the end of 2005. A diagram illustrating the data flow of the ground and space segments of the COSMIC system is shown in Figure 1. Each spacecraft will carry three payloads: 1) a GPS Radio Occultation (RO) receiver designed by JPL 2) a Tiny Ionospheric Photometer (TIP) designed by the US Naval Research Laboratory (NRL) and 3) a Tri-Band Beacon (TBB) designed by NRL. The payload data will be transmitted after each 100-minute orbit in near real time (via two northern Earth stations) to the CDAAC at UCAR in Boulder, CO. The CDAAC will analyze the GPS RO data and make it available to operational weather centers, universities and research institutions. In preparation for the upcoming COSMIC mission, UCAR is developing the CDAAC. A CDAAC prototype (version 1.0) has been finished and is now processing RO data from available missions such as GPS/MET, CHAMP and SAC-C in a post processing mode. This paper describes

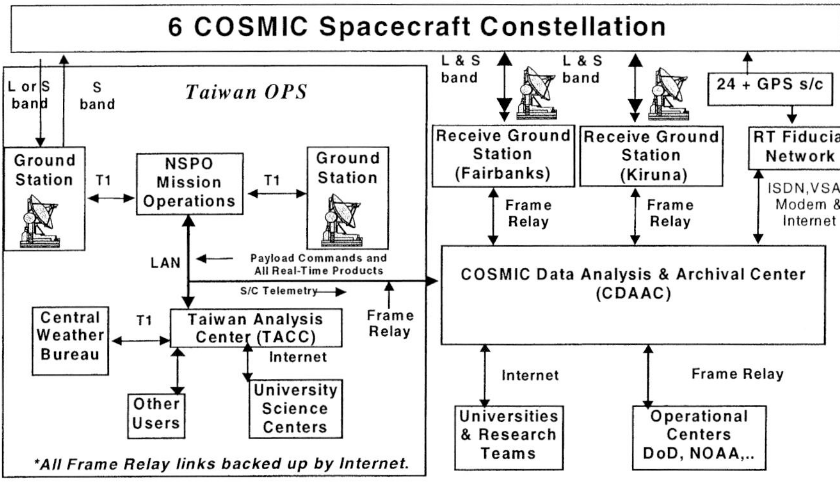

Fig. 1. The CDAAC within the COSMIC System

the CDAAC, gives an overview of its architecture, data processing software, and initial results that were obtained from neutral atmospheric RO data processing with emphasis on CHAMP and SAC-C.

2 The CDAAC System

The main responsibilities of the CDAAC are: payload monitoring, incoming data quality checking, RO data processing, validation, and data archiving and distribution. The CDAAC RO data processing will be performed in two modes: 1) a quasi-operational near real time mode where data are continuously analyzed and results fed to operational weather centers within 3 hours of observation and 2) a more accurate and better validated post processed mode to be used for climate and other research applications. Initially, Level 0 data consisting of a raw dump of low Earth orbit (LEO) data, IGS GPS orbits (predicted or final), and GPS ground fiducial data flow into the CDAAC and are quality checked and reformatted into a mission independent set of level 1a data files. The level 1a to level 1b processing consists of LEO precise orbit determination (POD), generation of an occultation table, and the calibration of excess atmospheric phase delay for each RO. The level 1b excess phase and amplitude files are lastly inverted into level 2 profiles of refractivity with the Abel inversion software. The CDAAC system is controlled by an event driven task scheduler, where tasks are automatically put on a processing queue when their corresponding dependencies (usually files) arrive on disk. This task scheduler is well suited for real time processing. A separate post-processing subsystem can also be run when all dependency files are present. All RO geometry as well as analysis information from each process are stored in a database. A web-based interface to this database is used to analyze and validate all CDAAC data products. The CDAAC software runs in a UNIX operating system, is written predominantly in Perl and Fortran, and is version controlled using CVS software. The current CDAAC development hardware consists of 2 dual Pentium-III 700 MHz machines connected to a RAID. More details regarding the CDAAC can be found at http://www.cosmic.ucar.edu/cdaac/index.html. Some details regarding level 1a-to-level 2 processing will now be discussed.

2.1 Level 1a to Level 1b Processing: Calibration of Excess Phase Delay

The most critical step in calibrating the atmospheric excess phase delay for a RO event is the POD of the LEO spacecraft. The goal for CDAAC POD is to determine the relative LEO GPS velocity to about 0.1 mm/sec. This corresponds to a requirement to estimate the LEO position to 10 cm 3-D root mean square (rms). The CDAAC is using a beta version of the Bernese software to perform the LEO POD. This software uses a combination of kinematic point positioning and subsequent dynamic smoothing over a defined arc length to provide precise LEO orbits (Bock et al. 2000). The kinematic point positioning requires precise high rate GPS clocks that are estimated from ground GPS data while holding the GPS orbits, zenith troposphere delay and ground station coordinates fixed (Bock et al. 2001). The dynamic model that is used to smooth the kinematic LEO positions includes the EGM96 gravity field, a cannon ball drag model with MSISE90 air density, and constant and 1 cycle per orbit revolution empirical accelerations.

Once the GPS and LEO orbits are available, the CDAAC computes an occultation table that defines the date, time and location of all possible RO events. The excess phase delay is then computed for each event by removing the GPS and LEO orbital motion (known) and clock errors (via double differencing with 1 Hz ground data) from the raw RO observations. The excess phase delay can also be computed with single differencing (i.e. no ground data), which eliminates potential error sources from the ground GPS data (Rocken et al. 2000, Wickert et al. 2002). This method of processing has been coded, but not tested.

2.2 Level 1b to Level 2 Processing: Abel Inversion

The CDAAC Abel inversion software computes the level 2 profile data from the level 1b excess phase files. The inversion software first filters the excess phase and computes a Doppler profile, and then computes a bending angle versus impact parameter profile that is inverted to refractivity versus altitude with the classical Abel integral transform (Eshleman 1973). Because there is little signal at high altitudes, the bending angle near the top of a RO is combined with a climate-derived (CIRA+Q, (Kirchengast et al. 1999)) bending angle in a statistical optimization procedure (Sokolovskiy and Hunt 1996). In the lower troposphere, the bending angle profile derived from Doppler is frequently multi-valued because of multipath created by sharp moisture-induced refractivity gradients. The CDAAC has tested three radio holographic methods (Back Propagation (Gorbunov et al. 1996),

Sliding Spectral (Sokolovskiy 2001), and Canonical Transform (Gorbunov 2001,2002a,2002b)) that use both phase and amplitude data in an effort to create single-valued bending angle profiles that are required by the Abel transform. These tests were performed by forward modeling and inverting of RO signals using high resolution radiosonde data, and it was found that the Canonical Transform performed the best. Another necessary task that is performed by the software is to find and remove erroneous tracking data near the bottom of the RO, which is needed when the receiver tracks in either the closed loop or so-called "fly wheeling" mode. This objective data screening is done by comparing the observational Doppler with Doppler predicted from orbital motion and climatology, and discarding data with large deviation. The definition of "large deviation" is based on the magnitude of the weather-induced spread of Doppler in LEO estimated for open loop tracking (Sokolovskiy 2001a). Finally, all level 2 data files are passed through a quality control procedure to remove corrupted profiles.

3 Early CDAAC Results

UCAR has now post processed a significant amount of RO data from the GPS/MET (1995.170-190), CHAMP (2001.148-309) and SAC-C (2001.213-299) missions with version 1.0 of the CDAAC software system. Processing data from different missions has helped discover and correct many software inconsistencies and has also provided an opportunity to compare each missions' POD and RO data quality. The remainder of the paper presents CDAAC POD quality, RO data quality issues, and validation statistics with the use of independent data from atmospheric dynamic models.

3.1 POD Quality

The POD quality results for each mission are shown in Table 1. Internal orbit overlaps are computed as the rms difference (over 1 orbit) between two adjacent CDAAC orbit solutions and are a good indicator of orbit quality, but they can under-estimate orbit quality if the orbit errors in adjacent solutions are correlated. A better indicator of orbit quality is to compute the rms difference between orbits from different processing centers, called an external orbit overlap. The average CHAMP external overlap of 1.5 m 3D rms with the JPL IGSLEO orbits is large due to gravity and drag forces that are not yet modeled correctly at CDAAC for an orbit altitude of 430 km. The JPL orbits are believed to be accurate at ~15 cm 3d rms. The last column in Table 1 gives an estimate of the retrieval temperature error at 30 km altitude due to the orbit error. As stated above the CDAAC goal for post processed orbit position quality is 10 cm 3D rms (~0.1 mm/sec in velocity).

Mission	Altitude [km]	Date Range	Internal Orbit Overlaps 3D rms [m]	External Orbit Overlaps with JPL (IGSLEO) 3D rms [m]	Resulting Temperature Error at 30 km [K]
GPS/MET	735	1995.170-192	0.5	-	0.4
CHAMP	430	2001.140-150	1.0	1.5	0.8
SAC-C	700	2001.213-226	0.3	-	0.3

Table 1. CDAAC LEO POD Quality Results

3.2 Radio Occultation Data Quality Issues

We have noticed several repeatable problems with the CHAMP and SAC-C RO data. (1) Very large noise on L2 Doppler in the troposphere. (2) 5 sec spikes on L2 and L1 Doppler. (3) Bias in Doppler. This bias can sometimes be seen above the neutral atmosphere, where the ratio of L2 and L1 Dopplers differs substantially from that expected due to ionospheric dispersion. In most cases (but not always) this error is accompanied by near-zero L2 SNR. An example: CHAMP, 2001, doy 141, 21:33 UTC, 47.6N, 15.6E. (4) Strong perturbations of L2 Doppler (while no perturbations on L1) correlated with ionospheric scintillation. An example: CHAMP, 2001, doy 151, 4:13 UTC, 40.1S, 107.9W. (5) Strong perturbation (highly correlated) on both the L1 and L2 Dopplers. In most cases (but not always) the source of this problem is in the reference link. An example: CHAMP, 2001, doy 143, 4:42 UTC, 31.6N, 85.9W. (6) Large L1 and L2 tracking errors at the beginning of each occultation. Problem (1) makes it impossible to apply the normal ionospheric calibration in the troposphere. Problem (2) makes it difficult to dynamically switch between the ionosphere calibration modes based on L2 noise (unless a matching filter is applied). Problems (3-5) can result in large retrieval errors at any altitude. We do not know the reasons for these problems. The potential sources could be related to ionospheric maximum, tracking under A/S on conditions, and receiver hardware/software problems.

3.3 Validation Statistics

Computing profile statistics with independent correlative data is a good approach to validate RO data from a particular mission. The CDAAC database interface web page has been used to generate refractivity statistics for CHAMP, SAC-C, and GPS/MET as shown in Figures 2-5, respectively. All RO profiles have been computed in the lower troposphere with the Canonical Transform method. The correlative data profiles are interpolated to the time and location of the RO. The GPS/MET profile data are compared to ECMWF, and the CHAMP and SAC-C data are compared to NCEP's low resolution Aviation Weather Model (AVN) because the ECMWF or high resolution NCEP data were not yet available. Figures 2-4 show the global mean and rms deviation of the differences between the RO's and model refractivity profiles, along with the number of compared profiles as a function of altitude for each mission. One can see from the figures that the mean percent difference in refractivity is less than ~1 % from the surface to ~17 km (the maximum altitude for AVN) for all three missions. The rms deviations of the percent differences for the three missions are ~1% except in the lower troposphere. The number of retrieved RO's that penetrate below 1 km altitude is ~45% for CHAMP and SAC-C and ~35% for GPS/MET. To evaluate the performance of the RO data in the tropics, the CDAAC database interface was used to compute statistics for CHAMP in the region defined by –30 deg < latitude < 30 deg. These results are shown in Figure 6 and exhibit a clear degradation due to tropical moisture. The mean percent difference in refractivity in the lower troposphere increases to ~1.7% and the percentage of RO's that penetrate to 1 km altitude is approximately 20%.

4 Conclusions and Future Work

UCAR has successfully inverted and catalogued over 15,000 RO's from the CHAMP (9,500), SAC-C (5,400), and GPS/MET (1,300) missions with version

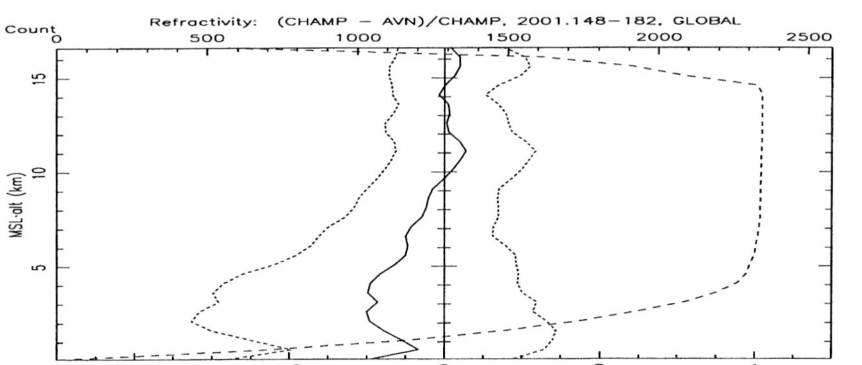

Fig. 2. Mean (solid), rms (short dash), and count (long dash) global comparison statistics between CHAMP occultations and AVN model. Approximately 45% of occultations penetrate to below 1 km

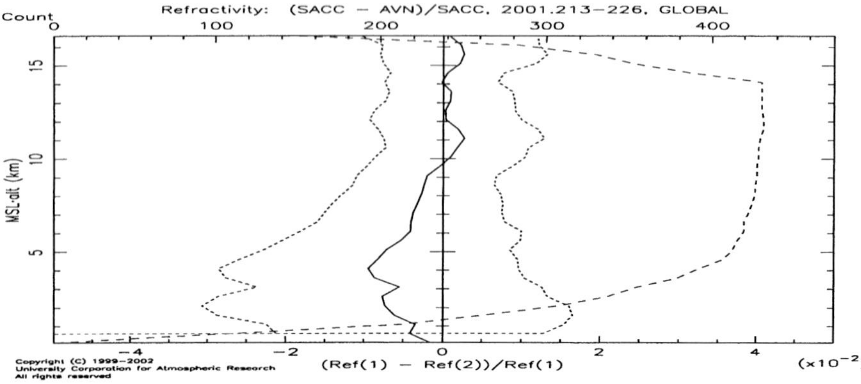

Fig. 3. Global comparison statistics between SAC-C occultations and AVN model. Approximately 45% of occultations penetrate to below 1 km

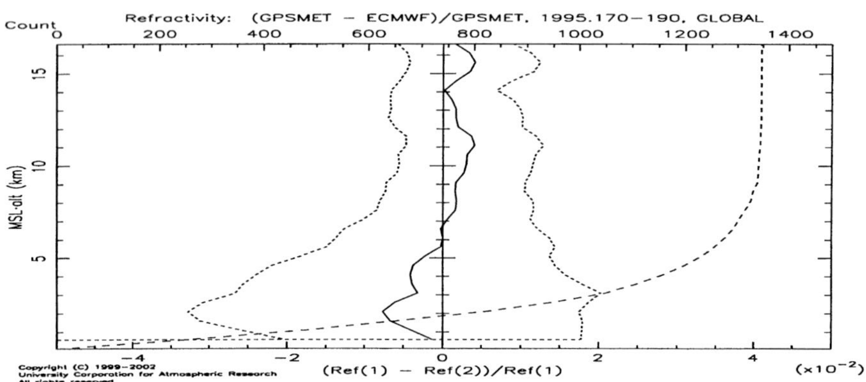

Fig. 4. Global comparison statistics between GPS/MET occultations and ECMWF model. Approximately 35% of occultations penetrate to below 1 km

1.0 of the CDAAC software. This analysis has enabled UCAR to identify problems and issues that should be studied and hopefully resolved before the COSMIC launch in 2005. The validation statistics presented above show a lack of penetration in the lower troposphere, especially in the tropics where only ~20% of the RO profiles penetrate to 1 km altitude. In addition, the tropical refractivity statistics for CHAMP show a significant bias below 10 km altitude with respect to the AVN model (and to radiosondes, not shown) that approaches 1.7% near the Earth's surface. It is not yet clear to what extent these lower troposphere problems are caused by receiver tracking errors, insufficient antenna gain, and atmospheric super-refraction. Open loop tracking will be tested by JPL in the near future with the SAC-C receiver and may provide some insight into this problem. A final issue that must be addressed before the COSMIC launch is the CDAAC POD quality. The post-processed results presented above using a preliminary version of the Bernese

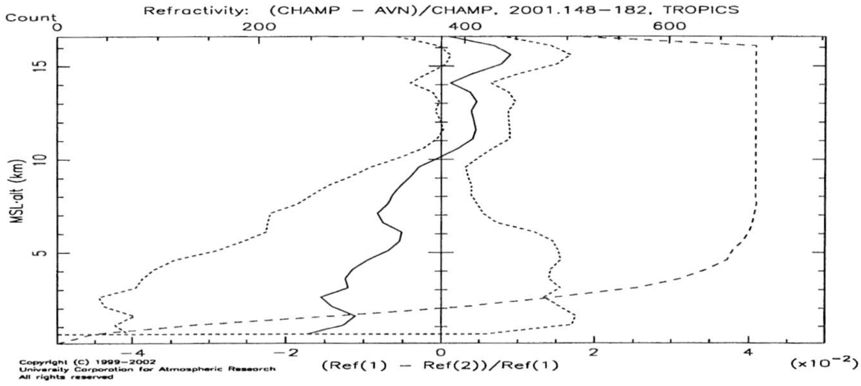

Fig. 5. Tropical comparison statistics between CHAMP occultations and AVN model. Approximately 20% of occultations penetrate to below 1 km

LEO POD software (i.e. 1.5 m 3D rms difference with respect to JPL CHAMP orbits), have already been improved. Recent CDAAC CHAMP POD results obtained with the new gravity model EIGEN (Reigber et al. 2001) show agreement with the JPL orbits at the 30 cm 3D rms level for the period 2001.140-150. The approach to be used for the near real time orbit determination mode is still being studied, but is not expected to significantly degrade the higher level data products. GeoForschungsZentrum (GFZ), JPL, and UCAR have just initiated a study effort called ROSE (Radio Occultation Sensor Evaluation) to compare RO results between agencies. This effort should lead to a better understanding of RO issues and will provide the best possible GPS receiver payload for the COSMIC mission.

Acknowledgements

The authors would like to thank GFZ for release of CHAMP data and JPL for the release of SAC-C data. The authors would also like to thank Jay Fein at the U.S. National Science Foundation for his continued support of the UCAR CDAAC.

References

Bock H, Beutler G, Schaer S, Springer TA, Rothacher M (2000) Processing aspects related to permanent GPS arrays. Earth Planets Space 52: 657-662

Bock H, Beutler G, Hugentobler U (2001) Kinematic Orbit Determination for Low Earth Orbiters (LEOs). paper presented a the IAG Scientific Assembly in Budapest, Hungary, Sept 2-7

Eshleman VR (1973) The radio occultation method for the study of planetary atmospheres. Planet Space Sci 21: 1521-1531

Gorbunov ME, Gurvich AS, Bengtsson L (1996) Advanced algorithms of inversion of GPS/MET satellite data and their application to reconstruction of temperature and humidity. MPI report 211, Hamburg

Gorbunov ME (2001) Radioholographic methods for processing radio occultation data in multipath regions. DMI report 01-02, Copenhagen, 32 pp

Gorbunov ME, (2002a) Canonical transform method for processing radio occultation data in the lower troposphere. Radio Sci (in press)

Gorbunov ME, (2002b) Radio-holographic analysis of MircroLab-1 radio occultation data in the lower troposphere. JGR (in press)

Kirchengast G, Hafner J, Poetzi W (1999) The CIRA86aQ_UoG model: An extension of the CIRA-86 monthly tables including humidity tables and a Fortran95 global moist air climatology model. IMG/UoG Techn Rep 8, Eur Space Agency, Paris, France

Reigber C, Schwintzer P, Koenig R, Neumayer KH, Bode A, Barthelmes F, Foerste Ch (GFZ Potsdam), Balmino G, Biancale R, Lemoine JM, Loyer S, Perosanz F(GRGS, Toulouse) (2001) Earth Gravity Field Solutions from Several Months of CHAMP Satellite Data. Eos Trans AGU 82 (47), Fall Meeting Suppl G4IC-02

Rocken C, Kuo Y-H, Schreiner W, Hunt D, Sokolovskiy SV, McCormick C (2000) COSMIC System Description. TAO 11 (1): 21-52

Sokolovskiy SV, Hunt D (1996) Statistical optimization approach for GPS/MET data inversions. URSI GPS/MET Workshop, Tucson, AZ

Sokolovskiy SV (2001) Modeling and inverting radio occultation signals in the moist troposphere. Radio Sci 36 (3): 441-458

Sokolovskiy SV (2001a) Tracking tropospheric radio occultation signals from low Earth orbit. Radio Sci 36 (3): 483-498

Wickert J, Beyerle G, Hajj GA, Schwieger V, Reigber C (2002) GPS radio occultation with CHAMP: Atmospheric profiling utilizing the space-based single difference technique. (in press)

Verification of CHAMP Radio-Occultation Observations in the Ionosphere Using MIDAS

Paolo Spalla[1], Norbert Jakowski[2], Andreas Wehrenpfennig[2], Paul S J Spencer[3], Cathryn N Mitchell[3]

[1]IFAC-CNR, Florence, Italy
[2]DLR, Neustrelitz, Germany
[3]University of Bath, Bath, UK.

Summary. The MIDAS (Multi-Instrument Data Analysis System) algorithm performs a 4-dimensional mathematical inversion. It is designed to assimilate a number of different measurement techniques, thus allowing the spatial and temporal factors to be accounted for in the inversion. In this study ionospheric data have been collected together from a number of observations in Europe. The MIDAS program incorporates these data into a single 4-D inversion and produces electron concentration maps.

CHAMP occultations over Europe have been identified for comparison purposes and the data have been inverted using a newly developed technique. First results indicate excellent agreement between the specification of the ionospheric electron concentration using MIDAS and that calculated from CHAMP.

Key words: Ionosphere, occultations, tomography, electron density, linear inversion

1 Introduction

The purpose of this short paper is to demonstrate the first results of a validation of the ionospheric radio occultation data of the CHAMP satellite. Comparisons are made between vertical electron-density profiles generated with CHAMP occultation data and those obtained using the MIDAS algorithm from a large number of observations collected all over the northern hemisphere.

2 Data analysis

To derive electron density profiles from CHAMP radio occultation data over Europe, a tomographic approach with a spherically layered voxel structure has been applied. The linear equation system describes the measured total electron content along numerous ray paths obtained during the occultation process, as a function of the sum of electron along the ray path elements within the crossed voxels. The solution provides the mean electron density within the spherically layered voxel. Due to the rather low orbit height of CHAMP (about 430 km) that is very close to the F2 layer peak, a model assisted technique has been applied to overcome the upper boundary problem. For more details of this technique see Jakowski et al. in this issue.

The MIDAS (Multi-Instrument Data Analysis System) algorithm performs a 4-dimensional mathematical inversion. It is designed to assimilate a number of different ionospheric measurement types, thus allowing the spatial and temporal factors to be accounted for in the inversion. The data used for the present work are:

- GPS data from the International GPS Service
- Peak density and height from ionosondes distributed over Europe
- NIMS observations from Italy

The MIDAS program incorporates these data into a single 4-D inversion and produces electron concentration maps. Another paper describing MIDAS in more detail is provided in this issue by C.N. Mitchell The comparisons of CHAMP and MIDAS profiles at the location of the midpoint defined by the CHAMP occultation are shown in the following figures. The date, time of occultation and the coordinates of each profile are given in the title. Each plot shows the vertical profile from the MIDAS inversion as a solid line and the profile from the CHAMP data as a broken line. These plots also show the peak densities and heights as given by observations from neighboring ionosondes. A key to these is given in the final figure. Different peak density values are due to different locations and local time.

3 Conclusions

The CHAMP occultations are often located between 25°E and 30°E, which is on the edge of the European region, where other ionospheric information used in the inversion is sparse. This could be the most important reason for any differences. Ionosonde data should (ideally) straddle the image area, which is not generally the case. In addition, ionosonde data give an information about peak heights via ionogram inversions that are always questionable: this is expected to lead to a worse agreement between peak heights than between peak densities.

Two very different measurement techniques have been compared here. The MIDAS inversion has used satellite-to-ground data in conjunction with ionosonde data to create maps of electron concentration over Europe. The CHAMP data has been inverted in isolation from this procedure. Despite these differences the agreement is remarkably good, providing a validation of the excellent ionospheric information now provided by CHAMP.

The results encourage further studies that will help to validate CHAMP derived electron density profiles. In future work the radio occultation data should also improve the accuracy of the MIDAS reconstruction results.

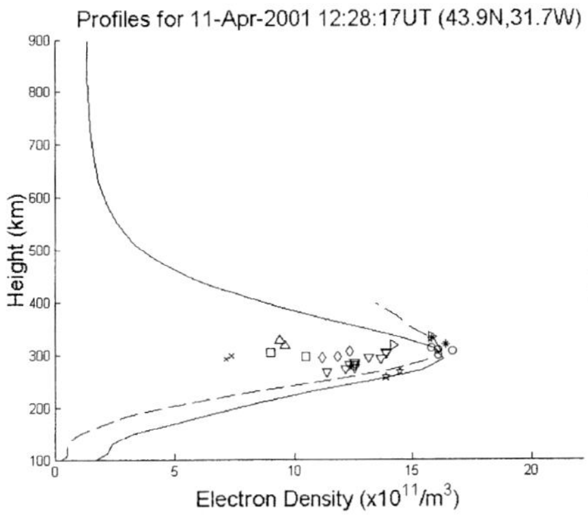
Profiles for 11-Apr-2001 12:28:17UT (43.9N,31.7W)
Height (km)
Electron Density (x10^11/m^3)

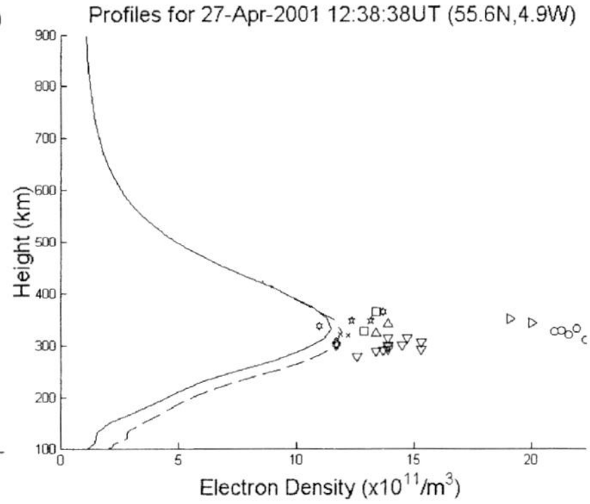
Profiles for 27-Apr-2001 12:38:38UT (55.6N,4.9W)
Height (km)
Electron Density (x10^11/m^3)

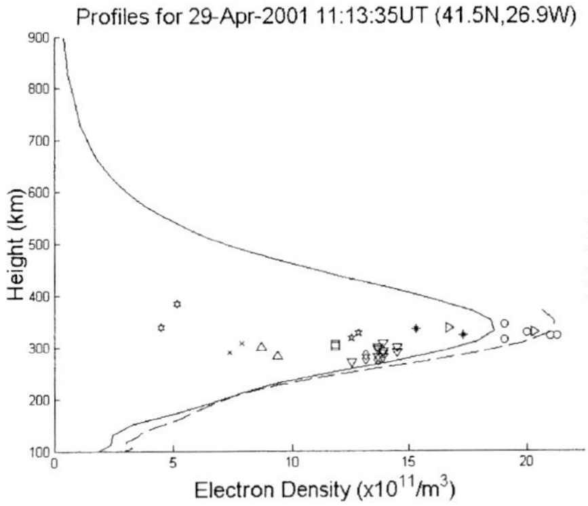
Profiles for 29-Apr-2001 11:13:35UT (41.5N,26.9W)
Height (km)
Electron Density (x10^11/m^3)

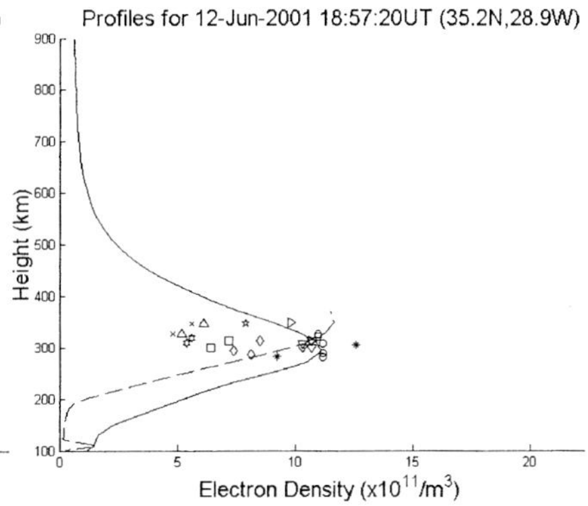
Profiles for 12-Jun-2001 18:57:20UT (35.2N,28.9W)
Height (km)
Electron Density (x10^11/m^3)

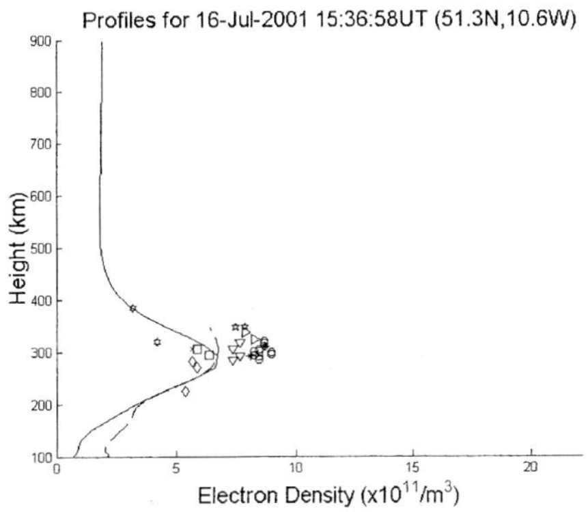
Profiles for 16-Jul-2001 15:36:58UT (51.3N,10.6W)
Height (km)
Electron Density (x10^11/m^3)

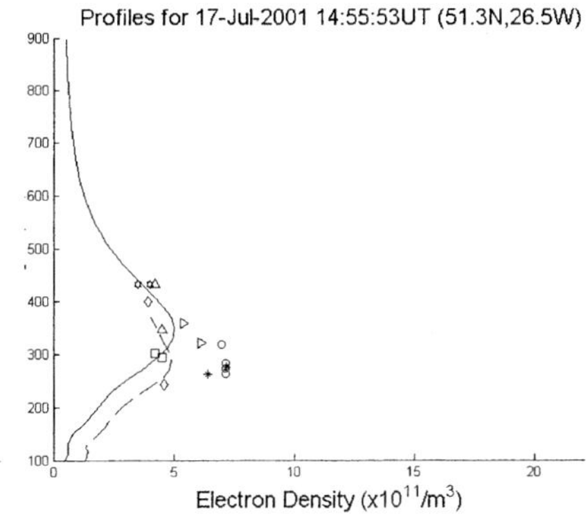
Profiles for 17-Jul-2001 14:55:53UT (51.3N,26.5W)
Electron Density (x10^11/m^3)

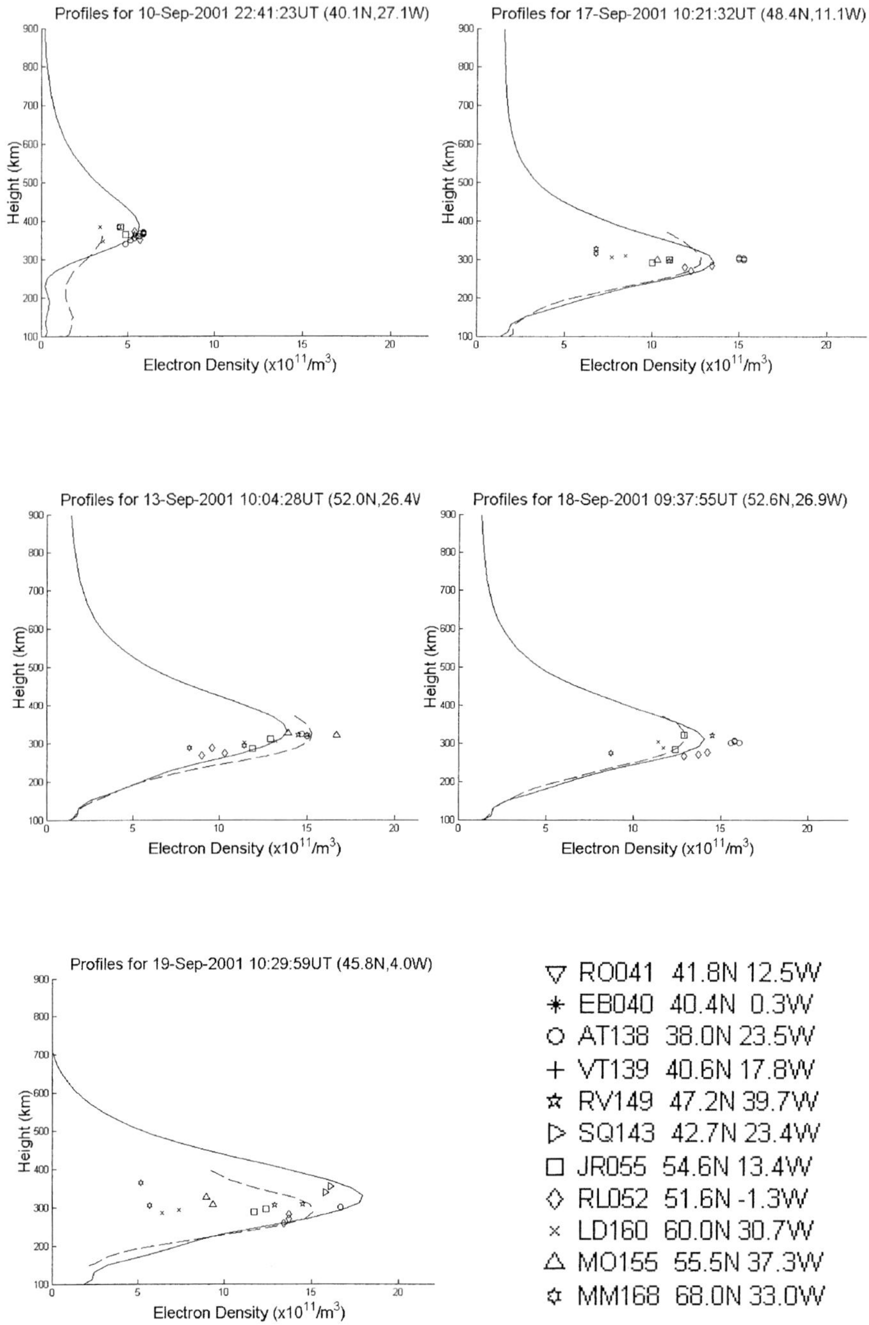

Profiles for 10-Sep-2001 22:41:23UT (40.1N,27.1W)
Profiles for 17-Sep-2001 10:21:32UT (48.4N,11.1W)
Profiles for 13-Sep-2001 10:04:28UT (52.0N,26.4V
Profiles for 18-Sep-2001 09:37:55UT (52.6N,26.9W)
Profiles for 19-Sep-2001 10:29:59UT (45.8N,4.0W)
Height (km)
Electron Density (x10¹¹/m³)
RO041 41.8N 12.5W
EB040 40.4N 0.3W
AT138 38.0N 23.5W
VT139 40.6N 17.8W
RV149 47.2N 39.7W
SQ143 42.7N 23.4W
JR055 54.6N 13.4W
RL052 51.6N -1.3W
LD160 60.0N 30.7W
MO155 55.5N 37.3W
MM168 68.0N 33.0W

Approach to the Cross-Validation of MIPAS and CHAMP Temperature and Water Vapour Profiles

Gabriele Petra Stiller, Tilman Steck, Mathias Milz, Thomas von Clarmann, Udo Grabowski, and Herbert Fischer

Institut für Meteorologie und Klimaforschung (IMK), Forschungszentrum Karlsruhe GmbH / Universität Karlsruhe, Postfach 3640, 76021 Karlsruhe, Germany *gabriele.stiller@imk.fzk.de*

Summary. The Michelson Interferometer for Passive Atmospheric Sounding (MIPAS) is a space-borne limb viewing mid-infrared high-resolution spectrometer launched on ENVISAT into polar orbit on 1 March 2002. Water vapour and temperature profiles are retrievable from spectral data in an altitude range from approximately 5 km up to the mesosphere. Complementing the operational analysis of ESA, a non-operational MIPAS level 2 processor has been developed at IMK to derive geophysical data from the MIPAS observations. The IMK MIPAS level 2 data processing scheme supports a multitude of regularization approaches. Full retrieval diagnostics in terms of retrieval covariance matrices, averaging kernel matrices etc. are available. Vertical profiles of temperature and humidity have been identified to be candidate quantities for comparison with corresponding CHAMP data as part of commissioning phase as well as long-term cross-validation activities. The accuracy of MIPAS water vapour and temperature measurements is investigated by test retrievals based on simulated data. The approach to cross-validate vertical profiles of temperature and humidity from CHAMP and MIPAS, respectively, is based on the χ^2 of the difference of MIPAS and CHAMP retrieved profiles under consideration of error correlations, averaging kernels, and spatial and temporal mis-matching of the observations.

Key words: Remote sensing, atmospheric observation, MIPAS, water vapour

1 Introduction

On 1 March 2002, the high-resolution mid-IR FTS limb emission sounder MIPAS (Michelson Interferometer for Passive Atmospheric Sounding) has successfully been brought into space on board of ENVISAT. ENVISAT is a sun-synchronous polar orbiter with 98.55° inclination in approximately 800 km altitude; subsequent orbits will provide global coverage about every 3 days with 14.4 orbits per day. MIPAS is a limb-viewing Fourier transform infrared (FTIR) emission spectrometer with 0.035 cm^{-1} spectral resolution (unapodised), covering the mid infrared from 685 cm^{-1} to 2410 cm^{-1} (14.6 – 4.15 μm). Tangent heights from about 5 to 150 km are supported. The field of view (FOV) of MIPAS corresponds to a width of 30 km in horizontal direction and approximately 3 km at the tangent point in vertical direction. The standard limb observation grid covers 6 to 68 km with a step width of 3 km. The sampling rate is about 500 km along-track and about 2800 km across-track at the equator. The standard observations cover the complete globe for day, night, or twilight conditions from the upper troposphere to the mesosphere.

2 IMK level-2 data processing

Besides the operational data processing under ESA responsibility, non-operational science-oriented data analysis will be performed at IMK within which vertical profiles of temperature, water vapour, and some 30 other species will be retrieved. The IMK level-2 data analysis (i.e. retrieval of atmospheric parameters) is based on the operational ESA level-1 data (i.e. calibrated and geo-located radiance spectra). As part of extensive validation efforts during commissioning phase as well as long-term activities, intercomparison and cross-validation of the vertical profiles of temperature and humidity as derived from MIPAS and CHAMP for the altitude range covered by both instruments is planned to be done. These activities will cover data amounts of several thousand limb sequences per year, an amount equivalent to several days of global MIPAS data.

2.1 Retrieval strategy

Inferring vertical profiles of atmospheric parameters from a limb sequence of infrared emission spectra involves the inverse solution of the radiative transfer equation. In order to account for nonlinearities, the linear estimation of the solution is embedded in a Newtonian iteration scheme. Following Rodgers' terminology (Rodgers 1976), the improved guess $\hat{\mathbf{x}}_{i+1}$ of the solution is calculated from the current guess $\hat{\mathbf{x}}_i$ as

$$\begin{aligned}\hat{\mathbf{x}}_{i+1} = \hat{\mathbf{x}}_i &+ (\mathbf{K}_i^T \mathbf{S}_y^{-1} \mathbf{K}_i + \mathbf{R} + \lambda \mathbf{I})^{-1} \\ &(\mathbf{K}_i^T \mathbf{S}_y^{-1} (\mathbf{y} - \mathbf{F}(\hat{\mathbf{x}}_i)) + \mathbf{R}(\mathbf{x}_a - \hat{\mathbf{x}}_i)), \qquad (1)\end{aligned}$$

where $\mathbf{K}_i$ is the Jacobian matrix $\partial \mathbf{y}/\partial \hat{\mathbf{x}}$ of the ith iteration, $\mathbf{S}_y$ is the covariance matrix of the measurement error, and $\mathbf{R}$ is a regularization term. The damping term $\lambda \mathbf{I}$ shall improve convergence in nonlinear cases and consists of the scalar Marquardt parameter λ and unity matrix $\mathbf{I}$ (Marquardt 1963; Levenberg 1944). $\mathbf{F}$ is the forward radiative transfer model which provides simulated measurements as a function of $\hat{\mathbf{x}}_i$, and $\mathbf{y}$ contains the measured spectral radiances; $\mathbf{x}_a$ is the a priori information the solution shall be constrained to.

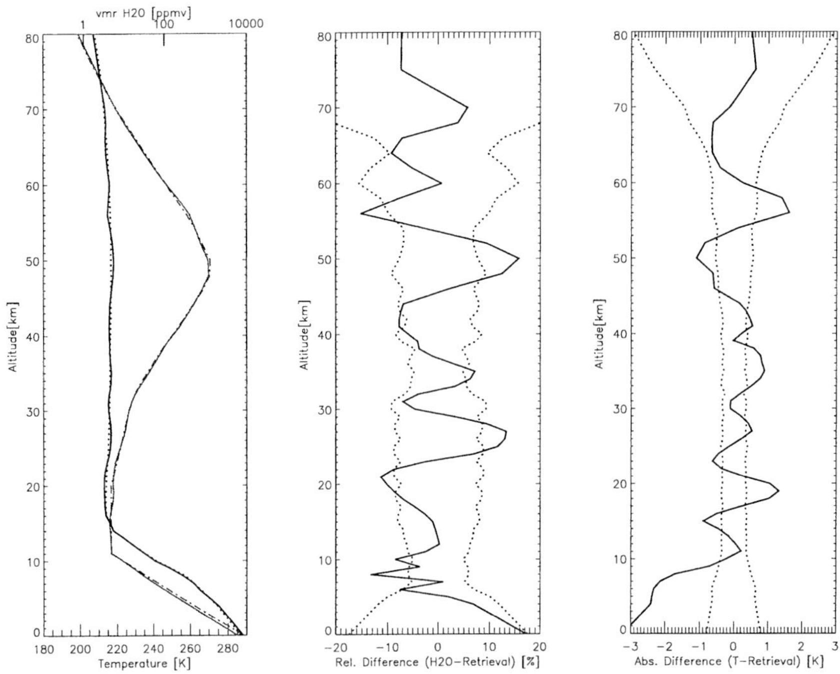

Fig. 1. Left: retrieval results (solid lines) and true profiles (dashed lines) of test retrievals, for H_2O (black) and temperature (grey). Middle: relative difference to true H_2O-profile (solid) and estimated noise error (dotted). Right: as middle but absolute difference of T-profiles. The spectra have been perturbed by detector noise as specified for MIPAS.

Atmospheric state parameters are retrieved on a fixed, i.e. tangent height-independent, altitude grid finer than the tangent altitude spacing. Typical gridwidth is 1 km. Retrieval of quantities where independent measurements are available, such as line of sight, are constrained by optimal estimation (Rodgers, 1976). Retrieval of quantities where only climatological a priori knowledge is available are constrained by Tikhonov's 1st and 2nd derivative operators (Tikhonov 1963).

Further details of the IMK retrieval strategy are published in Clarmann et al. (2001). In Fig. 1 examples for both, a simulated water vapour and a simulated temperature retrieval are given. The simulated measurement spectra are representative for midlatitudinal atmospheric conditions with a tropopause altitude of 11 km and a hygropause altitude of 16 km. The estimated error due to measurement noise is in the order of 10 % for H_2O, which is confirmed by the test retrieval. For the temperature, the estimated random error is in the order of 1 K for the complete stratospheric as well as the upper tropospheric region. Except for the middle and lower troposphere the temperature retrieval results show good agreement with the reference. The vertical resolution is high enough to reproduce the tropopause and the hygropause with good accuracy. Test retrievals based on more realistic scenarios including all relevant error sources in the simulated reference spectra confirm these estimates of achievable accuracies.

2.2 Retrieval Diagnostics

Besides the retrieved vertical profiles, retrieval diagnostics to evaluate the data quality and further properties of the data products are available. Here we focus on the ones relevant to validation.

The averaging kernel matrix $\mathbf{A}$ is given as

$$\mathbf{A} = \frac{\partial \hat{\mathbf{x}}}{\partial \mathbf{x}} = (\mathbf{K}^T\mathbf{S}_y^{-1}\mathbf{K} + \mathbf{R})^{-1}\mathbf{K}^T\mathbf{S}_y^{-1}\mathbf{K} = \mathbf{G}_y\mathbf{K}, \tag{2}$$

where $\mathbf{G}_y = \partial\hat{\mathbf{x}}/\partial\mathbf{y}$ is the gain matrix. $\mathbf{A}$ provides the dependence of the retrieved profile $\hat{\mathbf{x}}$ on the true profile $\mathbf{x}$. The width of the maxima in the rows of $\mathbf{A}$ gives an estimate of the vertical resolution (Carli 1998; Steck 2001).

The covariance matrix $\mathbf{S}_x$ of the retrieval, which includes the error component due to measurement noise is given as

$$\mathbf{S}_{noise} = \mathbf{G}_y\mathbf{S}_y\mathbf{G}_y^T \tag{3}$$

The systematic or forward model parameter covariance matrix $\mathbf{S}_f$ representing the error components due to uncertain model parameters represented by their covariance matrix $\mathbf{S}_b$ is given as

$$\mathbf{S}_f = \mathbf{G}_y\mathbf{K}_b\mathbf{S}_b\mathbf{K}_b^T\mathbf{G}_y^T. \tag{4}$$

3 Cross validation of different measurements

For the intercomparison of retrieved quantities from two instruments, the quantities have to be made comparable, e.g. by interpolation on a common grid. Then, by application of the formulation of Rodgers and Connor (1999) and Rodgers (2000) the mean expected difference between the two retrieved profiles $\hat{\mathbf{x}}_1$ and $\hat{\mathbf{x}}_2$ from different instruments can be expressed in terms of covariances $\mathbf{S}_\delta$ as

$$\mathbf{S}_\delta = (\mathbf{A}_1 - \mathbf{A}_2)^T\mathbf{S}_c(\mathbf{A}_1 - \mathbf{A}_2) + \mathbf{S}_{x1} + \mathbf{S}_{x2}, \tag{5}$$

where $\mathbf{S}_c$ is the climatological covariance of a comparison ensemble, which is needed to estimate the effect of different smoothing characteristics on the retrievals. Spatial-temporal mismatch can be considered by adding an additional term $\mathbf{S}_{mm}$ to $\mathbf{S}_\delta$. Finally the χ^2 of the difference between the retrieved profiles is

$$\chi^2 = (\hat{\mathbf{x}}_1 - \hat{\mathbf{x}}_2)^T(\mathbf{S}_\delta + \mathbf{S}_{mm})^{-1}(\hat{\mathbf{x}}_1 - \hat{\mathbf{x}}_2). \tag{6}$$

This quantity indicates if the two retrieved profiles are in agreement or not: The χ^2 expectation value for interconsistency measurements is equal to the number of degrees of freedom.

4 Future application to MIPAS and CHAMP

CHAMP currently provides temperature profiles in the troposphere up to the middle stratosphere, but water vapour profiles in the troposphere only. While the planned observation scenarios for MIPAS allow coverage of the troposphere down to about 5 km, retrievability of water vapour still needs to be confirmed with real MIPAS data. Studies based on simulated data show that, in the cloud-free case, meaningful water vapour volume mixing ratios can be retrieved down to about 11 km for the tropics; for midlatitude observations, the full altitude range observed can be exploited. Retrievability of mid-tropospheric water vapour depends strongly on the presence of (thin) clouds and aerosol layers.

Limitations in the altitude range where meaningful MIPAS retrievals can be performed could result in a reduced set of observations available for cross-validation. Crucial for the intercomparison is the availability of data sets with sufficiently small temporal and spatial mis-match. What sufficiently small is, however, currently is under investigation. If it should turn out that there are not sufficient observations available to perform any significant intercomparison ensemble mean data have to be compared instead of individual profiles. In this case the temporal and local scatter are expected to average out while biases remain visible.

Once successfully cross-validated, both observation systems will benefit from their consistency and can provide synergistic information. Tropopause and tropospheric water vapor information from CHAMP could be used as initial guess or a priori information for the MIPAS retrievals, or temperature information from MIPAS could be used by CHAMP within the water vapour retrieval. Finally, the retrieved profiles from both observing systems could be merged to provide consistently increased altitude coverage.

5 Conclusion and outlook

The IMK level 2 retrieval processor provides good results and useful retrieval diagnostics. The retrieval errors for H_2O and temperature in the stratospheric and upper tropospheric regions are in the order of 10 % and 1 K, respectively for MIPAS. Especially during the commissioning phase the cross validation with CHAMP will provide insight in the data quality and data characteristics of both instruments.

6 Acknowledgements

The main part of this work has been done under BMBF contract no. 01 SF9953/8. T.S. and M.M. have been supported by the EU project AMIL2DA (EVG1-CT-1999-00015) and BMBF contract no. 07 ATF43.

References

1. Carli B, Ridolfi M, Raspollini P, Dinelli BM, Dudhia A, Echle G (1998) Study of the retrieval of atmospheric trace gas profiles from infrared spectra. Technical report, European Space Agency, Final Report of ESA Contract 12055/96/NL/CN
2. Levenberg A (1944) A method for the solution of certain non–linear problems in least squares. Quart Appl Math, **2**, 164–168
3. Marquardt DW (1963) An algorithm for least–squares estimation of nonlinear parameters. J Soc Indust Appl Math, **11**(2), 431–441
4. Rodgers CD (1976) Retrieval of atmospheric temperature and composition from remote measurements of thermal radiation. Rev Geophys Space Phys, **14**(4), 609–624
5. Rodgers CD, Connor BJ (1999) Intercomparison of remote sounding instruments. In: Optical Remote Sensing of the Atmosphere, OSA Technical Digest, pages 46–48, Optical Society of America, Washington DC, 1999
6. Rodgers CD (2000) Inverse Methods for Atmospheric Sounding: Theory and Practice. World Scientific, Series on Atmospheric, Oceanic and Planetary Physics, FW Taylor, ed, Vol 2
7. Steck T, von Clarmann T (2001) Constrained profile retrieval applied to the observation mode of the Michelson interferometer for passive atmospheric sounding. Appl Opt, **40**(21), 3559–3571
8. Tikhonov A (1963) On the solution of incorrectly stated problems and method of regularization. Dokl Akad Nauk SSSR, 151:501
9. von Clarmann T, Fischer H, Funke B, Glatthor N, Grabowski U, Höpfner M, Kiefer M, Martín-Torres FJ, Milz M, Stiller GP (2001) MIPAS interactive semi–operational level–2 data processing. In: Smith WL, Timofeyev Yu M (eds) IRS 2000 Current Problems in Atmospheric Radiation, pages 785–788. A Deepak Publishing, Hampton, Va, USA

AUTHOR INDEX

KEYWORD INDEX

Printing: Mercedes-Druck, Berlin
Binding: Stein+Lehmann, Berlin